RODALE'S
VEGETABLE
GARDEN

PROBLEM
SOLVER

THE BEST AND LATEST ADVICE
FOR BEATING PESTS, DISEASES, AND
WEEDS AND STAYING A STEP AHEAD
OF TROUBLE IN THE GARDEN

FERN MARSHALL BRADLEY

RODALE

D1289345

Rodale books may be purchased for business or promotional use or for special sales. For information, please write to:
Special Markets Department, Rodale Inc., 733 Third Avenue, New York, NY 10017

Printed in the United States of America

Rodale Inc. makes every effort to use acid-free ♾, recycled paper ♲.

Book design by Tara Long

Illustrations by Lizzie Harper

Library of Congress Cataloging-in-Publication Data

Bradley, Fern Marshall.
    Rodale's vegetable garden problem solver : the best and latest
advice for beating pests, diseases, and weeds and staying a step
ahead of trouble in the garden / Fern Marshall Bradley.
        p.   cm.
    Includes bibliographical references and index.
    ISBN-13 978–1–59486–307–3   direct hardcover
    ISBN-10 1–59486–307–5   direct hardcover
    ISBN-13 978–1–59486–308–0   trade paperback
    ISBN-10 1–59486–308–3   trade paperback
    1. Vegetables—Diseases and pests.   2. Vegetable gardening.
3. Garden pests—Control.   4. Weeds.   5. Organic gardening.
I. Rodale (Firm)   II. Title.
SB608.V4B73   2006
635'.049—dc22   2006024238

Distributed to the book trade by Holtzbrinck Publishers

2   4   6   8   10 9 7 5 3 1   hardcover
2   4   6   8   10 9 7 5 3 1   paperback

RODALE
LIVE YOUR WHOLE LIFE™

We inspire and enable people to improve their lives and the world around them
For more of our products visit **rodalestore.com** or call 800-848-4735

CONTENTS

Acknowledgments.................................vii

INTRODUCTION: Healthy Soil,
    Healthy Plants, Healthy Harvest..............1

ALTERNARIA BLIGHT.........................9

ANGULAR LEAF SPOT.......................10

ANIMAL PESTS.................................11

ANTHRACNOSE................................ 23

APHIDS...25

ARMYWORMS.................................. 28

ARTICHOKES....................................30

ASPARAGUS.....................................34

ASTER YELLOWS...............................39

BEANS..40

BEET CURLY TOP.............................46

BEETS..47

BENEFICIAL INSECTS.........................51

BIRDS..62

BLISTER BEETLES.............................65

BLOSSOM-END ROT..........................67

BROCCOLI......................................68

BRUSSELS SPROUTS..........................74

CABBAGE.......................................79

CABBAGE LOOPERS..........................86

CABBAGE MAGGOTS.........................89

CANTALOUPE..................................92

CARROTS..93

CAULIFLOWER...............................100

CELERY..104

CERCOSPORA LEAF SPOT..................109

CLUB ROOT..................................110

COLDFRAMES.................................111

COLLARDS.....................................111

COLORADO POTATO BEETLES..............112

COMPANION PLANTING.....................115

COMPOSTING..................................116

Corn 124

Corn Earworms 131

Cover Crops 133

Crop Rotation 142

Cucumbers 151

Cucumber Beetles 158

Cutworms 161

Damping-Off 163

Deer 164

Disease Control 172

Downy Mildew 187

Early Blight 188

Earwigs 189

Eggplant 191

Endive 196

Fertilizers 199

Flea Beetles 205

Fleahoppers 208

Fusarium Wilt 210

Garlic 212

Gourds 215

Grasshoppers 216

Gray Mold (Botrytis) 218

Harlequin Bugs 220

Horseradish 222

Imported Cabbageworms 224

Japanese Beetles 228

Kale 231

Kohlrabi 234

Late Blight 237

Leafhoppers 239

Leafminers 241

Leeks 244

Lettuce 247

Melons 253

Mexican Bean Beetles 258

Mites 261

Mosaic 263

Mulch 265

Nematodes 271

Okra 276

Onions 278

Parsnips 284

Peas 286

Peppers 291

Pest Control 299

Pickleworms 321

Planting and Transplanting 323

POTATOES 330

POWDERY MILDEW 337

PUMPKINS 338

RADISHES 339

RHUBARB 342

ROTS 345

ROW COVERS 347

RUST 350

SALAD GREENS 352

SCAB 355

SEASON EXTENSION 356

SEED STARTING 359

SHALLOTS 363

SLUGS AND SNAILS 365

SOIL 372

SOUTHERN BLIGHT 378

SPIDER MITES 379

SPINACH 380

SQUASH 385

SQUASH BUGS 391

SQUASH VINE BORERS 393

STINK BUGS 396

SWEET POTATOES 397

SWISS CHARD 401

TARNISHED PLANT BUGS 406

THRIPS 408

TOMATOES 411

TOMATO HORNWORM 422

TURNIPS 424

VERTICILLIUM WILT 427

WATERING 428

WEATHER 431

WEEDS 435

WHITEFLIES 446

WHITE MOLD 448

WIREWORMS 449

Illustrated Key to Insects 452

Resources for Gardeners 456

Recommended Reading 458

Index 460

USDA Plant Hardiness Zone Map 472

ACKNOWLEDGMENTS

Thanks first to my family for their help and support. Barry Marshall and Jeri Burns welcomed me to share their garden, while Zachary patiently acquiesced whenever Aunt Fern needed to stop playing and write. Rory Bradley lifted my spirits by proclaiming his pride and excitement about my debut as an author. Tom Cole kept house and home together throughout months of research and writing and offered loving encouragement at every stage.

I'm grateful to Barbara Ellis and Ellen Phillips for teaching me how to be a good garden writer and editor. Thanks to Chris Bucks and Karen Bolesta at Rodale for offering me the opportunity to write this book. My editor, Anne Halpin, was a terrific partner in shaping the manuscript. Lizzie Harper provided beautiful illustrations. Tara Long, Keith Biery, and Nancy N. Bailey at Rodale ensured that the pages look elegant and read well from start to finish.

Many thanks to gardening consultant and author Sally Cunningham for creating the excellent design for a garden to attract beneficials, shown on pages 54 and 55.

I appreciate the insightful comments and corrections offered by technical reviewers Janet Bachmann, of Riverbend Gardens in Fayetteville, Arkansas; Dr. Linda Gilkeson, entomologist; and Beth Gugino, postdoctoral research associate at the New York State Agricultural Experiment Station, Cornell University.

For offering great gardening tips and answers to questions, I owe thanks to many fine home gardeners, farmers, university staff, and technical experts, including Joel Allen (Cornell Cooperative Extension of Columbia County), Brian Caldwell (Cornell University), Julie Callahan (Massachusetts Department of Agricultural Resources), Bev and Bud Cole, Leslie Cole, Dr. Scott Craven (University of Wisconsin), Don DeVault, Melanie DeVault (Pheasant Hill Farm), Mark Dunau (Mountain Dell Farm), Bill Foster (BioWorks, Inc.), Bruce Frasier (Dixondale Farms), Brian Goldman, Miguel Guerrero (Organic Materials Review Institute), Craig Harmer (Gardens Alive! Inc.), Diane Hartzell, Shawn Hill, Norbert Lazar (The Phantom Gardener), the late Bob Love, Sharon Lovejoy, Pam Marrone (AgraQuest), Bing McClellan (Green Light Company), Rose Marie Nichols McGee (Nichols Garden Nursery), Scott Meyer (*Organic Gardening* magazine), Kevin Moran (Troy Biosciences, Inc.), Jean Nick, Nancy Ondra, Barbara Pleasant, Dr. Steve Reiners (Cornell University), Pam Ruch (*Organic Gardening* magazine), Mary Lou Shillinger, Katie Smith (The Farm at Miller's Crossing), Nan Sterman, Heidi Stonehill, Richard Vento (St. Gabriel Laboratories), Eileen Weinsteiger (Rodale Institute), Carolyn Kay Wheeler, and the Columbia County Master Gardeners. I also learned a great deal from conferences, books, and newsletters of the Natural Organic Farming Association (NOFA) and NOFA–New York, as well as publications from Cooperative Extension in many states and from the National Sustainable Agriculture Information Service.

HEALTHY SOIL, HEALTHY PLANTS, HEALTHY HARVEST

Visualize your vegetable garden exactly as you wish it could be. All of your favorite crops are there, in just the right amounts to supply you with fresh produce every day. The soil is dark and crumbly and moist. The paths are well mulched. There's not a weed in sight, but cheerful flowers bloom here and there among the beds. Birds sing in the shrubs and trees nearby, but no rabbits or woodchucks ever dare show their furry heads. Slugs, beetles, and caterpillars have produced nary a single hole in a leaf. Every plant is the picture of health, without a yellow leaf or blemished fruit to be seen.

It's a paradise, isn't it? And although this paradise is a fantasy, a healthy organic garden can be a paradise of life and plenty. That's not to say it will be completely pest- and problem-free, but the soil will be rich and healthy, the plants will be vigorous, and pest infestations will rarely reach a crisis stage. Creating a healthy organic garden isn't a one-weekend project. It may take several years to develop a flourishing chemical-free garden, depending on the quality of the soil you started with, the resources you have available, and the amount of time you can devote to it.

REAL-LIFE GARDENING

Is your garden still on that multiyear road to achieving a healthy natural balance? Perhaps you're struggling to improve soil that was a rocky wasteland or a dry sandlot when you first started gardening. Maybe rabbits eat up half your crop of lettuce and salad greens, and you can't even consider growing corn because the local raccoons or deer would get to it before you could harvest it. Or perhaps the problems in your garden stem from the demands of the world beyond the backyard—there's just too little time to balance work, family, and community obligations with the needs of the garden.

Whatever the reason, if your vegetable garden is less than a paradise, this book is for you. It's filled with practical solutions to the gardening problems we all face. Advice for controlling insects, diseases, weeds, and animal pests abounds, but that's not all. You'll find practical ideas and helpful hints for all the logistical problems of gardening, too: where to find space for

1

starting seeds indoors, how to plant small seeds efficiently, how to set up a crop rotation plan, how to decide when to fertilize and what to use, how to reduce the time you spend weeding and watering, and lots more.

HEALTHY GARDEN BASICS

Healthy plants are more resistant to problems than stressed plants. The more you can do to promote a healthy garden environment, the less you'll have to struggle with fighting pest and disease problems. In undisturbed natural environments such as meadows and forests, it's rare for pest problems to reach epidemic proportions. For the most part, plants and their pests stay in balance. However, our vegetable gardens are not undisturbed natural environments. We dig in the soil; we plant seeds and transplants and pull out unwanted plants; we put up fencing; and we harvest leaves, fruits, and roots. We choose the plants we want to grow—and some of them aren't well adapted to our local climate and conditions. We plant in neat rows and beds. These are a lot of changes from a "natural" state. Even so, we can use gardening practices that mimic features of natural environments to help keep the balance and prevent disease or insect invasions.

TEND THE SOIL

The most important part of a healthy garden environment is biologically active soil—soil full of the living organisms that process nutrients, that help plant roots absorb nutrients and water, and that protect plants from pathogens.

The best way to tend garden soil is to enrich it with organic matter to stimulate the food web. The food web is the community of animals, insects, and microorganisms that live in the soil. Earthworms, nematodes, bacteria, fungi, and mycorrhizae (a specialized type of fungi) are part of this web, and their activity produces soil that is ideal for plant growth. Unfortunately, tilling and digging can kill some types of soil organisms and destroy the habitat for others. Synthetic fertilizers, harsh pesticides, and other materials put into the soil can kill even more. But good garden practices can support and enhance the underground community and let it work for the benefit of your plants. The more you can do to encourage living things in your garden soil, the better your crops will grow and produce. The Soil entry in this book explains how to overcome soil problems and build a diverse subterranean food web in your garden.

COVER THE SOIL

Bare soil is the exception to the rule in natural systems. You can see that in your own yard—if you dig a garden bed but then leave it alone, it won't stay bare for long. If it's spring or summer, plants will sprout. If it's fall, leaves will drop from the trees and cover the bed. Only persistent effort keeps garden beds clear of vegetation or leaf litter. Don't fight nature's tendency here; instead, work with it. After you dig a garden bed, if you don't plan to plant it right away, cover it up. Spread a layer of organic mulch or plant a cover crop. Or if you want to warm the soil, cover it tightly with black plastic.

Throughout the gardening season, continue to use mulch and cover crops to the fullest to avoid bare soil. Mulch plays many helpful roles. It moderates soil temperature and moisture levels, and that's beneficial for all the helpful organisms in your soil, as well as for your plants. Mulch provides a habitat for beneficial ground beetles and spiders. There are a few special circumstances when you'll leave soil uncovered to solve a problem, but for the most part, covering the soil is the best approach. To learn how to manage mulch and cover crops successfully in your garden, turn to the Cover Crops and Mulch entries in this book.

STRIVE FOR DIVERSITY

Natural environments include a mix of plant species of different sizes, shapes, and soil needs. Diversity is helpful in gardens, too. Although it seems counterintuitive, including some nonvegetables in your vegetable garden can *improve* yields because it tends to reduce pest problems. Interplanting crops among plants that aren't hosts to particular veggie pests can make it harder for the pests to find the crop plants. Even more important, a mixed planting provides a range of habitats for beneficial insects that help to control pests.

Outside your vegetable garden, plantings in other parts of your yard can help to supply habitat for these beneficial insects, too. Beneficial insects can't survive just by eating other insects; many also feed on pollen or nectar, and they need sheltered areas where they can hide to avoid damaging weather conditions and their own predators. Perennial plants, including shrubs and flowering perennials, are often important habitat plants. If you can't plant the perennials

in your vegetable garden, find other places in your yard for them, such as a foundation planting or a streetside garden. The beneficials attracted to these flowering islands will also forage for insect prey in your vegetable garden. Consult the Beneficial Insects entry for specific advice and recommended plants to encourage beneficials.

NEW-GARDEN WOES

Few home gardens start out with naturally rich, fertile soil. Contractors often strip topsoil and compact the remaining soil around new homes during construction. Or past owners may have treated the soil with synthetic herbicides and fertilizers, killing beneficial soil organisms that help keep plants healthy. In any case, vegetable crops growing in poor soil are likely to suffer stress and develop pest problems. If your soil is a rocky or compacted mess, you may feel overwhelmed by the double burden of improving the soil and fighting pest problems. But take heart! If you continue feeding the soil and following practices to attract beneficial insects and reduce pests, your soil quality will get better and you'll see pest problems decline from year to year. Your harvests will get bigger, too. You will find a wealth of soil-building solutions in the Soil and Composting entries. Organic methods really do work. Years of scientific research and the hands-on experience of millions of gardeners prove it.

TOO BIG TO MANAGE?

Another way gardeners tend to create problems for themselves is by planting a garden that's bigger than they can manage. It's easy to become overenthusiastic, especially when you page through the enticing seed catalogs that flood your mailbox each winter, or browse through local nurseries and garden centers in spring. You may end up planting lots of seeds and transplants, but the result can be the opposite of what you expect. Weeds shoot up in all your beds, competing for water and nutrients and stressing your crops. You don't thin all of the rows of seedlings, so the crops themselves put stress on one another. You can't keep up with watering, so plants wilt and struggle to survive in summer heat. Instead of harvesting a bumper crop of vegetables, you end up fighting a bumper crop of pests, simply because you didn't have enough time to keep your plants healthy.

If this description sounds like what happens in your garden, here's an idea for changing the pattern: Plant one-fourth of your garden with a cover crop instead of vegetables. Cover crops require less time and attention than vegetable crops, and they offer lots of great benefits, including improving the soil and providing habitat for beneficial insects. Try it and see. Even though you plant fewer veggies, you may end up with a better harvest. You may decide that cover cropping should be part of your garden routine, or you may gradually be able to increase the size of your garden as your gardening skills improve and your garden habitat matures. See the Cover Crops entry for information.

If you're starting a new garden, what's the best way to decide how much to plant? Start by assessing how much time you have available for gardening. Do you work full time? Do you have children to take care of and a household to run?

Do outside activities take you away from home on a daily basis? Write down your schedule during a typical week and see where you can fit in blocks of time for gardening. You may discover that you can devote only 4 hours per week to your vegetable garden, for example. If that's the case, think small. One rule of thumb is that you'll need 2 hours per week for each 100 square feet of garden you plant. Four hours per week equals a garden that's 10 feet by 20 feet.

SOLVING PROBLEMS ORGANICALLY

During a typical growing season, you may need to take steps to prevent or control roughly a dozen species of pests. Contrast that with the thousands of species of nonharmful organisms present in your garden. Many of the insects, animals, and microorganisms in our gardens are helpful, from lady beetles that prey on aphids to beneficial nematodes that protect root crops from root-knot nematodes. Since there are so few pests in our gardens, it makes sense to avoid spraying a pesticide or herbicide that kills a broad spectrum of living things, including ones that could be helpful to our gardening efforts. There are plenty of other important reasons to avoid using pesticides, and as an organic gardener, you will want to choose more benign measures. Organic gardeners deal with pest problems by setting up barriers to block pests, using repellents to keep pests at bay, encouraging pest predators and parasites in the garden, and resorting to organically acceptable natural pesticides only as a last resort.

Every year new kinds of organic gardening and pest-control products appear on the shelves of garden centers. It's hard to keep up with all the options, and you may wonder how to best educate yourself about this expanding world of organic gardening methods and products.

WHAT DOES ORGANIC MEAN TODAY?

Twenty years ago, organic gardeners had limited choices of commercial products for controlling pests. Most garden centers carried few or no organic fertilizers and pest-control products. Home gardeners who wanted to learn about organic gardening relied mainly on *Organic Gardening* magazine, Rodale gardening books, and their fellow gardeners for information. In the 1990s, interest in organic food, farming, and gardening snowballed.

The United States government became involved by passing the Organic Foods Production Act (OFPA), which directed the USDA to set standards for commercial organic food production, from vegetable crops to dairy products to processed foods. The purpose of the legislation is to protect consumers. Ideally, because of the OFPA, we can be confident that organic produce is commercially grown according to specific, detailed standards.

The USDA responded to the legislation by creating the National Organic Program (NOP), which wrote detailed rules governing organic food production and established a certification program. Farmers who want to label their products with the USDA Organic seal must create a written plan for their farm businesses detailing how they will meet government standards, and

WHO DECIDES WHAT'S ORGANIC?

The creation of national standards for organic farming and organic food products generated fierce debates among farmers, consumer groups, food processing companies, scientists, and environmentalists. Amendments to the standards are ongoing, and there is continuing controversy as to whether the standards uphold the high ideals that underlie the philosophy of organic agriculture. If you're concerned about the broader issues of organic food production and marketing, join an organization that supports organic farmers and gardeners. These organizations participate in the debate about additional amendments to the national standards for organic food production. Most publish newsletters that report on controversial changes, and inform their members about how they can respond to and comment on government decisions. See Recommended Reading on page 458 for a listing of some of these organizations.

their farms must be inspected once a year by a certification agency.

The regulations define which substances organic farmers can use as fertilizers and pest- and disease-control products. In general, synthetic substances are prohibited, and the OFPA defines synthetic as "formulated or manufactured by a chemical process or by a process that chemically changes a substance extracted from naturally occurring plant, animal, or mineral sources." Sewage sludge, ionizing radiation, and genetically engineered seeds also are prohibited. A limited number of synthetic substances are allowed, and they are listed in detail in the regulations. For example, copper nitrate is a type of synthetic chemical fertilizer, and it is prohibited for organic crop production. However, copper sulfate, a different synthetic form of copper fertilizer, is allowed if the farmer can prove (via soil testing) that there's a copper deficiency on his or her farm. The regulations are long and complicated, but if you are interested in reviewing them, you can find them online at www.ams.usda.gov/nop.

MAKING CHOICES ABOUT GARDENING PRODUCTS

As interest in organic food and farming increases, companies are researching and developing new products for organic farmers. Many of these products also become available to home gardeners. For example, there is an almost overwhelming variety of organic blended fertilizers, liquid fertilizers, and foliar fertilizers to choose from. Virtually all garden and home centers now offer some organic fertilizers and pest-control products. Information on organic gardening is available from many magazines, Web sites, and books. You'll find a list of good gardening books and magazines in Resources for Gardeners on page 456. *Organic Gardening* magazine maintains a searchable database of Web sites of interest to organic gardeners; to use that database, go to www.organicgardening.com/gardeninglinks/. The net result is that organic home gardeners today have a wide range of choices of products and methods, and sometimes it's confusing to know which options match up to your philosophy of organic gardening.

 FOR ORGANIC PRODUCTION

The three-leaf logo and the Organic Materials Review Institute logo indicate that a gardening product meets the standards of the national legislation governing organic food production.

Home gardeners don't have to follow the NOP standards, of course, but one way to check whether a product is approved for organic farming is to look for the three-leaf logo or the OMRI logo on the label. The three-leaf logo indicates that the Environmental Protection Agency has approved the product as meeting NOP standards. OMRI stands for Organic Materials Review Institute, which is a nonprofit organization that reviews applications from product manufacturers to determine whether a product meets NOP standards. (To learn more about OMRI, visit the institute's Web site at www.omri.org.)

EPA approval and OMRI listing of products are very important to commercial organic farmers because they may face penalties and lose their certification if they use products that don't meet NOP standards. Of course, this isn't the case in home gardens. OMRI listing and EPA approval are helpful guides for home gardeners but not requirements. Sometimes, you may choose to use a product that isn't OMRI listed, and you may also choose not to use a product that is listed by OMRI. Why choose a product that isn't OMRI listed? Some companies that make organic products for home gardeners don't seek OMRI listing because the review process is expensive and time-consuming. Also, new products come on to the market each year. Sometimes they are released for sale while still under review and aren't yet OMRI listed.

When choosing products, it's very helpful to find a local garden center or nursery staff person who's knowledgeable about organic products. Failing that, find a mail-order or online garden supplier that offers detailed information about its products (see Resources for Gardeners on page 456 for a list of suppliers). You can always check directly with a product manufacturer, too, and ask whether its product meets NOP standards.

On the flip side, why would you turn down an OMRI-listed product? One consideration in choosing pest- and disease-control products is how environmentally friendly they are. Some OMRI-listed products can have harmful environmental side effects. For example, many botanical pesticides—products that contain rotenone, neem, and pyrethrins—are OMRI listed. However, most botanical pesticides have some harmful effect on beneficial insects, and a few are also harmful to fish and other organisms. Organic farmers cannot use these pesticides except as a last resort—which in their case may mean the difference between harvesting a marketable crop and having to plow the crop under. In a home garden, thankfully, we rarely face such critical consequences if we lose some of our harvest. Thus, we can afford to be very conservative about the use of botanical pesticides, choosing only those that have the fewest harmful environmental effects and saving their use for severe pest problems only. (To learn more about botanical pesticides, see the Pest Control entry.)

KEEPING YOUR PERSPECTIVE

Even when a tough problem comes up in your vegetable garden, remember that it's not the end of the world. It can be heartbreaking to lose a crop that you've nurtured carefully through the season, but don't let that ruin your enjoyment of the crops that do succeed. Remember how delicious homegrown produce tastes and how good it is for you. Even if the broccoli heads from your garden are small and less than perfect, they're loads higher in vitamin content than the oversized broccoli at the grocery store that's been shipped hundreds or possibly thousands of miles. And growing your own organic vegetables is more environmentally sustainable than buying packaged frozen organic produce from places as far away as New Zealand or Asia.

Despite rows that aren't precisely even, tomatoes planted a few weeks late, weeds in the pathways, or some holes in the cabbage leaves, your organic vegetable garden really is a paradise of vitamin-rich, sustainably produced food. So take a new view of your gardening problems, and think of it this way: Every mistake you make or problem you face can lead to a discovery that will make next year's garden better.

HOW TO USE THIS BOOK

Rodale's Vegetable Garden Problem Solver is divided into entries that cover crops, insects and other pests, diseases, and gardening techniques. The entries are arranged from A to Z.

Let's say you've discovered a problem on your squash plants but you don't know what's causing the problem. Turn to the Squash entry and look for "Troubleshooting Problems." You'll find a complete listing of the common insect, disease, and cultural problems of squash. Browse the list to find the probable cause.

Sometimes you will know what's causing the problem. For example, you may know at a glance that the yellow bugs on your bean plants are Mexican bean beetles. If so, turn to the Mexican Bean Beetles entry for a full listing of control measures and find a control.

If you've found an insect in your garden but you don't know what it is, turn to the Illustrated Key to Insects on page 452 to quickly scan illustrations of common garden pests and beneficial insects. All of the insects covered in this book are included, and the key will direct you to the specific page where you can find information on that insect.

Sometimes a problem may be broader in scope: Perhaps your crops seem to be growing slowly and you suspect that your soil isn't supplying enough nutrients. You'll want to consult one of the gardening technique entries, such as Fertilizers or Soil, to find solutions to the problem.

Gardening isn't just about solving problems; it's also about having fun, enjoying the outdoors, and reaping a great harvest. And there's plenty of information in this book to help you there, too! Every crop entry includes a "Secrets of Success" section that you'll find helpful for enjoying greater success in your garden. And the entries on techniques such as seed starting and watering are full of great tips and ideas for more efficient and effective gardens. If you are an experienced gardener, look for "Beyond the Basics" boxes for tips on advanced methods.

If you'd like an overview of what's in the book, turn to the Contents page in the front, where you'll find a quick page reference guide to all the entries. And last but not least, if you can't seem to find an entry that addresses the problem, turn to the comprehensive index at the back of the book and look it up there.

ALTERNARIA BLIGHT

also called Alternaria leaf spot

It pays to act quickly if leaf spots appear on blight-susceptible crops because the Alternaria fungi that cause blight can spread rapidly in warm to hot, wet weather. A light infection won't do much damage, but if leaves end up blighted and dead, you may lose some of the harvest of roots, flowerheads, or fruits.

Alternaria solani causes early blight of potatoes and tomatoes; see page 188 for the description and control recommendations for this disease.

RANGE: United States; southern Canada. Rarely a problem in dry regions

CROPS AT RISK: Carrots, parsley, squash-family crops (especially muskmelons and watermelons), cabbage-family crops (including horseradish), beans

DESCRIPTION: Symptoms vary from crop to crop, but many crops develop targetlike spots that enlarge and merge. In carrots, leaf spots are dark brown or black and irregularly shaped. Elongated spots on leaf stems look like dark streaks. The disease develops rapidly in carrots, and most of the foliage may end up looking burned, collapsing to the ground.

In squash crops, small spots appear on older leaves. As the spots enlarge on muskmelon leaves, the pattern of concentric rings becomes clear. Infected leaves may curl and shrivel up. Often, fruits aren't directly infected but suffer from sunscald due to lack of shading after leaves die. Summer squash may develop moldy spots due to direct infection by the blight fungus.

In cabbage-family crops, spots mainly appear on older leaves, so damage is not serious. Spots may start as specks and grow quite large, showing concentric rings. Spores also can germinate on broccoli and cauliflower heads, creating dark lesions. Also, the spots can penetrate several layers deep in sprouts on Brussels sprout plants, ruining the harvest. Turnip leaves can be severely infected, and the roots of infected plants may develop symptoms in storage.

In warm conditions, Alternaria fungi can cause damping-off of seedlings (see page 163). However, infected seedlings often develop normally, showing no symptoms until warm, wet conditions prevail.

FIGHTING INFECTION: Remove leaves and stems that show symptoms, or pull out infected plants. Start a spray regimen of *Bacillus subtilis*, baking soda solution, or compost tea, but apply sprays early in the day so foliage will dry quickly. Stop using sprinkler irrigation.

Although symptoms spread rapidly through a carrot planting, the infection often doesn't occur until later in the season. Even if carrot tops die back suddenly, dig up the crop. Roots probably will be salvageable.

GARDEN CLEANUP: The fungus survives winter on seeds and in crop debris. After harvest, remove and discard noticeably diseased crop debris; compost or turn under the rest. Also uproot and remove weeds that belong to the same plant families as susceptible crops.

NEXT TIME YOU PLANT: Add compost to bolster beneficial soil microorganisms. Space carrot- and cabbage-family crop rows widely. Sow carrot seed in raised soil ridges. Plant clean seed; choose resistant varieties when possible. Spray *B. subtilis* or compost tea before symptoms appear.

CROP ROTATION: Different species of Alternaria cause blight in different crop families, and in general, Alternaria fungi can't survive more than 2 years in the soil. A simple 3-year rotation by crop family may break the disease cycle.

∽ ANGULAR LEAF SPOT

Cucumbers and their cousins are the only crops affected by this bacterial disease. Angular leaf spot spreads during wet conditions: Splashing rain or people handling wet plants spread the bacteria, which enters through wounds or pores in leaves, stems, and fruits.

In home gardens, even if leaves and stems show symptoms, fruits may escape infection. If the foliage is badly infected, though, yields will be poor.

RANGE: United States; southern Canada

CROPS AT RISK: All squash-family crops

DESCRIPTION: Water-soaked spots appear on leaves and stems. The spots don't enlarge past the leaf veins, so they end up with irregular geometric outlines. On wet mornings, small drops of bacterial ooze form on these areas (usually on leaf undersides), leaving a trace of white when they dry. The spots turn yellow and crisp and may tear out of the leaves, leaving ragged holes. White spots form on fruits, and the tissue underneath can rot, especially if soft rot sets in.

FIGHTING INFECTION: Prune off infected leaves and stems. If plants are small, uproot and discard them; start another planting elsewhere in your garden. Stop using sprinkler irrigation. Harvest fruits on the young side from infected plants. If the disease has been a severe problem in the past, spray with copper during wet weather to slow the spread of the disease. Stop spraying when dry weather returns.

GARDEN CLEANUP: The bacteria survive winter in seeds and plant debris.

After harvest, remove and discard any noticeably diseased plant debris. Compost or turn under the rest.

NEXT TIME YOU PLANT: Improve soil drainage before planting. Plant clean seed. Choose one of the many resistant cucumber varieties available.

CROP ROTATION: The bacteria persist up to 2 years in the soil in crop debris. A 2-year (or longer) rotation of squash-family plants should break the disease cycle.

ANIMAL PESTS

There's no simple solution for animal pest problems because furry marauders are much more complex creatures than insects are. Animals will behave differently depending on the weather, the amount of food available, the population level, and other factors. Our vegetable gardens are tempting targets because so much of the plant growth is young and tender, and because we water the plants during dry periods.

To cope with animal pests, start by deciding if you will protect only your vegetable garden or if you also want to protect flower gardens, trees, and shrubs. Note that if you take steps to stop animal pests from eating your vegetable crops, they may cause even more damage to your landscape plantings. Once you make the decision, act boldly. While insect pests or diseases can reduce yield of one crop or another, it's rare that they wipe out an entire crop overnight. With animal pests, on the other hand, such a scenario is increasingly common.

ASSESSING THE PROBLEM

The most common furry creatures that bother vegetable gardens from coast to coast are rabbits, woodchucks, deer, and raccoons. Rabbits, woodchucks, and deer eat a variety of crops. Raccoons are corn specialists but also will dig into ripening melons to feast on the sweet flesh. Several other kinds of wildlife, from armadillos to voles, sometimes feed in vegetable gardens in certain regions.

If you're new to gardening, ask your gardening neighbors or your Cooperative Extension office what animal pests raid gardens in your area. Keep a watchful eye out around your yard and your neighborhood, especially in the early morning and as dusk approaches. You'll probably spot rabbits and woodchucks if they're local residents, and in many areas, deer feed in full view of roads and houses.

DIAGNOSING DAMAGE

In your garden itself, look for rabbit droppings—small piles of brown pellets the size of peas. Rabbits clip off branches neatly. You may find a few wilted branch tips on the ground around damaged plants, leftovers that the rabbit didn't finish.

Deer droppings are larger than those of rabbits. Deer also tend to trample crops, and they feed roughly, tearing at plants.

You may never see a raccoon in your neighborhood, but if you grow corn and don't protect it, the raccoon will find it just as it ripens. If raccoons are reported as pests in your area, make the assumption that you'll need to protect your corn and melon patch from them. Otherwise, you'll wake one morning to find corn ears ripped open and stalks toppled over, as well as gaping holes in ripening melons.

Birds are garden helpers because they eat large quantities of pest insects, but birds can also be garden pests. To learn about the damage birds cause and how to deal with it, consult Birds. See "More Pests" on page 20 for signs of damage by small rodents and regional animal pests.

SOLUTIONS TO ANIMAL PEST PROBLEMS

Coping with animal pests can include four categories of action: Scare them away, repel them with repugnant odors and flavors, plant crops especially for them to eat, or fence them out.

A fence is the most secure option. If you decide against a fence, plan on launching a campaign with multiple products and gimmicks, and be prepared to switch them frequently. There's no magic repellent or scare tactic that works all the time every time. In fact, you'll find amazingly contradictory claims about animal repellents and scare devices. One gardener will swear that spraying plants with a hot pepper spray protected them from animals beautifully, while another will scoff that spraying hot pepper seemed to attract wildlife to the garden. Both could be telling the truth. Perhaps one gardener sprayed the repellent faithfully every 3 days, and it kept away the rabbits from her town garden. The other gardener may live at the edge of a wooded area overfilled with deer.

Unfortunately, the simple animal repellents devised from household items that protected our grandparents' and parents' gardens may no longer work for us. Decades ago, animals simply had more wild places in which to forage for food. Higher populations of animals that are less wary of feeding near our homes and roads make our gardens prime targets, and the ani-

THE PRESSURE FACTOR

"Wildlife pressure" and "deer pressure" are terms that refer to the density of the wildlife population in an area relative to amount of food and natural habitat available. If there are lots of wild animals in your neighborhood but also lots of appropriate habitat and food sources, wildlife may not visit your yard often—you live in a "low pressure" area. Using repellents and scare devices may be enough to prevent any significant damage.

Unfortunately, more and more gardeners find that wildlife pressure—especially deer pressure—in their area is high and getting higher as development encroaches on wildlife habitat. If you live in a high pressure area, a sturdy fence designed especially to keep out the pests that frequent your area will be your only hope of having a successful vegetable garden.

As you assess the wildlife pressure in your area, gather as much information as you can. Are new large developments planned nearby? Are any of your neighbors fencing their yards? Either of these factors could increase the pressure on your yard and garden. Also, keep in mind that pressure one year may be worse than another due to weather conditions (droughts will increase wildlife pressure) and cycles in local wildlife populations.

mals are learning faster that scare devices and repellents don't have to deter them from a night-time garden feast.

WHEN YOU NEED AN INSTANT DEFENSE

It's happened to most of us—you head out to your garden one fine spring morning and find unmistakable signs that animal pests have been chewing on young seedlings and transplants. There's no time to put up a fence until the weekend. What can you do until then? Try some or all of the following quick fixes. None are guaranteed, but with luck, they'll protect your plants until Saturday arrives.

APPLY A REPELLENT. Apply an odor-based repellent, such as a product made with rotten eggs, around the perimeter of your garden, and sprinkle or spray a taste-based repellent, such as hot pepper spray, directly on crops that were attacked.

SPREAD ROW COVERS. If you have it on hand, cover plants with floating row cover and fasten it down tightly along all sides. See Row Covers for tips on securing row covers.

Before you cover your plants, though, inspect them for insect pests. If young caterpillars, grubs, aphids, or flea beetles have found the plants, row covers won't work. You may lose your crop to the insects as they thrive in the protective space of the row cover.

BARRICADE PLANTS. Cloches used for protecting plants from cold will serve to keep out animal pests for a few days. See Season Extension for suggested cloches. Don't use cloches if temperatures are predicted to reach 80°F or more, or you'll cook your plants.

For an instant any-season animal barricade, check your shed for extra pieces of wire mesh fencing. Use wire cutters to cut sections just long enough to make arches over your crop rows, as shown on page 166. These barriers can be effective all season to protect crops from deer as long as the plants don't outgrow the arches. Caution: Animal pests that dig, such as woodchucks and rabbits, can tunnel under these kinds of barriers.

SCARE DEVICES

Items that move, make noise, resemble predators, or otherwise disturb the peaceful setting of your garden can spook animal pests for one day or several, but probably not all season. To get the best from scare devices, keep swapping one for another, and put them in different places around your garden.

FAKE SNAKES AND OWLS

Commercial versions of these decoys vary in design. Some are quite lifelike; others are more abstract, with surfaces that catch and reflect light. You can make your own "snakes" by cutting 4- to 6-foot sections of old garden hose and positioning them in the pathways of your garden, where they're easily seen. These may scare birds, rabbits, and some other small animals.

WEIRD SURFACES

Laying chicken wire on the ground around a corn patch could discourage raccoons because they won't like trying to walk across the wire. Planting corn as an island in a sea of prickly

To scare rabbits, birds, and other animals, hang aluminum pie pans and CDs on strings tied to sturdy twine stretched taut between stakes along garden rows. Drape shiny Christmas garland on tomato cages or trellises.

pumpkin or winter squash vines is another approach.

The chicken-wire tactic may deter deer, too, but they could jump across it—and it's awkward to try to surround a whole garden with a barrier of chicken wire "carpeting."

For rabbits, try surrounding crop rows or individual plants with spiny plant materials such as sweet gum balls or spiny holly trimmings.

SPRINKLERS

Motion-activated sprinklers send out a jet of drenching water when a marauding animal comes close enough to trip a heat sensor in the sprinkler unit. If you want to try a sprinkler like

this, keep in mind that it needs to be attached to a hose at all times. It takes more than one for full coverage because the motion detector has a range of only about 100 degrees (a full circle is 360 degrees). Thus, if animals can sneak into your garden from any direction, you'd need three sprinklers (and thus three hoses or a system of hose splitters to supply the three units). The units cost $50 to $75 each, and they include batteries that need to be replaced occasionally.

SHINY ITEMS

The flashes of light cast by aluminum pie pans, CDs, and metallic Christmas garland alarm animals and prevent them from investigating your garden. Arrange them as shown in the illustration on the opposite page.

NOISE

Commercial devices that make loud noises at random intervals work well for keeping pests out of farm fields, but if you have neighbors nearby, using one of these is a good way to ruin a friendship. One tolerable source of noise that's reputed to work well is a portable radio for scaring raccoons. Set it to an all-talk station and put it in the garden as corn approaches ripeness.

REPELLENTS

A foolproof animal repellent has yet to be invented. However, there are some repellents that work a majority of the time, especially if the animal pest pressure in your area isn't severe. Don't be disappointed if you see a little damage, because the animals may eat a bit here and there until they become turned off by the taste or smell.

Common odor-based repellents include bloodmeal (dried blood), predator urine, garlic, human or animal hair, rotten eggs, or synthetic chemicals. Taste-based repellents use capsaicin (the substance that makes hot peppers hot), essential oils, garlic oil, or various synthetic substances. Many studies show that overall, odor-based repellents work better than taste-based repellents. As with every general rule about repellents, though, there are exceptions.

Repellents may smell bad to you, too, at the time you apply them. You won't want to get them on your skin or inhale them. If they contain capsaicin, hot peppers, or garlic, it's important to protect your eyes while applying them. Wear goggles and a mask and gloves. Wash your hands thoroughly after applying.

COMMERCIAL REPELLENTS

Garden centers, nurseries, and mail-order suppliers offer a wide range of animal repellent products. Check the label to find one with the active ingredient of your choice. Avoid products that use synthetic substances as the repellent. When shopping for repellents, keep these tips in mind.

START SMALL. When you first try a repellent, buy the smallest container you can. This is more expensive on a per-unit basis but can be the least costly approach because it's impossible to predict the effectiveness of a particular repellent until you've tested it in your yard. After you find a product that works well, then invest in the larger container. And if you have success with more than one, buy both. Use one for a while, and then switch to the other.

DON'T DUPLICATE. Some companies package one repellent for deer, another for rabbits and woodchucks, and still another for pets. Ask a salesperson or check product labels to learn whether the products are significantly different. Common repellent ingredients such as putrefied egg solids, garlic juice, and capsaicin are repellent to all of these animal pests, so there may not be any significant differences in the formulations. If there isn't, then buy only one type.

CHECK THE SHELF LIFE. Some product labels state that one application will continue repelling pests up to 2 months. But in practice, most gardeners find that repellents work best when reapplied every 10 to 14 days and even more often when it is rainy or animal pressure is very high.

Check product expiration dates carefully. Some of the natural products have a limited shelf life and won't be as effective once they start to break down.

REPELLENTS FROM
THE HOUSE AND GARAGE

Items from around your house or garage sometimes work as deterrents, but limit your expectations. Sometimes the best use for these repellents is as a supplement to a garden fence. Spray the repellents around the perimeter of the fence or hang them from the fence wire or posts. This may be enough to keep animals from challenging the fence.

EGG SPRAY. Beat two eggs and add to 1 gallon water in a backpack sprayer (be sure no eggshell goes into the sprayer—it will block the spray hose). Mix well and spray on the ground around the perimeter of your garden.

CAYENNE PEPPER. Capsaicin extracted from hot peppers is the active ingredient in some commercial pest repellents, and cayenne pepper off the spice shelf may have the same power. Sprinkle pepper directly on foliage (but not on plant parts you plan to harvest).

STRONG-SMELLING SOAP. Use an ice pick to poke a hole through a bar of scented deodorant soap and stick a length of flexible wire through the hole. Hang the soap from a stake. Or collect odds and ends of leftover soap and put them in small bags made from cheesecloth, panty hose or old socks, or net bags from gro-cery store produce. Fasten the bags with twist ties or twine. Some gardeners even grate bars of soap directly onto the surface of garden beds.

OLD CLOTHES. Well-worn clothing and leather shoes supposedly carry enough human scent to scare away rabbits. Hang items on short stakes around the perimeter of your garden.

HAIR. Collect hair from a local barber or animal groomer. Spread it over the soil surface and work it lightly into the top inch of soil. Or put it in bags as you would soap.

BLOODMEAL. You may have a bag of bloodmeal in the garage that you use as fertilizer. It works as an animal repellent, too. Sprinkle it directly on foliage or put it in bags.

FENCING

Building a fence is the surest way to stop animals from damaging your garden. For rabbits, construct a 2-foot-tall fence of 1-inch chicken wire. Use metal or sturdy, rot-resistant wood stakes, and hammer them into the soil at least 12 inches deep. Use landscape pins to secure the bottom edge of the fence so the rabbits can't push underneath.

If woodchucks are potential raiders, increase fence height to 4 feet, but leave the top 18 inches floppy so that if the woodchuck climbs, its weight will pull the fence outward and over. Suspended in midair, the woodchuck will have no choice but to let go and drop back to the ground—outside your garden. Raccoons can climb taut vertical wire fences, but a floppy fence like this will probably stop them also.

See Deer to learn about fence designs to keep out deer.

UNDERGROUND BARRIERS

Rabbits and woodchucks sometimes tunnel under a fence, especially one that's put up after they've already sampled the treats growing in your vegetable patch. There are two design options to discourage burrowing: an underground barrier or an electric fence positioned a few inches outside the conventional fence.

Adding an underground feature to a fence requires a lot of digging, but it is effective. Some sources suggest extending wire 6 to 12 inches deep to keep out rabbits or woodchucks. Others say that the wire should curve or fold outward to create a three-dimensional barrier, as shown in the illustration below.

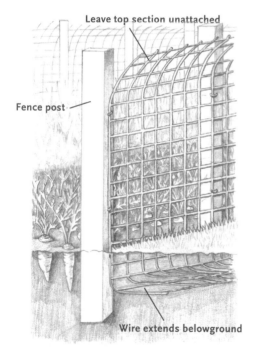

Extending a wire barrier belowground prevents digging animals from finding their way under the fence. The floppy top portion foils groundhogs and raccoons that try to scale the fencing.

ELECTRIC FENCES

Electric fences provide a psychological barrier, not a physical one. These fences scare animals by sending a pulse of electricity through the animal's body. The type of electricity is different than that in our household electric lines, so there's no danger of electrocution by an electric fence. Each pulse from the fence lasts only a fraction of a second, with enough time between for muscles to relax. Even if a raccoon, for example, gripped an electric fence wire in its paw, it would not become "frozen" to the wire. Instead, it would immediately pull its paw away and retreat in fear.

To learn how to set up electric fencing and attach it to a charger (which supplies the power to the fence), see Deer.

ELECTRIC FENCE DESIGNS. A fence about 2 feet high will keep out groundhogs, raccoons, and rabbits. The fencing is available as polypropylene netting with rods already attached to the netting (see the illustration on page 171). This type of fencing is easy to set up. For rabbits, the minimum voltage needed is 3,500 volts.

If raccoons are the sole animal pest troubling your garden, you can use one or two strands of electric wire (or polytape) rather than netting. String the wire on insulators attached to metal posts, setting one wire about 6 inches above ground level and the other at 18 inches.

IF YOUR FENCE FAILS. There are two common reasons why animals sometimes find their way through your fence into the garden. First, you turned off the fence so that you could work in the garden and forgot to turn it back on afterward. What's the solution for this?

If your regular garden fence isn't designed to keep out raccoons, set up a temporary electric fence connected to a battery-powered fence charger around your corn or melons as they approach ripeness. Two strands of polytape will do the trick.

Perhaps an old-fashioned string tied around your finger!

The second cause is that weed growth is interfering with the flow of current through the wires so the animals aren't getting much or any shock when they touch the fence. If that's the case, they'll easily worm their way under or through the flexible netting or between the wires. To prevent this, occasionally shut off the fence, move it aside, and use a mower or string trimmer to cut down weeds. (Remember to turn it back on.) Or, when you set up an electric net fence, lay flat stones or strips of heavy cardboard along the fence line to block weeds.

COMBINATION DESIGNS

Supplementing a chicken wire or woven wire barrier with electrified fence wire will keep out rabbits, woodchucks, raccoons, and some other pests, with no underground barrier required. One design calls for running two electric wires outside the nonelectric wire fence, a few inches away from the fence, as shown in the illustration on the opposite page. If rabbits or woodchucks try to push through or climb over the fence, they inevitably come in significant contact with one or both electric wires. This type of fence will keep out squirrels, too.

With this approach, it's unlikely that animals

will ever try digging, but if they do, it's certain that they will be zapped by the lower wire before their excavation hits pay dirt.

You can assemble a fence like this yourself or buy a kit that has all the components and instructions on how to assemble them. As with any electric fence, be sure to attach the wires properly to a charger and ground rod.

TO TRAP OR NOT TO TRAP

Bunny or woodchuck raiding the garden? One great solution would seem to be catching it in a humane trap and taking it off to the woods or open fields so it can live in nature and stop eating your plants. However, live trapping isn't that simple. Often, it's not one animal feeding in your garden; it's two or three, or eight or ten. If you expect to solve the problem by trapping, it may take weeks of repeated effort.

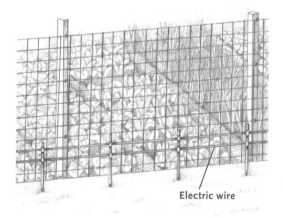

Electric wire

Electric wires set a few inches outside of a conventional garden fence will shock any small mammals that come close to investigate.

In some states, regulations limit the release of live-trapped nuisance animals to the property of the person who trapped them. So, it's legal to trap an animal only if you release it in your own yard. The only other option is to euthanize it.

To find out whether and where you can safely and legally release trapped animals in your area, contact your Cooperative Extension office, your state department of natural resources, or your state's USDA Office of Wildlife Services.

Be aware that even if you find an appropriate place to release the animal, you may be releasing it to a hard life or a quick death. Wild animals that have become accustomed to feeding in suburban areas often don't readapt to a wilder setting. Other resident animals may chase it out or kill it. If it does survive, chances are it will be because it finds some other yard or cultivated area to live in.

Finding solutions to these problems should be enough to discourage you. However, if you still decide to try trapping, educate yourself before you act. Find out which size trap you need. Learn how to bait traps and how to handle the traps when an animal is inside. Practice using the gate latch and release in advance so you won't fumble or panic once an animal is in the trap. Wear gloves and be cautious. Once a trap is open, the trapped animal will most likely run away from you as fast as possible. But while it's in the trap, it may try to bite or scratch you in self-defense if you get too close. If you live in an area where rabies among raccoons is a problem, be aware of the risks in handling a trap with a raccoon in it.

Hiring a professional wildlife contractor to set traps and remove the animals from your property is a safer, albeit more expensive, option.

For most gardeners, the only time a live trap comes in handy is when a pest animal finds its way inside a fenced garden and takes up residence inside. When that happens, set up the trap inside the garden. Lure the animal with the freshest and most alluring bait you can, setting a trail of food from outside the trap leading in. Once you've caught the animal, take the trap to the edge of your yard and release it. With luck, it will be so traumatized that it will head off to find a new home elsewhere.

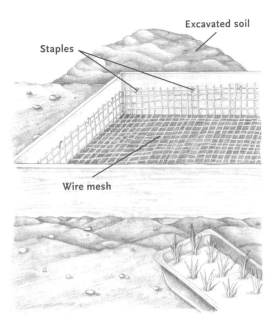

Line the bottom and lower sides of raised beds with wire mesh to foil hungry gophers.

MORE PESTS

ARMADILLOS

These unusual animals eat white grubs and other insects, earthworms, slugs, and carrion. They dig in the soil to find their food, and it's their digging that annoys gardeners in the southern states where these animals live. A standard mesh fence or electrified net fence should be effective for armadillos.

GOPHERS

With oversize heads and protruding teeth, gophers are well equipped to chew through soil and to decimate a tasty crop of carrots. Gophers also will eat other kinds of plant roots, seeds, and bulbs and sometimes devour entire plants— roots, stems, and leaves. Gophers can trouble gardens almost anywhere from the western United States east to the Great Plains and south into Texas and Oklahoma.

Gophers spend most of their lives in underground burrows, tunneling around in search of food. To protect crops from gophers, trench around your garden and install a barrier of $\frac{1}{4}$- or $\frac{1}{2}$-inch wire mesh fencing, or build raised beds with solid sides and line the bottom of the beds with wire mesh.

For regular garden beds, dig down 1 foot and line the bottom and sides of the bed with wire mesh. If you don't want to dig out an entire bed, trench around it 2 to 3 feet deep. Set fence posts along the trench and install a wire mesh fence (openings no larger than 1 inch; $\frac{1}{2}$ inch is better) to the bottom of the trench, extending up to 12 inches above soil level, too.

Special wire or box traps can be set in gopher burrows to kill them. Follow the instructions that come with the trap to position it and bait it properly. It may take time and repeated effort to lure gophers into traps.

GROUND SQUIRRELS

Ground squirrels are pests of the West and Midwest, and they have an appetite for leafy crops as well as seeds, grasshoppers, lizards, and sometimes other rodents. Their burrows are a nuisance around yards and gardens. They're most likely to trouble vegetable gardens in the spring. They can climb over or tunnel under most fences. A combination fence of wire mesh with an exterior electrified wire will keep ground squirrels out of gardens. Another design that may work is a 4-foot-high wire mesh fence with a 2-foot-wide skirt of smooth sheet metal at the top so the squirrels can't climb over. An underground skirt may help, but ground squirrels can burrow several feet deep.

Ground squirrels prefer a covered area to retreat to. Discourage them from bothering your garden by moving brush piles and wood piles at least 50 feet away from the garden.

Large-size snap traps (mousetraps) will work for trapping ground squirrels. Bait the traps with nuts or seeds and put them near the entrance to the squirrels' burrow.

HARES

Hares look like oversize rabbits, and they are related. Snowshoe hares and black-tailed and white-tailed jackrabbits are the most damaging to gardens. A sturdy fence of 3-inch wire mesh will keep out hares. It doesn't need an underground portion because hares don't dig. Repellents also may stop hares from feeding.

SKUNKS

A skunk's diet consists mainly of insects, including grubs and beetles, so they may frequent your vegetable garden. Occasionally, skunks will raid a corn patch. If you can, tolerate skunks around the garden, and stay out of their way. Fences that keep out rabbits and woodchucks usually are effective for skunks, too.

SNAKES

Snakes won't hurt your garden; in fact, they eat rodents, and some eat insect pests. So whether snakes are a pest depends on your attitude. If you're scared of snakes and don't want to encounter them in your garden, clear away brush, wood, or rock piles around your yard. Only about 1 percent of the snakes found in home gardens are harmful to people, but it's wise to find out whether poisonous snakes live in your area. If any do, be sure you know how to identify them. Encountering a poisonous snake in your yard *is* a problem, especially if you have children or dogs. Consult your local Cooperative Extension office for recommendations about dealing with the snake safely and appropriately (it may not be legal to kill the snake).

VOLES

Voles are secretive, hiding out of view in mole tunnels or under mulch. They eat a wide range of plants, and in vegetable gardens, voles nibble on plant roots or consume entire plants. They reproduce rapidly, bearing several small litters each year. Several different species can be garden pests, and each species has slightly different habits (such as how deeply they dig). If you can determine the particular species that is troubling your garden, you can tailor your control efforts to its habits.

Voles and moles look similar, but moles have slitted eyes, big padded feet, and a pointy snout.

Voles have open eyes and a short snout. If you catch a flash of movement from the corner of your eye as you work in your garden, it's likely a vole scurrying away from the disturbance.

Spilled birdseed attracts voles, so move your birdfeeders far from your vegetable garden. Some kinds of voles don't burrow, and you can discourage them by removing mulch from your garden. In this case, you'll have to balance whether the risk of weed problems or soil drying out is more harmful to your garden than the vole damage is.

To fence out voles, install metal flashing or ¼-inch mesh around the perimeter of your garden (you can position it flush with the inside or outside of a conventional garden fence). Use 2-foot-wide flashing or mesh, with up to 1 foot below soil level and 1 foot above. The barrier must be continuous. If you leave an opening for a gate, the voles could get through there.

Standard snap traps (mousetraps) baited with pieces of apple will work for catching voles. Place the trap close to the opening of a vole tunnel. Unfortunately, voles aren't very interested in bait when food is plentiful, so the best time for trap-

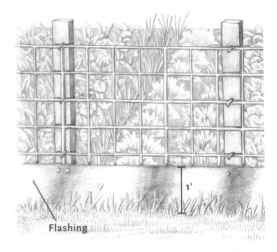

Flashing

A barrier of metal flashing added to your garden fence will make it vole-proof. Be sure the flashing extends 1 foot underground, too.

ping is in fall, but if you have a vole problem in spring or summer, you probably won't want to wait that long! If you have a serious infestation, your only choice may be to remove all mulch from your garden until you have time to put up a barrier. In fall, also set traps to reduce the resident population.

PETS

Defending a vegetable garden against hungry wildlife is tough enough, so we may reach the breaking point when we discover dogs and cats (especially our own) digging in the garden. In many cases, a garden fence will keep out dogs and cats, too. Or try some of these tricks to prevent pets from bothering your garden.

TEMPORARY BARRIERS

Loose soil is often what attracts dogs and cats to vegetable gardens. Once plants are well

MUCH ADO ABOUT MOLES

Homeowners complain about moles in their yards, primarily because of the raised tunnels that spoil a stretch of green lawn. Many gardeners also assume that moles feed on plants, but that's not true. Moles feed on insects (including many plant pests) and earthworms. Most likely, moles will be more interested in the grubs under your lawn than your garden.

established, the garden will be less attractive. Row covers or a screen cage like the one on page 64 will protect seedbeds from digging pets. Or simply lay chicken wire on the ground around the freshly worked soil. You can walk right across the chicken wire, but your pets will avoid it, not liking the feeling of the wire under their feet.

REPELLENTS

Try commercial repellents to keep dogs and cats out of gardens. Some home remedies may work, too:

- Sprinkle bloodmeal on the soil surface to repel cats.
- Spread orange and grapefruit rinds around the garden—they're reputed to repel cats.
- Spread crushed eggshells over freshly worked soil; the animals won't want to walk on the crunchy shells.
- Stick plastic forks into the soil with the tines protruding above the soil surface.
- Buy mats with protruding plastic teeth (available at garden centers) and spread them around the garden.

ANTHRACNOSE

Anthracnose shows up on a range of crops from beans to tomatoes, starting out as spots on leaves or stems. This fungal disease usually isn't severe in home gardens. However, if the fungus attacks fruits or pods, the infected areas are easy targets for soft rot pathogens.

Anthracnose spores are spread in wind-driven or splashing rain, by insects, or by people working among wet plants; the spores can enter through wounds or penetrate the leaf surface during wet conditions.

RANGE: Eastern and central United States; southeastern Canada; rare in arid climates

CROPS AT RISK: Beans, tomatoes, potatoes, peppers, cucumbers, melons, watermelon, gourds, and some other crops; rarely a problem on peas, squash, or pumpkins

DESCRIPTION: Dark spots or lesions, often sunken and wet-looking, are characteristic of anthracnose, followed by formation of pinkish or salmon-colored spores at the center of the lesions. Bean plants develop dark red or black spots first on the underside of the leaves, but these often go unnoticed. The elongated dark brown spots later show up on upper surfaces of leaves and on pods. Lesions on pods are more prominent and sometimes filled with pink masses of spores. Inside the pods, the seeds may show dark spots, too.

In melons, cucumbers, and watermelons, yellow water-soaked spots form. These spots turn black on watermelon leaves, brown on leaves of other vine crops. These spots may tear out of the leaves or spread so completely that the leaves die and fall off. Stems may develop long lesions, too, with a pinkish jelly present during wet weather. When stems become infected, the vines may lose all their leaves. Infected fruits will develop dark sunken areas and salmon-colored spores. Infected fruits may be flavorless or bitter.

On tomatoes, fruits develop sunken spots with concentric rings; spores may form at the center of the spots. Secondary rots may set in and ruin the fruit.

Seedlings that sprout from seeds infected with anthracnose may show symptoms similar to damping-off (see Damping-Off).

FIGHTING INFECTION: Remove and compost infected plant parts or plants. Stop using overhead sprinklers to water the garden. Spraying with sulfur will prevent the spread of anthracnose, but since the disease spreads mainly during wet weather, it's difficult to apply the spray effectively. Check tomato plants daily and harvest fruits as soon as they ripen. Tomato and pepper fruits can develop symptoms after harvest due to infection by spores on the surface of the fruit at harvest. To prevent this, wash and dry all healthy-looking fruits as soon as you bring them inside. Keep infected fruit separate from healthy-looking fruit.

GARDEN CLEANUP: Anthracnose fungi survive winter on seeds, weeds, and crop debris. When harvest is complete, compost, dig in, or discard all crop remains.

NEXT TIME YOU PLANT: Choose resistant varieties when possible; for beans, there are different races of anthracnose, so even resistant varieties may become infected. Plant clean seed. Watch for volunteer seedlings that sprout from infected seed (left in the soil when tomatoes or melons rotted in the garden the previous year). Uproot and discard them as soon as you spot them. Stake tomatoes and peppers to improve air circulation. Try preventive sprays of *Bacillus subtilis*.

CROP ROTATION: Anthracnose fungi can't survive very well in soil, and different species cause infection in different plant families. A simple 2-year rotation of crop families should break the disease cycle in your soil. However, new infections can arise from windblown spores (perhaps from a neighbor's garden), so don't be disappointed if it seems as though your crop rotation isn't "working." The disease would probably be more severe if you had planted muskmelons, for example, in the same bed year after year.

APHIDS

In early spring, aphid eggs hatch on weeds or perennial plants around your yard. The aphids that hatch immediately begin feeding and giving birth to live nymphs. Adult aphids are usually wingless, and they feed in clusters on the undersides of leaves or at the young growing tips of plants. A single aphid can produce dozens of "babies" per week. The nymphs start feeding right away and molt three times before reaching the adult phase, when they, too, produce live young.

When aphids sense that they're becoming too crowded or depleting their food source, they switch to birthing live, winged offspring. These aphids aren't strong fliers, but if they reach your garden, they'll settle on young crop plants and begin producing a new colony of wingless nymphs and adults.

In mild climates, the phenomenon of live aphid birth continues year-round. Each generation lives for about 1 month, and there can be more than 15 generations per year. In areas with hot summers, aphid activity will slack off when daily temperatures are in the 80s, but during cool summers, aphids may remain active all season. In places like the desert Southwest, aphids may be most troublesome on winter crops such as cabbage.

In fall in cold climates, winged aphids appear again, and some of them will be males. The males mate with females, and the mated winged females lay eggs on host plants.

There are thousands of species of aphids, and virtually all crops are susceptible to at least one species. Common aphids in vegetable gardens include the pea aphid, green peach aphid, cabbage aphid, and melon aphid.

Predatory insects such as aphid midges, lady beetles, and lady beetle larvae seek out aphid colonies on plants and feed on them. Parasitic wasps lay eggs on the aphids, and the wasp larvae tunnel into the aphids and consume them from inside. In humid conditions, fungi also may attack aphids.

Often these natural biological controls are very efficient. Look for signs of natural biocontrol such as aphid mummies (victims of parasitic wasps) or bloated, off-color aphids (killed by fungi).

PEST PROFILE

Many genera and species

RANGE: Throughout North America

HOW TO SPOT THEM: Check the growing tip of plants such as broccoli and the young shoots of bushy plants like peas for what look like clusters of colored dots. Or you may notice ants crawling around on some crops. Ants are attracted to the sticky honeydew that aphids produce as they feed; watch where the ants go and you'll find the aphids.

ADULTS: $\frac{1}{16}$- to $\frac{1}{8}$-inch-long, pear-shaped insects with long slender mouth parts, long legs, and long antennae. Two tubelike structures (called cornicles) stick out from the back end of their bodies and are unique to aphids. Color including gray, black, green, yellowish, or pink, varies by species. Some have a waxy coating. Most adults are wingless, but winged adults appear under certain conditions. Wingless aphids move little, even when disturbed.

LARVAE: Nymphs are similar to wingless adults but may vary in color.

EGGS: Laid on a wide range of overwintering host plants

CROPS AT RISK: Most vegetable crops

DAMAGE: Aphids suck plant fluids and excrete a sticky, saplike substance called honeydew. As leaves become coated with honeydew, a fungus may grow in it, producing a sooty mold that interferes with photosynthesis. Low populations of aphids usually don't harm the yield or quality of the harvest. But heavy feeding causes leaves to turn yellow, curl under, and become distorted. Some species also spread viruses as they feed.

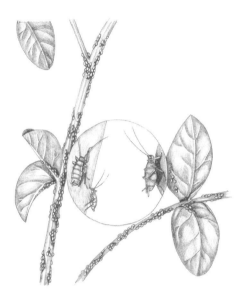

Clusters of feeding aphids

CONTROL METHODS

ENCOURAGE BENEFICIALS. Plant flowers that bloom in early spring to draw beneficial lady beetles and lacewings to your garden before aphids even begin feeding. The beneficial insects will feed on other kinds of insects but then will move to your crop plants if aphid populations are building up there. Plan to keep something flowering in or near your vegetable garden all season. See Beneficial Insects for a list of flowers that attract beneficials.

LURE BENEFICIALS WITH SUGAR. Another way to encourage beneficial insects to explore your crop plants is to spray the plants with a lightly sweet solution of sugar water (1 part sugar to 10 parts water).

COVER YOUNG PLANTS. If aphid-transmitted viruses are a concern, cover the crop with row

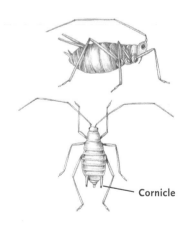

Cornicle

Aphid

cover at planting time, sealing the edges carefully. Remove the cover when temperatures warm up enough to reduce aphid pressure or when the crop needs pollination by insects. To learn how to seal row cover edges, consult Row Covers.

USE REFLECTIVE MULCH. A shiny mulch on the ground beneath plants may repel aphids. See Mulch.

WASH THEM OFF. If aphid populations seem to be increasing, use a strong stream of water to knock aphids off plants. Repeat the spray every 3 days for about 2 weeks. By then, beneficials should have arrived and will control the infestation.

TRY REPELLENTS. Applying a garlic or hot pepper spray may deter aphids. See page 308 for directions.

HOLD BACK ON NITROGEN. If aphids have been a regular problem in your garden, hold back on high-nitrogen materials when you enrich the soil. Too much nitrogen spurs lush foliage growth, and that's what aphids like best.

SPRAY AS A LAST RESORT. In an emergency, spray plants with insecticidal soap, oil spray, neem, or pyrethrins to reduce aphid numbers fast. If plants are severely infested, more than one spray will probably be needed because it's difficult to get sprays into sheltered areas in curled and distorted leaves.

CONTROL CALENDAR

BEFORE PLANTING

- Plant perennial and annual flowers to attract beneficial insects that prey on aphids.

AT PLANTING TIME

- Cover seed rows or transplants with floating row covers.
- Put reflective mulch under plants.

WHILE CROP DEVELOPS

- Spray infested plants with water twice a week to kill aphids and knock them off plants.
- Spray plants with garlic or hot pepper spray.
- If a serious problem develops, spray insecticidal soap, oil, neem, or pyrethrum.

AFTER HARVEST

- Pull weeds that may harbor eggs of aphid species that attack vegetable crops.

ARMYWORMS

Under cover of darkness in the spring, armyworm moths fly north from Central America and some southern states. When they reach gardens or farm fields, the moths lay eggs on plants. A few days later, the eggs hatch and the larvae begin to chew on plant leaves. When their numbers are high, armyworms can reduce foliage to skeletons or shreds. Fall armyworms and beet armyworms are the most common species that appear in home gardens.

Corn is a favorite host crop for fall armyworms, but they'll also eat foliage of other vegetables. Fall armyworms tunnel into corn ears, too, leaving a messy, dark brown opening (unlike European corn borers, which tunnel cleanly). The larvae feed for a few weeks before dropping to the soil to pupate.

Beet armyworms eat leaves, stems, and sometimes roots of a wide range of crops. They will bore into lettuce heads and tomato fruits. These armyworms live 3 to 4 weeks and then pupate. The next generation of adults emerges about a week later, ready to fly off in search of new host plants.

Fall armyworms overwinter as pupae in the soil, but they're hardy enough to survive winter only in the Gulf states (and points south). The strong-flying moths surge northward with each generation, though, so by midsummer fall armyworms are invading gardens as far north as New England. Northern gardeners have to contend with only one or two generations of armyworms, while southern gardeners must fight from three to six.

Beet armyworms are native to Southeast Asia, but they've adapted to the United States because they can overwinter in Arizona, Florida, and Texas. Beet armyworms also produce four genera-tions per year, moving northward to infiltrate the southern half of the country from coast to coast.

PEST PROFILE

Several genera and species

RANGE: Fall armyworm in most states east of the Rockies; beet armyworm in the southern half of the United States

HOW TO SPOT THEM: Search for cottony egg masses on leaf undersides or near plant tips; look for green or dark caterpillars on leaves, especially in the whorl of developing corn plants. Beet armyworms may spin fine webs around themselves.

ADULTS: Mottled gray moths with wingspan up to 1½ inches; fall armyworm moths have clear hind wings; beet armyworm moths have a pale yellow spot on the forewing.

Fall armyworm moth

Armyworm egg mass

Beet armyworm larva

Fall armyworm damage

LARVAE: Fall armyworms vary in color from tan to green to dark brown, depending on their stage of maturity; sometimes light-colored lateral stripes show on their backs; they have dark heads marked by a light upside-down Y mark.

Beet armyworms are smooth-skinned, light green caterpillars with narrow white lateral stripes; their appearance also changes as they grow and molt.

EGGS: Fall armyworm egg masses have a fuzzy gray covering; beet armyworm egg masses are covered with white hairs, so they look like fuzz or bits of cotton on leaf undersides.

CROPS AT RISK: Corn and many other vegetables

DAMAGE: Large, ragged holes in leaves; skeletonized leaves; larvae also bore into crowns, roots near soil surface, tomato fruits, corn ears.

CONTROL METHODS

TIMED PLANTING. Plant corn as early as you can and choose early-maturing varieties. If you're lucky, the crop will be harvested or close to maturity when armyworms arrive in your area.

FALL CULTIVATION. If you live in an area where armyworm pupae can overwinter, cultivate after harvest to expose the pupae to predators.

INSECT TRAPS. Early in the season, set up two commercial insect traps that use floral lures to attract the moths to the traps. This will not stop an infestation completely but can reduce the number of eggs laid in your garden.

HANDPICKING. Wipe egg masses off plants and crush them. Search for caterpillars in the whorl of corn plants or at the center of lettuce, spinach, and other rosette-forming crops.

KAOLIN. If you expect armyworms to show up, spray target crops with kaolin clay to discourage moths from laying eggs. See Pest Control for information on kaolin.

BIOCONTROL PRODUCTS. *Bacillus thuringiensis* (Bt) var. *kurstaki*, spinosad, and *Beauveria bassiana* are effective against armyworms. However, it's important to catch the infestation early, before the worms have started burrowing into buds, crowns, and ears of corn. See Pest Control for directions on use.

NEEM. As a last resort, if a large infestation of caterpillars is present, spray neem. Keep in mind that neem won't kill insects that have burrowed into plants unless they consume some of the spray.

CONTROL CALENDAR

BEFORE PLANTING

- Set up commercial insect traps.
- Research early-maturing corn varieties that will grow well in your area.

AT PLANTING TIME

- Plant as early as possible.

WHILE CROP DEVELOPS

- Spray plants with kaolin clay before pests appear.
- Handpick armyworms as you spot them.
- Apply *Bt* var. *kurstaki*, spinosad, or *B. bassiana* as soon as you spot larvae on plants.
- As a last resort, spray neem.

AFTER HARVEST

- In areas where armyworm pupae can overwinter, cultivate soil to expose pupae.

ARTICHOKES

Globe artichokes grow as perennials in pockets along the West Coast that offer the combination of mild summers and mild winters. Gardeners elsewhere in Zone 7 and north can succeed with artichokes by growing them as annuals. Artichokes generally are hard to grow in areas with very hot weather, such as Florida and Arizona.

Perennial artichokes will grow well in a vegetable garden, but they need lots of space—plants will spread 4 feet tall and wide (annuals won't get quite as big). Artichokes blend beautifully into an ornamental perennial garden; their foliage creates the impression of silvery fern fronds.

CROP BASICS

FAMILY: Lettuce family; relatives are chicory, endive, Jerusalem artichokes, lettuce, and radicchio.

SITE: Give artichokes a location in full sun. The plants benefit from light shading during periods of high heat.

SOIL: Artichokes need excellent drainage and fertile conditions. They will not do well in heavy or slow-draining soil.

TIMING OF PLANTING: Sow seeds for an annual crop indoors 2 months before your last expected spring frost; for perennial production, plant in fall or winter.

PLANTING METHOD: Start seeds indoors for an annual crop; for perennials, you can start

Harvest artichoke flower buds while the waxy petals are still tightly closed. Cut about 3 inches of stem along with the bud; the tender stem is good for eating, too.

temperature around the container from dropping below about 25°F.

HARVESTING CUES: With annual artichokes, plants usually begin to produce flower buds in early to midsummer. Cutting buds stimulates the plants to produce more. Perennials send up buds in spring.

SECRETS OF SUCCESS

PROVIDE PERENNIALS WITH A DORMANT SPELL. Perennial artichokes need a dormant period each year to trigger a new round of flowering the following year. In some areas, this dormant period is during the cool wintertime, but in warm Zones 9 and 10, cut back the plants to ground level after the spring harvest ends and the foliage dies back. Don't water or feed the plants while they're resting. After several weeks, mulch around plants with compost to provide a nutrient boost, and start watering. Foliage will resprout and the plants will produce flower buds again in fall. In general, thornless varieties are the most heat-resistant.

CHILL ANNUAL ARTICHOKES TO PROMOTE FLOWERING. Since annual artichokes don't receive a dormant period to promote flowering, gardeners have to trigger the flowering response by exposing the plants to a chilling period instead.

Temperatures between 60° and 70°F will encourage seedlings started indoors to grow vigorously. About 2 weeks before the last frost date, move the young plants to a coldframe for the chilling treatment. The plants need 8 to 10 days' worth of temperatures below 50°F. Plan to leave the plants in the coldframe for 2 weeks to ensure

plants from seed or use bare-root divisions or potted plants.

CRUCIAL CARE: For annual production, be sure plants are protected from cold once you plant them out in the garden. Don't rush the plants into cold soil and don't crowd them; set plants about 30 inches apart down the center of a 3-foot-wide bed. For perennial artichokes, be sure the planting site is free of perennial weeds and that the soil is well enriched with organic matter before planting.

GROWING IN CONTAINERS: Choose a durable container at least 12 inches deep. Mix potting soil and rich compost in equal proportions to provide long-lasting fertility. At the end of the growing season, move the container to an unheated garage or shed to overwinter, and insulate it as needed; your goal is to prevent the

ARTICHOKES 31

sufficient chilling (in case the temperature occasionally exceeds 50°F). The plants won't seem to grow during this period, and that's okay. Open the coldframe as needed during the day to prevent temperatures from climbing too much, and close it at night (cover it with a blanket, too, if needed, to be sure that the plants don't suffer frost damage). Watch the weather forecast for predictions of an extreme cold spell during this chilling period. If necessary, bring the plants back indoors, and keep them as cool as possible while they're inside. Resume the chilling treatment when the spell of dangerously low temperatures ends.

It's important to choose a variety that will respond to this chilling treatment and produce flower buds in time. 'Imperial Star' is a variety especially developed for growing as an annual.

After the chilling treatment, it's important to help the plants resume strong growth right away. Plant the artichokes into warm (60°F) soil and cover them with cloches or row covers if the weather is cold.

PREVENTING PROBLEMS

AT PLANTING TIME

- Put out slug and snail traps (see page 369).

WHILE CROP DEVELOPS

- Watch carefully for signs of aphids and wash them off plants immediately to prevent the spread of viral diseases.
- If earwigs are sometimes a problem in your garden, put out earwig traps before plants start to form flower buds (see Earwigs).

- If you're growing annual artichokes, mulch the soil if needed to prevent it from overheating.

AFTER HARVEST

- Bury cut foliage of perennial artichokes if it is infested with plume moth larvae.

TROUBLESHOOTING PROBLEMS

HOLES IN STEMS AND LEAVES. The larvae of artichoke plume moths feed on foliage, stems, and flower buds. The yellowish brown moths look like an old-fashioned biplane, with wings that jut out at right angles from the slim body. Moths lay eggs on plants, and the yellowish caterpillars eat tender emerging foliage, then tunnel into the plants. Spraying *Bacillus thuringiensis* (Bt) or a solution of *Steinernema* nematodes into the growing tips before the larvae tunnel into the plants can be effective. Cut off and destroy infested buds. When cleaning up foliage from damaged perennial plants before the summer dormant period, chop it well and cover it with 6 inches of soil or more so that new adults cannot emerge.

Artichoke plume moth

CURLED LEAVES. If curly dwarf virus is the cause, plants also will become stunted and mis-shapen. Remove and destroy infected plants. Replant using roots that are virus-free. Control aphids, which spread the virus as they feed.

LARGE HOLES IN LEAVES. Snails and slugs like to feed on artichoke foliage at night. There are many ways to control them; see Slugs and Snails.

PLANTS WILT SUDDENLY. Bacterial wilt and Verticillium wilt are possible causes. Remove and destroy the infected plants. Replant, but be sure the site is well drained and add organic matter to build soil fertility. Avoid overwatering, but don't let plants dry out too much. Space plants widely to allow adequate air movement even when they reach mature size.

PERENNIAL ARTICHOKE PLANTS DIE OUT. Globe artichokes aren't long-lived perennials. Generally, the plants may lose vigor and die away after about 5 years of production, even if they've been well cared for. Don't try to revive the plants. Instead, prepare a new planting area and plant new seedlings or root divisions.

AN ARTICHOKE ALTERNATE

Cardoon is a cousin to the globe artichoke. It has edible stems with an artichoke flavor. Cardoon tends to be problem-free, and there's no worry about a chilling requirement to promote flower bud formation. The secret to success with cardoon is knowing how to cook the stalks to bring out the best flavor.

Cardoon plants resemble globe artichoke plants, but the stems are covered with a waxy bloom that has a bitter taste. Use a two-step cooking process to vanquish the bitter coating.

After cutting the stems, slice off the winglike lateral edges of the stems, and also remove the leaves. Scrub the stems well with a vegetable brush—this removes some of the bloom. Cut the stems into 2-inch pieces and boil them in salted water for a full 20 minutes. Pour off the cooking water and drain well. Then bake the stems in a pan with Parmesan cheese or Italian seasonings for a delicious side dish with a Mediterranean flair.

Cardoon is a perennial hardy to Zone 7, and it grows well as an annual in colder areas. A cardoon plant, like an artichoke plant, is a handsome choice for a decorative container. As a garden perennial, cardoon will grow up to 8 feet tall—it won't reach such an imposing height in a container or as an annual.

Harvest takes place in early fall before frosts kill the tops. In mild-winter areas, harvest continues into winter and even spring.

Cardoon, a terrific artichoke alternative

BROWN SPOTS ON BUDS. If the spots are brown and develop a white or gray mold when the weather is wet, the problem is gray mold, a fungal disease. It spreads rapidly and will ruin the buds. If a spell of wet, cool weather is predicted, harvest as many uninfected buds as you can. For controls, see Gray Mold.

BLACK COATING ON FLOWER BUDS. Aphids suck sap from artichoke plants, and the honeydew the aphids produce allows mold to grow. Plants should still produce flower buds, but the aphids may infiltrate the buds, too, and they're hard to dislodge. While plants are in the garden, use a strong spray of water to knock aphids off plants. In the kitchen, soak harvested buds in a bowl of water with 2 tablespoons vinegar added to kill the aphids. They should float to the top of the water and you can pour off the water. See Aphids for other control measures.

HOLES IN FLOWER BUDS. Earwigs will bore into the base of the buds, which can ruin the quality of the bracts. Soak injured chokes in salt water before cooking to drive out the earwigs; cut away damaged portions before eating. To prevent earwig problems in the future, see Earwigs.

ASPARAGUS

This perennial crop is a spring taste treat. If you already have a healthy asparagus patch in your yard, consider yourself lucky. It will probably continue producing for many years and require little care. If the stand is unhealthy, though, don't struggle to revive it. Instead, dig out the old crowns and start fresh in a different location with young, healthy plants.

Ask your Cooperative Extension office or local organic vegetable growers to recommend varieties that will grow well in your area. Most gardeners choose varieties that produce only male plants (asparagus plants can be male or female; male plants yield better). Be sure to ask about disease resistance, too; many varieties are resistant to serious asparagus diseases such as rust and Fusarium wilt.

CROP BASICS

FAMILY: Onion family; relatives are garlic, onions, scallions, and shallots.

SITE: Asparagus does best in full sun, but it can tolerate light shade. If possible, choose a site sheltered from strong wind.

SOIL: Provide well-drained soil, with a pH near neutral, and enrich it well with organic matter and loosen it at least 1 foot deep (see "Start with Great Soil" on the opposite page).

TIMING OF PLANTING: Plant asparagus in spring as soon as the soil is dry enough to dig 6-inch-deep furrows.

PLANTING METHOD: Starting asparagus from seed is possible, but most home gardeners buy 1-year-old dormant crowns.

CRUCIAL CARE: Watering and weeding are essential while the stand becomes established. Never let perennial weeds become established in an asparagus patch.

GROWING IN CONTAINERS: Container culture isn't practical for asparagus.

HARVESTING CUES: Check your patch daily as soon as the young spears poke through the soil. At first, you may find harvestable spears (6 to 8 inches tall) only once every few days, but as the soil warms, the spears will grow faster. If a hot spell occurs, check the plot twice a day. Snap or cut the spears at soil level.

SECRETS OF SUCCESS

START WITH GREAT SOIL. Asparagus crowns can produce for 20 years or longer, but only when they start strong in the right environment. Give asparagus a home in the best soil your yard has to offer. Chances are that's in your vegetable garden, where you've been tending and building the soil for years. (If you're reluctant to give up space in your vegetable garden for asparagus, then expand your garden and plant beans or squash in the newly worked area—they can handle less-than-ideal soil.)

Site asparagus at the north or west side of your garden to minimize the shade it will cast on other crops. Start by clearing weeds from the bed. Dig a furrow 6 inches deep and at least 1 foot wide down the center of the bed (the bed should be 3 to 4 feet wide). Use a spading fork to check that the soil in the furrow is loose to a depth of 1 foot below the soil surface. Add a 1-inch layer of compost to the furrow along with 1 pound of bonemeal per 20 square feet of bed area. Mix the amendments into the soil at the bottom of the furrow. Also add compost to the soil you've removed from the trench.

Sometimes it's just not practical to plant asparagus in your existing garden. In that case, prepare the planting site the year before you plant the crowns. Remove sod or other plant cover from the area. If the soil is at all compacted, dig it out at least 1 foot deep, and add organic matter as you return the soil to the trench. Test the pH and adjust it if needed to between 6.5 and 7.5. Plant a dense cover crop, such as buckwheat, to prevent weeds from invading. Turn the crop under in

TRY THE MOTHERSTALK METHOD

In the northern half of the United States and in Canada, asparagus season ends all too soon. To enjoy a longer harvest, try the motherstalk method. This cultural technique is commonly used in Taiwan, where asparagus can be harvested year-round in the tropical climate.

At the beginning of the season, harvest spears for about 2 weeks. Then allow two or three spears from a few crowns to grow and produce ferny fronds. This will slow the production of additional spears from those crowns for a few weeks, but don't worry. After the small decline, spear production will pick up, fueled by the extra energy provided by the photosynthesizing motherstalks. Keep harvesting spears from these crowns for 10 weeks.

If this method works well for you, you can expand its use to your entire asparagus patch.

early fall so it has time to break down. Cover the bed for winter with straw or other protective mulch. Remove the mulch early in spring to allow the soil to dry out. Then proceed as described on page 35 to prepare a planting trench.

PLANT CROWNS WITH CARE. With your planting furrow prepared, soak the crowns in compost tea (see Composting for information) for 20 minutes. Then place the crowns in the furrow 18 to 24 inches apart, gently spreading the roots in all directions. Cover the crowns with 2 inches of soil and water well. After the young shoots emerge, keep adding more soil every week or two until you've filled the furrow. Continue to water when conditions are dry during the first year of growth. Mulch with a ½-inch layer of pine needles, straw, or well-rotted leaves to conserve moisture.

PHASE IN YOUR HARVEST. Don't cut any spears from your asparagus plot in the year you plant it. The following year, harvest for about 2 weeks, then let the plants grow undisturbed. The year after that, extend the harvest to 4 weeks. By the third year, you can enjoy a full harvest season up to 8 weeks long (or 12 weeks in warm areas of the West). Stop harvesting when spear thickness starts to decline and spears are thinner than a pencil.

STAY AHEAD OF WEEDS. A young asparagus planting can lose the battle with weeds if you don't help out. In the first year after planting, weed carefully by hand until the fronds are mature enough to shade the bed. In late winter, after cutting down last year's fronds, hand-weed or cultivate shallowly to remove winter annuals. You can also try flame weeding at this time of year. Flame weeding (see page 438) is most effective against annual weeds, not perennials.

Over time, asparagus crowns push their way up close to the soil surface. Don't try to dig them out or you'll damage the asparagus roots and possibly the crowns. Instead, cut off the perennial weeds at soil level repeatedly, about every 2 weeks.

REGIONAL NOTES

FLORIDA AND DEEP SOUTH

Asparagus is difficult to grow in the hot, humid conditions of Florida and the Gulf Coast area. Ten years may be a long life for asparagus in the South.

If you want to try growing asparagus, plant it as described under "Plant Crowns with Care" above. Don't harvest the first year. The ferns will keep growing and won't die. In December or January, cut all ferns down to ground level. If conditions are cool or dry enough, the crowns will produce new spears. Cut them for about 2 weeks, and then allow the plants to grow. Repeat the following January. Each year you can extend the harvest by a week or so.

You can try cutting down the ferns in August as well and allowing regrowth. The advantage to this method is that it reduces the threat of disease (fronds left to grow for an entire year are more likely to develop disease problems).

PREVENTING PROBLEMS

BEFORE PLANTING

• Prepare and improve soil thoroughly and remove weeds.

WHILE CROP DEVELOPS

- Cover your asparagus bed with row covers (see Row Covers) and seal the edges to keep out asparagus beetles.

AT HARVEST TIME

- Handpick and destroy any asparagus beetles or larvae that you find on spears.
- If you use a knife to harvest, don't cut below soil level or you may damage the crowns.

AFTER HARVEST

- If fronds show any signs of disease, remove them promptly at the end of the growing season and discard or destroy them.

TROUBLESHOOTING PROBLEMS

SPEARS CUT OFF OR DAMAGED. Cutworms and armyworms feed on spears near the soil line, sometimes before they even emerge from the soil. To control these pests, consult Cutworms and Armyworms.

CHEWED AREAS ON SPEARS. Slugs and snails and two species of small beetles chew on young spears and also on mature asparagus fronds. Tips of damaged spears sometimes curl

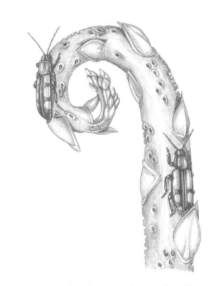

Asparagus beetles on a damaged stalk

Spotted asparagus beetle

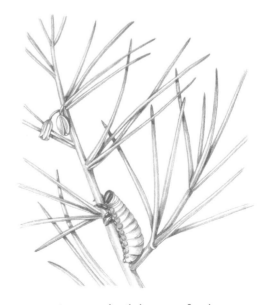

Asparagus beetle larva on a frond

over. The asparagus beetle (*Crioceris asparagi*) is metallic blue or dark green with white spots and reddish or yellowing markings on its back. Spotted asparagus beetles (*C. duodecimpunctata*) are orange with black spots. The larvae of both species are small gray grubs with black heads and legs. Asparagus beetles range throughout the United States and Canada. For slug and snail controls, see Slugs and Snails.

Spotted asparagus beetle larvae feed only on asparagus berries, but larvae of asparagus beetles feed on young spears and mature fronds. A healthy stand can withstand some damage. Damaged spears are still edible, although sometimes not attractive looking. Tips of damaged spears sometimes curl over.

Handpick beetles and larvae in spring. Handpick beetles in the morning, when they are less active; knock them into a container of soapy water. If an infestation threatens to devastate your stand, spray with insecticidal soap or pyrethrins as a last resort. To rid your stand of asparagus beetles, cut down the fronds as soon as they turn yellow in fall. Remove all debris from the patch and burn it or put it through a shredder and then compost it. Wait until late fall to mulch the stand; otherwise the beetles will take cover under the mulch.

To prevent asparagus beetles from finding a new stand of asparagus, cover the bed with floating row covers before spears begin to emerge. Pull back the cover daily to harvest; if beetles have penetrated the cover, then leave it off and start handpicking them.

WEAK, BROWN SPEARS. Fusarium wilt is probably to blame. Infected spears aren't edible. This soilborne fungus can't be eradicated from an asparagus bed. It usually doesn't reach serious levels in beds that are well drained and cared for. If Fusarium is ruining your harvest, your best course of action is to start a new bed in a different location. Be sure the bed is well drained, and choose a Fusarium-resistant variety.

BROWN OR GRAY FUZZY GROWTH ON SPEARS. Gray mold is a soilborne fungus that can infect spears in cool, wet weather. In wet weather, rake back mulch from plants to encourage the stand to dry more quickly. If gray mold is a continual problem in your stand, it could be related to poor drainage. Consider moving the crowns to a location with better drainage, and space the crowns farther apart.

TIPS OF SPEARS TURN PURPLE AND CURL. Frosty weather and cold winds can damage emerging spear tips. When the weather improves, normal spears will form.

FERNS LOSE "NEEDLES" EARLY. Rust or needle blight are probable causes. Check the fronds closely. If you find orange spots on the needles, the problem is asparagus rust. It's worst in wet weather. For controls, see Rust.

Needle blight also tends to occur during wet conditions. Gray spots with a purple margin appear on the ferns. In a severe case, ferns will lose all or nearly all of their needles, and the harvest the following year may be very small. Cut down infected ferns and discard or destroy them.

ASTER YELLOWS

Yellowed, twisted foliage is one sign of aster yellows, a disease caused by a phytoplasma, a single-celled organism that acts similarly to a bacterium. Leafhoppers pick up the phytoplasma as they feed on infected plants. After incubating 1 to 3 weeks in leafhopper salivary glands, the phytoplasma can be transmitted into new plants. Symptoms usually develop several weeks after transmission. Cool, wet weather favors the disease.

Many weeds are hosts for aster yellows, so weed control is important in minimizing problems.

RANGE: United States; southern Canada

CROPS AT RISK: A wide range of vegetables, but most serious in carrot-family crops and lettuce. Aster yellows also infects a wide range of annual and perennial flowers.

DESCRIPTION: Symptoms vary depending on the crop attacked, but stunting and leaf distortion are common. Young carrot leaves turn yellow and become twisted. Older leaves become reddish or bronzed and break off. New growth is pale and bushy, forming a witch's broom. Roots are distorted and covered with hairs. Celery stalks are yellowed and twisted; center growth may crack, leaving it susceptible to invasion by rot diseases.

In lettuce the disease is also called white heart because it causes the center of the heads to become very pale. Outer leaves don't develop fully, and plants become stunted. Infected onion plants develop yellow leaves. Bulbs may develop normally but won't keep well in storage.

In potatoes this disease is called purple top.

FIGHTING INFECTION: Pull up infected plants and compost them. If carrot roots are mature, sample them to see how they taste. Usually, infected roots are woody and bitter—you won't want to eat them.

GARDEN CLEANUP: The virus overwinters inside leafhoppers or in dormant perennial plants. Unfortunately, asters and many other garden perennials are host plants. Since you won't want to remove those plants from your garden, tackle this disease by protecting your crops from leafhoppers rather than by trying to clear your yard of potential host plants.

NEXT TIME YOU PLANT: Plant resistant or tolerant varieties; several resistant carrot varieties are available. Cover plants with row covers to prevent leafhoppers from reaching the plants.

CROP ROTATION: Not effective

BEANS

At their best, beans are an undemanding crop, producing more pods than you can eat. They generally do well in home gardens, but at their occasional worst, beans deteriorate into a mass of chewed or moldy foliage that's infested with Mexican bean beetles or ruined by diseases.

If you're raising beans for the first time or if you've had bean crops fail in the past, choose bush beans rather than pole beans. Bush beans produce pods quickly, increasing your chances of reaping a harvest before problems, should any arise, become severe. Sow small stands—perhaps a 6-foot row—of beans every couple of weeks to prolong the harvest period.

Once you've grown bush beans successfully and know which pest and disease problems to anticipate and prevent, graduate to pole beans if you like. Pole beans offer the advantage of greater productivity over a long season. They're more space efficient than bush beans, and you can harvest the crop standing up.

CROP BASICS

FAMILY: Legume family; relatives are peas.

SITE: Beans do best in full sun.

SOIL: Beans will grow fine in average soil as long as it's well drained. Avoid adding high-nitrogen materials to the soil before planting beans. Soil temperature is crucial; beans won't germinate or grow well in cold soil.

TIMING OF PLANTING: Plant beans in spring when the danger of frost has passed and soil temperature is about 60°F. Repeat sowings of bush bean varieties every 2 to 3 weeks until 60

days before the first expected fall frost. A single sowing of pole beans is all that's needed.

PLANTING METHOD: Sow seeds directly in garden beds; bean seedlings are fragile and difficult to transplant.

CRUCIAL CARE: Pole beans require strong support on a trellis. Both pole and bush beans need even soil moisture once flowering begins.

GROWING IN CONTAINERS: Choose pole bean varieties to maximize yield. Use at least a 10-gallon container and train the plants up sturdy stakes or a trellis fastened to a wall. Scarlet runner beans and purple beans are eye-catching choices for containers.

HARVESTING CUES: Harvesting cues vary depending on the type of bean you're growing. Pick filet beans before they reach pencil thickness. With snap beans, younger pods often have better flavor and texture. Begin harvesting as soon as you see and feel faint outlines of the seeds forming inside the pods. Bend a bean pod in the middle, and it should snap in half. Apply pressure near the tip, and the pod should bend but not break—if it breaks, it's past prime. Harvest green shell beans when the seeds have reached their full size but before the pods begin to dry. When harvesting dried beans, wait to pick until the pods are stiff and break open easily under pressure.

SECRETS OF SUCCESS

RIDGES FOR WARMTH, DEPTH FOR MOISTURE. Spring plantings of beans suffer if the soil is too cold. To help the soil warm up and stay warm, form raised ridges up to 6 inches high. Make a 1-inch-deep furrow along the top of the ridge and sow seeds. For summer plantings, soil moisture is the key to good germination. Skip the ridge-building step; instead, open a furrow 2 inches deep and sow seeds in the furrow. Treat the seeds with an inoculant (available from seed suppliers) before sowing.

SUPPORT BEFORE YOU SOW. Pole beans need strong support, and setting up the support structure should be step one in the planting process. Bean seedlings break easily and have shallow roots. If you try to set up a trellis around sprouted seedlings, chances are you'll damage both roots and shoots. Tepees of bamboo or wooden poles are a traditional support for beans, but you'll reap a better harvest if you use cylinders of wire fencing instead. Beans planted around an open cylinder benefit from better air circulation (reducing the potential for disease),

A cylindrical support for pole beans allows better air circulation and easier harvesting than a tepee-style support.

and the pods are easier to pick. Shape a length of wire mesh fencing about 5½ feet long and 6 feet tall (or taller) into a cylinder, cutting away the bottom piece of horizontal wire. Push the exposed vertical wires into the soil, and also anchor the cylinder with a sturdy stake. Plant

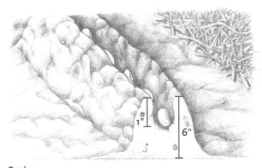

Spring

Summer

Sow beans in a raised ridge in spring and in a 2-inch-deep furrow during summer.

MESH TRELLIS FOR POLE BEANS

A classic stakes-and-string trellis works well for pole beans, but it's time-consuming to set up. Save time by using nylon mesh for a pole bean trellis. First, drive three 8-foot-long posts (wooden or metal fence posts or lengths of electrical conduit) along the planting row, 5 feet apart. Set the posts at least 1 foot deep—use a metal digging bar to open holes, or drive posts using a steel post driver. Tamp the soil firmly in place around the base of the posts.

Measure nylon mesh fencing to fit the trellis (use mesh with openings at least 5 inches square). The fencing should extend from about 4 inches above ground level to the top of the posts. Staple the mesh to furring strips along the top edge and fasten the furring strip to the posts with bolts or wire. At the end of the season, unfasten the furring strips and roll up the mesh around them for use in another season.

Nylon mesh   Furring strip

two or three seeds per hill, 6 to 8 inches apart around the base of the cylinder; thin to the strongest seedling in each hill.

Beans also grow well on a sturdy trellis like the one shown above.

TREAT SEEDS WITH CARE. Bean seed doesn't keep well from season to season. For best results, buy fresh bean seed each year. Moistening bean seeds before sowing can speed germination, but it's important to do it gently. If you place the seeds directly in water, they sometimes swell too quickly, and the cotyledons (seed leaves) may crack. Instead, wet a paper towel with warm water and squeeze out excess moisture. Lay the beans on the towel, roll it up, and set the towel on end in a jar. The beans will gradually absorb moisture and soften; they will be ready to plant the following day.

MULCH FOR EVEN MOISTURE. Beans don't need heavy watering, but they do need consistent soil moisture once flowers start to form. After bush bean plants have formed their first true leaves, cultivate shallowly to kill weeds, and then spread 1 to 2 inches of organic mulch alongside the plants. For pole beans, wait until the plants are about 1 foot tall and then mulch.

REGIONAL NOTES

FAR NORTH

If you garden in Zone 3 or Zone 4, try growing cool-season beans. Fava beans, for example, are more cold-hardy than peas. Plant them as soon as the soil is workable in spring. Scarlet runner beans, which produce lush vines and colorful flowers, also do well in cool conditions.

PREVENTING PROBLEMS

BEFORE PLANTING

- Cover planting area with black plastic to help speed soil warming.

AT PLANTING TIME

- Apply parasitic nematodes to planting furrows to reduce cutworm and seedcorn maggot populations (see Nematodes).
- Cover seedbeds with floating row covers and seal the edges (as described in Row Covers) to keep out a variety of insect pests.
- Put out wireworm traps if wireworms have been a problem in your garden (see Wireworms).

WHILE CROP DEVELOPS

- Thin seedlings promptly to promote good air circulation.
- If past crops suffered disease problems, spray young plants with a plant health booster (see Disease Control for information).
- Check soil moisture frequently once flowers start to form, and don't allow the soil to dry out.
- Inspect plants frequently for pests. Use a strong water spray to remove aphids and spider mites; handpick larger pests. Take other action as needed if pest populations climb.

AT HARVEST TIME

- Use two hands when harvesting (one to hold the vine, the other to grasp the pod) to prevent plants from breaking as you pull off pods.

AFTER HARVEST

- Clear out all residues that suffered pest or disease problems and put them at the center of a compost pile. Chop and turn under healthy plant residues to boost soil organic matter.

TROUBLESHOOTING PROBLEMS

SEEDLINGS NEVER EMERGE OR NEW SEEDLINGS DIE. Damping-off disease or seedcorn maggots are to blame. Both problems are worst in cool, wet conditions. If seedlings have dark, water-soaked stems, damping-off is the cause. Try replanting in a spot with better drainage. See Damping-Off for more information.

If seedlings that do emerge are deformed or have no leaves, it's a sign of seedcorn maggots, which are the larvae of small gray flies (similar to cabbage maggot flies). Dig around in the soil, and you may find the small yellowish white maggots. Remove the seedlings and treat the soil with parasitic nematodes as described in Nematodes. Replant when the soil is warmer and the weather is warm and dry.

NEW SEEDLINGS EATEN. A single cutworm can destroy several plants in one night. These pests hide just below soil level. See Cutworms for information.

SEEDLINGS STUNTED. Cold soil may be to blame. Bean seedlings that germinate in cool soil get a poor start in life, and they'll probably never grow out of their weakened condition. If you can't find any pest or disease symptoms in a new planting that's growing poorly, check soil temperature. If it's below 60°F, compost these seedlings and replant in a bed prewarmed with black plastic or in raised soil ridges (as shown on page 41).

LARGE HOLES IN LEAVES. Bean leaf beetles and cucumber beetles chew holes in bean leaves (see Cucumber Beetles for information). Bean leaf beetles are reddish orange with black edges and black spots on the wing covers. They feed only on beans and peas; the larvae bore into the roots. Handpicking bean leaf beetles is usually sufficient to avoid severe damage. If these pests show up in your garden regularly, cover seedlings with floating row covers until they start to flower, at which point bean leaf beetle feeding shouldn't hurt yields. For a serious infestation, spray pyrethrin as a last resort.

SKELETONIZED LEAVES. Mexican bean beetles and Japanese beetles feed on leaves between the veins, producing this pattern of damage.

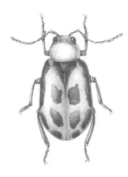

Bean leaf beetle

Check leaves for the yellow woolly larvae of the bean beetles, which also chew on foliage. Also look for iridescent beetles with green wing covers; these are Japanese beetles. Handpick any pests that you find. See Mexican Bean Beetles and Japanese Beetles for complete control information.

SMALL HOLES IN LEAVES. Flea beetles produce these small holes, and they can ruin a stand of young seedlings. See Flea Beetles.

YELLOW LEAVES. When aphid colonies form on bean plants, leaves first turn yellow and then shrivel up. Check leaf undersides and growth tips for clusters of black aphids. Spray plants with water to dislodge aphids. For more control options, see Aphids.

LEAF MARGINS YELLOW AND CURLED. Potato leafhoppers suck sap from plants, and when their populations are high, plants may drop their flowers and wilt. Spray insecticidal soap to kill leafhoppers. See Leafhoppers for more information.

WHITE STIPPLING ON LEAVES. This is a sign of spider mite infestation. Check leaf undersides for gray webbing. Spray plants with water to dislodge mites; for more control options, see Mites.

GRAY GROWTH ON LEAVES. Gray mold is a fungal disease that develops in warm, wet weather. Pods may end up covered with the gray mold as well. Remove infected plants; see Gray Mold for other controls.

WHITE GROWTH OR WATER-SOAKED SPOTS ON LEAVES. Four diseases are potential causes: powdery mildew, downy mildew, white mold, and bacterial blight. Powdery mildew leads to a thin white or grayish coating that eventually covers the topside of all leaves. If you catch powdery mildew early, a baking soda spray may prevent the spread of the disease (see Powdery Mildew). Downy mildew usually infects lima beans only, with white growth on flowers and young leaves (see Downy Mildew).

White mold causes both water-soaked leaves and a fluffy white mold on leaves and pods. Remove infected plants; see White Mold for more controls.

With bacterial blight, no mold will appear, but as the problem grows worse, the water-soaked spots turn reddish or brown and a yellow halo appears around the spots. If you catch the symptoms very early, spraying copper fungicide may control the problem. Otherwise, uproot the plants and compost them. Replant in a different area of the garden and take steps to reduce humidity and increase air circulation around the new planting.

RED BLISTERS ON LEAVES. Rust is a fungal disease that can ruin bean crops. Infected plants will drop their leaves, and blisters may also appear on pods. If you catch symptoms very early, spraying sulfur may prevent the problem from growing worse. See Rust for more controls.

DARK STREAKS ON LEAVES. Anthracnose is a fungal disease that can damage bean pods as well as leaves. Uproot young plants that show symptoms and compost them. See Anthracnose for prevention tips.

PALE OR YELLOWING LEAVES, STUNTED PLANTS. Viral diseases, wireworms, root-knot nematodes, and root rot are possible causes of yellowing and stunting. Aphids spread mosaic virus as they feed; plants with mosaic tend to be stunted and have mottled leaves (see Mosaic). Pull out diseased plants and destroy them.

To check for wireworms and nematodes, uproot a couple of affected plants. If you find segmented brownish grubs around the roots, the problem is wireworms. Destroy the worms. See Wireworms for information, then replant.

If you find knotty areas or dark swellings on roots (different in appearance from nodules, which are normal on bean roots), it's a sign of root-knot nematodes, and the crop probably won't produce a harvest (see Nematodes).

If the roots are brown and there are sunken, red spots at the base of the stem near soil level, the problem is root rot. Remove the plants and replant in a different part of your garden. See Rots for other controls.

BEAN PODS STOP FORMING. Check around the base of the plant and you'll probably find flowers that have dropped off. This is due to high heat or tarnished plant bug feeding.

Use a hand lens to look for tarnished plant bugs and nymphs on the plants. Pods that have already formed may be misshapen. See Tarnished Plant Bugs for control measures for future plantings. If you find no plant bugs, wait for the weather to change and new pods will probably start to form. Check soil moisture and water well if needed, because moisture stress also can cause flowers to drop.

WATER-SOAKED SPOTS ON PODS. These spots result from stink bug feeding. Pods may darken inside and secondary infections set in at the feeding injury sites (see Stink Bugs).

DARK, SUNKEN SPOTS ON PODS. These spots are due to anthracnose, a fungal disease. The pods may or may not be salvageable for eating (see Anthracnose).

BEET CURLY TOP

Beet leafhoppers spread curly top virus as they feed. The leafhoppers prefer warm, dry conditions, so that's when curly top is likely to be most severe. Some races of the virus produce more severe symptoms than others.

Curly top virus is very common in sugar beets. If you live near commercial sugar beet fields, you can expect that virus-infected leafhoppers will find your garden.

RANGE: Western United States; southern Canada

CROPS AT RISK: Beet-family crops, tomato-family crops, beans, and squash-family crops

DESCRIPTION: Symptoms vary by crop but generally include distorted growth, color changes, and stunting. Beet leaves are small, twisted, and curled or crinkled. Beet roots are deformed and covered with matted side roots. Tomato leaves become thick and crisp, and veins may turn purple. Young fruits will be dull and wrinkled. In squash-family plants, older leaves are yellowed but young leaves are very dark green. Bean plants produce tufts of abnormal leaves, flower prematurely, and produce a poor harvest. In general, if seedlings become infected, they will die fairly quickly.

FIGHTING INFECTION: Remove infected plants and compost them. If you replant immediately, try to separate the new planting from other susceptible crops, and cover the seeded area with row covers.

GARDEN CLEANUP: The virus survives in leafhoppers and weeds or ornamental plants around your garden. Weed your garden thoroughly in the fall to remove as many potential sources of the virus as you can. Mustards and tumbleweed are two of the many weeds that host the virus.

NEXT TIME YOU PLANT: Choose resistant or tolerant varieties when possible. Plant spring beets and spinach early to beat the arrival of infected leafhoppers. Leafhoppers like to feed in sunny spots, so plant susceptible crops in a part of the garden that receives light shade. Cover tomato plants growing in cages with fine netting to provide light shade; the netting may also keep leafhoppers away from the plants. Remove the netting when fruit begins developing. For smaller plants, cover them with row covers beginning at planting and leave them in place as long as possible.

BEETS

Greens for cooking and flavorful roots for baking or pickling are the rewards when you grow beets. Spring and fall are the prime growing times for beets because the roots develop the best flavor during cool weather. Also, early and late plantings suffer from fewer pest and disease problems than summer crops.

Although the swollen part of a beet root occupies only the top few inches of the soil, its taproot delves down more than a foot, allowing the crop to benefit from a large reserve of nutrients and moisture. Due to this rooting habit, beets are happiest in loamy or sandy soil and may struggle in heavy clay soil.

CROP BASICS

FAMILY: Beet family; relatives include spinach and Swiss chard.

SITE: Plant in full sun to produce healthy roots; if you're growing beets for greens only, partial shade is fine.

SOIL: Loosen the top 12 inches of soil before planting; remove as many rocks and clods as possible. Spread a 1-inch layer of compost and work it in. If you have heavy soil, grow beets in a raised bed if possible.

TIMING OF PLANTING: Begin sowing seed 4 weeks before last expected spring frost. In cool-summer areas, continue sowing every 2 to 3 weeks as desired. In most areas, stop planting from mid-spring through midsummer. About 2 months before the first expected fall frost, sow a crop for fall harvest. In mild-winter areas, repeat sowings are possible until mid- or late

fall. In the Deep South, beets will continue growing throughout winter.

PLANTING METHOD: Sow seeds directly in the garden; starting seeds indoors is possible but not usually necessary.

CRUCIAL CARE: Thinning seedlings and maintaining soil moisture are essential for producing tasty beet greens and roots.

GROWING IN CONTAINERS: Beets will do well in containers at least 1 foot deep.

HARVESTING CUES: Cut greens when young and tender; begin checking roots about 6 weeks after sowing; storage types take longer to mature.

SECRETS OF SUCCESS

THIN SEEDLINGS IN STAGES. Beet seeds are dried fruit husks; each husk contains several true seeds. Inspect beet seedlings shortly after they emerge and you'll find small clusters of seedlings. When the first true leaves have formed, use a small pair of scissors to snip off all but the strongest seedling in each cluster at soil level. Don't pull out seedlings at this stage or you'll disturb the tender roots of the seedlings you want to save.

About 2 weeks later, thin again so that plants are at least 2 inches apart. Use the young greens in a salad or sauté. Wait another 2 weeks, and you should be able to harvest a first crop of baby beets (about 1 inch in diameter). The remaining plants, now about 4 inches apart, will provide your main harvest in another 2 to 3 weeks (longer for large storage beets).

AVOID MIDSUMMER HEAT. Beets are fairly heat-tolerant, but root quality is poor if the roots mature during hot weather. For best flavor and texture, aim to pick spring beets before daytime temperatures exceed 70°F. Wait until temperatures are consistently in the 50s in fall to harvest roots of late-season plantings.

KEEP AHEAD OF WEEDS. Weed competition can ruin the flavor of beets. When weeds shade or crowd out beet foliage, the plants can't produce enough sugars to sweeten the roots. Stifle weeds from the start to allow your beet crop to produce a strong stand of greens. For spring crops, use a hoe to cultivate shallowly alongside rows of seed-lings (no deeper than 1 inch to avoid harming the beet roots). Once the beet tops are about 4 inches tall, mulch the soil between rows to suppress further weed growth. For fall crops, mulching between rows right at planting time is best—the mulch will both suppress weeds and cool the soil.

REGIONAL NOTES

WESTERN UNITED STATES AND CANADA

Beet webworms range throughout the United States, but they're usually problematic on beets only in the West. The damage they cause is similar to armyworm feeding—webworms chew at leaf surfaces, leaving a leaf skeleton behind. Newly hatched caterpillars are pale green. As they grow, they change to olive green and then nearly black; they also develop a black stripe and many black spots on their backs. You may not

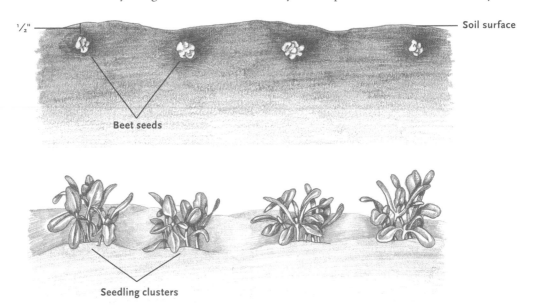

½" — Soil surface

Beet seeds

Seedling clusters

For every beet "seed" you sow, a cluster of several seedlings will sprout. Thin the excess seedlings so your crop will get off to a strong start.

see these caterpillars unless you look inside the webs they produce around a rolled or folded section of a beet leaf. Natural parasites help to control these caterpillars; include small-flowered annuals and perennials in or near your vegetable garden to attract the beneficials. Pigweed and lamb's quarters are host plants for webworms; if webworms are infesting your beets, seek out and destroy these weeds, too. Controls for armyworms also help for webworms; consult Armyworms for details.

PREVENTING PROBLEMS

AT PLANTING TIME

- Set up a slug barrier if slugs or snails are problematic in your garden (see page 368).
- Cover seedbeds with row covers and seal the edges to keep out aphids, leafminers, and flea beetles if you want to harvest beet greens (see Row Covers for details).
- Plant early or late to avoid blister beetle infestations.

WHILE CROP DEVELOPS

- Cover young plants with row covers if a long spell of cold weather is predicted.
- After the soil warms, mulch well to maintain even soil moisture, and check soil moisture regularly.
- Thin seedlings promptly to ensure good air circulation and adequate space for root development.
- If you suspect soil problems, apply foliar kelp spray once every 2 weeks to promote healthy roots.

- If past crops have suffered disease problems, spray young plants with a plant health booster.

AT HARVEST TIME

- To prevent damage to roots, insert a spade into the soil beside the row to loosen the roots before pulling them out.

AFTER HARVEST

- Remove any remaining beet roots (even tiny ones) and leaf debris and compost them.

TROUBLESHOOTING PROBLEMS

SEEDLINGS NEVER EMERGE OR NEW SEEDLINGS DIE. If the weather is wet and cool, damping-off is the probable cause; try replanting in a spot with better drainage. In hot weather, the problem may be lack of moisture. Resow seeds and water them daily to maintain moist conditions. See Damping-Off for more information.

NEW SEEDLINGS EATEN. A single cutworm can destroy several plants in one night. These pests hide just below soil level (see Cutworms for control information). Slugs and snails eat seedlings, too (see Slugs and Snails for controls).

TUNNELS IN LEAVES. Leafminer flies lay eggs and their larvae tunnel through leaves. You can still eat the damaged greens; cut away the tunneled areas and eat the rest. To prevent future damage, cover seeded beds with row covers in spring. See Leafminers for more controls.

SMALL HOLES IN LEAVES. Flea beetles chew on leaves; a heavy infestation can destroy seedlings (see Flea Beetles).

RAGGED HOLES IN LEAVES. Slugs, snails, cabbage loopers, and blister beetles are possible culprits; seek and find which pest is damaging your plants. Cut away damaged areas of leaves and use the rest. See Slugs and Snails, Cabbage Loopers, and Blister Beetles for control information.

SKELETONIZED LEAVES. Armyworm feeding produces this damage pattern. Look for webbing on leaves and search inside it for armyworms (see Armyworms for controls).

CURLED LEAVES. Aphids suck sap, causing leaves to curl and twist. Look for colonies on young leaves and undersides of larger leaves. Spray leaves with a strong stream of water to dislodge aphids; if the problem persists, see Aphids for more controls.

Curly top virus also causes leaves to curl upward and become crinkled. Check the roots, too; infected plants often sprout excess side roots. Leafhoppers transmit this disease as they feed. Pull out diseased plants and destroy them. See Beet Curly Top for more information.

LEAVES TURN RED. Exposure to temperatures below freezing can cause beet greens to change color. This isn't harmful, and frost actually improves the flavor of beet roots. Some varieties have naturally deep red foliage.

TAN SPOTS ON LEAVES. Cercospora leaf spot is the cause; the spots usually have red or purplish margins. Wet, warm weather can cause a flare-up of this fungus. Remove infected leaves and watch for new growth—it should be healthy. If the problem persists, see Cercospora Leaf Spot for more controls.

YELLOW PATCHES ON LEAVES. Downy mildew first shows up as yellow spots on leaves. In cool, wet conditions, white or gray fuzz appears on leaf undersides. Remove seriously infected plants. See Downy Mildew for more information.

PALE OR YELLOWING LEAVES, STUNTED PLANTS. Nematodes and root rot are possible causes of these symptoms. To check, uproot a couple of affected plants. Nematodes will cause the formation of galls or cysts on roots. See Nematodes for controls.

If you find dead roots or roots that are dry and black at the center, root rot is the problem (see Rots). Replant in a new area with enriched, well-drained soil. Water regularly (but don't overwater) to stimulate rapid growth.

ROUGH, CRACKED SKIN ON ROOTS. Infection by downy mildew can affect beet roots. Infected roots may be abnormally shaped, too. See Downy Mildew to prevent this problem in the future.

BLACK SPOTS INSIDE ROOTS. Boron deficiency leads to spots of black or brown tissue both inside roots and on root surfaces. Young leaves may turn brown and die; the center of the root may start to rot. Soil that is too acid or too alkaline is the usual cause because plants can't absorb boron well in extreme soil conditions. Dry conditions can worsen the problem. If the spots are small, try cutting around them and using the rest of the beet. For future plantings, check soil pH and adjust it to between 6.2 and 7.0. Keep beets well watered. For extra insurance, sprinkle $\frac{1}{3}$ cup of dried seaweed per 10 feet of row before planting and work it into the top few inches of soil, or apply a foliar spray of seaweed extract every 2 weeks as the crop develops.

WHITE RINGS INSIDE ROOTS. These rings develop when roots are exposed to high heat or uneven watering. Flavor of affected beets may be unpleasantly strong. To prevent this problem, mulch beets, water regularly, and harvest roots before hot weather arrives.

CORKY AREAS ON ROOTS. Scab fungus causes this problem (see Scab). The roots won't last in storage, but you should be able to cut out the damaged areas and use the rest of the roots.

WOODY ROOTS. Leaving roots in the soil too long can cause them to become woody, especially if the weather turns hot. Woody roots covered with hairy side roots are probably the result of curly top virus, which is spread by leafhoppers (see Beet Curly Top).

PLANTS FORM FLOWERING STALKS. Beets are biennials, and overwintered plants may naturally form a flowerstalk in spring. Harvest roots before the stalk forms or the roots may become stringy. Young beet plants exposed to a long period of cold weather (below 50°F) sometimes bolt, too. Uproot these plants and compost them. To prevent this problem in the future, cover young beets with row covers if a long cold spell is predicted.

BENEFICIAL INSECTS

Beneficial insects are the best supporting actors in the drama of garden pest management. Predators and parasites of insect pests work quietly and often unseen, but their drive to find nourishment helps protect your garden from serious pest flare-ups that can ruin a crop.

It's easy to take steps to shelter and attract beneficial insects, no matter whether you garden in the country, suburbs, or city. Beneficial insects are widespread in the environment. As long as they find shelter and food, many kinds of beneficials will be content to live out their lives right in your yard, even if you have only a small city plot.

WHAT BENEFICIALS NEED

When they're not out feeding, they need sheltered spots out of the sun and wind to rest. Also, beneficials expend lots of energy seeking out their prey. Many species of beneficial insects need to supplement their diet of insects with pollen (for protein) and nectar (for carbohydrates). These insects spend only one phase of their lives—usually the larval stage—consuming other insects; the adults are exclusively nectar feeders. For example, the larvae of parasitic wasps feed inside the bodies of caterpillars or insect eggs, but the wasps need flowers to feed on.

In general, if beneficial insects can't find nectar in or near your vegetable garden, they won't stay in the vicinity long. Your goal should be to maximize the sites, both in your garden and around your yard, where beneficials can rest, feed, and overwinter.

HABITAT IN THE GARDEN

Attracting beneficial insects is a desirable goal, but it's tough to sacrifice precious garden space to plants other than vegetables. The good news is that you can provide the diverse habitats and food sources that beneficials need without giving up more than a tiny fraction of garden space. Try some of the following space-efficient strategies for including shelter and food plants for beneficials.

LIVING MULCHES

Garden beds planted in sweet clover and buckwheat offer excellent shelter for beneficials. See Cover Crops for more information.

NECTAR FROM CROPS

Allowing some broccoli, mustard, or kale plants to go to flower in spring and fall offers beneficials that prey on cabbage-eating caterpillars an excellent source of late-season nectar.

SHELTER IN THE PATHS

Garden pathways can double as habitat for beneficials. Sod pathways provide shelter for ground beetles, which eat weed seeds as well as insects and slug eggs. The ground beetles will live in the soil under the sod but will venture as far as 10 feet over into your garden beds. Unlike your lawn, keep pathways mowed short so they won't offer refuge for voles or grasshoppers.

Straw is another good habitat for beneficial insects as long as you renew it once or twice during the growing season. Bark mulch will provide habitat for spiders.

ADD SOME HERBS

Most vegetable gardens include at least one perennial crop such as rhubarb or asparagus. Extend your perennial crop plot a bit and add some perennial herbs, such as rosemary, thyme, or mint (inside a root barrier). These bushy herbs offer excellent shelter for beneficials, and mint flowers are suitable feeding sites for tiny parasitic wasps.

If you grow dill, let one plant go to flower and set seed. Take the seed head and shake it over a garden bed or at the margins of the garden. It's easy to weed out most of the seedlings, but leave a stray plant here and there. It will unfurl flowers in the spaces above and around the crop foliage, bringing many beneficials right to your crops.

SWEETEN IT UP

Beneficials that prey on aphids, such as lady beetles, lady beetle larvae, and lacewing larvae, are attracted by the sugary honeydew that aphids produce. You can trick them into thinking aphids are present by spraying some of your garden plants with a sugar-water solution to attract the adults. Dissolve 5 tablespoons sugar in 1 quart warm water to make the solution. Spray the solution on the foliage of plants where you've noticed some aphid activity or where you expect aphids to show up based on past experience.

Plant an annual flower or herb from the list in "Flowers for Beneficials" on this page in place of every 10th or 15th vegetable transplant in a bed. This lowers potential yield from the bed a bit, but reduced pest problems can more than counterbalance the loss of harvest. If you don't want to sacrifice even a little yield, use a low-growing annual such as sweet alyssum as a border for vegetable beds. If you sow seeds about the same time that you set out transplants, the flowers won't compete with your vegetables.

MANAGE THE GOOD WEEDS

Some weeds have flowers that beneficial insects love. For example, dandelions are one of the best sources of early nectar for beneficial insects. The trick is to let the good weeds flower but then to uproot them, or at least remove the flowerstalks, before the weeds set seed. See Weeds for more information.

BEYOND THE GARDEN

Remember beneficial insects as you plan and plant other areas of your home landscape, too. An insectary garden close to your vegetable garden offers excellent year-round shelter for beneficials—a permanent refuge that will never be disturbed by harvesting or tilling.

DOUBLE-DUTY GARDEN. Your insectary garden can serve more than one purpose. For example, if your vegetable garden is near a street, plant a border of flowering shrubs and trees, perennials, ornamental grasses, plus some dwarf fruit trees (an excellent source of early-season

FLOWERS FOR BENEFICIALS

Not all nectar-producing flowers are equal in the view of beneficials. Because most beneficials are so small, they need to feed from small flowers that have a shallow, open structure—they could drown in the nectar pooled in a tubular flower such as a daylily. Daisy blooms and sunflowers are actually complex flowerheads consisting of hundreds of tiny individual flowers arranged in tight groups. The flat center of a sunflower or daisy is a gold mine of nectar and pollen for tiny insects. Flowers in the carrot family, such as those of dill and Queen Anne's lace, are also flowerheads made up of flat clusters of tiny flowers.

Many of the annual and perennial flowers listed below belong to the carrot or daisy family, and they're some of the best for attracting beneficial insects. For even more plant choices, see "A Hideout for Your Helpers" on page 54.

Bishop's weed (*Ammi majus*)
Dill (*Anethum graveolens*)
Chervil (*Anthriscus cerefolium*)
Astrantia, masterwort (*Astrantia major*)
Caraway (*Carum carvi*)
Coreopsis (*Coreopsis* spp.)
Coriander, cilantro (*Coriandrum sativum*)
Queen Anne's lace (*Daucus carota*)
Coneflowers (*Echinacea* spp.)
Fennel (*Foeniculum vulgare*)
Blanketflowers (*Gaillardia* spp.)
Sunflowers (*Helianthus* spp.)
Candytufts (*Iberis* spp.)
Lovage (*Levisticum officinale*)
Sweet alyssum (*Lobularia maritima*)
Parsley (*Petroselinum crispum*)
Anise (*Pimpinella anisum*)
Goldenrods (*Solidago* spp.)
Blue lace flower (*Trachymene coerulea*)

A HIDEOUT FOR YOUR HELPERS

This mini-hedgerow is a long-flowering "bed-and-breakfast" for many beneficial insects, birds, toads, and other welcome friends. Birds and toads may eat some of your beneficials, but they will eat more pests, and the beneficials will develop a population balance they can maintain. The garden includes annuals, perennials, herbs, and a few shrubs. It offers nectar, pollen, water, shelter, and prey insects for the beneficials. For the birds there are water, berries, seeds, insects, a perch, and sheltering shrubs. A toad will find cool shelter, water, and insects.

Some annuals may be available as seeds or plants in your area, and either are fine, but if you hope for flowers sooner, buy plants rather than propagating them yourself. Either way, many annuals will reseed on their own.

For a toad shelter, just space some small rocks 8 inches or so apart and place a board over them; then anchor them with more rocks. If you can, dig a little depression into the soil to keep the temperature in the toad shelter a bit cooler, and put the shelter near water.

The approximate number of plants you'll need for this garden is listed behind each plant name. Close planting is recommended so there's minimal bare soil. If using seeds, one packet is sufficient if you follow the sowing instructions on the packet.

In the West, use California buckwheat (*Eriogonum fasciculatum*) instead of *Fagopyrum esculentum*. Even though it's listed as a perennial, giant angelica is a biennial that often reseeds. If you live in a warm climate, black-eyed Susan may be an annual instead of a perennial. For the serviceberry or shadbush, choose a plant that's the right size for your space; in the West, substitute coffeeberry (*Rhamnus californica*).

SHRUBS

Serviceberries or shadbushes (*Amelanchier* spp.): 1 plant
Butterfly bush (*Buddleia davidii*): 1 plant
Common or spreading juniper (*Juniperus communis* or *J. occidentalis*): 1 plant
Snowberry (*Symphoricarpos albus*): 1 plant

ANNUALS

Dill (*Anethum graveolens*): 1 packet of seeds
Borage (*Borago officinalis*): 1 packet of seeds or 3 plants
Calendula (*Calendula officinalis*): 9 plants
Cosmos (*Cosmos bipinnatus*): 1 packet of seeds or 12 plants
Buckwheat (*Fagopyrum esculentum*): Less than 1/4 pound of seed
Alyssum (*Lobularia maritima*): 1 packet of seeds or 24 to 36 plants

---

pollen and nectar) between your garden and the roadway. Your neighbors will enjoy the flowers, and the border will protect you from noise and exhaust fumes as you garden.

LEAVE SOME LITTER. Small changes in your gardening practices can help beneficials, too.

Stop clearing away leaf litter and debris under shrubs around your yard. This debris is a perfect refuge for beneficials when the weather turns hot and dry.

BAN BUG ZAPPERS. If you have a bug zapper, unplug it—for good. Studies show that bug

PERENNIALS

Yarrow (*Achillea ptarmica* or *A. millefolium*):
2 plants

Giant angelica (*Angelica gigas* or *A. archangelica*): 1 plant

Swamp milkweed (*Asclepias incarnata*):
1 to 3 plants

Butterfly weed (*Asclepias tuberosa*):
2 plants

New England aster (*Aster novae-angliae*):
3 plants

Boltonia (*Boltonia asteroides*): 1 plant

Joe Pye weeds (*Eupatorium* spp.): 1 to 3 plants

Gayfeather (*Liatris spicata*): 6 plants

Black-eyed Susan (*Rudbeckia hirta*): 1 to 3 plants

Tansy (*Tanacetum vulgare*): 1 plant

New York ironweed (*Vernonia noveboracensis*): 1 plant

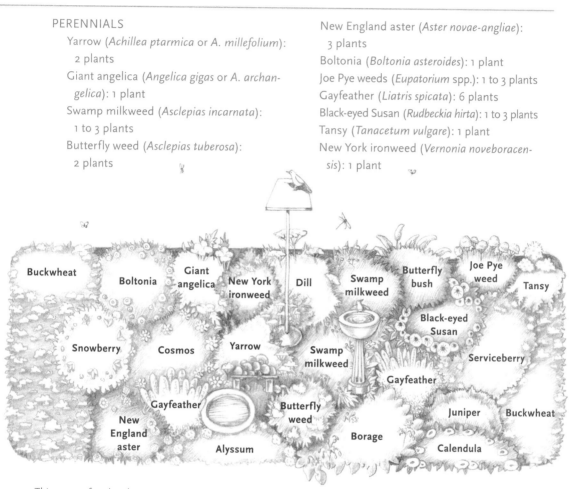

This 4 × 20-foot border can serve to screen your vegetable garden from the street and to shelter and attract beneficial insects.

zappers kill more beneficials than they do garden pests or mosquitoes.

SUPPLY WATER. It's easy to forget, but insects need water in order to survive. Because many beneficial insects are tiny, they can drink safely only from very shallow water. Supply water for beneficials by setting out near host plants a large clay plant saucer with a glazed bottom, with 1 to 2 inches of water in the bottom. Add a few flat rocks that stick up above the water surface as landing platforms for the insects. Check this insect watering hole every few days and refill as needed.

BENEFICIALS TO KNOW AND PROTECT

Being able to identify beneficial insects is important so that you don't mistake them for pests and handpick them or spray them. Here's a rundown of 10 of the most common types of beneficials that frequent vegetable gardens. Most of these insects are found throughout the United States and Canada. The spined soldier bug is found only in the eastern United States and the Midwest.

ASSASSIN BUGS

Many genera and species

HOW TO SPOT THEM: You may notice assassin bugs and their nymphs on crop plants or weeds, hunting for prey. Be careful about touching them because they may bite you.

ADULTS: Color, shape, and size vary by species; these bugs also have a long segmented beak curled below the head (or extended to attack prey). Up to 1 inch long; many are dark-colored, but some are brightly colored; crossed wings lie flat over abdomen when at rest.

NYMPHS: Similar to adults but wingless

EGGS: Some species lay eggs in soil or protected sites; others lay clusters of barrel-shaped eggs on leaves.

WHAT THEY EAT: Caterpillars, beetles, aphids, flies, leafhoppers, and more. A few species bite mammals and suck blood for nourishment.

GROUND BEETLES

Many genera; thousands of species

HOW TO SPOT THEM: These beetles hide under mulch and other protected spots during the day. You may be able to find them if you dig through surface mulch or check under stones or low-lying foliage of perennial groundcovers.

ADULTS: Long-legged beetles up to 1 inch long, with long antennae; most are black, reddish brown, or iridescent; scurry quickly when disturbed; may live as long as 3 years.

LARVAE: Wingless black insects with segmented, shiny, and elongated bodies up to 1 inch long

EGGS: Laid in the soil

Assassin bug

Assassin bug nymph

Ground beetle

Ground beetle larva

WHAT THEY EAT: Slugs, snails, cutworms, root maggots, Colorado potato beetle larvae, insect eggs, and some types of insect pupae. They occasionally eat earthworms, too, but don't cause significant loss of these beneficial creatures.

LACEWINGS

Chrysopa spp., *Chrysoperla* spp., *Hemerobius* spp.

HOW TO SPOT THEM: Look for the larvae around aphid colonies, and for adults on flat-topped flowers such as Queen Anne's lace. With luck, you'll also spot lacewing eggs, which appear suspended in air above a leaf (they're attached by a thin stalk). Lacewings do best when temperatures range from 65°F to the upper 80s; they will feed little during a long spell of cool weather. They do not like arid conditions.

ADULTS: Slim, pale green insects about ¾ inch long, with large transparent, veined wings and long antennae

NYMPHS: Fast-moving, wingless tan or gray insects resembling alligators in miniature, with pincerlike curving jaws. Nymphs live 15 to 20 days before pupating.

EGGS: White or green oblong eggs laid singly, attached to leaves by a threadlike stalk

WHAT THEY EAT: Lacewing adults feed mainly on nectar. Lacewing nymphs, also called aphid lions, eat aphids, insect eggs, mites, thrips, and other soft-bodied insects. The nymphs will also eat their fellow nymphs.

LADY BEETLES

Many genera; hundreds of species

HOW TO SPOT THEM: Watch for egg clusters on the top surfaces of leaves; larvae feed in and around aphid colonies. The colorful adults are easy to see around your garden and may land on you while you garden. Lady beetles are most active when temperatures range from 65°F to the upper 80s. They do not like arid conditions.

ADULTS: Small round beetles, usually red or orange, but may be other colors as well. Many species have black spots on the wings.

LARVAE: Blue or black wingless insects with an alligator-like shape, orange or yellow markings, and short spines along their bodies.

EGGS: Orange oval eggs laid in clusters on foliage

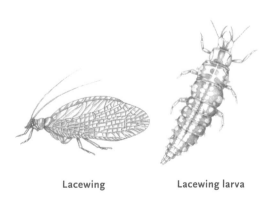

Lacewing   Lacewing larva

Lady beetle   Lady beetle larva

Lady beetle

WHAT THEY EAT: Adults and larvae eat insect eggs, aphids, and other soft-bodied insects. Adults also eat pollen, nectar, and honeydew. Two nonbeneficial lady beetles are the Mexican bean beetle (see that entry) and the squash beetle (a minor pest of squash plants).

PARASITIC WASPS

Many genera; thousands of species

HOW TO SPOT THEM: Many species are too small to see without a hand lens, but giant ichneumon wasps are as large as 3 inches (they parasitize tree pests, not vegetable garden pests). Look for small wasps on flowers such as dill, tansy, buckwheat, and yarrow. Also check crop plants for pest caterpillars (such as tomato hornworms) that have

Parasitic wasp

white or tan seedlike projections on their backs; these are the pupae that form after the wasp larvae finish feeding inside the caterpillar body (see page 423 for an illustration). Parasitized aphids turn dark and are called aphid "mummies."

ADULTS: Narrow-waisted wasps ranging from the size of a speck up to 3 inches long; some have long ovipositors (egg-laying tubes) that look like long stingers.

LARVAE: White grubs that feed inside the bodies of host insects or their pupae

EGGS: Laid in the bodies of host insects

WHAT THEY EAT: Pollen and nectar. The larvae feed inside caterpillars, beetle grubs, aphids, fly maggots, and other types of host insects. Some species of larvae parasitize insect eggs only.

SPIDERS

Arachnids; many genera and thousands of species

HOW TO SPOT THEM: Garden spiders aren't poisonous to humans. They like straw mulch and permanent plantings where they can seek shelter during unfavorable weather. Garden spiders often build their webs in full view. You may spot the spider resting in the web before you notice the web itself.

If poisonous spiders are native to your region, be sure you know how to identify them in case they do show up in your garden.

ADULTS: Spiders are arachnids, not insects. These eight-legged creatures vary widely in size, shape, and color. Most have a prominent abdomen.

EGGS: Laid in egg sacs

WHAT THEY EAT: Insects and other spiders

Garden spider

SPINED SOLDIER BUGS

Podisus maculiventris

HOW TO SPOT THEM: Put out a commercial pheromone lure in your garden and check plants close to it; inspect bean plants under attack by Mexican bean beetles; set a trap for the bugs as described on page 60.

ADULTS: Shield-shaped, grayish brown bugs with pointy "shoulders" and a brown spot near the tip of each wing. The spined soldier bug is part of a group known as stink bugs, some of which are plant pests (see Stink Bugs for a comparison).

NYMPHS: Similar to adults but without wings; young nymphs are more rounded and body shape develops as they grow.

EGGS: Laid in clusters on leaves

WHAT THEY EAT: Caterpillars, including armyworms and cabbage loopers; beetle grubs, including Mexican bean and Colorado potato beetles; flea beetles; also plant sap (but not harmful to plants). Soldier bug nymphs will eat each other as well.

SYRPHID FLIES

Many genera; hundreds of species

HOW TO SPOT THEM: Look for these bee-like flies hovering around flower blossoms. Bees have two pairs of wings, but syrphid flies have only one.

ADULTS: Shiny flies with yellow or white stripes on their abdomens; reddish eyes

LARVAE: Sluglike greenish or brown maggots up to ½ inch long

EGGS: Laid among aphid colonies

Spined soldier bug

Syrphid fly

PEST CONTROL ON DEMAND

Spined soldier bugs are excellent predators of caterpillars and beetles, but in nature there are usually not enough soldier bugs in any one spot to provide adequate control on their own. However, if you live in the eastern United States or the Midwest, you can rely more on spined soldier bugs if you put out a commercial pheromone lure designed especially to attract the bugs. Both male and female bugs will respond to the lure. Follow the instructions that come with the lure for mounting it in your garden; it will last for about 2 months.

For even greater benefit, incorporate the lure into a simple trap, and then cage the trapped bugs in the particular part of the garden where you want them to patrol for pests. To make the trap, you'll need a utility knife, two 2-liter soda bottles, a wooden stake about 3 feet long (an old tool handle may work well), and some tape.

With the utility knife, slice off the upper third of one bottle. Cut off the screw-top neck at the top, too. This will be the bottom section of your trap. Thread this part of the trap onto the stake. It should slide on easily, with about $1/2$ inch of wiggle room around the opening (so that bugs can crawl up the stake into the trap).

Cut off the lower third of the other bottle, but don't cut off the neck. Cut some narrow slits near the top of this bottle (to allow the pheromone to waft out of the trap and lure the beetles to it).

Pound the stake into the ground at the site where you plan to set up the trap (a spot near trees with lots of brush and leaf litter is ideal). Put the lure in the bottom portion of the trap.

Next, slip the top part of the trap onto the stake. Push the bottom portion up to connect with the top part and tape the overlapping edges. Don't leave any gaps.

At about the time that shade trees such as maples are blooming, spined soldier bugs overwintering in leaf litter will become active and look for mates. Attracted by the pheromone, they will reach the trap and crawl inside to be as close to the lure as possible.

When you've trapped a bevy of bugs, move them to a cage of $1/8$-inch hardware cloth or window screening set up over a garden bed. They will feed on insects in the caged area. Put some apple slices on the soil surface in the cage, too, as a supplement. Move the cage around your garden. Or if a particular crop is infested with a pest, move the cage to that part of the garden, let it sit a few hours until the bugs become acclimated, and then remove the trap and allow the bugs to roam free.

Slits

Tape

Pheromone lure

Wooden stake

Soda bottle pheromone trap

WHAT THEY EAT: Syrphid flies feed on nectar and help to pollinate strawberries and other crops. Also called hover flies or flower flies, these beneficials become active in early spring, providing pest control while other beneficial species are still dormant. The maggots prey on aphids. Some species also feed on spider mites, fungi, or decaying organic matter.

TACHINID FLIES

Many genera; thousands of species

HOW TO SPOT THEM: You can mistake tachinid flies for houseflies, but tachinid flies will frequent flower gardens.

ADULTS: Hairy black, brown, or gray flies, about ½ inch long, with reddish eyes and large bristles at the end of the abdomen

LARVAE: White maggots that feed inside the bodies of host insects

EGGS: Laid on host insects or on leaves

WHAT THEY EAT: Tachinid flies feed on nectar and occasionally on honeydew produced by aphids. The larvae feed inside the bodies of larvae or nymphs of beetles, true bugs, moths, butterflies, and grasshoppers. Tachinid flies parasitize some of the most troublesome garden pests, including squash bugs, cabbage loopers, and Japanese beetles.

Tachinid fly

RELEASING BENEFICIALS

Check garden-supply catalogs and you'll find that you can buy lady beetles, lacewings, and several other kinds of beneficial insects to release in your garden. Cost varies, depending on the insect, but a ballpark figure with shipping included is about $25. Is it worth the cost?

Many experts on biological pest control advise that releasing beneficials in your garden won't have a dramatic effect. And with widespread species such as lady beetles and lacewings, chances are that they're already present in your garden—especially if you've planted some of the nectar-bearing small flowers they prefer and provided permanent refuge such as a small perennial garden.

If you have a home greenhouse, releasing predators can be effective. For example, predatory mites released in a greenhouse can do an excellent job of reducing spider mite problems to a very low level. For more information on predatory mites, turn to Mites. *Encarsia formosa* is a parasitic wasp that provides excellent control of whiteflies in greenhouses.

Some gardeners also have success with releases of *Pediobius foveolatus*, a wasp that parasitizes Mexican bean beetles (see Mexican Bean Beetles for information).

Buying and releasing beneficial insects in your garden certainly won't be harmful, so if you want to experiment, go ahead. However, if you haven't yet set up an insectary garden in your yard, spending $25 on plants that attract beneficials is a better investment than buying lady beetles.

BIRDS

Birds in the vegetable garden are a big benefit because they feed on caterpillars, aphids, beetle grubs, and other pests, but occasionally the birds become a pest problem themselves. Although it's tough to quantify how much birds reduce pest problems, it's worthwhile to take simple steps to attract insect-eating songbirds to your garden.

If birds are visiting your garden to uproot seedlings or to attack tender transplants, you can try to scare them away, but a better strategy is to put a barrier between those plants and the birds.

ATTRACTING BIRDS

Birds desire insects most at nesting season, when they have babies to feed. Insects are protein-rich food to help nestlings grow. Birds nesting elsewhere will occasionally visit your garden to hunt for insects, but for a steady crowd of pest hunters, put up some nest boxes in your yard.

SHELTER AND WATER

All birds need water and a place to take cover from predators. Dense shrubs and evergreen trees offer shelter, and a simple birdbath provides the water. Even if you don't supply water for birds during winter, putting out a birdbath in summer will increase the number of birds that visit your yard during gardening season.

BIRDHOUSES

It's easy to make a simple wren nesting shelf or bluebird house (plans are available in books and on the Internet, and kits are sold at building supply stores and garden centers). Or buy one of the dozens of cute birdhouses available at garden centers or stores that sell bird-feeding supplies and equipment. If you live in the city, select nesting boxes for chickadees, sparrows, or wrens because these are the birds most likely to nest in your yard. Choose birdhouses without a perch, because perches can allow predators easy access to nestlings.

Mount the birdhouses on sturdy posts in a quiet part of your yard. (Birds don't like commotion near their nesting sites.) If there are raccoons or outdoor cats in your neighborhood, mount a predator guard on the post.

PERCHES

Birds like a tall vantage point where they can perch between pest-hunting forays. Set up a few near your garden to accommodate the birds. Don't put perches in the garden, though, or bird droppings on your plants can become a nuisance.

SOLVING BIRD PROBLEMS

City gardens and gardens in mild-winter areas tend to suffer most from bird damage. Starlings, grackles, and other blackbirds will attack lettuce and cabbage-family transplants, sometimes reducing leaves to shreds. In winter these birds travel in large flocks, so they can decimate winter plantings of cabbage-family crops. Protect plants by covering them with floating row covers. If you're worried that insect pests have already found the plants or that slugs may feast under the row cover, use bird netting instead, supporting it with branches as shown below. Substitute 1 × 1 wooden stakes for the sticks for a neater appearance, or use lightweight supports sold by garden centers and mail-order suppliers.

SCARE TACTICS

Another option is to use scare tactics. These are best as a preventive, and it's important to keep changing them so the birds don't become acclimated. Here are some possibilities:

- Stretch reflective tape directly over rows of seeds and seedlings. You can buy this tape at some garden centers and from online garden suppliers. (See the illustration on page 14 for more ideas for shiny scare devices.)
- Set stakes at the corners of the bed and weave fishing line back and forth to create a

Coat hanger pin

Bird netting draped over branched sticks will protect seeded areas, transplants, and fruiting crops from bird damage. Sections of wire coat hangers bent in a V and pushed through the edges of the netting and into the soil secure the setup.

protective spiderweb above the plants. Make this webbing low enough to prevent the birds from ducking underneath but high enough so that the plants don't protrude between the strands. A web of fishing line is a temporary setup, which you'll want to dismantle before the plants grow too large.

- Put out fake snakes and owl decoys at random spots in the garden.

CROWS IN THE CORN

Corn seedlings and sunflower seedlings are favorite foods of crows and jays. They'll spot the seedlings just as they emerge, then tear out the seedlings to eat the tasty seed at the base. To prevent these attacks, plant corn and sunflower seed in shallow trenches and cover each trench

with an arch of chicken wire or gutter guard. Leave it in place until the tips of the seedlings touch the guard; then remove it. By that point, the seedlings will have outgrown the tempting new-sprout stage.

Another option is to sow corn seed and sunflowers indoors in peat pots and plant

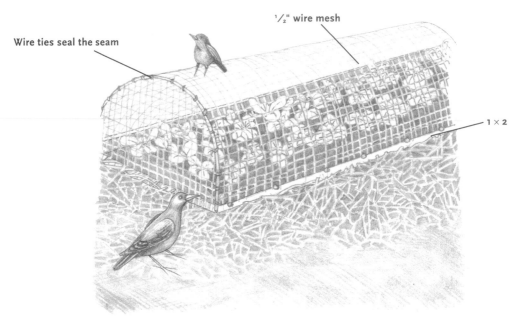

Create a portable cage to temporarily bar birds from a section of your garden. Move it from place to place in spring as needed.

sturdy transplants outdoors. Be sure you harden off the seedlings before transplanting them; handle the roots as little as possible while planting.

For corn, you may be able to fool birds by planting garlic cloves in spring where you plan to plant corn later. Birds may mistake the sprouting garlic for corn seedlings and uproot it, only to be disappointed by the false "corn." Try to time this so your real corn sprouts shortly after the birds have given up on the garlic sprouts. With luck, they won't return.

Crows sometimes visit cornstalks later in the season, too, to feast on the developing corn ears. You can discourage them by jacketing each ear in a paper lunch bag.

BLISTER BEETLES

This Jekyll-and-Hyde insect begins the growing season as an annoying garden pest but then is reborn as a beneficial predator. In spring, blister beetle larvae are present in the soil of gardens and farm fields, having overwintered in a dormant state. They pupate, and in early summer, adult beetles dig to the surface and then crawl or fly off in search of food. The beetles feed on foliage of a wide variety of vegetable crops, and a heavy infestation can nearly defoliate plants.

By midsummer, blister beetles lose their appetite and they seek out sites where grasshoppers have laid eggs. The blister beetles deposit their eggs in the soil nearby. When the larvae hatch, they troll through the soil, searching out and devouring grasshopper eggs for the rest of the summer.

Because of their beneficial side, it's best to tolerate blister beetles when their populations are low, especially if grasshoppers are a problem in your garden. However, if a large crowd is at work on your tomato plants, beans, or other crop, you'll need to take action. Just be careful when you do, because blister beetles earn their name by producing an irritating oil that causes skin blisters.

PEST PROFILE

Epicauta spp.

RANGE: Throughout the United States and Canada

HOW TO SPOT THEM: In early summer, look for black droppings on leaves; this may be the first sign of blister beetle activity. Shake plants and you'll see the beetles drop to the ground or scurry for cover under leaves.

ADULTS: $\frac{1}{2}$- to $\frac{3}{4}$-inch-long beetles with slender bodies and wing covers that don't reach quite to the end of the abdomen. May be gray, black, or metallic; some species have stripes or spots on wing covers.

Adult blister beetle

LARVAE: Soil-dwelling grubs that hatch with straight, narrow bodies and small legs. As they grow, they become fat C-shaped grubs.

EGGS: Laid in the soil

CROPS AT RISK: Tomatoes, potatoes, peppers, spinach-family crops, beans, and other vegetables

DAMAGE: Leaves chewed; in heavy infestations, plants are stripped of foliage.

CONTROL METHODS

HANDPICKING. Often, handpicking works well to control blister beetles. Always wear rubber gloves, though, to avoid the risk of blisters caused by the oil the beetles produce. It's fairly easy to knock the beetles directly off plants into a bucket of soapy water.

VACUUMING. Try using a handheld vacuum to suck up the bugs; then dump them into soapy water.

ROW COVERS. Row covers will keep blister beetles from damaging plants unless the beetles emerge from the soil under the row covers. Check covered plants periodically to catch a sneak attack like this.

OIL SPRAYS. Essential oil sprays may be effective for killing blister beetles, but using oil sprays in midsummer is risky if temperatures are high. See Pest Control for more information.

PYRETHRINS. If blister beetles are so numerous you can't handpick them, spraying pyrethrins may save your crop, but use it only for a severe infestation.

CONTROL CALENDAR

AT PLANTING TIME

* Cover plants with row covers.

WHILE CROP DEVELOPS

* Monitor damage. If it is mild, take no action, because when beetles finish feeding, they lay eggs that will hatch into beneficial predatory larvae.
* Handpick beetles (wear gloves) as you spot them.
* Vacuum beetles off plants.
* Spray essential oil to kill beetles when weather conditions allow.
* As a last resort, spray pyrethrins.

BLOSSOM-END ROT

Brown leathery patches at the blossom end of tomatoes and other fruits are a telltale symptom of blossom-end rot. The spots aren't due to a disease; instead, this disorder develops when there's too little calcium in young fruits. Plants take up calcium from the soil, and it moves through plants dissolved in water. When plants can take up only a limited amount of water, they send that water to the leaves rather than fruits. Without a steady supply of water (and calcium), the cell membranes at the blossom end of the fruit break down and the cells there die.

Water stress due to lack of soil moisture is the most common cause of blossom-end rot, but other factors that interfere with root function also can be to blame: too much water, rough cultivation that destroys crop roots, and cold soil. Soil pH that is too acidic or too alkaline also can interfere with root function and thus lead to blossom-end rot.

Applying too much nitrogen fertilizer may cause blossom-end rot by stimulating so much leaf growth that there's too much competition for calcium. It's also possible for soils to be deficient in calcium, but such deficiency is unusual in an organic garden if you've enriched the soil with a variety of types of organic matter.

RANGE: Throughout the United States and Canada

CROPS AT RISK: Tomatoes, peppers, eggplant, squash, zucchini, melons, watermelon

DESCRIPTION: A small, water-soaked area or tan area forms at the blossom end of an enlarging fruit. As the fruit grows, the affected area widens. The tissues dry and shrink, and the bottom of the fruit may become flat or sunken. Fruits may drop off plants, or if they cling to the vines, pathogens can invade and cause rot (turning the affected area black and soft). On peppers the spots usually are tan rather than brown.

Sometimes the symptoms develop internally even though the brown spot doesn't form on the outside of the fruit.

FIGHTING INFECTION: Periodically inspect developing fruit for symptoms, especially if the weather has been dry or there's been a long period of heavy rain. Remove fruits that show symptoms so that plants can put their energy into new, healthy fruits. Compost the immature fruits because they probably won't ripen properly. If the soil is dry, water well, then mulch to conserve soil moisture.

GARDEN CLEANUP: Remove fallen fruit at the end of the season, as it may harbor secondary rot organisms.

NEXT TIME YOU PLANT: For early plantings, take steps to encourage soil warming, such as shaping a low raised bed or covering soil with plastic mulch before planting. Harden off transplants gradually to promote good root system development. Plant susceptible crops in loose, fertile soil to encourage large root systems. Cultivate carefully to remove weeds; if weeds are growing close to the base of crop plants, gently hand-pull them rather than hoeing. Maintain even soil moisture by mulching and watering as needed. For beds covered with plastic mulch, monitor soil conditions frequently during hot weather to avoid soil overheating. Avoid nitrogen-rich fertilizers. During hot, windy weather, try shading plants during the hottest part of the day. If blossom-end rot has been a serious problem, check soil pH. If it measures below 6.5, add lime as recommended to raise the pH to 6.5 to 6.8. Choosing varieties that are resistant to blossom-end rot can help prevent the problem as well.

CROP ROTATION: A roots-shoots-leaves rotation as described in Crop Rotation may help prevent blossom-end rot by balancing the bank of nutrients in the soil.

BROCCOLI

Broccoli plants that grow quickly in rich, well-drained soil have the greatest chance of producing a big, beautiful flowerhead. Crop timing is important also, especially in spring. Plant too early and cold temperatures will spur plants to form a tiny flowerhead prematurely. Plant too late and heat will cause big plants to bolt suddenly.

Don't be surprised if your first spring crop isn't a great success, and don't give up. Fresh broccoli is worth the trouble. Be sure to plant a fall crop of broccoli, too, because pest problems tend to be less troublesome then, and flowerheads are usually bigger in the fall.

CROP BASICS

FAMILY: Cabbage family; relatives are arugula, broccoli raab, Brussels sprouts, cabbage, cauliflower, collards, horseradish, kale, kohlrabi, mustard, radishes, rutabagas, and turnips.

SITE: Broccoli does best in full sun. However, if you plant a crop in midsummer for fall harvest, provide protection from sun on very hot afternoons.

SOIL: Plant broccoli in well-drained soil that's been steadily enriched with organic matter. Add one shovelful of compost to each planting hole.

TIMING OF PLANTING: For spring crops, start seeds indoors 6 to 8 weeks before the last spring frost. For fall crops, start sowing seeds 10 to 12 weeks before the first expected fall frost. In mild-winter areas, gardeners can sow succession crops and harvest throughout winter.

PLANTING METHOD: It's possible to direct-seed broccoli in the garden, but it's usually safer to start seedlings indoors to protect young plants from insect pests and temperature extremes.

CRUCIAL CARE: Feed plants with compost tea about 2 weeks after transplanting. Side-dress plants with compost or a balanced organic fertilizer when the main flowerhead starts to form.

GROWING IN CONTAINERS: Choose a container at least 12 inches deep and enrich the planting mix with compost. Choose a variety that's known for producing plenty of side shoots.

HARVESTING CUES: Check plants frequently as the expected date of maturity approaches. When the central head reaches about 1 inch in diameter, start checking every other day. The head should be ready to harvest within 1 week's time, depending on the weather.

SECRETS OF SUCCESS

GROW MORE THAN ONE VARIETY. Some broccoli varieties tolerate wet weather well, while others are heat tolerant or cold tolerant (and a few claim to be both). You may need to choose a disease-resistant variety if downy mildew or another cabbage-family disease is widespread in your garden. Choosing a variety to match your conditions and the time of year can double your chance of success before you even sow a seed. To extend the harvest, try growing at least one non-hybrid, open-pollinated variety. These tend to mature less uniformly, and some are excellent side-shoot producers.

PROTECT SPRING CROPS FROM THE COLD. It's important to plant spring broccoli as early as you safely can because hot weather in late spring and early summer can push plants to bolt. However, you also need to be careful about chilling young broccoli plants too much. Tiny seedlings are fairly cold resistant, but once the main stem reaches the thickness of a pencil, exposure to cold can trick the plants into thinking it's time to reproduce. They form a tiny buttonlike flowerhead that never expands.

How cold is too cold? A few days with temperatures below 50°F is all it takes to trigger buttoning. The recommended time to plant broccoli transplants in the garden is about 2 weeks before the last spring frost. However, if temperatures are very unsettled then, it's wise to delay planting. Set the potted plants outside during the day on warm

If small broccoli plants form a "button" head *(left)*, there's no point in trying to stimulate them to grow bigger. Broccoli plants must be mature height, with a thick central stem, in order to produce a sizable head *(right)*.

To encourage production of side shoots (with heads 1 to 2 inches across), cut the main flowerhead with only 6 inches of stem. When harvesting side shoots, cut them as close to the main stem as possible.

den, a broccoli plant can potentially produce a head 2 inches bigger than the same variety might in spring. The trick is to get the fall crop off to a good start during summer heat. Try interplanting broccoli among tall crops such as tomatoes or corn, which will provide some shade from intense summer sun. Apply a thick mulch of straw or chopped leaves to cool the soil. At the opposite end of the season, be ready to drape row cover or newspapers over your broccoli plants if overnight temperatures are expected to fall below the high 20s.

In areas with long, mild fall seasons, well-managed broccoli plants will continue to produce side shoots for as long as 3 months. Space broccoli plants about 2 feet apart if you plan on an extended harvest. This will ensure that the plants won't outcompete one another for water and nutrients.

days, but keep them indoors under lights on cold days. Broccoli transplants can handle this extra time in pots, even if they're becoming rootbound. Just be sure they don't suffer from water stress.

If you've already planted your broccoli plants when a cold spell hits, protect the transplants under a double, or even a triple, layer of row covers. Remove the extra row covers as soon as the weather warms up. It's smart to leave the base layer of row cover in place to protect the young plants from flea beetles and other insect pests. If your plants aren't covered, be mindful of the wind. Chilling winds are hard on young plants, too, so set up a windbreak, such as a section of snow fencing, on the side facing into the wind.

LEARN TO LOVE FALL BROCCOLI. In the temperate conditions and warm soil of the fall gar-

REGIONAL NOTES

NORTHEAST AND EASTERN CANADA

The Swede midge is a small insect pest native to Europe that recently made the jump to North America and has become a problem on broccoli and related crops. It appeared first in Ontario in 2000 and has spread within Ontario and into Quebec, New York, and Massachusetts.

As the larvae of these tiny flies feed on growing tips of broccoli, they secrete an enzyme that causes plant tissue to break down. Infested plants produce a flush of small shoots and heads. Leaf stems swell, and leaves may be crinkled and puckered. At feeding sites, brown, corky scars appear, especially at the growing tips and on leaf stems. The pests are too small to see easily, so it's possible to mistake the symptoms for a nutrient deficiency or other reaction to stress.

In Europe, the pest is relatively minor because natural predators and parasites prevent population explosion. But here in North America, no beneficial insects seem to be attacking the midges, allowing them to invade commercial fields and rapidly reach epidemic levels.

Swede midges cannot fly very far; it's suspected that the pest will spread as pupae in soil or on infested transplants. If you live in eastern Canada or New England and find suspicious symptoms on broccoli plants (or other cabbage-family crops), put samples of damaged plants into sealable plastic bags and take them to your local Cooperative Extension office for an official diagnosis. Be careful not to spread soil from potentially infested beds to other parts of your garden or outside your garden.

PREVENTING PROBLEMS

BEFORE PLANTING

- Set up commercial insect traps to lure and trap looper and armyworm moths.
- Plant annuals and perennials with small flowers to attract aphid predators to your garden.
- Remove wild mustard and other cabbage-family weeds in and around your garden.

AT PLANTING TIME

- Put a cutworm collar around each transplant in spring.
- Cover seedbeds or transplants with row covers and seal the edges to keep out pests (see Row Covers).

- If you're not covering plants with row covers, use barriers or screen cones to protect individual transplants from cabbage maggot flies.
- For summer plantings, water well and mulch the soil immediately to keep it cool.

WHILE CROP DEVELOPS

- Handpick eggs and larvae of leaf-eating caterpillars frequently.
- For plants that aren't covered, spray a garlic or hot pepper spray to deter aphids and other pests (but not on leafy crops that are close to harvest).

AFTER HARVEST

- Clear out crop debris, including roots, from the broccoli bed. Compost healthy residues. Destroy or discard diseased residues.
- After the spring harvest is complete, renew the soil with compost before replanting in that spot.

TROUBLESHOOTING PROBLEMS

Although there's a long list of potential pest and disease problems that can trouble broccoli, some of the most widespread problems are caused by weather and soil conditions. The pest, disease, and cultural problems listed here are the most common in home gardens. If your plants show symptoms that aren't described here, refer to Cabbage for more symptom descriptions, causes, and solutions.

CURLED LEAVES. Aphids suck sap from leaves, causing leaves to curl and become deformed. Look for clusters of small, gray-green insects on leaf undersides and at the growing tip of the plant. See Aphids for controls.

NEW SEEDLINGS OR TRANSPLANTS EATEN. Cutworms or slugs and snails are responsible. Cutworms often cut transplants off cleanly near soil level; see Cutworms for preventive measures. Slugs and snails leave behind a slime trail; see Slugs and Snails for controls.

MANY SMALL HOLES IN LEAVES. Flea beetles chew small holes in leaves, creating a shot-hole appearance. Look closely and you'll see the tiny black beetles on the foliage. They can cause serious damage to young transplants. See Flea Beetles for controls.

LARGE HOLES IN LEAVES. Cabbage loopers, imported cabbageworms, or slugs and snails chew on broccoli foliage and florets. Look for green caterpillars with white stripes on plants; these are cabbage loopers. Imported cabbageworms are velvety green caterpillars. Handpick these pests as you find them. For more controls, see Cabbage Loopers and Imported Cabbageworms.

Since slugs and snails feed at night, you may not find them on plants, but you probably will find silvery slime trails on the plants. See Slugs and Snails for controls.

LEAVES TURNING YELLOW. Downy mildew and cabbage yellows are fungal diseases that first show up as yellow areas of leaves. If downy mildew is the cause, you'll find a white coating on leaf undersides (see Downy Mildew for controls).

Cabbage yellows is caused by a Fusarium fungus. Infected leaves may be twisted and eventually drop off the plant. See Fusarium Wilt to prevent future problems.

PLANTS FORM TINY FLOWERHEADS PREMATURELY. Cold temperatures, water stress, weed competition, or nutrient stress can cause "buttoning" of broccoli heads. Once it happens, there's no way to force or persuade the tiny flowerheads to enlarge. To prevent buttoning, avoid planting broccoli transplants too early in spring; if temperatures below 50°F are expected, keep the plants warm by covering them with cloches, a plastic tunnel, or medium-weight row cover. Keep plants growing rapidly throughout their development.

FLOWERING STEM RAPIDLY ELONGATES. Exposure to heat or drought can cause broccoli plants to bolt—the flower stem shoots up fast, bypassing the harvestable stage of tightly closed flower buds. This is a problem of spring broccoli crops, and once it happens, the best course of action is to pull out the plants. Even if you cut out the main flower stem, chances are that side shoots will also flower rapidly. To avoid this problem in the future, start your spring crop as early as possible and choose bolt-resistant varieties.

STEM OF FLOWERHEAD IS HOLLOW INSIDE. When broccoli grows too fast or can't take up enough boron, the main stem develops a hollow opening. As long as rot hasn't set in, the flowerhead is still usable. To prevent hollow stem in the future, enrich your garden soil with compost and cover crops (boron deficiency is rare in soils with sufficient organic matter content). Avoid applying too much nitrogen fertilizer to prevent sudden rapid growth spurts.

BROWN, ROTTING AREAS ON FLOWERHEADS. This type of soft rot occurs when bacteria in the soil splash up onto the plants, and it can spread rapidly in warm, wet weather, ruining the heads. There's no treatment for infected heads. If you catch soft rot early, you can cut

GROWING "TURNIP BROCCOLI"

Broccoli raab is more closely related to turnips than to regular broccoli; it's also called rapini, Italian turnip, and turnip broccoli. Adjust your expectations when you grow broccoli raab. These plants won't produce a large flowerhead like regular broccoli does. Instead, the harvest from broccoli raab is young flowering shoots, with heads only 1 inch or so across. Flower buds, stems, and leaves—all are tasty when steamed or sautéed and mixed with pasta and Parmesan cheese.

If you like broccoli raab, try Chinese broccoli (gai lohn), too. It's a broccoli relative that's traditional in Asian cooking and needs the same treatment in the garden as broccoli raab.

Broccoli raab isn't hard to grow. One advantage is that it's fast growing, sometimes ready to harvest as soon as 6 weeks from sowing. That means there's less time for pests and diseases to create problems that will reduce yields. Broccoli raab's major requirements are rich soil, cool weather, and a steady supply of moisture. In hot-summer areas, sow the spring crop as soon as the soil can be worked and soil temperatures rise into the 40s. Sow seeds 1 to 2 inches apart and thin the seedlings to about 6 inches apart (you can cook and eat the thinnings, too). The plants will resprout after the first cutting, and if the weather remains cool long enough, you can expect a second and even a third harvest. Harvest broccoli raab and Chinese broccoli at the same stage of growth, when flower buds are about 1 inch across. Cut the stems 10 to 12 inches long. Once the weather turns hot, though, uproot the plants and compost them. In the Far North, you can try succession planting broccoli raab through summer, but a surge in temperatures (above 80°F) can turn the flavor bitter.

Fall is a wonderful season for broccoli raab. For the fall crop, look for varieties listed as heat tolerant or intended for fall planting. Once the worst of the summer heat is past, sow a small patch every 2 weeks until about 3 weeks before the first expected fall frost. Light frost will improve the flavor. In mild-winter areas, a mulched late fall planting will overwinter and produce a crop in early spring. Covering the crop with mulch is good insurance because if temperatures drop much below 20°F, the plants will die.

Chinese broccoli

Broccoli raab

the head, excise the brown areas, and eat the other sections of the head. Don't try to store them, even in the refrigerator—the rot may continue to appear and spread. To avoid this problem, choose varieties with dome-shaped heads with tight florets. Plant broccoli in well-drained soil, and avoid wetting the heads when you water the plants. Cut stems on an angle when you harvest so water runs off—the rot organisms can also invade cut stems and spread

from there to the flowerheads. Also, apply mulch a few inches thick underneath the plants to stop spore movement.

WORMS IN HEADS. Cabbageworms and cabbage loopers sometimes hide in broccoli heads even after harvest. To avoid surprises in the kitchen, soak cut heads in salt water or warm water with a bit of vinegar added for 15 minutes before cooking, which will drive out any lurking cabbageworms.

BRUSSELS SPROUTS

Brussels sprouts are large, slow-growing plants that take up a lot of garden space, which means that most gardeners grow them as a treat, not as a mainstay of their vegetable supply. If you've never eaten homegrown Brussels sprouts, though, please try growing this crop once—you're in for a happy surprise.

Most commercially produced Brussels sprouts (primarily grown in California) are never exposed to the frosty weather that triggers the development of a pleasingly sweet and nutty flavor. The sprouts you harvest on a chilly fall afternoon will be unlike anything you've ever eaten before.

CROP BASICS

FAMILY: Cabbage family; relatives are arugula, broccoli, broccoli raab, cabbage, cauliflower,

collards, horseradish, kale, kohlrabi, mustard, radishes, turnips.

SITE: In the garden, Brussels sprouts need full sun. But seedlings in pots will grow fine outdoors in partial shade (bring them inside if temperatures will exceed 80°F).

SOIL: Provide a well-drained planting area that's been steadily fed with organic matter. Add a shovelful of compost to each planting hole.

TIMING OF PLANTING: Timing of planting is linked to when your first fall frost occurs.

See "Aim for Post-Frost Harvest" on this page for details.

PLANTING METHOD: You can sow seed directly in the garden, but most people start seeds indoors and plant transplants outside.

CRUCIAL CARE: Mulch the soil and monitor soil moisture carefully, especially during hot weather; don't let the soil dry out more than a few inches deep. Side-dress plants with bloodmeal or cottonseed meal when sprouts first form.

GROWING IN CONTAINERS: Brussels sprouts aren't a good choice for containers.

HARVESTING CUES: Sprouts develop from the bottom of the plant up. Begin harvesting as soon as sprouts reach 1 inch in diameter; don't leave sprouts on the stem too long or the tightly wrapped leaves begin to loosen, and aphids may invade. Brussels sprouts leaves are edible, too, but they're a bit thicker and tougher than other cabbage-family greens. Chop them up small before steaming them.

SECRETS OF SUCCESS

INTERPLANT TO START. Brussels sprouts plants grow tall (3 to 5 feet) and wide (up to 2 feet), but it takes them a while to reach that size. When you first set out transplants 18 to 24 inches apart, you'll be dismayed at how much open space is left between the plants. Make productive use of the open area between the transplants by sowing a fast-growing crop of lettuce or other greens. (Choose heat-tolerant varieties if you live in an area with hot summers.) The greens will reach cutting size while the Brussels sprouts plants are still adolescents. Cut the greens back to ground level when you harvest; water the bed well; and then cover the soil with a thick, cooling mulch of straw or shredded leaves.

AIM FOR POST-FROST HARVEST. There's nothing to gain in rushing to plant Brussels sprouts in spring. In most areas, Brussels sprouts are one of the last crops planted, after tomatoes, peppers, and melons. That's because you want to delay harvest until after frost. It's not that Brussels sprouts are inedible before frost, but you'll be disappointed with the flavor. Your goal is to allow the plants enough time to form firm,

Preserve your Brussels sprouts by stripping off all leaves and cutting the stalk at ground level. Put these Brussels sprouts logs in your root cellar or garage, and use up the sprouts over time.

tender sprouts by the time frost hits, but not to allow so much time that the sprouts become loose and overly mature before the temperature dips below freezing.

To figure out when to plant, start with the number of days to maturity (the time from transplanting to harvest) of the variety you're growing. Count back that many days from your first expected fall frost date. That's the time to plant transplants in the garden. If you're starting plants from seed, count back another 30 to 40 days from your planned planting-out date; that's the day to sow seeds (either indoors or in a garden bed).

It's a bit tricky to establish Brussels sprouts transplants, because they prefer cool weather (60° to 70°F). If you're planting them in July or August in a hot-summer area, transplant seedlings on a cloudy day. Pamper the transplants for the first couple of weeks by shading them on hot afternoons and watering daily if needed.

BOOST SPROUT SIZE IF DESIRED. It's your choice: Do you want your plants to produce the maximum number of sprouts, or do you want fewer but fatter sprouts? Brussels sprouts plants keep growing taller and producing more leaves until temperatures stay below 40°F. A sprout forms in each leaf axil, so the more

For maximum productivity, sow lettuce seed between transplants on the same day you set out your Brussels sprouts. Harvest the lettuce about 1 month later.

leaves, the more sprouts. If you want lots of sprouts, let the plants grow as tall as they can (tie them to a sturdy stake if they seem in danger of falling over). Or to channel the plant's energy totally into feeding the sprouts, you can remove the growing point to prevent more leaf growth. Once sprouts have formed on the bottom 12 inches of the plant, cut off the top few inches, including the growing point. It's important to continue feeding and watering your crop or the plants will stop all activity, including sprout development. If lower leaves are turning yellow, remove them to expose the sprouts to more sunlight.

PREVENTING PROBLEMS

BEFORE PLANTING

- Plant annuals and perennials with small flowers to attract aphid predators to your garden.

AT PLANTING TIME

- If the weather is moderate enough, cover new transplants with row covers and seal the edges to keep out flea beetles and maggot flies (see Row Covers).
- If the weather is hot, spray new transplants with kaolin clay to deter flea beetles. Use collars or screen cones to block maggot flies (see page 91).

WHILE CROP DEVELOPS

- Handpick cabbage loopers and other caterpillars that feed on foliage.

AT HARVEST TIME

- Check sprouts frequently for aphids. If aphids appear, harvest sprouts on the young side so aphids can't infiltrate.

AFTER HARVEST

- Remove and compost crop debris, including roots. Discard or destroy diseased plants.

TROUBLESHOOTING PROBLEMS

Brussels sprouts can suffer many problems that trouble cabbage-family crops. The ones listed below are the most common for this crop; refer to Cabbage for a complete listing of pest and disease symptoms and causes.

STICKY SUBSTANCE AND SMALL INSECTS ON SPROUTS. These are cabbage aphids, which suck sap and excrete a sugary honeydew. The aphids also crawl between layers of leaves if the sprouts are not tight. Badly infested sprouts may not be usable. Wash aphids off sprouts with a strong stream of water (see Aphids).

MANY SMALL HOLES IN LEAVES. Flea beetles chew small holes in leaves, creating a shot-hole appearance. Look closely and you'll see the tiny black beetles on the foliage. See Flea Beetles for controls.

LARGE HOLES IN LEAVES. Cabbage loopers and imported cabbageworms chew on foliage. Look for green caterpillars with white stripes on plants; these are cabbage loopers. Imported cabbageworms are velvety green caterpillars.

Handpick these pests as you find them. For more controls, see Cabbage Loopers and Imported Cabbageworms.

PLANTS WILT DURING HOT WEATHER. Cabbage maggots and club root are possible causes.

Plants infested with cabbage maggots sometimes look purplish and stunted. To confirm the presence of maggots, uproot an affected plant and you'll see winding tunnels through the roots; the roots may also be rotting. Badly infected plants won't produce usable sprouts. See Cabbage Maggots to prevent future problems.

If club root is the cause, you'll find swollen galls rather than tunnels. Infected roots also suffer from secondary rot. See Club Root for information.

CABBAGE

Cabbage is exciting when you grow a full range of Asian cabbage crops along with standard heading cabbage. Pac choi, napa cabbage, choy sum, and tatsoi are some of the most popular Asian cabbages. They're generally easier to grow than regular cabbage, and you'll enjoy their subtly different flavors and textures. Botanically speaking, Asian cabbages are more closely related to turnips than to cabbage, but they're grown and used similarly to cabbage.

Cabbages are cool-weather crops that grow best in spring and fall in most regions. In the Deep South, cabbage grows well through winter, and gardeners in cool-summer areas can enjoy cabbages in summer, too. It's the fall crop of cabbage that truly shines because pest problems are less severe and the heads will develop steadily in cool fall conditions, tolerating light frosts.

If you grow standard heading cabbage to make sauerkraut, then you'll need to harvest several heads at one time. Otherwise, planting even a single six-pack of cabbage can be too much. Remember, though, that overplanting isn't a problem as long as you've established a connection with a local food bank in advance. Share the extra harvest if several heads happen to mature at once. Or avoid a cabbage glut by

You'll never grow bored with cabbage if you plant a mixed bed of regular cabbage, Asian cabbages and greens, and lettuces. Flowering annuals help to draw beneficials.

starting your own seedlings. Sow just two or three containers; repeat 2 or 3 weeks later for an extended harvest if your season allows.

CROP BASICS

FAMILY: Cabbage family; relatives are arugula, broccoli, broccoli raab, Brussels sprouts, cauliflower, collards, horseradish, kale, kohlrabi, mustard, radishes, turnips.

SITE: Plant cabbage in full sun.

SOIL: Rich soil is important for cabbage. Plant in well-drained beds that have been steadily enriched with compost or cover crops. Add one shovelful of compost to each planting hole as you plant transplants. If you're working with poor soil, amend the planting area with about 3 pounds of alfalfa meal per 100 square feet at planting time as well.

TIMING OF PLANTING: Timing of planting varies depending on the type of cabbage or Asian cabbage you're growing, so check seed packet labels for specific timing. To grow a spring crop of regular cabbage or napa cabbage, it's typical to sow seeds indoors 7 to 9 weeks before last spring frost and then transplant to the garden 1 to 3 weeks before last expected frost. You can direct-seed Asian cabbages outside as soon as the soil can be worked in spring.

For fall crops, set transplants outdoors at least 6 weeks before the first expected hard frost in fall (timing will vary depending on the days to maturity of the variety you're growing). Or sow seeds about 12 weeks before the first hard frost. In the South, several succession plantings are possible. Shade seedbeds or young plants on hot

days until they become adapted to the garden, as shown on page 328.

Sow seeds of leafy crops such as tatsoi directly in the garden as soon as soil can be worked in spring. Make succession plantings every 3 weeks. Take a summer break, then resume succession sowing in late summer.

PLANTING METHOD: You can sow seeds directly in the garden or plant transplants—transplants generally withstand pest and disease problems better than seedlings. Some Asian cabbage crops grow better when direct-seeded, but only if you protect the seedlings from insect damage.

CRUCIAL CARE: Cabbage needs a steady supply of nitrogen and other nutrients for strong growth and head formation, especially if you use a close spacing. Apply 1 cup of liquid fish/kelp solution per plant weekly at planting and for 2 weeks afterward. Side-dress with blended organic fertilizer when heads start to form. When planting cabbage in summer for a fall crop, pay close attention to soil moisture while plants are becoming established. Moisture stress, especially on hot summer days, can shock young plants, resulting in slow growth and poor head production.

GROWING IN CONTAINERS: Heading cabbages require too much space and time to be worth growing in containers; fast-growing types like choy sum and tatsoi are a better choice. Add extra compost to create a rich, moisture-holding mix.

HARVESTING CUES: When regular cabbage heads reach the size of a softball, squeeze them to test firmness. Check napa cabbage when heads reach about 12 inches tall. A firm head is ready to be cut. Begin cutting individual outer leaves of leafy types about 1 month after planting, or use the cut-and-come-again technique.

SECRETS OF SUCCESS

FIND THE RIGHT WINDOW IN SPRING. Cabbages need at least 60 days from transplanting date to harvest, but they grow best when temperatures are in the 60s.

Cabbage plants can withstand some temperature fluctuations, but there are two key temperature exposures to avoid. The first is to protect cabbage transplants from any long stretch of temperatures below 50°F. If the transplant stems are thicker than a pencil, the plants will be mature enough to interpret the cold weather as a signal to send up a flowerstalk. You also need to avoid exposing enlarging cabbage heads to temperatures above 90°F. Heads are more likely to split or develop strong flavor when the weather is so hot. Watch the forecast. If a surge in temperatures is predicted, harvest heads on the young side rather than risk losing them.

MIX UP YOUR CABBAGES. Rotating cabbage-family crops from place to place in your garden each year is a good practice because of soilborne diseases and because cabbage and its relatives are heavy-feeding crops. Planting a mixed bed of Asian cabbages and regular cabbage is a great way to maintain your rotation and supply your culinary efforts with a beautiful variety of flavors, colors, and forms. Some Asian cabbage crops, such as tatsoi, are fast-growing, ready to pick in 45 days from seed, which also extends your harvest. A densely planted bed shades the soil, which is good for the plants and stifles weeds.

Cabbages do well in a 2-1-2 or 3-2-3 planting scheme. In this type of planting, imagine your bed as a series of triangles, as shown below. In a 2-1-2 scheme, you begin planting at one end of the bed, planting two plants across from one another and about 15 inches apart. Then you'll set a single plant in place at the center of the bed and about 15 inches away from each of the first two plants. In the open space left at the edges of

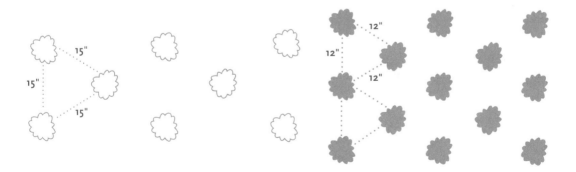

Planting cabbages closely in a triangular pattern uses space well, shades the soil, and helps prevent weed problems.

the bed, you can sow seeds of small, fast-growing cabbage-family crops, such as tatsoi, mizuna (Japanese mustard), or radishes. Or for greater diversity, sow lettuce seed, dill, or a low-growing annual flower in those gaps.

Setting cabbages 12 inches apart results in smaller heads that are ready to harvest sooner (and provides a more manageable amount of food to use at one time). If you want some large cabbage heads, increase the distance to 18 inches or even 2 feet for large-headed varieties. With the wider spacing, mulching between plants becomes more important to stifle weed competition.

REGIONAL NOTES

DEEP SOUTH

Winter is a great time to grow Asian cabbage crops from Florida west to Texas. But keep an eye out for damage by the yellowmargined leaf beetle, a pest native to South America that was accidentally brought to the United States in the 1940s. This beetle is a potential threat to any cabbage-family crops, but it seems to like turnip greens, pac choi, napa cabbage, mizuna, and other close turnip relatives the best. The yellow-margined leaf beetle hasn't become a major pest because many chemical insecticides that commercial growers apply to kill cabbage loopers and other foliage pests also kill the yellowmargined leaf beetle. However, in an organic home garden, beetle infestations can suddenly take hold, possibly because there are no natural predators or parasites that are likely to attack it in the United States.

The beetles and their larvae feed low on the plants and near the center—they can cause considerable damage before you know it unless you inspect plants carefully. The dark beetles are less than ½ inch long, and their backs have a yellow or brown edging. The brownish or gray larvae are also small, with a dark head and pointed tail end. They chew many small holes in leaves. Feeding in groups, the larvae can strip stems.

It's thought that this pest is active throughout the winter and dormant in summer. If you find damage from these beetles, be sure to plant your next cabbage-family crop in a different part of the garden. Cover the new planting with row covers so that emerging beetles can't feed or lay eggs there.

SOUTHEAST

If you find caterpillars with a distinctive pattern of black crosswise stripes and a yellow stripe up the sides on your cabbage plants, they're not mutant cabbage loopers. They're cross-striped cabbageworms, and they tend to feed on young leaves and growing tips of cabbage-family crops. These caterpillars are the larvae of yellowish brown moths that lay masses of flat yellow eggs on foliage in spring and summer. The larvae feed for 2 to 3 weeks, and there can be four generations per year. Control them as you would cabbage loopers (see Cabbage Loopers).

PREVENTING PROBLEMS

BEFORE PLANTING

- Set up commercial insect traps to lure and trap looper and armyworm moths.

- Plant annuals and perennials with small flowers to attract aphid predators to your garden.
- Remove cabbage-family weeds that are growing in and around your garden.
- Monitor temperatures to calculate degree days and predict the time of emergence of cabbage maggot flies.

AT PLANTING TIME

- Put a cutworm collar around each transplant.
- Cover seedbeds or transplants with row covers and seal the edges to keep out pests (see Row Covers).
- If you're not covering plants with row cover, use barriers or screen cones to protect individual transplants from cabbage maggot flies.

WHILE CROP DEVELOPS

- Handpick eggs and larvae of leaf-eating caterpillars frequently.
- For plants that aren't covered, spray a garlic or hot pepper spray to deter aphids and other pests (but not on leafy crops that are close to harvest). See Pest Control for information.
- If temperatures are rising into the 70s, remove row covers when crops are large enough to withstand some pest damage.

AFTER HARVEST

- Clear out all cabbage crop debris, including roots, from the garden. Compost healthy residues. Destroy or discard noticeably diseased residues.

TROUBLESHOOTING PROBLEMS

FEW OR NO SEEDLINGS SPROUT. Damping-off can be a problem when you start cabbage crops from seed in the garden. Seed will be dark or rotted, or shoots that germinated will have died before emerging. See Damping-Off to prevent future problems.

NEW SEEDLINGS OR TRANSPLANTS EATEN. Cutworms or slugs and snails are responsible. Cutworms often cut transplants off cleanly near soil level; see Cutworms for preventive measures. Slugs and snails leave behind a slime trail; see Slugs and Snails for controls.

YOUNG PLANTS FORM A FLOWERING STALK. When young cabbage plants are exposed to several days of chilly weather (below 50°F), they will prematurely produce flowers and seeds without ever forming a harvestable head. It's usually a problem only with the spring crop. Uproot and compost plants that bolt. To prevent bolting in the future, protect young transplants with cloches or other covers and/or plant them a little later in spring.

WATER-SOAKED SPOTS ON LEAVES. This is an early sign of white mold, a fungal disease that can ruin whole heads (see White Mold).

DARK SPOTS ON OLDER LEAVES. Alternaria leaf blight is a fungus that spreads fast in hot, wet weather. Check the spots closely for targetlike rings, a confirming sign of Alternaria. Remove infected leaves. See Alternaria Blight for controls.

BLUISH OR PALE SPOTS ON LEAVES OR STEMS. Blackleg is a fungal disease that can kill young plants. Black, sunken areas form on stems, and roots decay, causing plants to topple over. The fungus infects only cabbage-family crops and weeds and some ornamentals. Remove diseased plants. To prevent future problems, start a rotation for cabbage-family crops (a 4-year rotation is ideal). Consistently remove cabbage-family weeds such as shepherd's purse from your garden. Plant only disease-free seed.

LEAVES TURNING YELLOW. Downy mildew and cabbage yellows are possible causes.

Look for a white coating on leaf undersides to confirm downy mildew. See Downy Mildew for controls.

Cabbage yellows is caused by a Fusarium fungus. Infected leaves may be twisted and eventually drop off the plant. See Fusarium Wilt to prevent future problems.

V-SHAPED YELLOW AREAS ON LEAVES. Black rot is a bacterial disease of cabbage-family crops. It spreads easily; uproot and destroy infected plants. See Rots for controls.

BRONZING OF LEAVES. Onion thrips are the probable cause of this damage. Spray plants with water to wash off thrips (unless heads have already started to form). See Thrips for more controls.

MANY HOLES IN LEAVES. Diamondback moth larvae chew on leaves and can damage growing tips of young plants. Look for very small (about ¼ inch) greenish yellow caterpillars on foliage that wiggle rapidly when disturbed. They are the larvae of night-flying moths with a distinctive diamond shape on their wings. Handpick the caterpillars when you spot them. This pest is usually serious only on small plants. Applying *Bacillus thuringiensis* var. *kurstaki* is effective, although some diamondback moth

populations have developed resistance to Bt. To avoid future problems, cover seedbeds and new transplants with row covers and seal the edges (see Row Covers). You can remove the covers when plants are established; the plants should be able to withstand the larvae.

MANY SMALL HOLES IN LEAVES. Flea beetles chew small holes in leaves, creating a shot-hole appearance. Look closely and you'll see the tiny black beetles on the foliage. See Flea Beetles for controls.

LARGE HOLES IN LEAVES. Cabbage loopers, imported cabbageworms, or slugs and snails chew on cabbage foliage. Look for green caterpillars with white stripes on plants; these are cabbage loopers. Imported cabbageworms are velvety green caterpillars. Handpick these pests as you find them. For more controls, see Cabbage Loopers and Imported Cabbageworms.

Since slugs and snails feed at night, you may not find them on plants, but you probably will find silvery slime trails on the plants. See Slugs and Snails for controls.

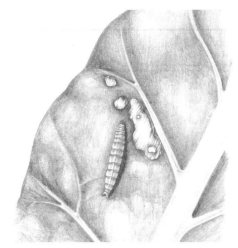

Diamondback moth caterpillar

SKELETONIZED LEAVES. Beet armyworms are a common pest in the southern United States. Check plants for green or dark caterpillars, which may be hidden inside fine webs. Handpick the caterpillars. See Armyworms for more controls.

CURLED OR CUPPED LEAVES. Cabbage aphids suck plant sap and can stunt growth of plants. Check leaf undersides for colonies of pale green aphids. See Aphids for controls.

TWISTED, BROWN LEAVES. Harlequin bugs are small orange-and-black insects that suck sap from leaves. Symptoms tend to show up during hot weather. See Harlequin Bugs for controls.

PLANTS WILT DURING HOT WEATHER. Harlequin bugs, cabbage maggots, and club root are possible causes. Look for splotchy areas on leaves (a sign of harlequin bug feeding), or small orange-and-black insects on leaves. Plants may recover if you get rid of the bugs. See Harlequin Bugs for controls.

Plants infested with cabbage maggots sometimes look purplish and stunted. To confirm the presence of maggots, uproot an affected plant. Roots will be filled with winding tunnels. Rot can set in due to the maggot injury. Plants won't produce usable heads. See Cabbage Maggots to prevent future problems.

If club root is the cause, you'll find swollen galls rather than tunnels. Fungi and bacteria that cause rotting can invade and make things even worse. See Club Root to prevent future problems.

STUNTED PLANTS. Viruses, cabbage yellows, wirestem, club root, or water stress can cause stunting.

The sweetpotato whitefly can spread cabbage viruses. Older leaves often show mottling. To prevent viruses, concentrate on controlling the whiteflies (see Whiteflies).

Cabbage yellows is caused by a Fusarium fungus. Infected leaves may be warped or drop off the plants. For controls see Fusarium Wilt.

Wirestem is caused by one of the fungi (*Rhizoctonia*) that also causes damping-off of seedlings and root rots. The main stem becomes dark, tough, and wiry. Infected plants won't produce good heads. See Damping-Off for prevention.

To check for club root, uproot a stunted plant and inspect roots for galls. See Club Root to prevent future problems.

If there's no sign of cabbage yellows or club root, the problem could be too little water. The plants probably won't yield well once they've become stunted. For the future, mulch around plants well and check soil moisture regularly.

HEADS CRACK OPEN. Cabbage crops that form tight heads suffer from this problem when the leaves inside the heads grow too fast. The pressure of the enlarging leaves causes the heads to split. This can happen if plants that are developing heads suddenly take up a lot of water or have an oversupply of nutrients (especially nitrogen). Harvest cracked heads immediately. They are still usable as long as they haven't rotted, but they won't store well. To prevent splitting in the future, monitor soil moisture frequently. If you find that the soil has dried out several inches deep, rewet the soil gradually over the course of a few days instead of all at once. If the soil is saturated (such as after a heavy rain) or you suspect that too much nitrogen is fueling growth, break some of the plant's roots to slow the uptake of water and nutrients. Do this by grasping the head with both hands and twisting it about 180 degrees, or by pushing a shovel blade into the soil on an angle near the base of the plant to slice through some of the roots.

MARGINS OF LEAVES INSIDE HEADS TURN BROWN. Internal tipburn can occur in any type of cabbage, and often you won't discover it until you harvest a head and cut into it. Tipburn happens when plants can't take up enough water, which creates a deficiency of calcium within the plants. The brown areas may be narrow strips along leaf edges or cover substantial areas of a leaf. These areas are subject to rot. The phenomenon is similar to blossom-end rot of tomato, and prevention is similar (see Blossom-End Rot).

DEFORMED HEADS. If aphids feed on young cabbage plants, they can damage the growing tip of the plants, which causes distortion of the heads when they form. To prevent future problems, cover young transplants with row covers. See Aphids for more controls.

HEADS ROT. Various bacteria and fungi can cause rot, especially if the heads have been damaged by insect feeding or other diseases. If you find any rotten heads, check the rest of your crop and harvest it as soon as possible. Remove all crop debris from the garden. See Rots and White Mold to prevent future problems.

CABBAGE LOOPERS

Loopers are the most destructive of the chewing "worms" that may feed on crops in the cabbage family. Cabbage looper moths emerge from silky cocoons in spring across the southernmost United States and fly north, eventually reaching throughout the United States and Canada. When they find suitable host plants, they lay eggs in small groups, which hatch a few days later into little green caterpillars.

The caterpillars first chew holes in leaf tissue between small leaf veins, creating an effect like a window in the leaf with many small panes. But as they grow, the caterpillars feed more heartily, reducing leaves to nothing more than a midrib.

After feeding for a few weeks to a month, the loopers spin a loose cocoon and attach it to a leaf or stem. The pupae rest for 1 to 2 weeks, and then moths emerge for another round of egg laying. In the South, there can be several overlapping generations per year; in the North, two generations is more typical. Pupae overwinter in crop debris but can't withstand much freezing weather.

Cabbage loopers leave piles of messy green excrement on leaves, as do imported cabbageworms (see Imported Cabbageworms). If these pests feed at the growing tip of cabbage, broccoli, or cauliflower plants, it can cause formation of multiple heads, which is disappointing but not disastrous. Worse is to cut into a head of cabbage to find that the worms have tunneled into the base of the head and left it filled with holes and excrement, or to cook broccoli and find in the cooking water worms that had been hidden in the openings between tightly packed florets. If loopers have been active in your garden, cut cabbage, broccoli, and lettuce heads on the early side to reduce the chance that loopers will breach the interior.

PEST PROFILE

Trichoplusia ni

RANGE: Throughout the United States and Canada

HOW TO SPOT THEM: Examine leaves of lettuce and cabbage-family plants, especially leaf undersides near the leaf margins, where loopers prefer to feed. Look carefully, as these green pests blend in well on leaf surfaces. Watch for their messy green droppings as well.

ADULTS: Night-flying gray or brown moths with a unique silvery white marking on each forewing; 1½-inch wingspan

Cabbage looper moth

Cabbage looper

Cabbage looper on damaged leaf

LARVAE: Larvae are creamy colored when they hatch but soon change to green caterpillars with narrow pale white stripes down the back and sides; they have three pairs of slender legs near the head and two pairs of stumpy legs near the rear end; up to 1½ inches long. They "loop" their bodies up as they move from place to place.

EGGS: White, dome-shaped eggs in small groups on leaf undersides

CROPS AT RISK: Cabbage-family crops, lettuce, spinach-family crops, peas, tomatoes, and some other crops

DAMAGE: Loopers eat leaves, produce messy droppings, and tunnel into heads of cabbage-family crops and lettuce.

CONTROL METHODS

ENCOURAGING BENEFICIALS. Many kinds of native beneficial insects eat looper eggs and larvae. To draw these beneficials to your garden, keep something flowering in or near your vegetable garden all season (see Beneficial Insects for suggestions).

CHOICE OF VARIETIES. If loopers are a standard pest problem in your garden, try growing red or purple cabbage so you can spot the worms more easily. Avoid savoy-type cabbages because the crinkled leaves provide even better camouflage for the loopers.

AVOIDING INFESTED TRANSPLANTS. Cabbage-family transplants at garden centers may be infested with eggs or even small loopers. To be safe, inspect plants carefully before buying or grow your own indoors or in a covered coldframe outdoors.

ROW COVERS. Cover plants at planting time and leave the covers in place until harvest, if possible, to prevent moths from laying eggs on the plants. To learn how to seal row cover edges, see Row Covers.

GARDEN CLEANUP. In spring, scout your yard and garden for cabbage-family weeds and pull them out to get rid of possible early egg-laying sites. As you harvest, remove all wrapper leaves or side leaves from the garden and put them at the center of a compost pile. (This will kill the pupae or moths as they try to emerge.) In southern states where cabbage loopers can survive over winter, dig or till in crop residues to bury the pupae.

HANDPICKING. Crush any eggs you spot on leaves. Pick off larvae and dump them in soapy water.

INSECT TRAPS. Early in the season, set up two commercial insect traps that include a floral lure to attract the moths. This will not eliminate loopers but can reduce the number of looper eggs laid in your garden.

GARLIC OR HOT PEPPER SPRAYS. At about the time when cabbage looper moths appear in your area, spray plants with garlic or hot pepper sprays. Repeat sprays weekly until plants are close to harvest. See page 308 for directions.

BIOCONTROLS. Apply spinosad or *Bt* var. *kurstaki* when small loopers are present. Biocontrol sprays are less effective when loopers are large. See Pest Control for directions for using these sprays. If you're applying Bt in fall, you'll get best results if you spray on a warm day so that caterpillars are sure to be feeding.

NEEM. If lots of large loopers are causing heavy damage to plants, apply a neem product. See Pest Control for more about neem.

CONTROL CALENDAR

BEFORE PLANTING

- Set up insect traps to lure and trap adult moths.
- Uproot cabbage-family weeds that can serve as egg-laying sites.
- Inspect commercially grown transplants for eggs or larvae before buying.

AT PLANTING TIME

- Cover plants with row cover.

WHILE CROP DEVELOPS

- Handpick loopers and crush eggs.
- Spray garlic and hot pepper sprays to repel moths and loopers.

- Spray *Bt* var. *kurstaki* or spinosad to kill young larvae.
- For an infestation of large loopers, apply neem.

AT HARVEST TIME

- If loopers have been active, cut cabbage, broccoli, and lettuce heads on the young side to minimize amount of damage.
- Remove all wrapper leaves and cuttings from the garden.

AFTER HARVEST

- In areas where loopers can overwinter, dig or till in crop residues to destroy pupae.

CABBAGE MAGGOTS

Inconspicuous cabbage maggot flies may escape your notice, but your crops won't escape the attention of these little flies. They zero in on cabbage-family plants, laying masses of eggs in cracks in the soil near the plants. When the eggs hatch, tiny white maggots head for nearby plants and tunnel down into the roots, where they feed for up to a month before pupating.

Meanwhile, your plants are suffering. They will look stunted and purplish and be prone to wilting. Unfortunately, there's no cure for a plant infested with root maggots.

Unlike many vegetable pests, cabbage maggots tend to be problematic only in the northern half of the United States and Canada. This is a pest that just can't do well in the South because the eggs can't survive well in hot soil. For the same reason, the second and third generations of cabbage maggots in a growing season usually cause less damage than the first.

The only reliable way to control root maggots is to prevent the flies from laying their eggs by your precious plants. In a home garden, that's easy to do with various types of barriers. Commercial radish growers are experimenting with fencing their fields with window screening to keep out flies. In general, cabbage maggot flies navigate near soil level searching for egg-laying sites. When they encounter the window screen barrier, they will try to fly up and over, but by leaving a loose overhang that flops outward at the top of the fence to baffle the flies, growers have been able to reduce maggot damage in fields by up to 80 percent. In a home garden, though, even a 20 percent loss of the radish crop could be a disappointment, so you'll probably want to rely instead on one of the methods described below.

PEST PROFILE

Hylemya brassicae

RANGE: Throughout the United States and Canada, but rarely a problem in the southern half of the United States

Cabbage maggot fly

HOW TO SPOT THEM: Flies and their eggs are difficult to spot. Usually, the first sign of infestation is plants that wilt suddenly. Or plants may look abnormally purplish or blue-gray.

ADULTS: ¼-inch-long gray flies with smoky-colored wings and black stripes down their bodies

LARVAE: ¼-inch-long, white, legless maggots

EGGS: Laid in the soil

CROPS AT RISK: Asian cabbages, arugula, broccoli, Brussels sprouts, cabbage, cauliflower, collards, horseradish, kale, kohlrabi, mustard, radishes, rutabagas, turnips

DAMAGE: Stunted, discolored, wilting plants; plants may die; roots of root crops have winding tunnels filled with frass and maggots; infested roots prone to disease

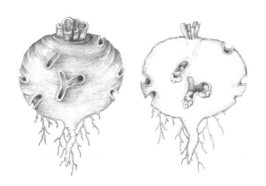

Cabbage maggots and damaged roots

CONTROL METHODS

TIMED PLANTING. Plant radishes early and plant cabbage late to avoid the emergence of the first wave of flies. Another timing tactic is to skip planting cabbage-family crops in spring altogether. Instead, plant them from midsummer on for fall harvest. Crops planted in the heat of summer are less prone to damage because the maggot eggs can't survive well in hot soil. Fall root crops such as turnips may need protection (such as row cover) from a late generation of flies.

CROP ROTATION. Rotating cabbage-family crops from one place to another in your garden won't end your cabbage-maggot problems, but it's very important if you plan to use row covers to protect your crops. If you plant cabbage-family crops in the same bed 2 years in a row, flies may emerge under the row covers, rendering them useless.

INDIVIDUAL BARRIERS. Fashion barriers about 6 inches in diameter from heavy cloth, old carpeting, foam rubber, or tar paper as shown in the illustration below. Slip a barrier in place at the base of each transplant at planting time. These barriers prevent flies from laying eggs in the soil near the plants.

A window screen cage bars cabbage maggot flies from the close contact they seek with cabbage-family plants for egg laying.

SCREEN CONES. Another barrier to cabbage maggot flies is a cone made of window screening. Cut a fan-shaped piece of window screening as shown above, and use a staple gun to fasten it to a piece of wood molding. Set the cone over a transplant and twist it to firm it into the soil as shown above.

ROW COVERS. Cover crops at planting and keep them covered until harvest. See Row Covers for details on sealing row cover edges. This is an especially good approach for root crops, which are hard to protect with individual barriers. Be sure that no cabbage-family crops were planted in the bed the preceding year, or flies could emerge under the cover and lay eggs.

An alternative to row covers is a framed cover made of window screening. Follow the pattern for a bird barrier on page 64, but

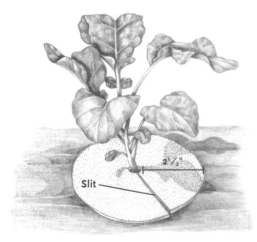

A circle (or square) of a sturdy material, such as tar paper or foam rubber, blocks cabbage maggot flies in search of egg-laying sites. Cut a slit to the center of the barrier so you can slip it around the base of a transplant.

substitute window screening for the wire mesh. Use needle and thread to sew the screening tightly closed along the seams.

REPELLENTS. Hot pepper dusted over the soil surface may repel flies and discourage them from laying eggs. Try combining repellents with a trap crop planted a week or so earlier and not treated with repellent. With luck, the flies will pass up the seedlings protected by repellent and lay eggs only around the trap crop. Be sure you uproot the trap crop, roots and all, and plunge the plants into soapy water to kill the larvae before they pupate. Also, don't plant cabbage-family crops where the trap crop was growing, or larvae left behind in the soil could infest the new crop.

GARDEN CLEANUP. After harvest, check beds to be sure no roots remain in the soil, or larvae in those roots may form pupae and overwinter. Remove all roots and destroy them.

CONTROL CALENDAR

BEFORE PLANTING

- Monitor temperatures to calculate degree days and predict emergence of cabbage maggot flies.
- Sow a trap crop (if you plan to try to protect seedlings with a repellent rather than with barriers).

AT PLANTING TIME

- Plant early or late to avoid damage from the first generation of maggots.
- Cover crops with row cover and seal the edges.
- Place squares or circles of sturdy material around the base of individual transplants.
- Cover individual transplants with screen cones.
- Sprinkle hot pepper or diatomaceous earth on the soil along seeded rows. Reapply frequently.

AFTER HARVEST

- Remove and destroy any remaining roots of cabbage-family crops.

CANTALOUPE

These tasty melons are a favorite of gardeners throughout the United States and Canada. Technically speaking, a cantaloupe is a particular type of muskmelon, and true cantaloupes are more common in Europe than in North America. Most gardeners don't worry about the fine distinctions; they just enjoy planting and eating the tasty, aromatic flesh of these wonderful fruits. To learn about growing cantaloupes (muskmelons) and protecting the plants from pest and disease problems, see Melons.

CARROTS

Pamper your carrots for the first 4 weeks after planting, and chances are good that you'll enjoy a successful harvest 1 or 2 months later. Carrot seeds germinate slowly (over a 3-week period); each seedling produces a tiny, thin taproot that stretches through the soil to full length before it begins to thicken. Anything that interferes with the growth of those delicate young taproots has a lasting effect that can't be overcome, and that's why the first month after planting is critical.

Pest and disease problems of carrots usually aren't serious; most carrot problems stem from poor seed-sowing practices, neglect of seedlings, or lack of proper thinning. The good news is that you can easily avoid these problems by learning the proper techniques—and a few tricks—for sowing carrot seeds and taking care of carrot seedlings.

CROP BASICS

FAMILY: Carrot family; relatives include celeriac, celery, cilantro, dill, and parsley.

SITE: Plant in full sun to light shade.

SOIL: Loose, debris-free soil not too rich in nitrogen is essential.

TIMING OF PLANTING: Sow seed from early spring through midsummer in most areas. In the Deep South and parts of the Southwest and Far West, plant in winter and/or in late summer. Time planting so roots will mature when temperatures are between 60° and 70°F.

PLANTING METHOD: Sow seeds directly in garden beds in narrow or wide rows. Do not transplant.

CRUCIAL CARE: Never allow seedlings to suffer moisture stress.

GROWING IN CONTAINERS: Short-rooted cultivars grow well in containers.

HARVESTING CUES: Bright orange color is a sign that roots are ready to harvest.

SECRETS OF SUCCESS

DIG, DIG. Carrot roots start out thread-thin. Only after the roots reach maximum length do they thicken and toughen up. In heavy or compacted soil, the young roots often get stuck or twisted: When such carrots mature, they'll be stunted and/or deformed. Even if you've built raised beds, always dig test holes to check the condition of the soil. Your goal is a garden bed that contains few or no rocks or pieces of plant debris and has soil that is loose at least as deep as your carrot crop will be long. For example, 'Danvers Half-Long' carrots, which grow 7 inches long, need soil that is loose and open at least 7 inches deep (9 inches would be even better). Take the time to hand-dig to the proper depth. While you work, remove rocks, soil clods, and debris.

EXPERIMENT WITH SOWING STYLE. Carrot seeds are very small; it's hard to sow them evenly. With great patience, you can place them one by

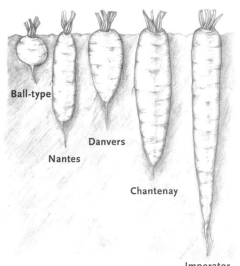

Read variety descriptions carefully when you buy carrot seeds. Pick carrots that will grow well in your soil conditions. Ball-type and Nantes carrots are best for shallow soil; Danvers and Chantenay can produce well in heavy soil; Imperator types need ideal soil conditions.

moisture better than plain soil. Another excellent option is to cover the seeded area with row covers, which protect seedlings from both drought and downpours. Try sowing some radish seeds along with the carrots (1 seed per inch). The radish seeds germinate quickly and help prevent soil crusting. About 3 weeks later, pull out the radishes so they won't compete with the carrots. Another option is to cover the seeded area lightly with grass clippings or row cover.

PLANT MORE THAN ONCE. Sometimes the weather foils your careful planting efforts. For example, a drenching downpour can wash out a seedbed before the seedlings take hold. The best insurance is to limit the size of each planting and to sow repeatedly. Sow some seeds every 2 to 3 weeks. If you're not sure what times of year are favorable for planting carrots in your area, check with your local Cooperative Extension office.

CHECK DAILY. Some vegetable seedlings can tolerate a few days of dryness while emerging, but not carrots: Let the seedbed dry out just once after seeding, and germination may fall as much as 50 percent. Check the seedbed daily after sowing, always with watering can or hose in hand to moisten the surface if it's dry. Use a rose-type nozzle that breaks the stream of water into small droplets. Avoid a heavy gush of water, which will wash seeds right out of the bed.

GRIT YOUR TEETH AND THIN. Most gardeners hate to pull out plants (unless they're weeds). But because carrot seeds are so small, it's nearly impossible to sow them at precisely the right spacing. Chances are, if you sow properly and water frequently, you'll have a prolific crop of carrot seedlings. If you're timid about thinning,

one, but most gardeners prefer to sprinkle the seeds, either from their fingers or direct from the packet. Try various methods to see which one works best for you (refer to page 327 for tricks for sowing small seeds). If you're not sure whether you're sowing enough seed, experiment: Sow lightly in one area, more heavily in the next, and even more thickly in the next. After the seeds germinate, you can decide which seeding rate is best.

COVER SEEDS CAREFULLY. The stems of carrot seedlings are just as delicate as their roots; they can't push through a soil crust or past rocks or soil clumps. Cover rows of carrot seeds with vermiculite, sawdust, or sifted compost. These materials won't form a crust, and they hold the roots will have no room to enlarge later on—and that means a poor harvest. Thin the stand when seedlings are about 2 inches tall, leaving seedlings 1 inch apart. Two weeks later, thin again to 2 to 4 inches apart (depending on the ultimate size and shape of the carrots). If you've never thinned seedlings before, check the suggestions on page 328 for thinning techniques that won't damage tender young plants.

MAGNIFY THE VIEW. Once a week, turn a magnifying glass on carrot foliage. Look for small dark brown, black, gray, or tan spots on the leaves. This is a symptom of leaf blight, and it may be treatable if you catch it early. Yellowing leaves may indicate a lack of nutrients (try spraying with kelp or using a water-soluble organic fertilizer). However, it may also be a symptom of aster yellows or root-knot nematodes. Pull up a couple of plants: If the roots show unusual symptoms such as galls or lots of side roots, diagnose the specific cause and take steps to deal with it.

PLANTING A GRID OF CARROTS

Frustrated with trying to sprinkle carrot seed evenly in rows? Forgetful about thinning the seedlings? Then try a grid-seeding technique. It takes more time than sprinkling seeds, but it eliminates the risk of underseeding or overseeding. Plus it minimizes the need to thin.

Use your fingers or a small dibble to poke shallow ($\frac{1}{4}$ inch deep) holes in a grid pattern, with holes 2 inches apart. Drop two seeds in each hole. Cover the whole grid area with sifted compost or vermiculite. Use a tool or the flat of your hand to firm the soil surface. Water gently to settle seeds in place. When seedlings are 1 inch tall, check for duplicate seedlings (in some planting holes, both seeds will have germinated). Use nail scissors to clip off the duplicate seedlings. If there are gaps in the grid, resow if desired to fill the space.

Making a grid pattern for seed sowing

Thinning seedlings

SPOT-CHECK MATURING ROOTS. As your crop nears maturity, pull up a couple of carrots every few days for inspection. If the roots show holes or tunnels, wireworms or carrot root maggots may have invaded. Taste the carrots. As long as the flavor is good, harvest the whole crop now before the insects cause any more damage. Even if the roots look perfect, it's still important to taste them each time you check. That way, you'll know when the crop has reached its peak of sweetness. When you're satisfied with crop flavor, begin harvesting. (Pick late in the day for optimum sweetness.) Harvesting your crop a little at a time is fine, but don't stop spot-checking: Roots left in the ground can still develop pest or disease problems. If a hot spell is predicted, you may want to harvest the whole crop because the heat could cause bitter flavor.

REGIONAL NOTES

FAR WEST

Motley dwarf is a disease problem that results when carrots are infected by two viruses: carrot redleaf virus and carrot mottle virus. Symptoms are stunting of plants and red to yellow foliage (if plants are infected late in development, they may not be stunted, but foliage will turn color). It's easy to mistake this condition for a nutrient deficiency. The disease may be a problem in northern California and parts of Oregon and Washington. The viruses are transmitted by the carrot willow aphid or by leafhoppers, and the problem is worst on spring carrots. If this problem shows up in your carrot patch, switch to growing cultivars that seem resistant (most Imperator carrots show resistance). Don't leave carrots in your garden over winter because already-infected carrots are the main source of virus when insects begin feeding in spring. Planting for fall harvest can help, but harvest all the carrots in fall at once.

HUMID SOUTHEAST

Most gardeners struggle to provide enough moisture for young carrot seedlings, but in the Southeast, frequent drenching rains can saturate the soil for hours at a time. The waterlogging can damage tiny carrot roots, causing them to fork. To avoid this, plant carrots in raised beds to encourage flooding rains to drain away quickly.

PREVENTING PROBLEMS

BEFORE PLANTING

- Loosen the soil deeply and prepare a fine seedbed to help ensure good germination.

AT PLANTING TIME

- Cover seeded areas with row covers and seal the edges (see Row Covers).
- Insert wireworm traps into the soil between rows (see Wireworms).

WHILE CROP DEVELOPS

- Check under row covers once a week for signs of pests or disease; remove weeds as needed.

- Monitor wireworm traps; if you find wireworms, take steps to control them.
- Handpick slugs and snails as needed.
- Check soil moisture twice a week and water whenever the top inch of soil has become dry.
- To prevent roots from developing green shoulders (the green flesh will have an unpleasant flavor), apply mulch along rows of carrots about 1 month after planting.

AFTER HARVEST

- If your crop suffered any disease problems, carefully remove all traces of foliage and roots from the bed. Discard noticeably diseased material.

TROUBLESHOOTING PROBLEMS

FEW OR NO SEEDLINGS APPEAR. A soil crust may have blocked seedlings, or seedlings may have died from moisture stress or been killed by damping-off fungi. If you suspect that a soil crust or lack of moisture is to blame, replant in the same spot, using tricks described under "Secrets of Success" on page 93. If you suspect damping-off, refer to Damping-Off for tips on preventing a recurrence.

WEEDS OVERWHELM SEEDLINGS. It's hard to establish carrots in a weed-prone garden. Sometimes the effort needed to remove the weeds isn't worth it, especially because pulling weeds can injure the young carrot roots, resulting in forked or deformed carrots. Instead, cut your losses early, and try the trick described on page 436 for reducing weeds in a seedbed before you replant.

CHEWED FOLIAGE. Slugs and snails sometimes feed on carrot leaves, and carrot leaves are a food source for swallowtail butterfly caterpillars. Many animal pests find carrot foliage enticing, too. Check your plants at night or first thing in the morning to spot slugs or snails. If you find them, try control methods described in Slugs and Snails. Swallowtail caterpillars are easy to spot, with their bold yellow and black stripes. Gently remove them from the plants and transport them to some carrot-family weeds away from your vegetable garden. If you don't find slugs, snails, or caterpillars on chewed plants, refer to Animal Pests for techniques to discourage animal pests.

SPOTS ON LEAVES. Leaf blight is the probable cause, and it may be due to either fungi or bacteria. Eventually, infected leaves shrivel and die. Turn to Alternaria Blight and Cercospora Leaf Spot for description and controls for these fungal diseases. With bacterial blight, leaf spots appear shiny on the underside of the leaves, and symptoms usually spread down the leaves and then into the leaf stems. Control measures are similar to those for the fungal leaf spots, but as a last resort, use copper sprays, if necessary, to prevent the spread of bacterial blight.

PALE LEAVES, SOMETIMES WILTED. Here's a signal that something's wrong belowground. Pull up one or two plants and check the roots. If the carrots are forked, have tufts of small roots, and/or have galls on them, the problem is root-knot nematodes. This planting probably won't yield a worthwhile harvest. Choose a new site for your next planting. To learn how to rid infested areas of root-knot nematodes, see Nematodes.

If, instead, the developing carrots are covered with lots of very thin roots, the problem is probably aster yellows. Dig and destroy infected plants. If timing allows, replant. For prevention, see Aster Yellows.

CHEWED AREAS WHERE STEMS AND ROOT MEET. Coppery colored carrot weevils munch on stems and crowns, and their feeding may stunt or kill plants. If you spot weevils on your carrots, handpick and kill them. Evaluate the level of damage; the crop may still be worth harvesting (cut away chewed areas of roots) or may not be fit for harvest. Replant in a different area of your garden; since weevils don't fly, this alone may be enough to protect your crop. Covering the seedbed with row cover will help, too. Carrot weevils lay eggs on plant stems, and their larvae also feed on roots. To kill the larvae, drench the soil with parasitic nematodes (see page 272 for instructions).

CARROTS HARD TO UPROOT. You grab those lush carrot tops and pull, but nothing happens. Or worse yet, the tops break off, leaving the roots stuck in the ground. The solution is to insert a digging fork into the ground parallel to the row of carrots and gently lever it back and forth to loosen the soil. The roots should then slide out when you pull on the tops. For topless roots, you may need to dig down on both sides and use the fork to pop the roots up out of the ground.

To prevent this problem in the future, plan ahead and wet down a section of your carrot bed the day before you want to harvest. As you pull on the tops, use a gentle twisting motion as well to help break side roots. (Avoid soaking the soil around roots that you plan to leave in the ground for later harvest, because excess moisture can cause the carrots to rot in the soil.)

ROTTED ROOTS AND LEAF STEMS. Southern blight, white mold, and other types of fungi can cause rot in carrots. Significant rotting can occur below the soil surface before aboveground symptoms (such as wilting) develop. There's no instant cure for this problem. To minimize rot problems in the future, follow a 3-year rotation with related crops. Planting in raised beds may lessen this problem. See Southern Blight and White Mold.

BROWN, WATER-SOAKED LESIONS ON ROOTS. Cavity spot is a soilborne fungus. It thrives in cool soil. You may be able to salvage some roots by cutting away damaged areas. To prevent the problem on future crops, follow a 3-year rotation with related crops. Harvest your crop promptly when it is mature, because older roots are more susceptible to this disease than younger roots.

SURFACE TUNNELS IN ROOTS. The larvae (maggots) of carrot rust flies form tunnels as they feed, and you may see brown material in the tunnels—it's the castings (excrement) of the larvae. The idea of maggots and their excrement in the roots may turn you off, but if the tunneling isn't severe, you can go ahead and harvest your crop. Use a vegetable peeler to strip away the damaged sections of roots. The leftovers may be useful only as cooking carrots, but that's better than nothing. Don't compost the damaged pieces of root because they may contain root maggots: Throw the debris in your household trash. Don't try to store carrots in the garden over winter if they are infested. If there's enough time remaining in your growing season, resow carrots in a different part of your garden. Cover the bed with row cover to prevent the flies from laying eggs around the plants. Leave the row cover on as the crop grows; pull back when needed to thin and weed.

To kill maggots in the soil, drench the bed with parasitic nematodes (see page 272 for instructions).

SMALL BLACK HOLES IN ROOTS. These holes are typical of wireworm damage; you may find leathery brown wireworms in the holes. Harvest your crop and cut away the damaged sections; the roots won't last in storage, though. Use them promptly or freeze them. For wireworm details and controls, consult Wireworms.

ROOTS WITH BITTER FLAVOR. When nights are warm, carrot roots burn up the sugars that they accumulated during the day, resulting in bitter flavor. But when nighttime temperatures drop below 60°F, the sugars remain intact, so the carrots build up sweetness as they mature. To avoid bitter carrots in the future, check the sowing date of your bitter crop. Chances are you sowed too late in spring, and the roots came to maturity when nights were warmer than 60°F.

Plan to sow your spring carrot crop 2 to 3 weeks earlier next year, or choose a variety that matures more quickly.

FORKED ROOTS. Rocks, soil clods, debris in the soil, moisture stress, and disease are possible causes of forked roots. If the roots taste good, there's no reason not to harvest and use the crop. Forked roots may not store as well as straight roots, though. To prevent this problem, work the soil well before planting your next crop. If possible, put the soil through a sifting screen to filter out rocks and debris. Water regularly. If you suspect that disease caused the problem, have a sample analyzed, and take measures to deal with the disease problem in future plantings. Adding fresh manure or other uncomposted organic matter to the soil just before planting can cause roots to fork. Forked roots can be a sign of root-knot nematode damage, but these roots probably also will show excess root hairs and galls on side roots. See Nematodes for control options.

CRACKED ROOTS. Excess water (from rainfall or irrigation) when roots are maturing causes cracking. The roots are still fine to eat, but they won't keep well in storage. For future crops, monitor the weather forecast. If heavy storms are predicted when your crop is close to maturity, harvest before the rain hits. Cracking can also be a sign that you left roots in the ground too long. Usually these overmature roots will be woody and/or tough as well. The best way to avoid this problem in the future is to note on your gardening calendar when the crop should reach harvest size, and begin spot-checking 2 weeks before that date.

ROOTS TWISTED TOGETHER. Overcrowded roots will intertwine as they grow. There's no reason not to harvest and eat these roots if they taste good, but they probably won't keep well in storage. Prevent this problem by thinning carrots to at least 2 inches apart by the time the plants are 1 month old.

HAIRY, MISSHAPEN ROOTS. Crowding, overwatering, or too much nitrogen can cause these symptoms. But if you've thinned your crop properly, checked soil moisture regularly, and avoided using nitrogen-rich soil amendments, the problem is probably root-knot nematodes. Consult Nematodes for control options for future plantings.

CAULIFLOWER

Cauliflower can confound gardeners because it's so sensitive to high temperature, low temperature, water stress, and nutrient stress. Stress leads to poor curd formation (curds is the term used for the immature flower stems that make up a head of cauliflower). Plus cauliflower needs to be blanched for the best white color and flavor.

If you decide to try cauliflower, give yourself permission to fail a couple of times before you succeed. Try to plant several varieties—in any garden, it's difficult to predict precisely which variety will grow well. Be sure to choose at least one that matures rapidly. Also choose one of the cauliflower varieties that produce a colorful head (orange, green, or purple). They don't need blanching and may be a little easier to grow.

CROP BASICS

FAMILY: Cabbage family; relatives are arugula, broccoli, broccoli raab, Brussels sprouts, cabbage, collards, horseradish, kale, kohlrabi, mustard, radishes, turnips.

SITE: Cauliflower needs full sun, but young plants planted for fall harvest benefit from afternoon shade on hot days.

SOIL: Cauliflower is not a crop to try growing in poor soil. Build the organic matter content of your soil with compost and cover crops, especially if you have sandy soil or heavy clay soil. Add a shovelful of compost to each planting hole.

TIMING OF PLANTING: Plant transplants outside 1 to 2 weeks before the last expected spring frost. If possible, wait until the soil temperature reaches 55°F. Sow seeds for a fall crop about 75 days before your first fall frost. In Zones 8 through 10, it's possible to grow cauliflower through the winter, too (choose varieties described as designed for overwintering).

THE BENEFITS OF BLANCHING

Pure white cauliflower heads look pretty, and they taste better, too. If you're growing a white variety of cauliflower, it's worth the effort to blanch the heads for a week or so before harvesting. It's easy to do. When a cauliflower head reaches the size of an egg, gather some leaves loosely over the head and secure them with twine, a long twist tie, a rubber band, or clothespins. Leave enough slack for the head to expand and to allow air circulation.

Securing leaves over developing heads has a secondary benefit, too. It provides a moderate amount of protection from frost. Light frost improves cauliflower flavor, but too much can hurt the plants. Thus, for fall crops, using the standard blanching technique even on orange and green cauliflower heads can shield the heads from cold.

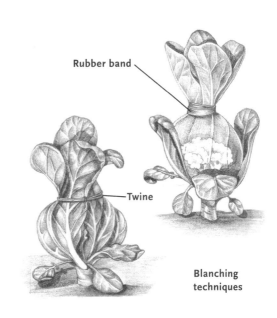

Rubber band

Twine

Blanching techniques

PLANTING METHOD: It's possible to grow cauliflower directly from seed in the garden, but it's safer to start seedlings in containers. Sow seeds indoors 5 to 6 weeks before your planned outdoor planting date.

CRUCIAL CARE: Check soil moisture at least weekly; twice weekly is better. Water whenever the top few inches of soil is dry. For fall crops, mulch the soil surface (3 inches of shredded leaves is a good choice) to conserve moisture and keep the soil cool.

Cauliflower usually benefits from fertilizing, although too much nitrogen can lead to poor head formation. Try feeding plants every 2 to 3 weeks with a dilute solution of liquid fish and/or seaweed fertilizer. Or spread a 1-inch layer over

the soil surface around the plants about 1 month after planting.

GROWING IN CONTAINERS: Cauliflower's not a space-efficient choice for container growing because once you cut the main head, the crop is done. Instead, plant broccoli, which will produce an extended harvest of side shoots.

HARVESTING CUES: Once cauliflower heads start to form, it should only be a week or two until harvest time. To start, cut heads when they reach about 6 inches across. They will seem very small compared to the giant heads you see in grocery stores, but it's wiser to harvest a small, high-quality head than to risk that the head will become overmature. As you gain experience with cauliflower, you can try allowing the heads to grow larger.

SECRETS OF SUCCESS

PLANT YOUNG, VIGOROUS SEEDLINGS. Stress can ruin cauliflower transplants. Even after transplanting into rich soil, a stressed or pot-bound cauliflower seedling will mope, grow poorly, and probably yield only a button-size head. To be sure you'll succeed with cauliflower, grow your own transplants, and plan to transfer them to the garden before they're 6 weeks old—planting 4- to 5-week-old transplants is ideal. In spring, harden transplants gently by exposing them to cool outdoor weather for a few hours each day. In summer, expose transplants to short periods of strong sun for a week before planting them outside. Don't harden cauliflower transplants by withholding water or nutrients—there's too much risk of causing a stress reaction.

GET IN TOUCH WITH THE 60s. No, not the decade, but the temperature. Cauliflower is happiest when temperatures are in the 60s, and moderate temperatures are essential when heads are forming (too much heat causes poor appearance and flavor). In a few areas of the country, such as coastal areas of the West, it's not hard to identify a 2-month-long period when temperatures will be in the 60s—or close to that—during the day.

For the rest of us, choosing when to plant presents a dilemma. In springtime, if you wait to plant until daytime highs are close to 60°F,

temperatures will probably be in the 80s by the time the cauliflower heads appear. On the other hand, if you delay planting until temperatures are in the 60s in early fall, a hard frost could hurt the plants before heads form. The better choice is to plant transplants in late summer, even though it's hotter than recommended for cauliflower. Young plants can take heat better than big ones would. Cool the soil with a thick organic mulch. Plant in the shade of tall crops or trellised cucumbers or beans, or drape shade cloth over the plants until the weather cools off. There's a better chance that plants will produce sizable, tasty heads in cool fall conditions than in spring, when temperatures may unexpectedly soar just when heads start to form.

PREVENTING PROBLEMS

BEFORE PLANTING
- Set up commercial insect traps to lure and trap looper and armyworm moths.
- Plant annuals and perennials with small flowers to attract aphid predators to your garden (see Beneficial Insects for a list of choices).
- Remove cabbage-family weeds in and around your garden.

AT PLANTING TIME
- Put a cutworm collar around each transplant in spring.
- Cover transplants with row covers and seal the edges to keep out pests (see Row Covers).

- If you're not using row covers, put a paper or foam barrier around the stem or cover each plant with a screen cone to prevent cabbage maggot flies from laying eggs at the base of the plant, as shown on page 91.
- For summer plantings, water well and mulch the soil immediately to prevent overheating.

WHILE CROP DEVELOPS

- Handpick eggs and larvae of leaf-eating caterpillars frequently.
- For plants that aren't covered, spray a garlic or hot pepper spray to deter aphids and other pests (but not on leafy crops that are close to harvest). See Pest Control for more information.
- Blanch white varieties, starting 1 to 2 weeks before harvest.

AFTER HARVEST

- Clear out crop debris, including roots. Compost healthy residues. Destroy or discard noticeably diseased residues.
- If you grow cauliflower for late-spring harvest, renew the soil with compost before replanting in that spot.

TROUBLESHOOTING PROBLEMS

On top of the cultural problems that complicate cauliflower production, the plants are also susceptible to the same pests and diseases as cabbage. Below is a listing of cultural problems for cauliflower. To diagnose pest and disease problems, refer to Cabbage.

YELLOWING OR BROWNING OF HEADS. Brown spots dotted across a cauliflower head may be a symptom of downy mildew. Infected heads often rot, too. Try cutting away brown areas to salvage the rest of the head for eating; don't try to store infected heads. See Downy Mildew for controls.

If the discoloration is diffuse, the problem is more likely sunburn or boron deficiency. Hollow areas inside the stem are a confirming sign of lack of boron. Try tasting the curds. They may have developed an off flavor, but if not, they're fine to eat.

To avoid this problem in the future, blanch the heads for about 1 week before harvest. Also, check soil pH. Boron deficiency is rare in good garden soil, but if the soil is too alkaline or acid, plants may not be able to absorb the boron that's present in the soil. Adding compost helps buffer pH and supplies a wide range of nutrients, including boron.

UNEVEN HEADS. Heads that have opened up, with gaps between curds, can result from an uneven moisture supply, such as if the soil dries out and then is saturated. Or you may simply have waited too long to harvest. Cut the heads; they should still be usable. In the future, mulch the soil, monitor soil moisture carefully, and water cauliflower whenever the top few inches of soil is dry. Check heads frequently for maturity, and cut them while they are still smooth and firm.

LEAVES GROWING THROUGH HEADS. This condition can happen if heads are overmature, if the soil is too rich in nitrogen, or if the weather has been hot and dry. Sample a head. If the flavor is still palatable, harvest the crop and enjoy it. Next time you plant, try a heat-tolerant variety, or plant at a different time of year so heads can develop during mild weather. Monitor soil moisture carefully, and don't overfertilize with nitrogen-rich amendments (such as bloodmeal).

CURDS TASTE BITTER. Heat and water stress can turn cauliflower bitter; it's usually a problem only with spring-planted cauliflower. Try growing cauliflower in fall or winter instead. Lack of blanching can result in curds with an off flavor (though usually not bitter).

∽ CELERY

Congratulate yourself if your celery plants are trouble-free. Celery can be challenging to grow because it needs steady moisture, fertile soil, and a long period of cool temperatures, plus it's susceptible to many pests and diseases.

Celery originated from a marsh plant. As a general rule, it needs twice as much water as the other crops in your garden. For ideal development, celery likes 4 months or longer of daytime temperatures in the 60s and nighttime temperatures of about 50°F. If you can't supply the ideal, you can still grow celery, but you may have to harvest plants before they reach their full size.

Some gardeners prefer blanched celery to unblanched celery. Unblanched stems will be bright green with stronger flavor (also higher vitamin content). Self-blanching varieties are also available, but they may be even more finicky about conditions than regular varieties. It's best to experiment by growing more than one variety—sometimes one will grow well in your soil and climate while another growing right alongside will do poorly.

CROP BASICS

FAMILY: Carrot family; relatives are carrots, celeriac, and parsnips.

SITE: Plant celery in full sun, but provide shading during sudden heat spells or on hot summer afternoons (for plants intended for fall harvest). Use shading devices like those shown on page 328.

SOIL: Celery needs rich soil that drains well but holds lots of moisture. See the opposite page for details on preparing soil for celery.

TIMING OF PLANTING: Sow seeds 10 to 11 weeks before the desired transplanting date. Transplant shortly before or after your last spring frost date. In many areas, a fall crop is possible, too, if you set out transplants in late summer when the worst heat has passed.

If you'd like to try a taste comparison of blanched and unblanched celery, try wrapping the stems of a few plants about 2 weeks before you plan to harvest. Use a washed paperboard juice or milk carton, a brown paper bag brushed with vegetable oil (to keep the bag from ripping or disintegrating) and with the bottom cut out, or a section of newspaper stapled to itself. Allow the leaves to extend above the wrapper so they can photosynthesize and continue to feed the plant. Be sure plants are dry when you put the wrapper in place, and water wrapped plants carefully at the base to avoid getting water inside the wrapping (too much moisture may lead to rot).

Hill up soil at bottom edge of bag

Blanching in a bag

PLANTING METHOD: It's possible to sow seeds directly in the garden, but most home gardeners buy or raise transplants instead.

CRUCIAL CARE: Starting seeds indoors requires patience and care because the seeds are very small and slow to germinate. See Seed Starting for tips on speeding germination and working with small seeds. Consistently moist soil is crucial; see below for watering guidelines. To ensure that the plants won't suffer nutrient stress, water them with a nutrient-rich tea or organic liquid fertilizer every 2 to 4 weeks (see Fertilizers for information).

GROWING IN CONTAINERS: In fall, try transplanting a couple of plants from your garden into pots and placing them on a sunny windowsill indoors. The plants won't grow very fast, but they will serve as a source of celery leaves and the occasional stalk for flavoring soups and casseroles. Use a compost-rich potting mix and never let the container dry out.

HARVESTING CUES: Begin harvesting individual stalks when they are 6 to 8 inches tall. Carefully cut outer stalks with a sharp knife; plants will continue producing new stalks from the center. If plants start to form a flowerstalk, cut the whole plant at once. If you catch it early enough, the flavor and texture of the stalks should be fine.

SECRETS OF SUCCESS

FOCUS ON SOIL AND WATER. To meet celery's demand for steady moisture and nutrients, prepare your soil beyond the norm for other crops. Even if you build your soil yearly with organic matter, plan to add extra to the bed where you'll grow celery. Spread 2 to 4 inches of compost (or a mix of compost and peat moss) over the bed and work it into the top several inches of soil. If you're concerned about soil nutrients, also add a balanced blended organic fertilizer, applying it according to label instructions.

Another approach is to concentrate the compost or compost/peat mix in a trench about 12 inches deep. Dig the trench, fill it with compost (or if your soil quality is good, with a mix of compost and the removed soil), and top that with 2 to 3 inches of soil. Water the trenches thoroughly the day before planting to settle the soil and compost. Set your celery plants 1 foot apart along this slightly raised planting ridge.

If possible, use drip irrigation, soaker hoses, or watering reservoirs (as shown on page 430) to supply water. Celery should never be allowed to dry out, and it may need daily watering during hot or dry weather.

Covering the soil with black plastic can cause overheating, which damages celery's shallow roots. Instead, cover the soil with a cooling layer of mulch, such as several thicknesses of newspaper or 3 to 4 inches of straw.

REGIONAL NOTES

CANADA AND NORTHERN UNITED STATES

Celery can withstand cold outdoor temperatures down to the high teens with protection, but in areas where temperatures drop lower, you'll need to harvest all plants sometime during fall. If you grew only a few plants, refrigerate them for up to 1 month. But to store several plants for a longer time, dig them with their roots intact and transfer them to a wooden crate lined with newspaper or a large planter filled with a blend of half compost and half garden soil. Plant them just a few inches apart, and store in a cool place, where they'll last for up to 3 months.

PREVENTING PROBLEMS

BEFORE PLANTING

- Plant flowering annuals and perennials that attract beneficial insects that prey on aphids, leafhoppers, leafminers, and other pests. See Beneficial Insects for information.

AT PLANTING TIME

- If earwigs have been a problem in your garden, put out earwig traps and empty them every morning.

WHILE CROP DEVELOPS

- Cover young plants with cloches or row covers if extended periods below 50°F are predicted. Remove covers when the weather warms.
- Wash aphids off plants with a strong spray of water.
- Water plants frequently (possibly even daily) so that the soil never dries out. Use drip irrigation or watering reservoirs to avoid wetting the leaves and stems.

AT HARVEST TIME

- Cut outer stems carefully to avoid injuring young, developing stems.

TROUBLESHOOTING PROBLEMS

YOUNG PLANTS FORM FLOWERSTALKS. Plants will bolt prematurely if they're exposed to too many days with temperatures below 55°F. In the future, don't plant celery as early in spring, and protect young plants with cloches when cold weather is predicted. Remove the protective covers as soon as the temperature rises, though, because celery plants are also susceptible to heat stress.

DISTORTED LEAVES WITH STICKY COATING. Aphids suck sap from celery leaves. Light feeding isn't a problem, but a large colony can cause leaves to grow abnormally. Wash aphids off plants with a water spray. See Aphids for other controls.

WATER-SOAKED YOUNG LEAVES. This is an early sign of a calcium deficiency, which is usually the result of water stress. If you notice this problem, increase watering and apply mulch to maintain even soil moisture. Don't apply any more nitrogen fertilizer because overfertilizing with nitrogen can also cause this problem. Plants may recover and go on to produce normal new growth.

YELLOW SPOTS ON LEAVES. Early blight of celery causes these spots, which usually appear on outer leaves first. As the disease worsens, the spots enlarge and turn grayish brown. Brown streaks appear on the stems, and overall growth is usually weak. Early blight of celery is not related to early blight of potatoes and tomatoes, but instead is a species of the Cercospora fungus. Other species of this fungus cause leaf spot disease in carrots and several

other crops. See Cercospora Leaf Spot for control measures.

BROWN SPOTS ON LEAVES. Brown spots are an early symptom of late blight, which is not the same disease as late blight of potatoes and tomatoes. However, it can cause serious damage to celery as infected areas die and plants lose leaves and become stunted. Spots appear on stems, too, and may ruin whole plants. Late blight of celery is caused by a Septoria fungus, which survives on seeds and in the soil. To minimize late blight problems in the future, use disease-free seed, keep plants growing rapidly (late blight is most problematic late in the growing season), start a 3-year rotation of celery and celeriac with other crops, and try spraying plants with a plant health booster (see Disease Control for information).

CENTRAL LEAVES TURN DARK. Tarnished plant bugs suck sap from plants, and the feeding injury causes plant tissues to break down. Injured leaves turn dark, then blacken and die. The injury can look similar to blackheart (caused by calcium deficiency). Use a hand lens to look for tarnished plant bugs on the foliage. See Tarnished Plant Bugs for controls for future plantings.

MOTTLED LEAVES. Mosaic virus causes mottling and stunting of plants. It's rarely serious. Aphids spread mosaic as they feed, so to limit mosaic problems, take steps to reduce aphids (see Aphids and Mosaic for more information).

HOLES IN LEAVES. Loopers and armyworms sometimes feed on celery. They don't usually cause extensive damage, but their excrement among the stalks can be messy and unpleasant to deal with. If you spot lots of young caterpillars on your plants, spray with *Bacillus thuringiensis* var. *kurstaki* to control them. See Cabbage Loopers and Armyworms for other controls.

YOUNG STEMS CHEWED. Check in between stems to find earwigs hiding. They feed at night and sometimes gather in large groups. Put out earwig traps as described in Earwigs.

WATER-SOAKED SPOTS ON STEMS. White mold is a fungal disease that infects many vegetable crops. In celery, infected plants may appear pink at the base of the stalks; the disease is also called pink rot. Uproot and destroy infected plants. See White Mold for preventive measures.

HOLES IN STALKS. Slugs and snails will feed on celery stalks during the night. Handpick them at night; see Slugs and Snails for more control measures.

DEFORMED OR TANGLED STALKS. This is a sign of a disease called aster yellows, which is spread by leafhoppers. The only way to prevent it is to prevent leafhoppers from feeding (see Leafhoppers).

CENTER OF PLANT TURNS BLACK. If the outer stalks are healthy, this condition is called blackheart, and it's due to calcium deficiency. It's similar to blossom-end rot of tomatoes and other fruits. To prevent it in the future, follow the recommendations in Blossom-End Rot.

STUNTED PLANTS. Fusarium yellows, nematodes, and carrot weevil damage can cause stunting. Fusarium yellows is a fungus that also causes yellowing of plants and internal discoloration of the crown and stems, which you can see if you break a plant open. Control is similar to that of other diseases caused by Fusarium fungi; see Fusarium Wilt.

If you suspect a nematode problem, uproot a stunted celery plant. Root-knot nematodes cause formation of galls and bunched roots. If root-lesion nematodes are the problem, you will see some roots that are dark and dying back. To reduce nematode problems in future crops, refer to Nematodes.

Carrot weevil larvae are white grubs with dark heads that feed at the crown of celery plants. Uproot a plant and look for the larvae. If you find them, see page 98 for controls.

ROTTEN AREAS ON STALKS. Various bacteria and fungi cause rotting of celery stalks. Sometimes the rotten areas turn brick red; other times the rot is soft and bad-smelling. Discard or destroy rotten stalks. Harvest remaining uninfected stalks right away and use them as quickly as possible. To prevent rot problems on future plants, see Rots.

CRACKS IN STALKS. When boron is deficient, stems develop crosswise cracks; they look as though they've been scratched. This problem is most likely to occur in new gardens where soil quality is poor. Foliar kelp sprays may help the plants produce normal stalks. For the future, increase soil organic matter content by adding compost or growing cover crops, and check soil pH.

TOUGH, STRINGY, OR STUNTED STALKS. Water stress and too little nitrogen can lead to poor stalk quality. In the future, space plants farther apart, and enrich the soil with more organic matter before planting rather than relying solely on supplemental fertilizer.

CERCOSPORA LEAF SPOT

Carrots, peanuts, and crops in the beet and squash families are the odd mix susceptible to Cercospora leaf spot. Different species of this fungus are responsible for the disease in each crop family. The disease is also called Cercospora blight, Cercospora leaf blight, and frog-eye leaf spot.

Symptoms can develop quickly in wet, hot weather, and the disease is most often a problem in humid regions, including the South and the Northeast. Spores can be blown by the wind or move in splashing rainwater.

RANGE: Some species throughout the United States and Canada; squash-family crops affected mainly in the southern United States

CROPS AT RISK: Carrots, beets, spinach, Swiss chard, peanuts, cucumbers, squash, melons, watermelons, pumpkins

DESCRIPTION: On carrots, spots often appear first at the edges of young leaves. The spots have a tan center and a well-defined dark margin. Leaves may curl, and as the spots spread, leaflets die and wither. When the disease is rampant, it may infect older leaves, too.

In other crops, symptoms appear first on older leaves. On beets, spots are brown or gray with a reddish purple margin. Tan streaks sometimes appear on leaf stems, and the leaf stems may be girdled.

On squash-family crops, spots are white or tan in the center, with a dark or yellowish halo.

The disease does not cause symptoms on fruits and usually isn't serious enough to reduce yield.

FIGHTING INFECTION: Clip off infected leaves if feasible. Feed and water crops regularly to reduce stress. Harvest infected crops as soon as possible.

GARDEN CLEANUP: The fungus survives winter in crop residue and infected seeds. Remove noticeably infected crop debris and discard it. If plants showed minor symptoms, be sure to turn under or compost all crop debris.

NEXT TIME YOU PLANT: For future crops, plant clean seed, and choose tolerant varieties when available. Space rows more widely than normal. Plant crops such as carrots and beets into raised soil ridges so that rainwater and dew will evaporate more quickly. Spray compost tea (see Composting for information) before symptoms appear.

CROP ROTATION: Because different species of Cercospora cause the disease in different crop families, practicing a 2-year rotation by crop family will help to break the disease cycle.

CLUB ROOT

The roots of cabbage and its relatives are the target of this fungus. When the fungus invades, galls form on the roots, interfering with water and nutrient uptake. If club root attacks a young seedling, the plant will probably die. When it infects older plants, club root doesn't kill, but it weakens the crop enough to ruin the harvest.

The most common way for club root fungus to reach a home garden is in infected transplants—the young plants may appear perfectly healthy even though the fungus has begun to attack the roots. Once club root reaches your garden, it's probably there to stay because the fungus produces resting spores that persist many years in the soil. When conditions are right, the resting spores transform into swimming spores that move through soil water to find plant roots. The fungus attacks cabbage-family weeds, too.

RANGE: United States and Canada

CROPS AT RISK: Broccoli, Brussels sprouts, cabbage, cauliflower, kale, kohlrabi, radishes, turnips; some varieties of turnips and radishes are less susceptible than others. Horseradish rarely suffers from club root.

DESCRIPTION: The first noticeable symptom is that plants wilt on warm days; the plants may be a little yellowed. Plants become stunted; young plants may die. Uproot some plants and check the roots. Infected roots show a range of symptoms, from small swellings to large, club-like galls. Secondary rot fungi may also invade, resulting in soft, dark roots. The symptoms may resemble those caused by cabbage maggots. If

you're unsure which problem you have, take root samples to your local Cooperative Extension office for a diagnosis.

FIGHTING INFECTION: If club root is a new problem in your garden, uproot all potentially infected plants, including the soil ball around the roots, and discard or destroy them. If you can remove the plants before the fungus has produced resting spores, you may escape future club root problems.

GARDEN CLEANUP: If club root is present in only one part of your garden, clean up that area very carefully, removing all crop residues and cabbage-family weeds, and also dock, bentgrass, orchardgrass, and ryegrass (all potential hosts). Discard or destroy these materials (and soil clinging to roots), or put them in a long-term refuse pile well away from your vegetable garden. Don't add the materials to compost, or you may end up spreading the resting spores to other areas of your garden. Also, clean all soil from tools after using them in the infected beds.

NEXT TIME YOU PLANT: Club root is more active in acid soil, and keeping pH at the proper level is one of the most important ways to suppress the disease after soil becomes infected. Before planting cabbage-family crops in soil that may contain club root fungus, test the pH. If needed, adjust the pH to at least 6.8 by adding ground limestone (do not raise above 7.5). Wait at least 6 weeks after adjusting the pH to plant.

To avoid infecting any additional areas of your garden, buy cabbage-family transplants only from a trusted supplier, or grow your own from seed. Build soil organic matter to improve drainage and increase populations of beneficial microorganisms.

CROP ROTATION: A long crop rotation (7 years or longer) of cabbage-family crops can reduce problems with club root. However, such a long rotation may not be practical in a home garden. Relying on rotation to fight club root is also difficult because the fungus can infect the roots of many common garden weeds, too.

COLDFRAMES

Building or buying a coldframe is an investment of time and money that will expand your vegetable gardening options. In spring, coldframes are helpful for handling the overflow of seedlings from your indoor seed-starting area. Even if you have enough room indoors, transferring pots or flats of seedlings to the coldframe is useful as part of the hardening-off process. In fall, the coldframe is a great place to grow salad greens and other cool-weather crops for an extended harvest after you've cleared out your regular garden for the season. To learn how to test whether you like using a coldframe and also how to manage coldframes to avoid problems, refer to Season Extension.

COLLARDS

South, north, east, and west, collards are a great cooking green for every garden. Traditionally a southern vegetable, collards are nutritious and tasty no matter where they're grown. Collards are closely related to kale, and the growing, harvesting, and cooking methods for these two crops are very similar. To learn how to grow collards and troubleshoot problems, see Kale.

COLORADO POTATO BEETLES

Sluggish but hungry, Colorado potato beetles emerge from the ground at about the same time that potato shoots push through the soil in spring. The beetles walk in search of host plants, then climb up the stalks to begin feeding on foliage. The adults mate, too, and females lay clusters of bright orange eggs on the plants.

Humpbacked red grubs hatch about 5 to 15 days later, depending on how warm the weather is, and begin chewing away on leaves. They feed in groups at first, then spread out to cover all areas of a plant. After 10 days to a month (also depending on temperature) of feeding, the larvae are full grown, and they drop to the ground to pupate in the soil. In most regions of the United States, a second (and sometimes a third) generation of adults emerges to repeat the cycle. By late summer, though, adult beetles go dormant in the soil to overwinter.

Colorado potato beetles are the worst insect pest of potatoes in most home gardens, and the beetles and larvae also can damage eggplant, and sometimes peppers and tomatoes. Damage to young plants or plants close to harvest may not affect yield much, but feeding at the time of flower production can be serious. Large larvae do a disproportionate amount of leaf damage, so the more young larvae you can kill, the less likely it is that your plants will suffer serious damage from Colorado potato beetles.

Fortunately, there are several cultural practices and organic sprays that will help keep these beetles from ruining any of your crops. Some of these measures take advantage of the beetles'

habit of walking in spring, but keep in mind that if the beetles can't find a suitable host plant, they will resort to flying; second-generation beetles are fairly strong fliers.

PEST PROFILE

Leptinotarsa decemlineata

RANGE: Throughout the United States and Canada except for parts of California, Florida, Nevada, and the maritime provinces of Canada

HOW TO SPOT THEM: Begin looking for orange-and-black striped beetles on plants as soon as potatoes sprout in your garden. Examine leaf undersides for clusters of bright orange eggs (these eggs look similar to the bright yellow egg clusters of beneficial lady beetles). Also watch for humpbacked grubs and dark globs of excrement on foliage.

Colorado potato beetle grub

ADULTS: Oval beetles with yellowish white wing covers marked with lengthwise black stripes; ⅜ inch long

Colorado potato beetle laying eggs

Colorado potato beetle grubs on damaged leaf

LARVAE: Humpbacked grubs ranging from ⅛ to ½ inch long; young grubs are reddish, older larvae tend to be pinkish or salmon-colored. All grubs have a black head and two rows of black spots along both sides of the body.

EGGS: Bright orange ovals laid in clusters of 20 to 60 on leaf undersides

CROPS AT RISK: Potatoes, eggplant, tomatoes, peppers; rarely found on cabbage-family crops

DAMAGE: Adults and grubs chew on leaves; grubs leave dark excrement on leaves; heavy infestation of grubs can defoliate plants.

CONTROL METHODS

TIMED PLANTING. Plant early-maturing varieties of potatoes about 1 month earlier or later than the "standard" planting time in your area. Early crops will be near harvest by the time beetles begin feeding. Late crops may suffer some damage when young, but by the time they are flowering, the last generation in your area should be winding down.

REMOVING ALTERNATE HOSTS. In spring, scout your garden and uproot any tomato-family weeds. They can serve as food sources and egg-laying sites (see Weed Control).

CROP ROTATION. Rotating tomato-family crops from one place to another in your garden won't end your Colorado potato beetle problems, but it is important. If you plant these crops in the same bed 2 years in a row, the emerging beetles will have an easy time finding the crop and beginning to feed. Crop rotation at least slows the start of the problem in spring.

STRAW MULCH. Studies show that mulching around crops with about 4 inches of straw significantly reduces Colorado potato beetle problems. It seems that the straw is difficult for the beetles to navigate, so they may not find the plants. The straw also creates a favorable environment for beneficial insects and spiders that prey on the pest.

ROW COVERS. Cover plants with row cover at planting time and seal the edges (see Row Covers for information). Check under the cover occasionally to be sure that no potato beetles emerged from the soil under the cover.

PROTECT BENEFICIALS. Spined soldier bugs prey on adults and larvae, and lady beetles eat Colorado potato beetle eggs. Be sure you can identify these beneficial insects, and don't handpick or otherwise harm them.

HANDPICKING. Pick beetles and grubs off plants at least twice a week and dump them in soapy water. Or find a helper to hold a large snow shovel or pan under the plants while you shake them or bat the foliage with a broom. Adult beetles will drop from the plants, and you can also dispense with the collected beetles in a bucket of soapy water.

VACUUMING. Collecting adult beetles with a handheld vacuum may work well. Do this when the weather is cool and the beetles are less likely to fly away.

BIOCONTROLS. Spinosad and *Beauveria bassiana* will kill Colorado potato beetle larvae. *Bacillus thuringiensis* var. *san diego* or *tenebrionis* also is effective against larvae, but it can be difficult to find an organic formulation of this variety of Bt. See Pest Control for more information about this and other biological controls.

NEEM. If biocontrols don't work to control Colorado potato beetles in your garden, spray neem as a last result. See Pest Control for more information about neem.

CONTROL CALENDAR

BEFORE PLANTING

- Choose an early-maturing potato variety.
- Pull out tomato-family weeds in spring before beetles emerge.

AT PLANTING TIME

- Plant potatoes 1 month early or 1 month late to avoid damage at critical times.
- Mulch plants with 4 inches of straw.
- Cover plants with row covers and seal the edges.

WHILE CROP DEVELOPS

- Collect and destroy eggs, larvae, and beetles, but be sure you don't destroy lady beetles or their eggs by mistake.
- Use spinosad, *Beauveria bassiana*, or Bt to kill larvae.
- As a last resort, spray pests with neem.

AFTER HARVEST

- Plan to plant tomato-family crops in a different bed next year to slow the progress of the beetles.

COMPANION PLANTING

Every organic garden benefits from the practice of companion planting. A mix of folk tradition and educated reasoning (now confirmed in some cases by controlled research studies), companion planting is the art of arranging plants in ways that improve your overall gardening success.

BEYOND THE OLD-FASHIONED APPROACH

We're most familiar with companion planting from old adages that recommend pairing specific crops—planting dill beside cabbage to improve the flavor of the cabbage is just one of many examples. Most of these recommendations are unproven, and a pairing that one book on companion planting praises may be advised against in another. Trying out old-fashioned companion planting duos can be fun, and you can keep track of your own impressions of the results.

HIGH-YIELDING COMPANIONS

The potential of companion planting goes far beyond the possibility of making one crop or another taste better, however. Planting different kinds of vegetables together in a garden bed can make better use of your garden space, which means higher yields overall. Gardeners call this interplanting, and it's a very practical form of companion planting (see page 325 for information).

INSECT-ATTRACTING COMPANIONS

Another form of companion planting is including specific flowering plants that attract beneficial insects. Not just any flower will do, however. Consult Beneficial Insects for information.

Sometimes you can use a companion crop as a trap crop to attract a particular pest, thereby protecting your desired crop from damage. This trap-cropping technique is a little tricky, but if you'd like to try it, turn to page 307 to learn more.

COMPANION PLANTING VS. CROP ROTATION

From the standpoint of pest control, separating crops that belong to the same crop family can be important—it's companion planting in reverse, so to speak. For example, planting a garden bed with cauliflower, broccoli, and kale could work well, but it could be a disaster because the pests and diseases that trouble cauliflower also infect broccoli and kale. A better approach would be to plan your garden to separate beds of cabbage-family plants with crops from other families. Also, plan over time to change the position of crops from one place to another, a process called crop rotation. As you develop companion planting schemes, you'll need to consider how they impact your plans to rotate crops from year to year. To figure out how to practice both companion planting and crop rotation, see Crop Rotation.

COMPOSTING

Maintaining a steady supply of mature compost is the most common composting problem for organic vegetable gardeners. We know the almost magical effects of compost—we just can't seem to make enough to meet the need. Whether your problem is lack of composting materials, a need for more composting space, or stagnant compost piles that never seem to mature, read on to learn how to produce more compost in less time without hassles.

FINDING AND STORING COMPOSTING MATERIALS

High-quality compost is the product of a mix of dry, brown materials and fresh, green material; for example, dry leaves mixed with green pulled weeds and kitchen waste. The trick is to collect enough browns and greens at one time to build a pile. Here are some helpful hints for stockpiling materials.

If you scavenge in your neighborhood for composting materials, avoid loose leaves or other materials along roadsides because they could be contaminated with motor oil, toxins, or trash.

LEAVES. Capture your fall leaves—and your neighbors'—for use next season. Set up a leaf corral near your compost pile using stakes and chicken wire or mesh fencing. Dump your leaves there, cover them with a tarp (so they'll stay dry), and weight the tarp with boards, bricks, or rocks. Or bag your leaves and stack the bags around your compost pile (they'll provide some insulation, keeping the pile active longer).

PAPER. Newspaper is easy to stockpile in your garage for use whenever it's needed. Inks used to print newspaper are not toxic; you can use both black-and-white and colored newspaper. Nonglossy newsprint is the best choice.

Shredded office paper and junk mail will compost, too, but it's not known whether or how synthetic toners used in photocopying and laser printing affect soil or soil organisms. Paper is very high in carbon, so add it to piles in small amounts and shred it first for best results.

GRASS CLIPPINGS. Grass clippings are difficult to stockpile; they tend to end up as a moldy, matted mass. Don't add moldy clippings to the compost pile—they'll be too heavy to mix well. Instead, spread them over a vacant garden bed and dig or till them in. In the future, to prevent clippings from matting, mix the fresh clippings immediately with dry hay or leaves. This mix may begin to compost before you add it to your compost pile, but it shouldn't end up matted.

KITCHEN WASTES. Gather kitchen wastes from the cafeteria of a local school by asking them to save the discards from their salad bar. Collect coffee grounds from your office or a local coffee joint.

MANURE. Manure is an excellent compost ingredient, but don't stockpile manure unless you're knowledgeable about how to handle it properly. If you find a local source, pick up only the amount you need for building one compost pile (or two if you keep two active piles running at once). It's fine to add manure and bedding from pet rabbits and rodents to a compost pile, but never use manure from dogs, cats, or reptiles.

Keep in mind that even if you compost manure in a hot compost pile, there's a small risk that the finished compost could contain pathogens (such as *E. coli*) that are harmful to humans. To be safe, be sure to allow at least 4 months between the time you mix manure-based compost into your soil and the time you harvest crops. Ideally, add manure-based compost to the soil at the end of your active gardening season. Always wear gloves when you work with manure or compost made with manure and wash your hands with soap afterward.

BINS THAT DON'T REQUIRE BUILDING SKILLS

Building a compost bin is no problem for people who enjoy do-it-yourself woodworking projects. But if you can't hammer a nail in straight (or don't want to learn how), it's still no problem to set up a compost bin. Constructing a bin is as simple as stacking up some concrete blocks or bales of hay. Or try one of these simple setups.

SAPLING BINS

You can assemble a compost frame using saplings or sunflower stalks as shown above. These simple frames allow good air circulation for a passive compost pile that you don't plan to turn. In the case of the sunflower stalks, the frame eventually breaks down into compost, too.

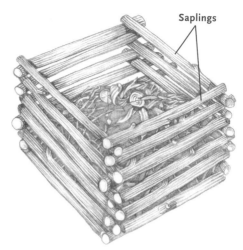

Saplings

After cutting down saplings, stack them to frame a compost pile. You can also harvest sunflower stalks— just cut off the heads and stack them log-cabin style.

A CIRCLE OF SNOW FENCE

Snow fencing is reasonably priced, easily available, and perfect for enclosing a compost pile. Simply drive a sturdy wooden or metal fence post into the soil about 12 inches deep as an anchor.

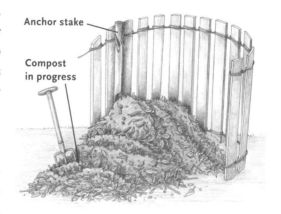

Anchor stake

Compost in progress

Turning compost is easy when you use a snow fence bin. Simply unfasten the ends of the fence, move the compost, then drag the fence to the new position and refasten the ends.

Yard waste recycling programs are a huge benefit to society, diverting tons of organic materials from landfills. You may live in a community that operates a municipal recycling center where you can pick up compost for free or for a small fee. The question about such compost is: Is it safe for organic gardens?

Because municipal compost sites accept yard waste materials from everyone, that means that many of the materials contain pesticide residues. Studies show that an active composting process will break down many pesticide and herbicide residues, but not all.

You may choose to avoid municipal compost altogether because of the risks. Or if you use municipal compost, apply it only to ornamentals, and save your own home compost for your vegetable garden. Pretest every batch of municipal compost by filling a flat with the compost and sowing seeds in it. If the seeds germinate well and the seedlings remain healthy, then that batch is safe to spread in your yard.

Attach the midpoint or one end of a 15-foot length of snow fence to the stake, and then draw the ends of the snow fence together to make a cylinder. Fasten the cylinder with wire or twine. Fill the cylinder with a varied mix of composting materials.

WORM COMPOST BIN

Composting in a worm bin is another great option for gardeners who have no room for outdoor compost or don't want to compost outdoors. A worm bin is a large plastic bin that you fill with a bedding mixture suitable for redworms, which you can buy from mail-order suppliers. These worms are voracious eaters of kitchen wastes, and their castings (excrement) are a wonderful soil amendment. You can make your own worm bin or buy a commercial kit. See Resources for Gardeners on page 456 for recommendations and detailed sources of information on vermicomposting.

FIXING COMPOST GONE "BAD"

WET, SMELLY COMPOST. When compost gets too wet, it turns anaerobic (has no oxygen), and that results in unpleasant smells. To remedy this, you'll need some rough organic material (such as cornstalks or dry branches), some large branches, and a bag or two of dried leaves or some straw or shredded paper. Spread a tarp beside the nasty compost. Use a shovel or pitchfork to move the compost onto the tarp (wear old clothes for this job). Next, use a spading fork to loosen the soil you've uncovered at the base of the pile. Spread 3 inches of rough organic stuff (such as branches or cornstalks) over the pile area. Then begin returning the compost to the pile, mixing it with your dry materials as you go. Occasionally lay one of the branches across the pile, which will help allow air to infiltrate the middle of the pile. Keep everything loose and break up packed clods of material. When you finish, spread the tarp over the top of the pile to prevent rainfall from drenching the pile; leave the sides of the pile exposed so air can penetrate. Ideally, turn the pile every 6 weeks to refresh the air supply, too.

COMPOST COVERED WITH WEED SEEDLINGS. If your compost pile is covered with weed seedlings, think twice before you spread it on your garden. Compost that's contaminated with weed seeds can do a garden more harm than good. There are ways to use weedy compost if you're careful. For example, you can use weedy compost to enrich the soil in a planting trench for a perennial crop such as asparagus. Mix the compost into the soil at the bottom of the trench only.

Another possibility is to mix weedy compost with potting soil for use in containers. The containers will sprout some weed seedlings, but it's much easier to catch such weed problems early and pull out the seedlings than it would be in a large garden bed.

ANIMALS ROOTING IN COMPOST. A compost pile that's pure vegetable matter usually won't attract animal pests, except perhaps mice, who may come to the pile in search of weed seeds. If you see evidence of other animals rooting in your compost, try to use that compost immediately, perhaps to start a new garden bed. Work the compost into the soil of the new bed, and then let the bed sit a few weeks before planting. Or dig a trench in an empty bed and bury the compost under a few inches of soil. Start a new pile, and never add meat, bones, or oils. If animals still come, you'll need to buy an enclosed compost bin.

COMPOST ISN'T COMPOSTING. Lack of moisture is the main reason that home compost piles don't break down. The materials in a compost pile should glisten with moisture. When you squeeze a handful, only a few drops of water should come out. If you can't force out a single drop, the pile needs moisture. If you prefer not to tear the pile apart, place an oscillating sprinkler on top of the pile and turn it on at low volume for about an hour. This slow but steady sprinkling may allow water to infiltrate the pile. Check the center of the pile when you're done to make sure that the water soaked in rather than running off. Check the moisture content of your compost pile every 2 weeks throughout the year. Even in winter, unless the pile is frozen solid, cold wind and frosty temperatures tend to dry out the pile. Keep the pile moist at all times. That way, whenever temperatures are warm enough, your pile will become active.

If the pile has the ideal moisture content,

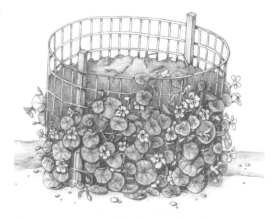

Screen a wire compost bin from view by planting nasturtiums or morning glories to twine up the fencing. Or plant sunflowers just outside the fence.

then a failure to heat up is probably due to lack of nitrogen. Open up the pile at spots and add small amounts of a nitrogen-rich material such as bloodmeal or alfalfa meal.

COMPOSTING WITHOUT A COMPOST PILE

If a traditional composting area isn't right for your yard, make compost in a compost tumbler or right in your garden beds instead. Commercial tumblers cost $100 and up, but they will last for years and take up very little space. Or you can transform a plastic trash can into a compost tumbler.

Sheet composting or trench composting is low-cost but may require that you take part of your garden out of production.

A third option is to plant cover crops and then turn them into the soil. The cover crops break down rapidly in the soil, enriching the soil in the same way that digging in mature compost would. (See Cover Crops for instructions.)

TRENCH COMPOSTING

Trench composting requires digging, but it can transform terrible soil into great soil as your plants grow. The technique is simple: Dig a trench 1 to 2 feet wide and as deep as you can down the middle of a garden bed, heaping up the soil along the edges of the trench. Pick a pleasant spring or fall day, and dig as much as you can without overdoing it. Then, as they become available, dump organic materials such as kitchen waste, garden trimmings, and leaves into the trench. Add the soil you dug out in

among the compost materials as you go. Don't pack the materials down tightly—they'll settle on their own. Between "feedings" of your trench, top the added materials with a couple of inches of soil and cover that with a scrap board, for appearance's sake and to keep out animals. Over time, earthworms and other soil organisms will digest the riches in the trench and distribute them through the bed. When a portion of the trench is near full, top it with about 3 inches of the soil. You can plant into this area right away.

SHEET COMPOSTING

To sheet-compost, you simply spread your composting materials on the surface of a garden bed, mix them together, and let them compost in place. For this method, it's best to use only leaves, grass clippings, straw, and manure. Kitchen scraps could attract animals. Avoid manure if it would tempt your pet to dig in the

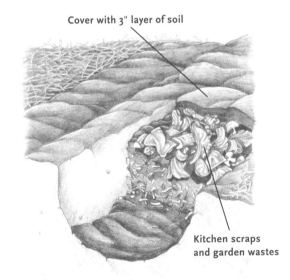

Cover with 3" layer of soil

Kitchen scraps and garden wastes

Composting below ground level, directly in your garden, is an alternative to building a compost bin.

bed. When the materials are fairly well broken down, spread a thin layer of soil or finished compost over the top, and plant right into the enriched bed.

You can use this method to make a new garden bed by first covering the site with cardboard or several thicknesses of wet newspaper.

BUYING COMPOST

Sometimes your best or only option is to buy bagged or bulk compost. Garden centers offer many types. How do you choose? One option is to ask for products that are OMRI listed (see page 6). Another good choice is to buy from a locally based supplier whom you trust, or to follow the recommendation of knowledgeable employees at a locally owned garden center.

COMPOSTED SEWAGE SLUDGE

Some commercial compost products contain composted sewage sludge. It's your choice whether to use these products. One argument in favor of composting sewage sludge is that it's a helpful solution to the immense problem of disposing of human waste. However, there are concerns that sewage sludge contains harmful contaminants, including industrial by-products, petroleum residues, antibiotic residues, and heavy metals. Although composted sewage sludge must be tested for contaminants, some people feel that the standards set are not stringent enough. Because of concern about contaminants in sewage sludge, certified organic farmers are not allowed to use sludge-based compost.

If you want to avoid sludge-based composts, check product labels carefully. These products may not say that they're made from sewage sludge. Instead, they may list "biosolids" as an ingredient, or they may not say anything to indicate that they contain composted sludge. Tell your supplier if you don't want to use sludge-based compost.

COMPOST TEA

Compost tea has fourfold benefits for your garden. It's a fertilizer, it speeds the breakdown of crop residues, it promotes long-term soil fertility, and it helps plants remain healthy and resist infection by diseases. Research on how compost tea works and how to use it for best results is a fascinating and active field. If composting and soil intrigue you, you'll enjoy experimenting with making and using compost tea.

In general, compost tea helps plants by supplying nutrients and substances that promote growth. It's also theorized that compost tea helps protect plants from disease because it's loaded with beneficial organisms. When you apply compost tea to the soil, you're supplying an army of these beneficial microbes to fill the zone that surrounds crop seeds and roots. These beneficial organisms act like bouncers, stopping pathogens from entering seeds, roots, or leaf openings. They may simply outcompete the pathogens for food and space, they may secrete antibiotics that kill the pathogens, or they may consume or parasitize the pathogens. The same kinds of interactions take place on the surface of leaves when you spray your crops with compost tea.

Compost tea can be a simple passive tea or an aerated brew. Passive tea is easier to make, but it takes as long as several weeks to brew. Also, if conditions in the tea become anaerobic, biological reactions take place that produce alcohols and other compounds that can be toxic to plants.

Aerated compost tea requires more effort and

BEYOND THE BASICS

MAKING HOT COMPOST

Most insect pests and disease organisms survive from one growing season to the next in crop debris, weeds, or the soil. One way to fight against pests is to incorporate crop debris and weeds into a hot compost pile. A hot pile is one in which the high level of microbial activity generates so much heat that the pile temperature climbs high above ambient air temperature. The internal heat of an active compost pile can kill insect pests, nematodes, most pathogenic fungi and bacteria (not heat-tolerant viruses such as mosaic), and most weed seeds.

Making hot compost can be hard work. The reward is an increased supply of compost and the satisfaction of dealing with multiple pest, disease, and weed problems without the fuss and cost of traps, cultivators, or sprays. The compost pile has to stay hot throughout the process to be effective. To ensure that, build a pile of minimum size, with the right mix of green and brown materials and sufficient moisture. The process has to remain aerobic, which means frequent turning. Here are the key factors for keeping a compost pile cooking.

MEET THE MINIMUM SIZE REQUIREMENT. Volume matters for hot composting. A pile should be 3 feet wide, deep, and high in order to generate enough heat to kill pests.

GO FOR 160°. Use a thermometer to check pile temperature; 160°F is the desired temperature. Compost thermometers are available for $15 and up, or you can use a meat thermometer (but don't use it for meat if you do). If you've built your pile properly, with the right moisture content, the internal temperature should reach 160°F within 48 hours.

SET UP A TURNING SCHEDULE. Turning a pile maintains the temperature in the right range. Turn it twice a week to prevent the temperature from rising too high or dropping too low. Try to move materials from the outside to the center as you turn it. If you turn the pile daily, the process should be complete in less than 2 months, perhaps in as little as 3 weeks.

You'll know your pile is cooking when you see white fungal growth among the materials when you turn it. The pile will smell good and earthy, not rank and sour, and you may see water vapor rising from the pile as you work.

TROUBLESHOOT PROBLEMS RIGHT AWAY. Check the pile daily. If the pile isn't heating up, it's a sign that the pile is too wet, too dry, or has too little nitrogen. Check pile moisture. If it feels soggy, spread out the materials for a day or so to dry. Then rebuild the pile.

If the pile seems dry inside, open up the pile, but rebuild it right away, sprinkling materials with water as you go. If the moisture content and size are right, the problem is probably too little nitrogen. Open up the pile, and as you rebuild, sprinkle layers of fresh grass clippings or alfalfa meal to seed the pile with nitrogen.

If pile temperature rises more than a few degrees over 160°F, the beneficial organisms that power the composting process start to die. To cool the pile, turn it.

You'll know that the composting process is complete when the pile stops producing heat even after turning. By then, it will be considerably smaller than when you built it, and the materials will have been transformed to a uniform brown color.

equipment to make, but it's ready to use in just 1 or 2 days. Some researchers claim that aerating the tea spurs development of beneficial bacteria and fungi, making a supercharged tea that's more effective in promoting plant health and suppressing disease organisms. The drawback of aerated teas is the concern that they're a breeding ground for potentially harmful bacteria.

MAKING A PASSIVE TEA

Brewing passive compost tea is simple: Place some finished compost in a burlap bag, immerse the bag in a bucket of water, and let it sit for 10 days or longer. Cover the tea so that it won't be contaminated. A suggested ratio is 1 part compost to 5 parts water. Soluble nutrients and growth substances will pass from the compost into the water, creating a dark tea. You can use the tea straight, or dilute it up to 50 percent with more water. Use a watering can or sprayer to apply the tea to soil or directly to plants.

AERATED COMPOST TEA

Many garden suppliers offer kits for making aerated compost tea. The kits include equipment for continuously bubbling air through the water as the tea brews. They cost from $40 on up.

You can set up a homemade brewer using a 5-gallon bucket, a couple of aquarium air pumps (look for them at garage sales or flea markets), plastic aquarium tubing, and two air stones (available from stores that sell tropical fish and aquariums). A more sophisticated setup uses a stronger pump that can supply air to four air stones via a gang valve, as shown in the illustration above.

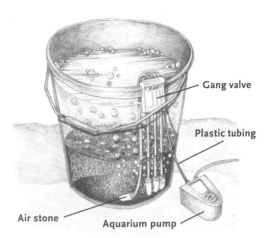

An aquarium pump attached to air stones supplies continuous aeration of compost tea as it brews, increasing the oxygen level of the water to encourage rapid growth of beneficial bacteria and fungi.

To make the tea, fill the bucket about half full of mature compost and add a sugar source (such as a tablespoon of molasses). Push the air stones to the bottom of the bucket and fill it nearly to the top with water. Then turn on the pump and let the air bubble through the mixture. After 1 to 2 days, disconnect the pump and remove the air stones, then strain the tea through an old pillowcase, a nylon stocking, or cheesecloth.

CONCERNS ABOUT AERATED COMPOST TEA

Some studies show that aerated compost teas do a great job of suppressing disease, while other studies don't. There's also a concern that these microbe-rich brews could contain bacteria that would cause disease in humans. Laboratory studies have shown that it's possible for *E. coli* bacteria to thrive in compost tea if a sugary substance such as molasses is added. (Some compost inoculants sold for aerated compost tea brewing include a sugar source.)

Is bacterial contamination something that you should be concerned about? There's no easy answer. No one has demonstrated that spraying compost tea containing *E. coli* actually leads to illness in humans. The risk for an individual home garden is probably very small. If you're concerned about possible health risks, buy a pre-made compost tea that has been tested for safety. If you make your own, follow these precautions:

- Only use fully mature compost for making compost tea.
- Be sure the water you use is drinking-water quality.

- Keep the tea covered while it's brewing to prevent harmful bacteria from contaminating the tea.
- Always clean your equipment thoroughly as soon as you finish brewing a batch of tea.
- Use aerated compost tea only as a soil drench before planting or to spray young transplants of long-season crops such as tomatoes, corn, and squash. Don't apply the tea to leafy crops such as salad greens or kale. As a rule of thumb, wait at least 3 months after applying compost tea to harvest crops that don't come in contact with soil, or 4 months for crops that do contact the soil (such as carrots and potatoes).

⚘ CORN

Corn may seem like a space-hogging luxury in the home garden, but there are some excellent reasons to grow corn, even if you can't fit it in every year. Corn works well for interplanting with vine crops, and the stalks provide light shade in midsummer for a neighboring bed of salad greens.

Corn is the only vegetable crop in the grass family, so it adds an extra dimension to a crop rotation. And it goes without saying that corn fresh from the garden is one of the most enjoyable eating experiences you'll ever have. Corn is not as hard to grow as you might think if you prepare the soil right and take steps to anticipate pest problems.

CROP BASICS

FAMILY: Grass family

SITE: Corn needs full sun and moist, well-drained soil.

SOIL: Add extra nitrogen and compost to the spot in your garden where you plan to plant corn, as described on the opposite page.

TIMING OF PLANTING: Begin planting about the time of the last expected spring frost. Succession plantings 1 to 2 weeks apart extend the harvest. In the North, you can sow until mid-June to early July. In the Deep South, mid-summer heat sometimes interferes with kernel formation. If so, stop planting about the end of April, but sow again in late summer for a fall crop.

PLANTING METHOD: Sow seeds directly in the garden. If you start seedlings indoors, treat the seedlings with great care. Use biodegradable pots. Transplant seedlings while they are young: 1 inch tall is large enough for transplanting.

CRUCIAL CARE: A steady supply of moisture is critical from the time that tassels appear until the ears are ready to pick. Depending on your climate and soil type, the corn patch may need as much as 2 inches of water per week. Your best bet is to use drip tape or soaker hoses, and mulch between the rows with 1 to 2 inches of straw or grass clippings to conserve moisture.

Unless you have excellent soil, feeding the plants periodically is also key to reaping a good harvest. Side-dress with compost or balanced organic fertilizer 1 month after planting and again when the tassels form.

GROWING IN CONTAINERS: Corn is difficult to grow well in containers.

HARVESTING CUES: Watch for the first silks to appear in your corn patch. When you spot them, go to your calendar or garden journal and mark a date 3 weeks in the future as a reminder. That's about how long it will take for the ears to ripen. To check ripeness, look for dry, brown silks and ears that are filled to the tip.

If you need a further test, peel back the husk at the top of one ear and pop a kernel with your fingernail. White milky juice will drip out if the ear is ripe. (If the ear fails the test, be sure to pull up the husk and wrap a rubber band around the top of the ear to keep it in place.)

If you grow popcorn, leave the ears on the plant until the husks are fully dried.

SECRETS OF SUCCESS

STOCK UP YOUR SOIL. Corn needs plenty of nitrogen, especially while the ears are developing. When you prepare your garden for planting in the spring, give your corn patch a little extra boost. Dust the surface of the planting area with cottonseed meal or soybean meal (about 3 pounds per 100 square feet), and spread 1 extra inch of compost over the soil. Another excellent preparation technique is to plant a legume cover crop in that part of your garden and work it in a few weeks before planting.

Raised beds are excellent for corn because the soil warms quickly and drainage is good. However, you can't plant a square block of corn if you have permanent beds only 3 feet wide (square beds are important for pollination, as explained on page 126). The best choice is to enrich the soil well, choose a fast-maturing variety that doesn't grow too tall, and set plants closer than the normal spacing. The risk with this type of spacing is that the center row may suffer a bit from lack of light, and all the plants run a greater risk of water and nutrient stress. Pay close attention to soil moisture and any signs of nitrogen deficiency.

GROW THE TYPE THAT'S RIGHT FOR YOU.
Sweet corn comes in three types: standard, sugary enhanced, and supersweet. Within each type, you can choose from yellow, white, or bicolor varieties. Standard corn is the "old-fashioned" corn with plenty of great corn flavor. Sugary enhanced types retain their sweetness longer after picking (about 3 days). Supersweet types are the sweetest of all and keep their sweetness longest after reaching maturity.

Choose standard corn if you want to plant early—standard varieties can tolerate colder soil (as cool as 55°F) than supersweet types do. Also,

BEYOND THE BASICS

PERFECTING POLLINATION

The tassel at the top of a cornstalk is a group of male flowers. Each individual silk emerging from the husks covering an ear of corn is a female flower. When conditions are right, pollen drops or is blown from the tassels and lands on the silks, and pollination occurs. If a silk is pollinated, it will produce a plump, sweet kernel at the spot where it connects to the ear. Gaps in an ear of corn represent silks that were never pollinated.

PLANTING FOR BETTER POLLINATION. Maximize pollination in your home corn patch by clustering the stalks in a square formation, with plants 2 feet apart in rows 2 feet apart. Or you can plant hills of corn in a square block, with three or four stalks per hill and hills about 3 feet apart.

HAND-POLLINATING. The second part of perfecting pollination is to hand-pollinate the silks. This can be as simple as visiting your corn patch daily to shake the tassel on each stalk. For even better results, take a paper bag along and shake the tassels over the open bag, catching the pollen inside. Then gently dispense the pollen from the bag.

PREVENTING UNWANTED POLLINATION. Supersweet corn has weaker pollen than other types. Thus, if pollen from a standard or sugary enhanced corn plant reaches silks of supersweet plants, it's likely to outcompete the supersweet pollen. The cross-pollination results in kernels that aren't supersweet. It's easy to avoid this problem. You have a choice: Either sow your supersweet corn at least 25 feet away from any other type of corn, or select varieties or time planting so that tassel formation of your other corn won't occur at the time when your supersweet corn is in silk. A difference of 12 days (in planting dates or dates to maturity) should be enough.

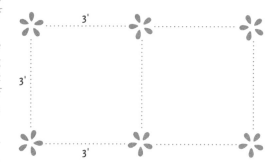

Hills of corn planted in a block pattern

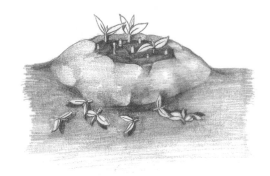

Thin each hill to three plants

if you want to grow heirlooms, then you'll be growing standard corn.

Sugary enhanced corn is a good choice for busy gardeners who don't have time to check ripeness of developing ears on a daily basis; the 3-day window allows some flexibility. Hybrid sugary enhanced corn is good for producing a large harvest all at once for canning and freezing (hybrids tend to ripen more uniformly than open-pollinated standard varieties).

Supersweet corn is for those who love their corn sweet—sweet enough to eat for dessert. It takes patience to grow supersweet corn, though, because it won't do well unless it's planted in soil that's at least 65°F. Also, it doesn't have the creamy texture that most people prefer for frozen or canned corn; it's primarily for eating off the cob. Another drawback of supersweet corn is that it's less vigorous than other types. It needs closer supervision to prevent problems, as well as special growing techniques such as prewarming the soil with black plastic and growing young plants under row covers.

BEWARE OF BIRDS. Birds love to visit the corn patch to uproot new seedlings and eat the tender seeds. Protect your seedlings by covering the seeded area with floating row cover (leave some slack) and leaving it in place until the seedlings are a few inches tall. Or try one of the other tricks described in Birds.

REGIONAL NOTES

EAST AND MIDWEST

Yellow or brown stripes on corn leaves is a sign that corn flea beetles have infected your crop with Stewart's wilt. This disease causes formation of bacterial slime in the vascular system; infected plants wilt, and they may become stunted or even die. It's hard to control flea beetles thoroughly enough to prevent the spread of the disease. If Stewart's wilt has become a problem in your garden, choose tolerant varieties, which usually grow and produce well even when they're infected. It's okay to compost diseased plant debris because the bacteria overwinters inside flea beetles hibernating in soil, not in crop debris.

PREVENTING PROBLEMS

BEFORE PLANTING

- Cover soil tightly with black plastic for 2 weeks before planting to prewarm it.

AT PLANTING TIME

- Cover seeded rows with row cover or other barrier to protect seedlings from birds.
- Put out earwig traps; empty and replenish them as needed (see Earwigs).

WHILE CROP DEVELOPS

- Side-dress corn with fish emulsion when young plants have about five leaves.
- Handpick leaf-feeding caterpillars and Japanese beetles as you spot them.
- Apply 5 drops of vegetable oil to the silks of each ear (just as the silks begin to turn brown) to prevent earworm damage to ears.
- Shake tassels daily to help disperse pollen.
- Protect ears from raccoons as they near ripeness by laying chicken wire on the

ground around your corn patch, leaving a radio playing in your garden overnight, or installing a floppy fence or electric fence.

AFTER HARVEST

- Chop crop debris well and compost it.

TROUBLESHOOTING PROBLEMS

NO OR FEW SEEDLINGS SPROUT. Cold soil, damping-off, or seedcorn maggots may be the problem. Dig around in the soil for the seeds. If they look healthy, soil temperature is the problem. Wait until the soil warms, then replant using new seed.

If the seed is dark and rotting, or has germinated but the shoot died, the problem is damping-off. See Damping-Off to prevent future problems.

You may find small yellowish white maggots in the soil—these are seedcorn maggots, which feed inside sprouting seeds. Remove any seedlings that have germinated there and treat the soil with parasitic nematodes as described in Nematodes. Replant in a different spot when the soil is warmer and the weather is warm and dry. Give organic matter time to break down before you sow seeds in the bed.

SEEDLINGS UPROOTED. Crows and other birds will pull up corn seedlings to feed on the seeds. Replant using one of the strategies described in Birds to protect the seedlings.

YOUNG PLANTS LOOK PALE. Nitrogen deficiency is the most likely cause. Give your plants a boost by drenching them with fish emulsion and kelp. Continue to feed plants throughout their development to keep them growing strongly. In the future, enrich the soil before planting corn.

YELLOW LEAVES. If lower leaves on cornstalks begin turning yellow about the same time as tassels form, it's a sign that the plants are running out of nitrogen. This problem occurs most often in sandy soil. Try watering the plants with fish emulsion. For the future, build soil organic matter and also side-dress plants proactively, when they have about five leaves.

MOTTLED LEAVES. Mosaic virus causes mottling. If plants develop symptoms when young, they probably won't produce healthy ears. See Mosaic to prevent this disease in the future.

LEAVES LOOK RUSTY. Rusty brown spots and blisters on leaves are signs of rust disease. Pick off and destroy infected leaves; see Rust for other controls.

YELLOW STRIPING OF LEAVES. This is a sign of Stewart's wilt; see "Regional Notes" on page 127 for controls.

SPOTS ON LEAVES. Leaf blight appears as grayish or tan oval spots. Two different fungi may be to blame: northern corn leaf blight or southern corn leaf blight. Both need wet conditions to thrive, so if conditions turn drier after you notice symptoms, the crop may continue to grow well. After harvest, chop and turn under all crop debris or chop it into a compost pile, because the fungus survives over winter on undecomposed plant debris. If the problem was serious, seek out resistant varieties to plant next year.

LARGE HOLES IN LEAVES. Fall armyworms and grasshoppers feed on corn foliage; handpick

Corn borer eggs

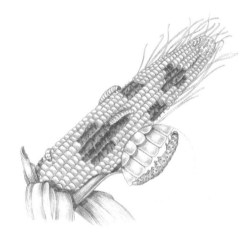

Corn borers and damaged ear

moths with a ¾-inch wingspan then emerge in late spring to mate. They lay masses of flat white eggs on leaf undersides. Larvae are light or brown caterpillars with dark brown heads and dark spots on their bodies. They hatch in about 1 week and begin feeding in the whorl; after a few days, they bore into the stalk to feed. As the tassel emerges, the larvae feed there, too. Larvae that hatch near corn ears bore directly through the husks or travel down the silk channel to reach the ears and feed on kernels. This pest is found in most areas of the United States and Canada east of the Rocky Mountains. There is one generation per year in the Far North, two generations in most of its range, and up to four generations in the Southeast.

Handpick larvae as you find them. If the problem grows worse, apply *Bacillus thuringiensis* (Bt) or spinosad to foliage before the borers tunnel into stalks. As tassels develop, check them frequently for frass—a sign of larvae feeding there as well. Treatments that prevent corn earworms from entering ears also will block corn borers that try to enter via silk channels; see Corn Earworms for details. At the end of the season, remove all infested stalks, chop them up, and compost them.

SOOTY COATING ON PLANTS. This dark coating shows up when corn leaf aphids are feeding. The aphids tend to build up late in the season during hot, dry weather. Natural enemies usually keep these aphids under control in organic gardens. See Aphids for more control options.

STUNTED PLANTS. A heavy infestation of corn rootworms or wireworms can sap the vigor from your corn crop. Dig into the soil around the roots of a stunted plant and look for white grubs with brown heads; these are rootworms. If

them and dump them in soapy water. See Armyworms and Grasshoppers for other controls.

SMALL HOLES IN LEAVES NEAR THE WHORL. European corn borers like to feed on foliage in the whorl of emerging leaves. Later the borers will tunnel through the husk into developing ears. Borers primarily attack corn, although occasionally they are found on bean plants and tomato-family crops. Larvae overwinter in old cornstalks and pupate in spring. Light brown you find them, try applying *Heterorhabditis* nematodes to the soil to reduce the grub population (see Nematodes). Plant corn in a different spot in your garden next year.

Wireworms are brown or yellow leathery worms. For controls, see Wireworms.

STALKS TOPPLE OVER. Usually this is a sign of European corn borer damage. Look inside damaged stalks for tan or brown caterpillars. Injured plants probably won't produce a good harvest. Chop and compost the damaged stalks; see other control measures for borers described on page 129.

STALKS ROT. Injury to stalks from insects or careless handling opens wounds where rot organisms can enter. Controlling insect pests and moving carefully among your corn plants is the best way to prevent this problem.

TASSELS BREAK. Check the tassels for tan or brown caterpillars feeding—these are European corn borer larvae. If you find them, take steps to control them as described on page 129.

SILKS ARE CHEWED SHORT OR CLIPPED OFF. Earwigs, Japanese beetles, and corn rootworm beetles can cause this damage. Ears will have lots of missing kernels if there's heavy damage to silks. As appropriate, consult Earwigs, Japanese Beetles, or Cucumber Beetles to pre-vent this problem in the future (southern corn rootworm is the same pest as the spotted cucumber beetle).

EARS NOT COVERED COMPLETELY WITH KERNELS. Moisture stress, hot weather, or high winds during pollination result in poor pollination. Another possibility is that insects chewed on the silks. For the future, use soaker hoses or drip irrigation to maintain a steady moisture supply; time plantings so pollen won't be shedding in the hottest part of summer; and be prepared to control Japanese beetles, earwigs, and/or cucumber beetles.

WHITE GALLS ON EARS. Corn smut is the fungus that causes these galls; as they mature, the galls become filled with dark spores. The galls are edible, and in Mexico they're used in a variety of dishes. If you'd prefer to avoid corn smut in the future, remove the galls while they are still white and firm and discard them. Choose resistant varieties for future plantings, though, because even if you remove all spores from your garden, the wind may carry in spores from other infected plantings.

PINKISH KERNELS. Insect damage to ears opens the way for rot organisms to invade. The best way to prevent this is to minimize insect feeding in the ears.

CORN EARWORMS

Aliases abound for this garden pest. Corn earworm is its most common name, but this caterpillar is also called the tomato fruitworm, cotton bollworm, and vetchworm. Adult moths emerge from the soil in the South in spring (they can't survive over winter farther north) and lay eggs on plants. The caterpillars feed on foliage or fruits (if available). After about a month, the larvae drop to the soil to pupate. After a brief rest, the second generation emerges.

The moths are good fliers and also are pushed north on strong winds that accompany storm fronts. By midsummer, the pest is widespread throughout its range, and that's when the major damage begins, as the next generation of caterpillars moves into corn ears as well as fruits and pods of various other crops to feed. In the South, there can be four generations yearly.

With good management, you'll be able to keep corn earworm damage low in your garden, but this pest is so widespread that it's unusual to avoid it completely. With that in mind, take a tolerant view: If you grow sweet corn, be prepared to cut off the tips of damaged ears. If the worm is a problem in your tomato and bean crops, plant a little extra so you can afford to throw out the damaged fruits and pods.

tomato foliage and developing pods and fruits frequently.

ADULTS: Night-flying moths with brownish or olive wings (1½-inch wingspan) with a dark spot at the center; the moths have bright green eyes.

LARVAE: Pale caterpillars with brown heads, lengthwise stripes, and short spines or bristles on their bodies. The larvae change color as they grow, generally becoming darker and reaching up to 2 inches in length.

EGGS: Laid singly on plants

CROPS AT RISK: Beans, corn, lettuce, peppers, tomatoes, and other crops

DAMAGE: Larvae eat or tunnel into leaves of many crops, but that's not the main problem. The worms cause serious damage when they eat

PEST PROFILE

Helicoverpa zea

RANGE: United States and Canada, except for Alaska and extreme northern Canada

HOW TO SPOT THEM: Since the moths fly at night and the worms feed mainly under cover, these pests aren't easy to spot. Check insect traps for the presence of moths. Inspect bean and

Corn earworm moth

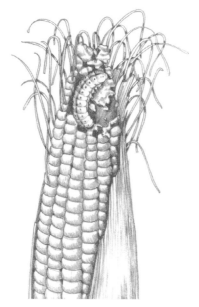

Corn earworm feeding in ear of corn

kernels at the tips of corn ears, bore into the stem end of fruits of tomatoes and other crops, or tunnel into bean pods and lettuce heads. Their feeding causes unsightly damage and opens pathways for rot organisms to invade.

CONTROL METHODS

ENCOURAGING BENEFICIALS. Plant annuals and perennials with small flowers, such as sweet alyssum, to attract parasitic wasps that parasitize earworm eggs. In the South, plan to keep something flowering in or near your garden all season so the wasps will be around to attack all four generations of the pest. See Beneficial Insects for a list of flowers that attract beneficials.

RESISTANT VARIETIES. To reach the tip of a developing ear of corn, earworms have to travel down the silks; they can't chew through the tough husk leaves. Choose varieties with tight husks that extend well beyond the tip of the ear. Earworms will be less successful in infiltrating these ears.

TIMED PLANTING. In the North, planting early and planting early-maturing varieties is a smart way to avoid serious corn earworm damage. With luck, your crop will mature and be harvested before or shortly after the earworm moths reach your area and lay eggs.

INSECT TRAPS. Earworm moths lay eggs daily over the course of their short life span. To reduce the number of eggs laid in your garden, set up two commercial insect traps that include a floral lure to attract the moths. Moths enter the trap and can't escape.

HANDPICKING. Early in the season, before fruits form, the worms also feed on foliage of some crops. Pick worms off plants and dump them in soapy water.

KAOLIN CLAY. Spraying tomatoes and other susceptible fruiting crops with kaolin will discourage the moths from laying eggs. If the clay coats fruits or pods, wash it off well before using them. Kaolin isn't a good choice for protecting lettuce or cabbage.

BLOCKING ENTRY. Since corn earworms have to enter the ear through the silk channel, you can block their path by clamping down on the husk tip. Wrap a rubber band around each husk tip after the silks lengthen, or clip each husk tip with a clothespin.

VEGETABLE OIL. Stop corn earworms from reaching corn kernels by coating the silks with vegetable oil (such as corn or soybean oil). When corn silks have reached full length and are just beginning to turn brown, use an eyedropper to apply five drops of oil to the silk of each ear. As the earworms crawl down the silks toward the ear, the oil smothers them. Timing is important with this technique. If you apply the oil too early, the oil interferes with pollination, resulting in poor kernel development at the ear tip. If you wait too long, earworms will already be safely through the channel, feeding on the ears. Apply the oil only once; it will be effective even if rain follows application.

BIOCONTROL PRODUCTS. As a last resort, if earworm/fruitworms are a serious problem in your garden, apply *Bacillus thuringiensis* var. *kurstaki* or spinosad. Keep in mind that the worms must ingest sprayed plant parts. You may have to reapply several times, and worms already hidden inside ears or fruits won't be killed.

CONTROL CALENDAR

BEFORE PLANTING

- Plant small-flowered annuals and perennials that will attract parasitic wasps.
- Seek out early-maturing varieties and corn varieties with tight husks.
- Put out insect traps to lure and catch adult moths.

AT PLANTING TIME

- In the North, plant early so crops will mature before pest populations rise.

WHILE CROP DEVELOPS

- Spray fruiting plants with kaolin clay to discourage egg-laying.
- Apply vegetable oil to corn silks to kill larvae before they reach the ear.
- As a last resort, apply Bt or spinosad to kill larvae.

COVER CROPS

Inexpensive and versatile, cover crops are an underused problem solver for home gardeners. Are compost and mulch in short supply? Cover crops can substitute for either. Does your garden produce more weeds than vegetables? Cover crops can beat out weeds. Is your soil less than optimal? Let cover crops build structure, tilth, and fertility.

Cover crops are also called green manure crops or catch crops. Some experts say that there are differences in the definition of these three terms, but from the home gardener's perspective, the names are interchangeable. Broadly speaking, cover crops are grains, grasses, or legumes planted primarily to cover and enrich the soil. What you call cover crops isn't important, but planting them in your vegetable garden is!

SOWING SEEDS FOR GARDEN SUCCESS

Some of the best cover crops for home gardens have names that you'll recognize, even if you've never grown them. For example, buckwheat and oats are great cover crops. Others are less familiar, such as Austrian peas and sudangrass. Different crops offer different benefits, and it's a good idea to experiment with several of them. (You'll find details on how to grow and use the best cover crops for home garden on the opposite page.)

When you first try cover cropping, it seems like a strange practice. You sow seeds, water the crop until it becomes established, let it grow until it flowers, and then cut it down and dig it back into the soil. You may wonder: What's the purpose?

REDUCE EROSION. One purpose of a cover crop is to protect your soil, especially during winter, when rain and melting snow can wash away precious soil. Prevent erosion by planting cover crops in late summer and early fall after you clear out vegetable crop remains.

SOLVE COMPACTION PROBLEMS. Planting cover crops opens up compacted soil, especially in a garden that doesn't have permanent raised beds. Simply walking on garden soil repeatedly gradually causes a compacted layer to form 6 to 12 inches below the soil surface. Some cover crops, such as oilseed radish, sweet clover, and rape, generate deep roots that punch down through the compacted subsoil layer, which will maintain good drainage.

MINE FOR NUTRIENTS. Those deep roots on cover crops can tap into a free source of nutrients for your garden—nutrients that are deep in the soil, beyond the reach of most vegetable crop roots. Cover crops bring up those nutrients and use them to fuel their own growth. Then, when you turn the crops into the top few inches of soil, those nutrients from way down deep are released right where your vegetable crop roots can absorb them.

BLOCK WEEDS. Weeds love an empty garden bed at almost any time of year. Cover crops fill the blank spaces early in the season, between crops during summer, and after harvest in fall, denying weeds some of their best opportunities to invade the garden.

SUBSTITUTE FOR COMPOST. Turning under cover crops provides a rich source of organic matter for your garden. It's a great way to reduce your garden's demand for compost, allowing you to avoid spending money on bagged compost and to save your homemade compost as a side-dressing and supplement for the vegetable crops that need it most.

REDUCE PESTS AND DISEASES. Growing cover crops helps to prevent pest and disease problems in two ways. First, cover crops provide habitat for beneficial insects, and the flowers of many cover crops provide food for beneficials, too. When you dig in cover crops, populations of beneficial microbes increase as they work to break down the rich new supply of organic matter. This is especially important in small gardens where it's difficult to plan for crop rotations. By keeping populations of beneficial microorganisms high, your crops have a better chance to succeed even when pathogenic fungi and bacteria are present in the soil. (See Soil for an explanation of how this works.)

All this for the cost of a few pounds of seed! And best of all, cover crops establish themselves quickly and suffer from few pest and disease problems, so they require very little of your time and attention.

COVER CROP HOW-TO

Lack of experience with cover crops may leave you hesitant to try them. If you're unsure when to plant, how to plant, or how to manage a cover crop, read on to learn the simple secrets of cover cropping.

KNOWING WHEN TO PLANT

Cover crops are part of the standard crop rotation pattern on many organic farms, planted in spring to grow through fall. But in home gardens, devoting garden space to cover crops for a whole growing season is the exception—unless the garden is very large. Instead, we tuck in cover crops during gaps between crops: after the harvest is complete in late summer, in spring before planting a warm-season vegetable crop, or in summer after picking a spring crop and before planting a fall crop. Another good option is to interplant cover crops and vegetables by giving the vegetables a head start and then undersowing the cover crop.

Half-Season Cover Cropping

As soon as you can work the soil in spring, sow a frost-hardy cover crop such as oats in beds designated for warm-season crops such as peppers, tomatoes, melons, or sweet potatoes. The cover crop will have 6 to 10 weeks to grow (depending

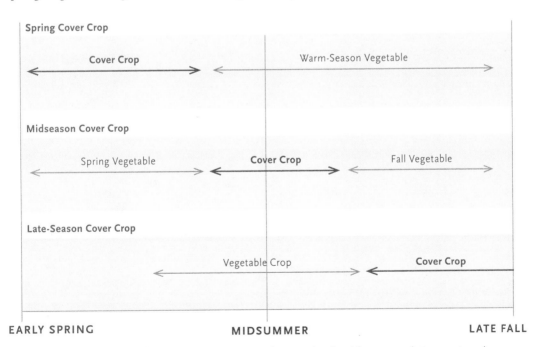

With a little planning, you can fit cover crops into your garden in spring, in midseason, and at season's end.

on your soil and climate). Chop and turn under the cover crop 1 week after it starts to flower for maximum nutrient benefit. However, even if the crop hasn't started to flower, turn it under at least 2 weeks before you expect to plant your vegetable crop.

Midseason Cover Cropping

Midseason sowing is great in hot-summer areas where it's tough to grow cool-loving crops once spring ends. After you harvest lettuce, spinach, and spring root crops, sow a heat-tolerant cover crop such as cowpeas. Cut it down and turn it under about 3 weeks before you plan to sow seeds for the fall garden. If it's been a dry summer, wet the soil a day or two before turning the crop under, because crop residues won't break down in dry soil.

In gardens farther north, where the gap between crops may be only a month long, choose a fast-growing cover crop such as buckwheat.

Winter Cover Crops

In cold-winter areas, sowing cover crop seed should be a standard part of putting the garden to bed. As soon as you harvest a crop, turn under or compost the crop remains. Then sow a cover crop on the vacant bed. If you start early enough (6 weeks before killing frost), you'll have a range of crop choices, some of which may survive winter and continue growing in spring, and others of which will die during winter but remain in place to cover the soil and prevent erosion. If killing frosts are only a few weeks away, the best choices are oats (which will winter-kill) or cereal rye.

Undersowing

You don't have to wait until you've harvested and pulled out a vegetable crop to plant a cover crop. Instead, you can sow the cover crop seed on the soil surface around crop plants to form a living mulch. The cover crop will sprout and begin to grow in the partial shade of the vegetable crop.

The benefits of undersowing are many. First, the living mulch suppresses weeds. Second, it buffers the soil from temperature and moisture extremes, which in turn can help protect crops such as tomatoes from fruit cracking and blossom-end rot. Earthworms and other beneficial soil organisms also benefit from the buffering of soil temperatures. Third, the cover crop gets an early start, affording you more choices of cover crops to use if you live in an area with cold winters.

HOW TO UNDERSOW. In general, wait 4 to 5 weeks after sowing or transplanting your vegetable crop before undersowing cover crop seeds in

A cover crop sown about 1 month after sowing the vegetable crop becomes a living mulch that carpets the bed, suppressing weeds and moderating soil temperature.

GREEN PATHWAYS

Garden pathways are a great place to grow cover crops. A pathway covered with low-growing clover is a pleasant place to walk and kneel. Cuttings from the crop make a nutrient-rich mulch for the garden. The deep clover roots help to ensure that the paths don't become compacted and poorly drained. And the crop provides hiding places for plenty of beneficial insects. You will need to weed out any excess that sprouts in garden beds.

It's a small inconvenience to avoid walking on the paths until the clover becomes established, but that takes only a few weeks. Sow 1 to 1.5 ounces per 100 square feet as early as possible in spring and rake to cover, as you would any cover crop. Mow the crop as needed (or cut it with a string trimmer) and use the clippings as mulch in the garden.

Clover planted in pathways around raised beds

the bed. If you sow too soon, the cover crop will compete too strongly with your vegetable crop. For fruiting crops, another general guideline is to undersow when the crop has started to produce flowers. With corn, sow the cover crop when the corn seedlings are about 15 inches tall.

The cover crop will germinate quickly and crowd out weed seedlings that germinate at the same time, but it can't suppress weeds that are already in place. Thus, be sure the bed is weed-free when you sow the cover crop. Ideally, weed about 2 weeks after planting the main crop,

again 1 week later, and then 1 to 2 days before you sow the cover crop seeds.

Full-Year Cover Cropping

Planting cover crops for a full season seems drastic, but it produces impressive results. For example, planting successive crops of buckwheat for a whole growing season can choke out some of the most persistent garden weeds. A season of cover crops also reduces root-knot nematode populations enough to allow a nematode-susceptible crop to produce well the next season. And if you need to improve a bed of infertile, compacted soil, planting a cover crop for a season and turning it under will do the job at low cost and with little effort compared to double-digging and adding store-bought amendments.

Even if your garden doesn't suffer from serious weed or disease problems, planting a full-season cover crop rejuvenates and revitalizes the soil. As a general garden improvement and problem prevention measure, treat each bed in your garden to a season-long cover crop once every 10 years.

PLANTING COVER CROPS

Cover crops are easy to plant. They don't need a fine seedbed. Simply remove weeds and crop debris, and rake off clumps from the soil surface or break them up. Measure the length and width of the area you want to plant, and multiply those numbers together to determine the total square footage (a bed 4 feet wide and 10 feet long, for example, has an area of 40 square feet). As a rule of thumb, sow 1 cup of seed per 50 square feet. For more precise seeding rates, see the table on page 140.

Broadcasting seed works reasonably well. If the crop has small seeds, such as clover, mix the seed with fine soil or sand and then broadcast it over the bed. Larger seed doesn't need to be mixed with anything. After broadcasting, rake the bed to cover the seed, and then use the back of the rake to tamp the soil surface.

For a more uniform cover crop, use a hoe to open furrows across the bed, 5 to 6 inches apart. For small seeds, use furrows ½ inch deep. For pea- or bean-size seed, make furrows 1 to 2 inches deep. Sow the seed in the furrows, using the same total amount per area as you would if broadcasting. Then use your rake or hoe to cover the furrows and smooth the surface of the bed and tamp the soil surface.

Legume cover crops benefit from the addition of an inoculant that contains the bacteria that work together with the crop roots to fix nitrogen. Be sure to add this the first time you use a legume cover crop in your garden. Soybeans, clovers, and vetch need different types of inoculants. Ask your seed supplier which kind you need.

BEYOND THE BASICS

WEEDS BEGONE WITH BUCKWHEAT

Buckwheat bursts out with so much growth so quickly that it can smother tough grassy weeds such as quackgrass. Although it will take a full season, you'll not only beat your weed problem but also boost the soil's organic matter and improve its structure dramatically.

As soon as the danger of frost is past in spring, clear existing weed growth from the infested bed and sow buckwheat at a rate of 3 to 5 ounces per 100 square feet. Once the buck-wheat flowers, cut it down, let it dry a day, and till or dig it in. Immediately after tilling, rake the soil surface and sow buckwheat again.

Let this second crop flower, cut it down, dig it in, and replant again right away. Repeat this process until the first frost kills the buckwheat. After digging in that crop, sow oats to provide a winter cover crop for the bed. The following spring, you should be able to plant vegetable crops in the bed with no weed worries.

PLANTING THROUGH A KILLED COVER CROP

In heavy soil, a cover crop that survives over winter can become a liability in spring because it is covering the soil so well. Sunlight can't penetrate to help the soil dry out, so you can't turn the crop under. That delays your whole spring planting routine.

One way to avoid this problem and still reap the benefit of a cover crop is to plant a crop that will die during winter, such as spring oats, Alsike clover, or sweet clover. You can plant spring crops such as cabbage or broccoli transplants straight into the killed crop in spring. That way, you don't have to worry about turning the soil and incorporating the crop.

As early in spring as you can, cut out blocks of the killed cover crop to expose the soil at each planting site for your planned vegetable crop. When you plant, mix a little compost into the soil of each planting hole. If the killed mulch seems to be a lair for slugs and snails, pull it back farther from the plants to expose more bare soil. Once the garden warms and dries and slugs are less of a worry, roll the carpet of killed mulch back into place.

This technique is also helpful for planting warm-season crops such as tomatoes and peppers in areas where summer heat tends to overheat the soil. Soil temperature is buffered by the layer of organic material covering the ground, so there's less risk of blossom-end rot or tomato fruits cracking due to wide swings in soil moisture levels.

When Things Go Wrong

If your cover crop doesn't come up, lack of moisture or hungry birds may be to blame. Although cover crops generally don't need watering once they're established, they do need a steady moisture supply to get started. Ideally, sow seeds just before rain is expected. But if that doesn't work out, then water the seed in well, and cover the seeded area with a light sprinkling of straw or grass clippings—enough to retain surface moisture but not block the emerging seedlings.

If you notice birds digging around in an area where you've sown a cover crop, resow the area and immediately cover it with row cover or bird netting. If you use row cover, leave some slack, but weight the edges with rocks, bricks, boards, or plastic bottles filled with water (you don't need to seal the edges with soil). With bird netting, use wooden stakes to support the netting a few inches above the soil surface. It's safe to remove the row cover or netting when the crop is 2 to 3 inches tall.

HOW TO KILL A COVER CROP

The fast and easy way to kill a cover crop is to till it into the top few inches of soil. Keep in mind, though, that tilling can kill earthworms and blast apart soil structure. If you do till a cover crop, be sure the soil is neither too wet nor too dry. Set the tiller at a shallow setting—if you incorporate the crop residue too deeply, it won't break down quickly. And till as gently as you can.

Tilling may be the only method for incorporating a vigorous crop such as sudangrass, but *(continued on page 142)*

COVER CROPS AS PROBLEM SOLVERS

Each cover crop has unique benefits for the garden, but no cover crop is perfect. Diversity is the goal: Use as many different cover crops as you can, as long as they're suited to your climate and purpose. The list below points out the needs, benefits, and potential pitfalls of 19 cover crops for home gardens.

CROP	PLANT TYPE	SEEDING RATE	GROWING TIPS	PROBLEM-SOLVING ROLES
Alsike clover (*Trifolium hybridum*)	Perennial legume	1.5 oz/100 sq ft	· Sow in spring or fall · Tolerates heavy, wet, acid soil · Winter-kills in North · Won't grow well in South	· Improves structure of heavy soil · Attracts bees
Annual ryegrass (*Lolium multiflorum*)	Annual grass	1.5 oz/100 sq ft	· Sow spring through fall · Grows in wide range of climates and soils · Winter-kills in Zone 5 and colder · Becomes weedy if allowed to set seed · Easy to till under	· Outcompetes weeds · Prevents winter erosion · Loosens compacted soil · In the North, good killed cover for spring transplants
Austrian winter pea, field pea (*Pisum sativum* var. *arvense*)	Annual legume	3–6 oz/100 sq ft	· Sow in fall or early spring · Survives over winter in many areas · Grows quickly in cool, moist weather · Tolerates heavy soil · Can become weedy	· Produces large amounts of organic matter · Winter cover · Attracts beneficials early in the season
Berseem clover (*Trifolium alexandrinum*)	Annual legume	1.5 oz/100 sq ft	· Sow in late summer and early fall · Tolerates many kinds of soil · Allow 4 weeks after turning under before planting a vegetable crop	· Produces very large quantities of organic matter · Winter cover for the South and California
Buckwheat (*Polygonum esculentum*, also sold as *Fagopyrum esculentum*)	Broadleaved annual	3–5 oz/100 sq ft	· Very frost tender · Does well in poor soil · Fast-growing · Sow spring to midsummer · Will reseed itself if flowers allowed to mature	· Smothers weeds · Releases potassium as it breaks down · Attracts parasitic and predatory wasps and syrphid flies
Cereal rye, winter rye (*Secale cereale*)	Annual grain	5–10 oz/100 sq ft	· Grows fast in spring and fall · Can be planted from spring until time of first killing frost · Very winter hardy · Tiller needed to incorporate into soil	· Excellent for winter cover · Improves soil structure · Remove undecomposed debris before planting a vegetable crop, or crop may grow poorly
Cowpea, southern pea (*Vigna unguiculata* ssp. *unguiculata*)	Annual legume	3–5 oz/100 sq ft	· Sow in spring · Grows well in hot weather · Tolerates some shade	· Summer cover crop for the South · Smothers weeds · Undersow around sweet corn · Suppresses nematodes if allowed to bloom before tilling under
Crimson clover (*Trifolium incarnatum*)	Annual legume	1–1.5 oz/100 sq ft	· Shade tolerant · Overwinters in areas with mild frosts · Does well in rich soil but not acid soil · Easy to turn under	· Provides winter cover · Good choice for Southeast gardens · Undersow around fall crops in Pacific Northwest · Attracts beneficial insects
Fava bean, bell bean (*Vicia faba*)	Annual legume	8–12 oz/100 sq ft	· Related to vetch · Plant as soon as soil can be worked in spring in the North · Plant in late summer or through fall in mild-winter areas · Tolerates acid soil	· Provides winter cover · Good choice for Western gardens · Attracts parasitic and predatory wasps · Mines the soil · Smothers weeds · Produces large amounts of organic matter

CROP	PLANT TYPE	SEEDING RATE	GROWING TIPS	PROBLEM-SOLVING ROLES
Hairy vetch (*Vicia villosa*)	Legume	1.5 oz/100 sq ft	• Sow from spring through late summer • Does best in well-drained soil • Can grow well in sandy soil; very hardy • Never let it go to seed or it becomes a persistent weed • Works well combined with ryegrass	• Winter-killed mulch for tomatoes, peppers, and other warm-season transplants • Undersow around potatoes
Kale (*Brassica oleracea* Acephala Group)	Broadleaved annual	.5 oz/100 sq ft	• Sow spring or fall • Harvest leaves as desired, then till under the crop remains	• Good source of organic matter and iron
Oats (*Avena sativa*)	Annual grain	5–6.5 oz/100 sq ft	• Plant as soon as soil can be worked in spring through early fall • Likes cool, moist conditions • Doesn't grow well in heavy clay but tolerates acid soil • Winter-kills in most areas	• Winter cover; fills gaps between vegetable crops • Combine with legumes to provide fast cover until legumes become established
Oilseed radish (*Raphanus sativus*)	Broadleaved annual	3 oz/100 sq ft	• Sow in fall • Don't plant just before or after cabbage-family crops • Winter-kills	• Suppresses sting nematodes and weeds • Breaks through compacted soil
Rape (*Brassica napus*)	Broadleaved annual	.5 oz/100 sq ft	• Plant in fall • Grows well in cool, moist weather • Dies back in harsh winters	• Winter cover crop for MidAtlantic and South • Breaks up compacted soil • Biofumigant
Red clover (*Trifolium pratense*)	Biennial legume	1 oz/100 sq ft	• Plant from spring through fall • Shade tolerant	• Combine with annual ryegrass as a living mulch • Smothers weeds • Winter cover
Soybean (*Glycine max*)	Annual legume	6 oz/100 sq ft	• Sow after last spring frost • Grows fast in summer heat	• Smothers weeds • Grow before potatoes to lessen scab problems
Sudangrass (*Sorghum bicolor* var. *sudanense*)	Annual grass	1.5–3 oz/100 sq ft	• Plant late spring or early summer • Cut back when it reaches 3 feet (mature height to 12 feet)	• Good summer cover crop for the South • Helps suppress some plant disease organisms • Suppresses weeds • Provides massive amounts of organic matter
Sweet clover (*Melilotus officinalis*)	Biennial legume	1–1.5 oz/100 sq ft	• Sow in spring and summer • Grow as annual Zone 5 and colder • Doesn't like acid soil	• Very attractive to beneficials • Opens up compacted soil • Produces very large amounts of organic matter if allowed to overwinter
White clover, white Dutch clover, New Zealand white clover (*Trifolium repens*)	Perennial legume	.5 oz/100 sq ft as living mulch; 1 oz/100 sq ft for pathways	• Shade tolerant • Undersow around vegetable crops • Turn it under in fall	• Living mulch • Good pathway crop • Attracts ground beetles, parasitic wasps, and tachinid flies

hand-digging works well for many crops. The trick is to cut off the topgrowth first, about 2 inches above the soil surface, and allow it to dry in place for a couple of days before you dig. A lawn mower, string trimmer, or hand sickle or scythe work well for cutting topgrowth. Even a sharp pair of hedge shears suffices for a small planting. Drenching the cut material with compost tea can encourage faster breakdown.

If a crop is already in flower or is more than 2 feet tall, there may be so much topgrowth that you can't work it into the soil easily. If that's the case, rake off the clippings and just dig in the stubble and roots. You can return the clippings to the bed as a surface mulch, or use it as mulch on another bed, or add it to a compost pile. This reduces the amount of organic matter that you're returning to the bed, so you may need to add a little fertilizer to the crop that you plant in that bed for optimal growth. You will see a benefit from the microbial activity stimulated by the cover crop. Plus you've generated some high-quality mulch or compost ingredients for your garden.

The same rule applies for a spring cover crop if the soil is wet when you want to turn under the crop. Cut down the topgrowth and remove it from the bed. The exposed stubble and soil beneath will dry out in a few days, and then you can dig in the remains of the crop.

CROP ROTATION

Think of crop rotation as supplemental insurance for your garden. The garden won't fail if you decide not to worry about crop rotation. However, you may discover that your crops grow a little better and suffer fewer problems if you follow a crop rotation scheme. That's because rotating crops from bed to bed in your garden over time optimizes the nutrient bank in your soil and provides extra insurance against pest and disease problems.

Crop rotation is an organized approach to deciding what to plant where in your garden each year. The basic principle is to change the location of crops from year to year following a cyclical pattern. For example, a simple crop rotation could call for planting tomatoes in a bed one year, salad greens and broccoli in that bed the following year, onions and carrots the year after that, and then starting the pattern again with tomatoes in the fourth year. Crop rotation patterns call for grouping crops either by the plant part harvested or by the plant family to which they belong.

HOW ROTATION WORKS

Over time, rotating crops improves soil structure, reduces weeds, and helps fight insect and disease problems. Practicing crop rotation can boost yields, too, even if you don't apply supplemental organic fertilizer.

IMPROVING SOIL

Some vegetable crops, such as beets and tomatoes, are naturally deep-rooted; others, like onions and lettuce, have shallow roots. When you rotate crops, your entire garden benefits from the ability of deep-rooted crops to delve down into the subsoil and promote better moisture drainage. Crops with deep roots also mine the lower layers of the soil for nutrients, reducing the drain of nutrients from the surface soil. Peas, beans, and some types of cover crops play a special role by transforming nitrogen found in the air in soil into a form that plants can take up through their roots (a process called nitrogen fixation).

REDUCING PEST PROBLEMS

Rotating crops won't magically solve pest and disease problems, but crop rotation can stop problems from growing worse year after year, and in some cases it can reduce the severity of a pest or disease problem.

Here's an example. Suppose that your tomato crop develops late blight symptoms in late summer, cutting short the harvest. You remove the diseased plants, but late blight spores will have contaminated the soil of that garden bed. If you plant tomatoes in the same bed the following year, the plants will become infected early in the season by those spores in the soil, possibly ruining the crop before it sets any fruit. However, if you plant tomatoes in another part of your garden, the plants may escape infection altogether or at least until late in the season, when spores blown by wind or rain from some other field or garden reach your yard.

In some cases, crop rotation very effectively breaks the disease cycle of soilborne fungi. One example is angular leaf spot, which infects only squash-family crops and has relatively short-lived spores (they survive in soil for only about 2 years). If angular leaf spot shows up in your garden one year, be sure to plant your squash, melons, and cucumbers in other areas of the garden for the following 2 years. With luck, no new spores will reach those crops, and they won't develop leaf spot symptoms. The leaf spot spores left behind in the soil of the original squash bed will die out, unable to infect the crops from other plant families that you plant there.

Farmers rely on crop rotation to reduce the impact of pest problems, too. For example, a farmer can plant his potato crops far away from the field where potatoes grew the previous year. The relocation delays the onset of Colorado potato beetle damage because when the beetles emerge from winter dormancy, they have to travel a mile or more to find the new potato planting. By the time they reach the crop, it will have grown large enough to tolerate some beetle damage without losing yield. In the confined space of a home garden, this type of delay isn't nearly as dramatic. But in the case of pests that don't fly well, crop rotation can contribute to your overall scheme for reducing pest problems.

CROP ROTATION RULES

The first rule to remember is that crop rotation is not a substitute for good garden management. You can develop an elaborate rotation scheme, but if you neglect building soil organic matter, tending to your crops, choosing resistant varieties where possible, and scouting for pests, your garden will suffer. On the other hand, some experienced organic gardeners find that they can plant crops in the same location year after year without any increases in pests or disease. Their excellent garden management minimizes problems even though they're not rotating crops. Still, crop rotation is a useful tool in a gardener's management kit, especially where soil is less than ideal or where pest and disease problems persist.

Creating a crop rotation plan is like working out a logic puzzle. It's a matter of applying one of the general rules to your unique garden layout and choices of crops. You must take into account how much space you want to devote to each crop. And if part of your garden is lightly shaded, you need to consider which crops can tolerate those conditions. In most regions, for example, it's important to plant peppers, eggplant, and other heat-loving crops in full sun, but lettuce and spinach can tolerate some shade.

The easiest way to rotate crops is to group them according to harvest category. As you gain experience, you may find that you can rotate crops by plant family as well. To whatever degree you use crop rotation, your garden is sure to benefit.

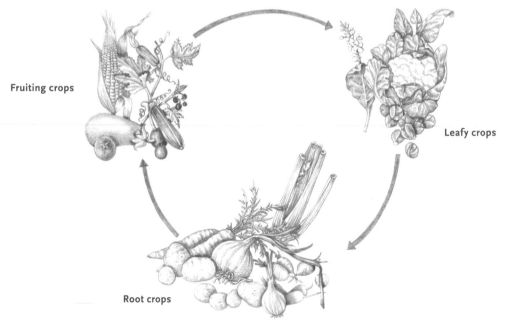

Fruiting crops

Leafy crops

Root crops

One simple approach to crop rotation divides crops into three groups: fruiting (flowering) crops, leafy crops (which includes crops such as broccoli, cauliflower, and celery), and root crops.

ROTATING BY HARVEST CATEGORY

The most basic way to organize a crop rotation scheme is to group crops by the part of the plant harvested: leaves, fruits (and flowers), or roots. The rotation goes like this:

1. Leafy crops follow fruiting crops.
2. Root crops follow leafy crops.
3. Fruiting crops follow root crops.

It's a simple circle (see the illustration on the opposite page). This type of rotation helps maintain balanced soil over time because each group of crops draws nutrients in different proportions from the soil.

You can take this rotation a step further by adding green manure crops into the rotation. The goal is to plant the green manure crop after the root crop, and then follow the green manure with the fruiting crop. The cycle becomes green manure, fruiting crop, leafy crop, root crop. The green manure crop helps to recharge the soil nutrient supply and encourage soil microorganisms to flourish. See Cover Crops for more information on green manures.

CREATING A FRUIT/LEAVES/ ROOTS ROTATION

Start by listing all the crops you'd like to grow in your garden this year. Next, figure out approximately how much space you'd like to devote to each (you can use a rough listing, such as half a bed or two rows). Sort the crops into fruit, leaf, and root groups. Tally up how much space you need for each group. If you're lucky, you'll find that you need the same amount of

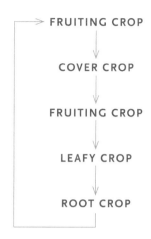

FRUITING CROP

COVER CROP

FRUITING CROP

LEAFY CROP

ROOT CROP

To build on a fruit/leaves/roots rotation, divide your garden into five areas. Each year, plant two areas in fruiting crops, one in a cover crop, one in root crops, and one in leafy crops. Move the crops from one year to the next as the arrows show.

space for each group. However, you'll probably discover that you need more space for fruiting crops than for the other two groups. (In general, fruiting crops are bigger and more sprawling than leaf and root crops.) If that's the case, you have some choices.

- Bend the rule and plant fruiting crops two years in a row in the same spot.
- Try one of the unconventional options described under "Less-Than-Perfect Rotation" on page 147 to create more "space" for fruiting crops.
- Include a cover crop in your rotation, and sandwich it between the years when you grow fruiting crops in a bed, as shown above.

If part of your garden is shaded, it will limit your options even more. Do the best you can with the space available.

ROTATING BY PLANT FAMILY

Following the fruit/leaves/roots rule won't, unfortunately, help to reduce pest and disease problems. If disease reduction is your goal, you need to rotate your crops by family instead of by edible part. In this kind of rotation you wait at least 2 years before replanting a bed with a crop from the same botanical family.

Crops that are closely related often are susceptible to the same disease problems. For example, tomatoes and potatoes are botanical cousins, and both can suffer from late blight, leaf spot, and other diseases. So if you plant tomatoes in a bed, you should wait at least 2 years before you plant tomatoes, potatoes, peppers, or eggplant (also cousins) in that bed again.

This rule is harder to apply than the fruit/leaves/roots rule because commonly grown vegetables fall into 11 different botanical families. You must juggle as many as 11 groups rather than only three when creating a crop rotation plan. Although complicated, rotating by botanical families can be an effective way to reduce the impact of troublesome pests and diseases.

ROTATING CROP FAMILIES

Many organic market gardeners devote serious study and deliberation to developing an intricate rotation plan. However, they have the advantage of lots of space and a wide variety of crops to rotate. They also are strongly motivated to plan: If a disease problem flares up, they could lose an acre or more of a crop.

For home gardeners, the stakes are lower, and you can take a more relaxed approach to rotating by crop family. One simple technique is to map your garden each year. Then, at planting time, refer to last year's map as you plant. Check the map to ensure you're not planting a crop from the same botanical family you planted there last year. This way, you'll allow at least one year between repeats of a crop family in any part of your garden.

If you like planning out your garden in detail, you could try creating a comprehensive 3- or 4-year rotation plan. To do this, follow the instructions in "Setting a Crop Rotation Pattern" on page 148.

HEAVY FEEDER, LIGHT FEEDER

Some crop rotation recommendations group crops by how heavily they draw nutrients from the soil. Crops are rated as heavy feeders, moderate feeders, or light feeders. In this type of rotation, begin with a heavy feeder; plant a moderate feeder the second year and a light feeder the third year. The difficulty with this type of rotation is deciding which group a particular crop belongs in. Different sources of information disagree about how to rate some crops. For example, you may find tomatoes ranked as a heavy feeder in some sources but as a moderate feeder in others. You'll do just as well with the fruit/leaves/roots approach or rotating by plant family as you will trying to figure out which crop is a heavier feeder than another. And no matter what your rotation pattern, you must continue to add organic matter to the soil regularly to maintain its quality.

LESS-THAN-PERFECT ROTATION

Following a perfect crop rotation may be impossible if you have a small garden or if you grow crops from only a few plant families. You can apply some of the principles—and reap some of the benefits—if you are creative.

ROTATE TO CONTAINERS

Perhaps you love tomatoes so much that you devote half your garden to tomatoes each year. To create a 3-year rotation plan, plant tomatoes in containers every third year instead of in your garden beds.

THE DRAWBACK. Your harvest from the containers will be smaller than from the garden, and your choice of varieties will be limited.

THE ADVANTAGE. You'll free up garden space for crops that there's no room for otherwise. Experiment with crops you've never tried—Asian vegetables or fresh herbs, perhaps. You may discover new favorites! Or you can plant half your in-ground garden to a soil-improving cover crop, which should bolster next year's tomato yield.

PLANT AMONG THE FLOWERS

Another option for gardeners with limited space is to rotate a pest-troubled crop out of the vegetable garden into a sunny flower garden. Most vegetable crops will fit in fine among annual or perennial flowers.

THE DRAWBACK. It may be hard to protect the veggies from hungry animal pests in an aesthetically pleasing way. Try a variety of repellents, and switch them frequently (see Animal Pests to

PLANT FAMILIES

Botanists use Latin names for plant families, but each family is also known by one or more common names. Use the listing below to find the vegetables you plan to grow, and group them by family.

SQUASH FAMILY: Cucumbers, gourds, melons, pumpkins, squash, watermelon, zucchini

NIGHTSHADE OR TOMATO FAMILY: Eggplant, peppers, potatoes, tomatoes

BRASSICA OR CABBAGE FAMILY: Asian cabbages (bok choy, tatsoi, and others), arugula, broccoli (and broccoli raab), Brussels sprouts, cabbage, cauliflower, collards, horseradish, kale, kohlrabi, mustard, radishes, turnips (and rutabagas)

LEGUME FAMILY: Beans, peas

GRASS FAMILY: Corn

LILY OR ONION FAMILY: Asparagus, garlic, leeks, onions, shallots

ASTER OR LETTUCE FAMILY: Artichoke, endive, lettuce, radicchio

GOOSEFOOT OR BEET FAMILY: Beets, spinach, Swiss chard

CARROT FAMILY: Carrots, celery, parsnips

MALLOW FAMILY: Okra

MORNING GLORY FAMILY: Sweet potatoes

---

learn more). Adding vegetables may interfere with the design and color scheme of the flower garden. If that's important to you, put the vegetables in corners instead of front and center.

THE BENEFIT. Surrounding a vegetable with flowering plants may help hide it from insect pests or offer pest predators that will keep your

(continued on page 151)

SETTING A CROP ROTATION PATTERN

To devise a structured rotation based on crop families, you'll need a pencil, paper, and several copies of a blank plan of your garden layout.

Start by listing the crops you want to grow, grouping them under the family headings (refer to the "Plant Families" table on page 147 if needed). Next, roughly list how much space you want to devote to each crop. Tally the space needs for each family. Now it's a numbers game. Can you combine families to end up with three to five groups that require approximately the same amount of growing space?

SQUASH FAMILY

CROP	SPACE NEEDED
Cucumber	$1/2$ bed
Zucchini	$1/2$ bed
Winter squash (buttercup, acorn)	1 bed
Melons	1 bed
Total	**3 beds**

CABBAGE FAMILY

CROP	SPACE NEEDED
Bok choy	$1/4$ bed
Arugula	$1/4$ bed
Broccoli	$1/2$ bed
Cabbage	$1/4$ bed
Collards	$1/4$ bed
Kale	$1/2$ bed
Total	**2 beds**

TOMATO FAMILY

CROP	SPACE NEEDED
Peppers (bell, Cubanelle)	$1/2$ bed
Tomatoes	$1/2$ bed
Total	**1 bed**

LETTUCE FAMILY

CROP	SPACE NEEDED
Lettuce mix	$1/2$ bed
Total	**$1/2$ bed**

LEGUME FAMILY

CROP	SPACE NEEDED
Snap peas	$1/2$ bed
Filet beans	$1/2$ bed
Total	**1 bed**

CARROT FAMILY

CROP	SPACE NEEDED
Carrots	$1/2$ bed
Total	**$1/2$ bed**

ONION FAMILY

CROP	SPACE NEEDED
Yellow & white onions	$1/2$ bed
Total	**$1/2$ bed**

SPINACH FAMILY

CROP	SPACE NEEDED
Spinach, beets, and chard	$1/2$ bed
Total	**$1/2$ bed**

For a multiyear plant-family rotation plan, list what you want to grow and how much space each crop will need. Look for ways to combine plant families so you end up with three (or more) groups with equivalent space needs.

Once you've done that, figure out how to divide your garden into the correct number of equal-size sections. Sometimes you'll need to compromise or improvise. In the example shown below, the garden has 10 beds, and the rotation calls for three groups of three beds and one group of one bed (sweet potatoes). The sweet potatoes will "jump" around the garden from year to year as needed to allow the other three groups to move to fresh territory.

In this garden, the rotation groups include three three-bed units (A, B, and C) plus a single one-bed unit (D).

(continued on page 150)

SETTING A CROP ROTATION PATTERN—Continued

As you fill in the base plans, consider whether shading of short crops by tall crops will be a problem. In this example, tomatoes, peas, and cucumbers (on a trellis) are the only tall crops. The peas will be harvested by early summer and replaced by bush beans, so it's not necessary to worry about the shade they cast on a neighboring bed. (If you grow pole beans, though, shading could be a problem.) Strategic positioning of cucumbers and tomatoes will provide beneficial shade to cool-season salad crops during the peak of summer (especially helpful in warm climates where the summer sun is intense).

Also consider substituting a season-long cover crop in place of sweet potatoes or one of your squash beds. If you can afford the space, allowing a cover crop such as clover or vetch to grow for a full season will work wonders for the soil, helping to ensure healthier and more productive crops in the future.

YEAR 2

Squash Family	Sweet Potatoes (or other crop)
Squash Family	Cabbage Family
Squash Family	Cabbage Family
Spinach Family/Lettuce	Carrots/Onions
Tomatoes	Legumes

YEAR 3

Cabbage Family	Carrots/Onions
Cabbage Family	Legumes
Sweet Potatoes (or other crop)	Tomatoes
Squash Family	Spinach Family/Lettuce
Squash Family	Squash Family

Each unit rotates to a new set of beds each year.

crop pest-free. Also, some vegetables are orna-mental in their own right—for instance, red-stemmed Swiss chard and purple snap beans.

SWAP CROPS

You love tomatoes, your friend loves squash. So, every third year, ask your friend to grow the tomatoes in her garden, and you grow the squash. Then split the harvest.

THE DRAWBACK. Your friend's gardening skills might be disappointing, and the harvest may suffer.

THE BENEFIT. You'll see your friend more often!

SKIP A YEAR

Sometimes the best approach is simply not to grow a crop for a year. For example, if squash bugs have become a serious problem in your garden, consider taking a squash sabbatical. Without any host plant available for a year, the local population of squash bugs in your garden will die out. The following year, squash bugs will have to fly in from other locations to reach your squash. If your next-door neighbor grows squash, this strategy won't help, but if your garden is isolated, this tactic may give you a chance to enjoy unsul-lied squash next year.

THE DRAWBACK. No homegrown squash for a year.

THE BENEFIT. This is a chance to try new crops. Or plant a cover crop where you planted squash last year and enjoy increased leisure because cover crops require less attention and time than most vegetable crops.

CUCUMBERS

Warm weather is great for growing cucumbers. Conversely, the vines can't take frost at all. Al-though cucumbers need warmth, gardeners in almost any zone can grow them well—some vari-eties go from seed to harvest in as little as 50 days.

Cucumbers are vulnerable to pests and diseases, but many varieties offer disease resistance. If you can avoid cucumber beetle damage when plants are young, keep fruits up off the soil, and pre-vent fungal diseases from spreading, you'll har-vest a good crop of juicy cucumbers.

CROP BASICS

FAMILY: Squash family; relatives are squash and melons.

SITE: Plant cucumbers in full sun.

SOIL: Cucumbers form an extensive root sys-tem if they grow in loose, rich soil that holds moisture well. If your soil isn't rich in organic matter already, spread up to 4 inches of compost over the planting area and work it in, loosening the top foot of soil as you work.

TIMING OF PLANTING: Sow seeds or set out transplants after danger of frost when the soil is warm (at least 60°F; 70°F is ideal); plant again 4 to 5 weeks later if your growing season is long enough to allow a second crop. In the Deep South, wait to plant until 80 to 90 days before first fall frost.

PLANTING METHOD: Sow seeds directly in the garden in raised hills or in rows. Start seedlings indoors in peat pots about 3 weeks before you plan to set them outside. Harden them off in a protected outdoor area (such as a coldframe) for the last week.

CRUCIAL CARE: Protect seedlings from cucumber beetles to prevent bacterial wilt from ruining your crop. Avoid any moisture stress once fruits begin to form.

GROWING IN CONTAINERS: Bush and compact varieties of cucumbers grow well on a trellis in a container at least 18 inches deep.

HARVESTING CUES: Watch for the cucumber flower to drop off the blossom end of the fruit. You can pick the fruit anytime after that. If the cucumber has spines, watch for the dimpled areas around the spines to fill out; that's a sign that the cucumbers are juicy and ready to pick. Harvest cucumbers for pickling when they are 2 to 4 inches long. Harvest cucumbers for slicing and salads when they are 6 to 8 inches long (many varieties will grow larger than this, but younger fruit will have better quality). Pick European cucumbers when they reach about 10 inches.

SECRETS OF SUCCESS

PICK VARIETIES KNOWLEDGEABLY. Cucumbers are prone to several serious diseases, including mildews, scab, anthracnose, and bacterial wilt. Fortunately, plenty of disease-resistant varieties are available; some show resistance to several different diseases. However, it's important to check what kind of cucumber you're buying as well as whether it's disease-resistant, because some cucumber varieties need special conditions or care in order to produce good fruit.

Many cucumber varieties bear flowers in the traditional way for squash-family crops: Each vine first produces male flowers, then female flowers. Insects carry pollen from male to female flowers, and then the female flowers produce

European cucumber   Pickling cucumber   Slicing cucumber

Pickling cucumbers are good for fresh eating as well as pickling. Some slicing varieties offer resistance to multiple diseases, while others are especially heat and stress tolerant. European cucumbers are long and thin, with tender skins, no seeds, and a mild flavor.

fruits. Special varieties known as gynoecious cucumbers and parthenocarpic cucumbers, however, work differently.

Gynoecious cucumber varieties produce female flowers only. Planting a second variety (one that produces male flowers, too) is required for pollination and fruit set. Seed suppliers usually add a small amount of pollinator seed (marked with a dye) to the packet of gynoecious seeds. You must plant some of the pollinator seeds (and not thin out those seedlings) in order to reap a harvest from the gynoecious variety.

Parthenocarpic varieties produce only female flowers, but they do not require pollination to produce fruit. In fact, some parthenocarpic varieties will produce distorted fruits if the flowers become pollinated. These varieties are intended for greenhouse growing, where insects can be excluded and there's no risk of pollination. Other parthenocarpic varieties are adapted to garden growing and will produce normal fruit even if the flowers are pollinated.

Why grow gynoecious and parthenocarpic varieties? As it happens, some of the varieties with the broadest range of disease resistance are gynoecious varieties, so they're a good choice if you've had trouble with cucumber diseases in your garden. Parthenocarpic cucumbers produce very tender-skinned crisp fruits (often called European cucumbers) that many people prefer to standard slicing cucumbers.

All of these details should be spelled out in the catalog description of a cucumber variety. If the description isn't clear to you, contact the seed supplier and ask for clarification before you buy.

TRELLIS TO SAVE SPACE. Cucumbers and trellises are a great combination. The fruits grow straight, they're easy to find and pick, they suffer fewer disease problems, and you can grow more in less space. Construct an A-frame trellis (shown on page 154) or a vertical trellis (like the one on page 42). The vines don't need any special support once they start to climb—the strong tendrils will cling to the trellis as the vines grow. Set up the trellis before you plant. Sow seeds 4 inches apart and thin the plants to 1 foot apart once they have 2 or 3 leaves. As the vines elongate, pinch off the first few side shoots that form on each vine. After that, allow the side shoots to sprout and spread.

DRIP IRRIGATION FOR FINE FRUIT. Cucumbers are more than 90 percent water. Because of this, the fruits will show the impact of moisture stress while they're forming. Cucumbers with a skinny portion are the result of a lack of water during their development. The appearance is sometimes comical, but the flavor won't be. Water-stressed fruits tend to be bitter. The best way to provide even soil moisture, especially for trellised cucumbers, is to run a soaker hose or drip irrigation line along the row of fruits. Trellised cucumbers are more prone to drought stress because the foliage is more exposed to sun and air (but they are less likely to suffer disease problems).

TRY FOR TWO CROPS. Cucumbers grow fast, some producing fruit in as little as 50 days from sowing seeds. A vine can produce fruits for a month or longer, but sometimes productivity declines—due to insect or disease problems, or simply because you didn't keep picking, and maturing fruit sent a signal to the vine to stop producing more young fruit. Rather than

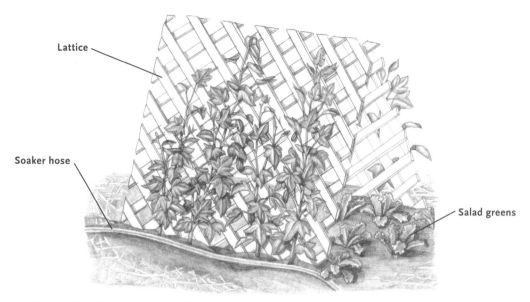

Lattice

Soaker hose

Salad greens

An A-frame trellis offers strong support for cucumber vines. In hot-summer areas, sow salad greens in the shaded area under the trellis, too.

counting on your vines to stay productive for several weeks, simply make a second planting (be sure to choose a short-season variety) about 4 weeks after the first. If you try this approach, don't double the size of your cucumber planting. Instead, plant half of your trellis early, and the other half late. Or plant the second crop in a space where you've harvested a spring planting of broccoli or other cool-season crop. If you don't want to bother with constructing another trellis, simply train each vine up a single sturdy stake.

REGIONAL NOTES

DRY WESTERN STATES

Spider mites and moisture stress are prime problems for cucumbers in arid regions. Spider mites feed more and reproduce faster in dry condi-tions, causing leaves to drop off. Any fruits that form will probably be stunted and bitter flavored. And in a dry climate, it's more difficult to maintain even soil moisture. Instead of trellising cucumbers on a tall trellis, train them over an arch of wire fencing. The vines will be less exposed to drying wind and hot sun. You'll be able to spray the vines lightly with water to promote humidity. Using a tunnel also results in the vines shading the garden bed, which helps preserve soil moisture. A drip line running down the center of the bed under the arch allows regular watering without a fuss.

Cover the arch with row cover at planting time to protect seedlings from cucumber beetles. Remove the covers when the weather becomes too hot, or once the vines are ready to twine through the wire mesh.

PREVENTING PROBLEMS

BEFORE PLANTING

- Sow a trap crop to attract cucumber beetles.

AT PLANTING TIME

- Cover seeded rows or seedlings with row cover and seal the edges as described in Row Covers to keep out insect pests.
- If possible, arrange a soaker hose or drip irrigation line along the row.

WHILE CROP DEVELOPS

- Spray exposed plants with kaolin clay to deter cucumber beetles (see Pest Control for information).
- Remove row covers when plants start to flower or when you need to start training them onto a trellis. If you're growing parthenocarpic cucumbers without a trellis, you can leave them covered until harvest.
- Spray vines with plant health boosters or compost tea to help prevent disease problems (see Disease Control).
- Check soil moisture frequently, and water deeply whenever the top few inches of soil are dry.
- Weed out squash-family weeds and volunteer seedlings.

AFTER HARVEST

- Remove and destroy any noticeably diseased crop residues, including damaged fruits.
- Turn under or compost all other crop debris.

TROUBLESHOOTING PROBLEMS

SEEDLINGS CHEWED. Cucumber beetles can wipe out young cucumber seedlings or infect them with bacterial wilt. Don't try to save damaged seedlings; replant instead and cover the seeded areas with row cover. See Cucumber Beetles for more controls.

YELLOW AND/OR WATER-SOAKED SPOTS ON LEAVES. Angular leaf spot, anthracnose, downy mildew, and scab are possible causes. As angular leaf spot develops, the spots become clearly geometric and the diseased tissue sometimes drops out, leaving holes in the leaves. See Angular Leaf Spot for controls.

Leaf spots that drop out are also a symptom of anthracnose. Check plant stems, too; if the problem is anthracnose, you find dark streaks and a pink jellylike substance (during wet weather). If the disease has progressed this far, you'll probably have to uproot the vines. See Anthracnose for prevention tips.

To confirm that the problem is downy mildew, check leaf undersides, where purplish mold will grow on the spots. Remove infected leaves. Applying a copper spray may stop the problem from spreading. See Downy Mildew for more information.

In wet weather, spots due to scab become covered with a velvety mold. Fruits will also develop spots that turn into tan dry scabs. Scab is not common, because most cucumber varieties are resistant. If you determine that scab is the problem on your plants, be sure to choose scab-resistant varieties in the future.

SPOTS ON OLDER LEAVES. Alternaria leaf blight causes these yellow spots, which later turn light brown. The spots show concentric rings, unlike the spots caused by other disease organisms. Remove infected leaves and take steps to improve air circulation and decrease moisture around the vines. See Alternaria Blight for more controls.

EDGES OF LEAVES DIE. Gummy stem blight fungus is a serious problem of cucumbers, especially in the Southeast. Infections of stems can girdle the stems, causing the vines to collapse suddenly. Feeding by cucumber beetles and aphids makes plants more susceptible to gummy stem blight, and so does infection by powdery mildew. The fungus is also transmitted by seed. Be sure to buy clean seed, and keep in mind that commercially grown transplants may be infected but not show symptoms. Also, grow powdery mildew resistant cultivars and cover young plants with row covers to prevent insect damage. Remove and destroy infected vines. Because the fungus persists in the soil, start a 2-year or longer rotation of squash-family crops.

POWDERY WHITE COATING ON LEAVES. Powdery mildew is a widespread fungus. The coating may cover stems and fruits, too, if not checked. Handpick infected leaves and start other control measures as described in Powdery Mildew.

MOTTLED LEAVES. Cucumber mosaic virus causes mottling. Leaves may curl under. There is no treatment, and the virus will infect and ruin the fruits, too, so uproot and destroy the infected vines. See Mosaic for prevention tips.

CURLED AND TWISTED LEAVES. Melon aphids suck sap, causing leaves to curl and twist.

Look for colonies of yellow to dark green aphids on leaf undersides. Spray leaves with a strong stream of water to dislodge aphids; if the problem persists, see Aphids.

LEAVES OR WHOLE VINES WILT. Bacterial wilt, Verticillium wilt, or gummy stem blight can cause vines to wilt suddenly. Cut through a wilted stem. If a thick ooze comes out, the problem is bacterial wilt, and there is no cure. Uproot and compost the diseased vines. Cucumber beetles spread bacterial wilt as they feed; to prevent future infections, see Cucumber Beetles.

With Verticillium wilt, older leaves usually wilt first. Split open stems and you'll find long dark streaks inside. For prevention and controls, see Verticillium Wilt.

Gummy stem blight is a fungus, and it also produces a gummy ooze, but on the surface of stems, not internally. See "Edges of Leaves Die," above, for more information.

FLOWERS DON'T FORM FRUITS. If you're growing a standard variety, the vines will produce male flowers first, then female. Inspect the vines; you may just need to wait until female flowers form. However, if you find female flowers (which have a tiny cucumber at the base) that aren't developing into fruits, the problem is a lack of pollination. Double-check the type of cucumber you're growing. If the variety is gynoecious, the problem may be that no vines of the pollinator variety survived to produce male flowers.

Insects pollinate cucumber blossoms, and if the weather is cloudy and cool, insects won't be active. As long as both male and female flowers are available, you can improve pollination by hand-pollinating, as shown on page 390.

MISSHAPEN FRUIT. Funny-looking cucumbers are the result of incomplete pollination or moisture stress. To prevent the problem as more fruits form, be sure to maintain even soil moisture, and try some hand-pollinating.

BITTER-TASTING FRUIT. Cool weather or drought can cause bitter flavor. Some varieties are more prone to bitterness than others, too. Take note of the variety that has bitter fruit, and choose a different one for your next planting. Try salvaging the bitter fruits by peeling them; remove some of the flesh under the peel at the stem end as well because that's where bitter flavor tends to concentrate.

MOLD ON BLOSSOMS OR SMALL FRUITS. Some fungal organisms can invade cucumber blossoms and then grow into the young fruits; the disease is commonly called blossom blight or wet-rot. Remove and destroy the infected blossoms and fruits. See Rots for more information.

HOLES IN BLOSSOMS AND FRUITS. Pickleworms feed on cucumber blossoms and fruits. Their tunneling in young cucumbers allows rot to set in. The fruits usually aren't salvageable. See Pickleworms for more information.

Slugs and snails sometimes feed on cucumber fruits, leaving holes or scars. Fruits may be stunted. Growing cucumbers on a trellis will prevent this problem in the future.

SCABBY AREAS ON UNDERSIDE OF FRUITS. Belly rot results when spores of fungal rots in the soil come in contact with cucumber fruits, especially those resting on the soil surface. It develops quickly in hot weather. To prevent this problem for future crops, grow cucumbers on a trellis. Plastic mulch also helps to prevent spores from coming in contact with the fruit.

MOLDY OR DRY ROTTEN SPOTS ON FRUITS. Many of the disease organisms that cause symptoms on leaves and stems of cucumber plants can also infect the fruits. These disease organisms can cause rotting or allow infection by secondary bacteria that produce wet or dry rots. The fruits probably won't be usable. Remove diseased fruits to a hot compost pile or bury them deeply so that the seeds inside won't sprout in the future (the seedlings could be infected with disease organisms). To prevent fruit rots in the future, improve air circulation around vines, use a support structure to keep vines and fruit off the ground, or place cans or other small supports under individual fruits.

CUCUMBER BEETLES

Little yellow beetles with stripes or spots are easy to overlook on cucumber and squash vines, but the damage they cause can't be ignored. The beetles emerge from dormancy before cucumber seedlings have even poked out of the ground. Biding their time, they feed on leaves and pollen of trees, shrubs, and weeds. But when squash-family plants sprout, the seedlings give out a chemical signal that cues the cucumber beetles to fly to your vegetable garden and chow down.

Young seedlings may die quickly as a result of beetle feeding. And even plants that aren't seriously damaged sometimes decline prematurely from disease because cucumber beetles spread bacterial wilt and cucumber mosaic virus as they feed (see Bacterial Wilt and Mosaic for information).

The beetles also lay eggs in soil cracks by plant stems. One to 2 weeks later, wormlike larvae hatch and feed on crop roots. Their feeding isn't too severe, and as long as plants are well cared for, the roots can rebound from the injury. After feeding for about a month, the larvae pupate, and a second generation of adults emerges in midsummer. These beetles feed on leaves and flowers but don't do as much harm unless they congregate on young cucumber and summer squash fruits, where their feeding can disfigure the crop. Second generation larvae can ruin fruits that are resting on the soil by tunneling into the rind. From year to year and region to region, there will be one to three overlapping generations per year.

Striped cucumber beetles can overwinter in most parts of the United States and Canada. A similar species, the western striped cucumber beetle, is found only in western states. It will feed on snap bean pods as well as seedlings of a variety of crops. Spotted cucumber beetles survive winter only in the southern United States, but they can migrate north in plenty of time to become a problem for northern gardeners as well. The larval stage of the spotted cucumber beetle is the southern corn rootworm, which can cause serious damage to corn crops. Western corn rootworm beetles look similar to cucumber beetles but have yellow abdomens and shorter stripes on the wings. These beetles will feed on squash-family plants but don't do much damage. Their larvae, though, are serious pests of corn (see Corn for details).

PEST PROFILE

Acalymma spp., *Diabrotica undecimpunctata*

RANGE: Striped cucumber beetle throughout the United States except for the Far West and central and eastern Canada; western striped cucumber beetle west of the Rocky Mountains; spotted cucumber beetle throughout the United States, in Canada from east of the Rocky Mountains into Quebec

HOW TO SPOT THEM: Inspect squash-family seedlings frequently, especially the underside of cotyledons, along the stems, and inside blossoms, because cucumber beetles prefer shaded spots.

Striped cucumber beetle   Spotted cucumber beetle

Damage from beetle feeding

ADULTS: Striped cucumber beetles are ¼ inch long, yellow with black heads and three black stripes down the back and rows of tiny indentations in the wing covers; spotted cucumber beetles are ¼ inch long, yellow-green with black heads and 12 black spots on their wings.

LARVAE: Cucumber beetle larvae are white, wormlike grubs with dark heads and tails, up to ¾ inch long.

EGGS: Laid in the soil

CROPS AT RISK: Cucumbers, melons, winter squash, summer squash, watermelons, corn, and occasionally other vegetable crops

DAMAGE: Beetles feed on plants, sometimes killing young seedlings; larvae feed on plant roots; even minor beetle activity can be a problem if their feeding spreads bacterial wilt or mosaic virus, because the diseases can kill the vines later in the season.

CONTROL METHODS

TIMED PLANTING. If your growing season is long enough to allow it, plant vine crops late to avoid the first round of beetle feeding and egg laying.

ROW COVERS. Cover seedbeds and transplants immediately after planting and seal the edges (see Row Covers for information). Remove the cover when plants start to flower to allow insect pollination.

STICKY TRAPS. Cucumber beetles are attracted to the color yellow. Put out yellow sticky traps by your vine crops to help monitor whether beetles have found your garden.

GARDEN CLEANUP. After harvest, remove all crop debris or dig it into the soil to destroy overwintering sites for adult beetles.

FALL CULTIVATION. Turning the soil just before frost in beds where squash-family crops grew will expose dormant beetles to killing cold.

KAOLIN CLAY. Spray as often as twice a week, especially the undersides of leaves, to deter cucumber beetles. Reapply after heavy rains as well.

HANDPICKING. As often as possible, seek out beetles and kill them by crushing them or dumping them in soapy water; or use a sticky bucket trap as shown on page 160 to nab beetles.

VACUUMING. Use a handheld vacuum to collect beetles and dump them into soapy water.

Paint a plastic bucket bright yellow, wrap plastic wrap around it, and coat the plastic with a sticky substance such as Tanglefoot. Walk beside your cucumber and squash plants, brushing the plants to disturb the beetles. They will fly up and get stuck on the sticky surface.

TRELLISING. If you don't use row covers, train cucumber and melon vines on trellises. Beetles will be easier to spot and handpick on trellised vines, and there's no danger that larvae will damage fruits.

TRAP CROPS. Some farmers deliberately plant one early squash-family crop, using yellow plastic mulch as an extra attractant for cucumber beetles. One variety that's very attractive to the beetles is 'Dark Green' zucchini. The farmers spray or vacuum the trap crop to kill off the beetles, and then plant their regular crop. This will work in a home garden only if there are no other vegetable gardens or farm fields nearby; otherwise, beetles will travel to your garden from those areas.

DRENCHING SOIL WITH NEMATODES. Applying *Heterorhabditis* nematodes to the soil helps kill the larvae; refer to Nematodes for application details.

PYRETHRINS. If you need to control an infestation fast, try spraying with a pyrethrin product. Pyrethrins kill beneficials, too, so limit your spray to plants you know are infested with cucumber beetles.

CONTROL CALENDAR

BEFORE PLANTING

- Plant a trap crop and plan to destroy the beetles attracted to it before your desired crop emerges.

AT PLANTING TIME

- Delay planting to avoid the first flush of beetle activity.
- Cover plants with row covers and seal the edges.
- If you don't cover plants, set up yellow sticky traps to monitor beetle activity.

WHILE CROP DEVELOPS

- Train cucumbers and melon vines on a trellis.
- Spray plants with kaolin clay to discourage beetles.
- Handpick or vacuum plants to remove beetles.
- Use a sticky bucket trap to catch beetles.
- Remove row covers when plants begin to flower.
- Drench soil with nematodes to reduce larvae.
- Spray pyrethrin as a last resort to kill beetles.

AFTER HARVEST

- Clean up or till under all crop debris.
- Just before fall frost, turn the soil to expose beetles to cold.

CUTWORMS

Seedlings look like they've been felled by a miniature chainsaw or even disappear completely when cutworms are active. These soil-dwelling caterpillars can be active nearly any time during the growing season, which is bad news when you're trying to establish seedlings or tender transplants.

Various species frequent gardens in different regions, and the variegated cutworm is active in most parts of the United States and Canada. The larvae become active in spring, feeding at night by chewing through stems near soil level or climbing up on plants to eat leaves and buds. By morning, the larvae have returned to the soil, safe from the gardener's wrath.

Larvae pupate in the soil, too. The moths emerge and fly at night, laying eggs on host plants. Newly hatched larvae will feed briefly on foliage before they head into the soil. One or two generations per year are usual in the North; up to six generations can plague gardens in the South.

With cutworms, prevention is vital because there's no help for seedlings once they're sawn through at the base. If cutworms are widespread in your garden, start as many crops as possible in seed flats, and protect all transplants with cutworm collars as shown on page 162.

PEST PROFILE

Several genera and species

RANGE: Throughout the United States and Canada

HOW TO SPOT THEM: In spring before planting, stir the soil to turn up pupae or dormant larvae. Unfortunately, the first noticeable sign of cutworm activity may be ruined seedlings or transplants.

ADULTS: Dull gray or brownish moths, sometimes mottled, some with markings on wings; wingspan up to 1½ inches

LARVAE: Greasy-looking, smooth, gray or brownish C-shaped caterpillars, up to 2 inches

Variegated cutworm

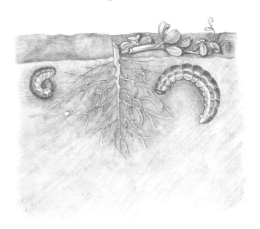

Cutworm feeding damage

long. Variegated cutworms have yellow spots down the center of the back.

EGGS: Clusters of small white eggs laid on plants

CROPS AT RISK: Most vegetable crops

DAMAGE: Cutworms eat tender young plants, sometimes chewing through transplant stems at soil level or consuming whole seedlings.

CONTROL METHODS

SPRING CULTIVATION. If cutworms were a problem in your garden last year, before you sow seeds, cultivate seedbeds twice to turn up cutworms or pupae for birds to eat.

PLANT COLLARS. Protect individual transplants with collars fashioned from toilet paper or paper towel tubes or tin cans with the ends cut out. Or save those paper or Styrofoam cups from your morning coffee run and punch out the bottom to make a wider collar, which may be easier to use for large transplants. Plastic food containers, such as yogurt cups, will work, too.

INSECT TRAPS. Putting out insect traps that contain floral lures (not sex pheromones) will help decrease cutworm populations by preventing egg laying. However, traps won't prevent damage from cutworms that are already present in the soil.

HANDPICKING. During the day, use a fork to sift the soil in beds where cutworms have been active. With luck, you'll turn up some cutworms; stab or crush them (or drop them in a cup of soapy water). Use a flashlight at night to seek cutworms feeding directly on plants.

PARASITIC NEMATODES. Apply *Steinernema* nematodes to kill cutworms in garden beds. See Nematodes for information.

After setting transplants in place in the garden, carefully thread each plant top through a cutworm collar and push the collar 1 inch into the soil. Leave collars in place all season or remove them once plants are too large and tough to interest cutworms.

CONTROL CALENDAR

BEFORE PLANTING

- Cultivate soil twice to expose cutworms to insect-eating birds.
- Set up insect traps to lure cutworm moths and prevent egg laying in your garden.

AT PLANTING TIME

- Cover each transplant with a cutworm collar.

WHILE CROP DEVELOPS

- Apply parasitic nematodes to kill cutworms in the soil.
- Handpick cutworms from soil or plants.

DAMPING-OFF

Gaps in seedling rows and new seedlings that suddenly collapse are telltale signs of damping-off. Several species of fungi can cause damping-off; Pythium and Rhizoctonia are two of the most notorious. Damping-off can be a problem in garden beds, coldframes, and indoor seed-starting areas. These common soilborne fungi draw nourishment from decomposing organic matter as well as living plants, and thus, they're not "all bad."

Your strategy should be to avoid the environmental conditions that favor damping-off, rather than trying to exterminate the fungi that cause it. In fact, in healthy soil, many other types of soil microorganisms suppress damping-off fungi. As you'd expect from the name of the disease, damp conditions favor damping-off. Some damping-off fungi thrive in cool soil; others do well in warm soil. The fungi can infect seeds and seedlings through natural plant openings. When conditions are right, the fungi begin to attack as soon as a seed swells and softens in moist soil.

RANGE: Throughout the United States and Canada

CROPS AT RISK: Most vegetable crops

DESCRIPTION: Germination may be spotty, with gaps in seed rows. Some seedlings will die before they emerge from the soil; other seedlings emerge normally but then collapse. Seedling stems have reddish brown constricted areas near the soil line or are dark and soft at the soil line, with mushy, rotted roots. Seedlings that don't die will turn into stunted plants that yield poorly if at all.

FIGHTING INFECTION: Outdoors, pull out infected seedlings and watch to see what happens. Don't water the bed for a few days. The remainder of the crop may not become infected, especially if the soil surface dries out. If the infection continues to spread, uproot all the seedlings and take steps to improve conditions before you replant.

Indoors, it's sometimes possible to save a flat of seedlings that has started to damp-off. First, pull out all the affected seedlings. Then use a fork to lightly rake the soil surface around the seedlings. Set up a small electric fan (a clip-on type works well) and leave it running continuously to promote air movement. If possible, increase the light intensity in your seed-starting area. If the infection continues to spread, compost all the soil and plants in the infected flats or containers. Replant in clean containers using fresh mix.

GARDEN CLEANUP: Remove infected seedlings and clear away rough, partially decomposed organic matter before you replant an area where damping-off has been a problem.

NEXT TIME YOU PLANT: Indoors and out, your goal overall is to help seedlings germinate fast and grow steadily. Use fresh seed, or test the viability of old seed before you use it. (If it germinates slowly, throw out that seed and buy a new packet.)

Outdoors, improve soil drainage and add compost to encourage populations of beneficial microorganisms. Check soil temperature before sowing seeds; wait until temperature is optimal to plant. Sow seeds at the correct depth—the deeper you sow, the longer it takes for seedlings to emerge, and that leaves the seedlings vulnerable to the fungi longer. Avoid sowing seeds too thickly. Seedbeds should be moist but not soaking wet; water in the morning so that the seedbed won't be damp overnight. Thin seedlings before they become overcrowded.

Indoors, take steps to prevent a stagnant, humid atmosphere in your seed-starting area. Also, provide lots of light and position seedlings close to the light source (1 to 2 inches away). Top off seeded flats or containers with a thin layer of milled sphagnum peat moss or coarse sand. Water containers from the bottom, but don't let containers or flats sit in water and become waterlogged. Also, watch for patterns in the development of damping-off. The fungi can be seed-borne. If damping-off seems to happen when you sow seed from a particular packet, no

matter how careful you are, chances are the fungi are in the seed. Buy fresh seed and try again.

It's also possible for the fungi to contaminate a soilless seed-starting mix, and it can survive in dried soil debris sticking to containers. If you've had any recent problems with damping-off, wash containers with a 10 percent bleach solution before reusing them. Don't reuse seed-starting mix. Instead, after transplanting, dump any leftover soil in your compost pile.

Run a small electric fan in your seed-starting area to encourage air circulation. If possible, direct it to blow across the surface of containers to keep the soil surface dry.

If you've had serious problems with damping-off, indoors or out, buy a beneficial fungal product such as *Trichoderma harzianum* and apply it according to package instructions.

CROP ROTATION: The fungi that cause damping-off are present in all soils; crop rotation is not a direct factor in controlling this problem.

⌒DEER

Predicting whether deer will damage your vegetable garden isn't difficult. If deer live in your neighborhood, chances are you'll spot them in fields or wooded areas as you drive by. In many places, deer have adapted so well to living near humans that they'll graze in suburban backyards in the early morning or at dusk.

It's estimated that more than 25 million deer roam the forests, fields, and suburbs of the United States, with especially dense populations

in the eastern states. Deer damage can be so devastating and discouraging that gardeners just give up. However, deer and gardens can coexist if you're willing to be persistent and invest some time and money in the effort.

REVIEWING THE OPTIONS

First, decide whether you want to protect only your vegetable garden or your entire landscape. Deer favorites include alfalfa, clover, apples, acorns, cabbage-family crops, tomatoes, corn, and certain ornamentals, but deer will eat almost any plant when they are very hungry. In summer, when food is plentiful, you may be satisfied to fence them out of your vegetable garden and tolerate some light deer damage to flowerbeds and tree buds. But if deer pressure is high in your area during winter and early spring, fencing your entire property to prevent destruction of valuable trees and shrubs as well as garden beds might be the best choice.

REPELLENTS AND SCARE DEVICES

Most of the repellents and scare devices described in Animal Pests will have an effect on deer as well. How much damage you can prevent by hanging soap bars on trees, spraying plants with repellents, or installing motion detector lights varies greatly from area to area, even more so than with small mammals. If you haven't tried repellents yet, it's worth the cost (about $25) to see how a couple of them work. If you've tried them already, with variable results, it's probably time to plan a budget for a garden fence.

FENCING OUT DEER

Fortunately, there are plenty of options for deer fences, and some are reasonably priced. For example, fencing a 40 × 40-foot garden should be possible for less than $200. A quality woven wire, plastic mesh, or electric fence will cost quite a bit more but should last more than 10 years. Think about what you've invested in buying soil amendments, tools, and plants for your garden and landscape over the years. It may turn out that the losses you'll suffer by not fencing your yard are greater than the cost of installing a fence.

A SIMPLE REPELLENT FENCE

If deer pressure in your area is low and you're willing to live dangerously, try setting up a simple fence that relies solely on repellent odor to discourage deer from trespassing in your garden. The fence consists of a single length of cotton cord or rope, tied to metal garden stakes. Pound 3- to 4-foot stakes around the garden perimeter and tie the cord to the stakes 30 inches above soil level. Spray an odor-based deer repellent on the rope. For extra protection, tie strips of cotton cloth to the cord, 3 feet apart, and spray the strips with repellent, too. (Renew the repellent at least monthly.) If you set up this fence before you start spring planting, it may be enough to keep deer away. It will not stop small mammals, though.

MESH TUNNEL FENCE

Another low-cost option for protecting shorter crops is to use sections of wire mesh fencing laid on the side to create protective arches over your crops. In essence, your crops will grow in the

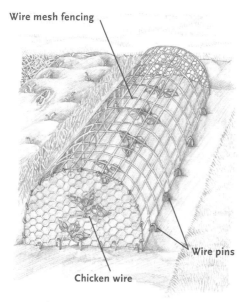

Wire mesh fencing

Chicken wire

Wire pins

Turn wire fencing on its side to create protective arches over crops that will prevent deer from nibbling.

protective space of a tunnel, as shown above. Use fencing with openings about 2 × 4 inches. A 6-foot length will create an arch of about the right height to span a 3-foot-wide bed. You can buy a 100-foot roll of this fencing for about $75. Anchor the fencing with metal pins and cover the ends of the tunnels with chicken wire.

SNOW FENCE SOLUTION

If your vegetable garden is less than 60 feet long and 40 wide, a barrier of snow fencing should be sufficient to keep out deer. Snow fence isn't attractive, but it's relatively cheap and easy to set up. Pound in metal fence stakes about 15 feet apart around the perimeter of your garden. Hook the fencing to the stakes, and rope it tightly closed where the ends of the fence overlap. It's easy to pull back one end of the fence

when you want to work in your garden. Although deer can easily jump snow fence, they will choose not to because they don't like to jump into a confined area. In the off season, roll up the fence and store it in a shed or outdoors covered by a tarp. Or if necessary, use it to block snowdrifts on your property. Wear gloves when working with snow fence to avoid splinters.

TALL WIRE FENCES

In areas with high deer pressure, an 8- to 10-foot woven wire fence is good insurance against deer damage in your vegetable garden. This type of fence is a major investment, costing as much as $8 per foot installed. Installing it yourself will reduce cost, but it's heavy work. You'll need to drive 4 × 4 wooden posts every 12 to 15 feet around the garden's perimeter to support the wire.

Flag the top of the fence with strips of white cloth to make it more visible to the deer. That way they'll be less likely to try to jump it.

If rabbits and other small mammals raid gardens in your area, add a chicken wire skirt along the bottom of the woven wire fence. It's possible that raccoons will scale an 8-foot fence if there's corn on the other side. To prevent this, mount a single electric fence wire to the outside of the fence, 4 inches above ground level. Connect it to a small charger as described on page 170.

A SHORTER SHORTCUT

A 3-foot welded wire fence topped by ranks of galvanized wire may be sufficient to protect your garden, especially if other food sources for deer are plentiful in your neighborhood. Use ½-inch welded wire. Sturdy 8-foot metal fence posts are an alternative to wooden fence posts with this design. Trench along the fence line after setting posts, so you can set the bottom edge of the wire at least 6 inches below soil level. The welded wire should rise at least 3 feet above soil level. Once the wire is in place, refill the trench. Above the welded wire, attach three strands of galvanized wire (not electrified), 18 inches apart. Flag the wires with repellent-sprayed cloth strips (as described on page 165) to make the wires visible to the deer and to add to the fence's effectiveness. This fence should also keep out rabbits and woodchucks.

PLASTIC MESH FENCING

Polypropylene mesh fences are a lightweight option for fencing a vegetable garden or an entire landscape. The mesh is available in different

A DEER FENCE WITH STYLE

If you'd prefer a garden fence with more elegance than plastic mesh or electric wire can provide, consider installing a wood fence about 4 feet high and topped with wire (nonelectric). The wooden barrier can be prefabricated garden fencing or 4 × 8 sections of wooden lattice. Use 10-foot rot-resistant wooden posts sunk at least 1 foot deep. Above the wooden fence, string galvanized wire or heavy-duty fishing line 1 foot apart to the top of the posts. You can use a staple gun or fence staples to attach the wire to the posts. Tie strips of white fabric at 3-foot intervals to the top two wires of the fence. The fluttering fabric signals the deer to the true height of the barrier (they may not be able to see fishing line or wire alone) so they won't try to jump it.

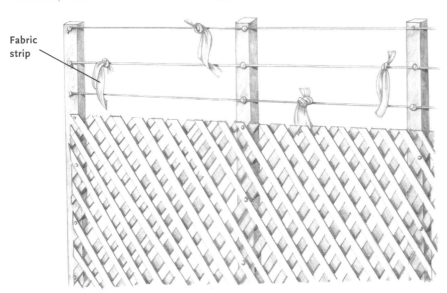

Fabric strip

Lattice fence topped with wire

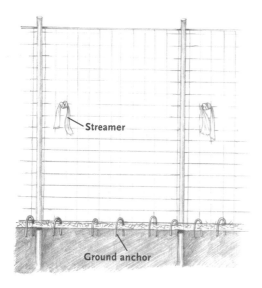

Plastic mesh fence looks flimsy, but it's remarkably effective as a deer barrier. When installing the fence, be sure the mesh is flush with the soil surface so deer can't push it up and crawl underneath. White streamers along the fence alert deer to its presence.

then make your own from strips of old sheets or T-shirts. Tie them to the fence 10 feet apart.

Lightweight plastic mesh 7 to 8 feet in height costs less than $1 per foot; heavier mesh costs close to $2 per foot. (Also add the cost of poles or posts, ground anchors, and heavy plastic ties for attaching the mesh to the poles.) The illustration on this page shows the components of a mesh fence setup.

When the growing season ends, roll up the mesh fencing and store it in a dark, dry place for winter. Although the mesh is manufactured to include protection against ultraviolet light, it will eventually break down. The more you can shelter it from UV light in the off season, the better. Generally, manufacturers claim the fencing will last 15 years, but 10 years is a more realistic estimate.

Small mammals such as woodchucks can tear or chew through plastic mesh fence. Fencing skirts made of metal grid are available for fortifying and repairing damage to the lower sections of plastic mesh fence.

strengths—the greater the deer pressure in your area, the stronger the mesh needs to be. There's even a type of flexible metal mesh fencing for areas with heavy deer pressure. The metal mesh will keep out rabbits and woodchucks, too.

Support the fencing with metal poles or 4 × 4 posts. (If you're fencing your whole landscape, you can attach the fencing to trees along your property lines, too.) It's critical to anchor the bottom edge of the fence into the soil or thread a taut wire horizontally through the bottom edge of the mesh so that deer can't push up the mesh and crawl underneath it. Some suppliers include white streamers to tie to the mesh to alert deer to its presence. If the streamers aren't included with the roll of wire you've purchased,

ELECTRIC FENCE

Electric fences to protect gardens from deer and other wildlife are increasingly popular, especially now that lightweight, easy-to-use kits and components are available. Standard electric fences use 12½-gauge steel wire, but if you're planning to fence your vegetable garden only, you should be able to use electrified netting or tape (polypropylene ribbon with fine steel wire woven through the ribbon) instead. These materials are easy to install and to take down and store for winter.

The first step in deciding whether you want to install an electric fence around your garden is to check local regulations by calling your local government office or zoning official. Some communities prohibit electric fencing, which will make the decision for you. Others may restrict what type you can use.

Every fence needs to be attached to a fence charger unit, which converts household current into a pulsing-type current. Battery-powered chargers are also available. Small chargers cost less than $100; battery-powered chargers may cost closer to $200. The larger the area you want to fence, the more powerful the charger needs to be. Also, whatever type of fence you choose, the minimum voltage to repel deer is 5,000 volts. Your fence supplier can help you decide what charger to buy. Check local regulations regarding charger installation; some municipalities require that a professional electrician install and connect the charger.

ULTRASONIC SOLUTIONS

Ultrasound deer repellent devices emit a frequency that humans can't hear at all, but manufacturers claim that deer can and will be frightened away from your yard by the sound. Visit Web sites for these devices and you'll find testimonials from satisfied customers, but there's little, if any, independent testing to demonstrate that these devices work. Some of these units cost less than $50, so you won't lose much if you want to experiment, but don't expect miracles.

Once you've installed a fence, you'll need to check it frequently with a voltage meter to be sure the charge is not declining. Battery-operated chargers need to be recharged regularly. Patrol the fence weekly to look for debris on the wires that could interfere with the charge.

A garden protected from deer by a tall fence needs a gate. If you're not handy at construction, though, simply use a sturdy stepladder to block deer from entering. Move the ladder aside when you want to work in the garden.

LARGE-SCALE ELECTRIC DEER FENCE

To fence a full home landscape with a quality electric fence, cost per foot will be $1.50 or more. If you plan on fencing your entire yard, you'll probably need a multistrand fence supported by sturdy posts. This job is one for a professional fence installer.

POLYTAPE FENCING

Polytape electric fences can be the easiest and least expensive method for fencing a moderate-size vegetable garden. The materials are lightweight and easy to work with. It's important to install the fence before deer get a taste of your garden, though; otherwise, they may challenge or jump over the fence to reach the tasty reward they know is on the other side. If deer have already discovered your garden, you'll need to take a more intensive approach, using multiple strands of wire attached to sturdy wooden posts. These fences are more costly, and you'll probably need to hire a professional installer to set up the fence.

A SIMPLE ELECTRIC FENCE

One strand of polytape is all you need to fence out deer in areas where deer pressure isn't high. This fence is simple to install and quite inexpensive. The key to this type of fence is to bait the fence with tasty morsels of peanut butter or to flag the fence with fabric strips soaked in deer repellent.

Setting up the Fence

To set up the fence, you'll need polytape, 4-foot fiberglass fence rods, four metal fence posts, about 16 feet of PVC pipe, a fence charger, a grounding rod, and a digital voltage meter.

CAGING SPECIALTY CROPS

Is your asparagus or herb bed separate from the rest of your food garden? If so, it's also vulnerable to deer damage. Cage it by fencing it with woven wire to whatever height the crop will reach (about 6 feet for asparagus, 4 feet for herbs), then make a cover of framed woven wire to place on top of the fence. Fully enclosed, your special garden will be safe from browsing deer.

Start by pounding in the metal fence stakes at the corners of your garden at least 1 foot deep. Cut the PVC pipe into four equal sections and thread one over each metal stake. (Wooden posts fitted with insulators also work for corner posts.) Then add fiberglass rods around the perimeter, 15 to 20 feet apart. Many of these rods are designed with a pointed end and a projecting step so that you simply press on the step with your foot and your weight drives the rod into the ground.

Tie off the polytape to one of the corner posts about 30 inches above ground level. Then work your way around the garden, threading the tape through each fiberglass rod in turn. At the corner posts, wrap the polytape around each sleeve twice (the PVC is an insulator). When you've encircled the whole garden, pull the tape tight and tie it off.

Connecting the Charger

A charger must be mounted on a sturdy wooden post or on the wall of a building near the fence (if that's possible). Consult the instructions that come with the charger and follow them care-

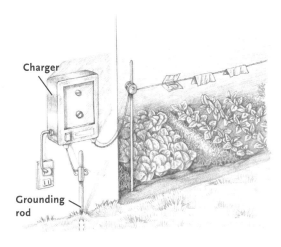

Charger

Grounding rod

A fence charger is the link between a source of electricity and the fence wire. A grounding rod is an important safety feature.

the grounding rod (usually a copper rod or galvanized pipe) into the soil near the charger and attach it to the charger as well.

If there's no source of electricity convenient to your garden, choose a battery-powered charger instead. These chargers do cost more than standard chargers; look for a type that uses solar panels to recharge the battery.

Adding Baits or Repellents

Once the fence is installed, plug it in and use the voltage meter to check that current is pulsing through the fence. Then unplug the fence and immediately add baits or repellents. These increase the fence's effectiveness significantly by drawing the deer's attention. Without baits or flags, deer may unwittingly charge through the fence. They'll get a shock, but that will only make

fully. The schematic above shows a typical charger setup. You'll use a length of steel electrical wire to connect the fence to the charger. Pound

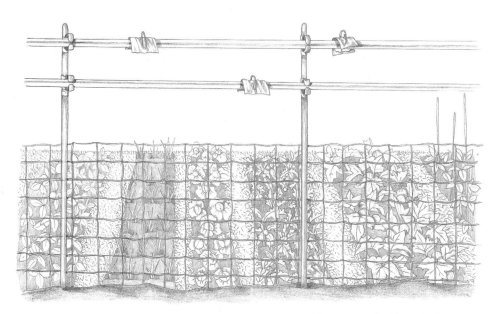

Electric netting keeps out rabbits, raccoons, and other small animals, while two strands of baited polytape ensure that deer won't cross the barrier.

them more skittish inside your garden—resulting in trampling damage as well as feeding damage.

Baits made of pieces of aluminum foil smeared with peanut butter lure deer to come close to sniff or lick the aluminum. When they do, the electric charge flows right through the aluminum, delivering an unpleasantly painful shock to the deer's tender nose or mouth. Usually, one shock is all it takes to train a deer to avoid your polytape barrier for the rest of the season.

To make bait tabs, cut pieces of aluminum foil 4 inches square. Fold them in half. Mix equal parts peanut butter and vegetable oil and smear the mixture on the inside of the folded aluminum tabs. Set the baits 3 to 4 feet apart. Use a strip of adhesive tape or a paper clip to secure each of the tabs to the polytape, or staple the tabs in place.

An alternative is to tag the fence with flags made from strips of old sheeting and sprayed with an odor-based repellent. Flags should be about 6 inches long. Make enough of them to tag the entire fence line at 3- to 4-foot intervals. Tie the flags to the polytape (be sure the power is turned off first), and spray them thoroughly with repellent. (Respray monthly.) Deer will notice the flags, and curiosity will drive them to come close to investigate. In many cases, the odor will repel them before they even touch the flags. But again, if they do, they'll receive the doubly negative experience of the nasty smell and the shock.

Whether you use bait or repellent, it's critical to leave the fence turned on all the time. The cost is minimal (less than $1 per day) considering the benefit of the protection you get. It's fine to turn off the fence when you want to work in your garden, but use some kind of signal—like a bright scarf draped on the charger—to remind yourself to plug it back in when you leave the garden.

DISEASE CONTROL

Blights, spots, and rots are the big three among vegetable garden diseases, with mosaic and rust following close behind. Chances are you'll have to contend with a few of these diseases in your garden each year. They may not be the most serious garden problems you face, but they often cause the greatest frustration. Unlike insect or weed problems, where you can fight back after the problem appears, the only course of action for some disease problems is to pull out the infected plants and start again. The sooner you spot symptoms, though, the greater the chance you can save your crop and enjoy a satisfying harvest.

MATCHING SYMPTOMS AND CAUSE

When you discover sick-looking plants in your vegetable garden, your first challenge is to figure out the nature of the problem. Fungi, bacteria, and viruses cause plant diseases, of course, but those aren't the only possibilities. Nutrient deficiencies, lack of water or light, or extreme weather conditions also can cause plants to grow poorly and look "sick." Problems not caused by living organisms are called *disorders* rather than diseases. That may seem like a technicality, but it's important to know the difference between diseases and disorders because they require different treatments. Even though symptoms of a disorder such as potassium deficiency may look a lot like symptoms of Fusarium wilt, the way to treat the deficiency is very different from the treatment for Fusarium wilt.

Diagnosing diseases and disorders can be more difficult than identifying insects. Disease symptoms vary from crop to crop and may develop differently depending on temperature and weather conditions. Only a few diseases have unique symptoms that are easy to diagnose, such as the white coating that appears on plants suffering from powdery mildew fungus. The right diagnosis can mean the difference between saving a crop and losing it, and also will ensure that you can take effective action to prevent the problem from recurring next season.

FUNGAL DISEASES

Pathogenic fungi cause many diseases of vegetable plants, including early and late blight of tomatoes and potatoes, powdery and downy mildew, and club root of cabbage-family plants. However, most of the fungi in gardens are beneficial. Soil fungi break down organic matter and enhance soil structure. They help to decompose dead plants, releasing their nutrients to the soil. Many kinds of fungi compete with plant pathogens to *reduce* disease—and scientists have figured out how to turn some of these fungi into sprays that you can use to prevent disease problems. Other types of fungi associate with plant roots to help with absorption of nutrients (this association of fungi and plant roots is called mycorrhizae).

Fungi are multicellular, but they don't contain chlorophyll. They reproduce by producing spores (their equivalent of seeds), which can be spread by wind or water or on soil or plant debris. The classic symptom of a fungal infection is mold. Mold is actually a network of microscopic tubes called hyphae that grow from spores. Hyphae tend to extend in all directions from a spore, which is why symptoms often show up as round spots. Hyphae produce enzymes that can break down plant tissue, eventually causing leaves or plants to die.

Most fungi like wet, dark conditions, and many fungi that cause vegetable diseases are soil dwelling. Some species can survive for years in the soil in a resting state. Club root fungus can persist in the soil for up to 20 years. Soilborne fungi are spread when rain splashes soil up onto plants, or when you till or dig in an infected area and then use that tool (covered with spore-laden soil) in another part of the garden. Soil that sticks to your shoes or garden gloves can also spread soilborne spores. Some fungi produce their spores inside a fruiting body. Fruiting bodies may be so small we can barely see them, or they may be quite large. Mushrooms are fruiting bodies, and so is the growth called corn smut.

BACTERIAL DISEASES

Bacteria usually consist of simple single cells that have a cell wall but no nucleus or chlorophyll. These cells are so small that you can see them only by viewing them through a powerful microscope. Bacteria enter plants through wounds or stomates (natural openings in the leaves). Inside the plants, the bacteria secrete toxins that cause plant cells to break down. The bacterial cells use the nutrients released from the plant cells as fuel for their own reproduction (which they do simply by splitting to produce more cells). Masses of bacteria cells can clog the xylem and phloem—the internal circulatory system of plants that moves water and food through roots, stems, and leaves. With circulation hampered, plants become stunted or wilt and die. As the disease worsens, the masses of cells become a visible slime or ooze that you can see when you cut into a plant stem.

As with fungi, most of the bacteria in soils and compost aren't harmful to plants. Bacteria are an important part of the organic food chain in the soil. Decomposer bacteria live and reproduce by breaking down soil organic matter. In turn, earthworms and other soil animals eat these bacteria, and the waste products of the earthworms produce a rich source of nutrients in forms that plant roots can absorb.

Pathogenic bacteria usually can't live long outside a host organism, but some species can form resting spores that persist in a dormant state. Also, bacteria may be present in plants but inactive. When conditions change to favor their reproduction, populations skyrocket, and symptoms of infection will appear within a few days.

Bacterial spores can spread on the wind or in water. Insects spread bacteria as they feed, and the spores can be spread in soil that clings to gardening equipment and shoes or clothing.

VIRAL DISEASES

Mosaic is a familiar viral garden disease, and the organisms that cause it are extremely small, even smaller than bacterial cells. In fact, scientists describe viruses as particles rather than cells, because viral organisms don't have cell walls or a nucleus. Viruses consist of DNA inside a protein coat, and they cannot survive long outside of a living plant or animal. Many viruses that cause plant disease are spread from plant to plant by insects, especially insects that suck plant sap. For example, an aphid may feed on an infected plant and ingest viral particles as it feeds. Then, when the aphid moves to a healthy plant, it introduces viral particles into the tissue of the healthy plant while feeding there.

Viruses cause disease by forcing plants to become virus factories, producing more and more viral particles. Because of this, the plants can't function properly, which results in symptoms such as stunting or distorted growth. You won't see fruiting bodies, slime, or any visible signs of the virus itself.

Symptoms caused by viruses look similar to those caused by some nutrient deficiencies, and in fact, infection by a virus can induce a nutrient deficiency. One way to help decide if symptoms are caused by viruses is to observe the pattern of the symptoms. If scattered plants within a crop row or a bed show symptoms, the problem is likely to be a virus. But if whole groups of plants show the same symptoms, the cause is more likely a deficiency.

The key to preventing viral disease problems is to control the insects that spread the viral particles and also to eliminate weeds that serve as reservoirs of the viruses.

ENVIRONMENTAL DISORDERS

A variety of environmental problems can cause symptoms similar to those caused by diseases. Here are some disorders to look for.

NUTRIENT DEFICIENCIES. When plants can't get enough nitrogen or other nutrients from the soil, they can't grow and develop properly. Plants suffering from nutrient deficiencies may be pale overall or have patchy pale areas. They may be stunted, or some plant parts may turn purplish or reddish or yellowish.

Nitrogen, potassium, and phosphorus are the major plant nutrients needed for proper growth, and there are a variety of necessary micronutrients, too. Sometimes nutrients are simply in short supply in the soil; or they may be present but plants can't absorb them because the soil is too dry or the soil pH is too high or too low. To learn about how to remedy nutrient deficiencies, turn to Fertilizers.

WEATHER-RELATED PROBLEMS. Drought, frost, high ozone levels, and other weather-related phenomena can interfere with plant growth, producing symptoms such as wilting, chlorosis, bleaching of plant surfaces, and dieback. In most cases, there are no cures for these problems, but if you diagnose them properly, at least you can save yourself the effort of fighting a disease organism that isn't there.

DEALING WITH PLANT DISEASES

In a healthy garden, disease problems are few and usually not serious. Vigorous plants have innate defense systems that help them fend off disease, and biologically rich soil provides other kinds of defenses against plant disease, too. Dealing with disease is a process. The first step is to avoid introducing diseases into the garden when you plant. Second, you can take steps to prevent disease organisms that are already present from infecting your plants. Finally, if disease symptoms do show up on your plants, take steps to break the cycle of the disease so it won't cause problems in future seasons.

PLANT SELF-DEFENSE

If a pathogen tries to invade a healthy plant, the plant fights back at the cellular level to prevent the pathogen from spreading or causing damage. This kind of response is called systemic acquired resistance (SAR). SAR is so effective that scientists have developed sprays that proactively trigger SAR in order to boost plant growth and prevent disease problems. (See "Plant Health Boosters" on page 178.)

HEALTHY SOIL HELPS

Garden soil is a bacteria-eat-bacteria world. In healthy soil rich in organic matter, specialized bacteria called myxobacteria may attack plant pathogens, destroying their cell walls and

When a crop is languishing but a diagnosis eludes you, one other cause to investigate is nematodes. You can't see these tiny, round-bodied worms on roots, but root-knot nematodes cause galls or swellings on roots. Root-knot nematode is a widespread garden pest, especially in the Southeast. Infected plants may be stunted or grow poorly, but it's the root symptoms that will clinch a diagnosis. For complete information on diagnosing and controlling harmful nematodes, see Nematodes.

consuming the cell contents as "food" to fuel their own reproduction. Healthy soils are also rich in beneficial fungi and bacteria that live in the zone immediately surrounding plant roots, where they outcompete plant pathogens for space and nutrients so that the pathogens never have a chance to invade plant roots.

What happens when the garden environment isn't healthy? Plants that are suffering from nutrient stress or moisture stress won't have a strong SAR response, and disease may spread quickly. Thus, if disease problems are appearing in most or all of your garden beds, it's a signal that you need to analyze your general garden management. Turn to page 2 for an overview of healthy gardening practices.

STARTING OUT DISEASE-FREE

How do disease organisms "find" your vegetable garden? Some travel as spores blown long dis-

tances on the wind. Insects carry viruses and bacterial diseases from plant to plant as they feed. And a few types of plant disease organisms may have been waiting, dormant in the soil in your yard, before you even started your vegetable garden.

We bring new disease problems into our gardens, too, when we plant diseased seeds and transplants. Many disease organisms can overwinter in seeds. If you plant diseased seeds, the seedlings may get sick and die before or just after they emerge. Or the seedlings may grow well for weeks before disease symptoms appear. If you don't spot the disease quickly, the fungus or bacterium may have time to produce spores that spread to other plants or enter the soil. The way to block this avenue of disease development is to plant only disease-free seeds and transplants.

CLEAN SEEDS

Of course, reputable seed companies don't want to sell diseased seed, and most companies offer a money-back guarantee on seed. But getting your money back won't solve a disease problem, so check whether the companies you buy from participate in screening programs for seed-borne diseases. Ask also whether they treat their seed with hot water to kill disease organisms.

If you save seeds from your garden, choose seeds only from plants that were vigorous and showed no disease symptoms. And keep in mind that seed from seed swaps or from a friend's garden can carry disease. That doesn't mean you should avoid these sources altogether, but be aware that it involves some risk.

HEALTHY TRANSPLANTS

Transplants sold at garden centers and home centers aren't guaranteed disease-free; in fact, they should be labeled "buyer beware." They may appear very healthy because they've been pampered in a greenhouse, but if they're grown from infected seeds, symptoms could develop after you transplant them into your garden.

One way to lessen the risk is to buy transplants from a local grower (preferably organic) who has a good reputation. Or grow your own transplants from clean seed.

CHOOSING RESISTANT VARIETIES

If a few tough diseases trouble your crops year after year no matter what you do, seek out disease-resistant or -tolerant varieties to save yourself the heartache of lost crops. Resistant varieties aren't immune to disease, but they have some ability to fight off specific pathogens. A disease-tolerant variety will show symptoms, but it will grow better in the presence of a disease problem than "regular" varieties would.

Breeding disease-resistant varieties is a high priority in the seed industry, but the choices are sometimes limited. For example, you'll find plenty of tomato varieties resistant to Fusarium and Verticillium wilt but few that are resistant to late blight.

To find a variety that's resistant to a particular disease and also suitable for your growing conditions, ask someone on staff at your favorite seed company or a good local garden center for recommendations. Or ask your local Cooperative Extension office for a list of recommended resistant varieties for that crop.

Planting resistant and tolerant varieties is also a smart choice if you know that a particular disease was rampant in your area the preceding year. And if you're still in the transitional years of building up the health and quality of your soil, planting as many disease-resistant varieties as possible will help you avoid problems and enjoy a better harvest.

STAYING DISEASE-FREE

Covering crops with row covers can block out insects that transmit disease. Cucumber beetles can infect squash and cucumber vines with bacterial wilt, which can devastate the vines just when they're bearing fruit. Leafhoppers transmit aster yellows and ruin a crop of lettuce or carrots. Covering plants keeps out the insects and prevents the disease. (To learn more about this technique, turn to Row Covers.)

Biological disease control is an active area of research, and new products are developed almost every year. One biological approach to keeping plants disease-free is to spray them preventively with substances called plant health boosters, which bolster their natural self-defense systems. Compost tea and products containing beneficial fungi can prevent pathogenic fungal spores—such as the fungi that cause damping-off—from germinating.

Most of the organic sprays that prevent plant diseases are nontoxic to people and pets. However, almost any substance can produce an allergic reaction in some people or can irritate lungs if inhaled. To protect against that possibility, it's a good idea to wear long sleeves, long pants, shoes, socks, eye protection, and a dust mask or respirator when spraying plants, even with something as harmless as a baking soda solution.

PLANT HEALTH BOOSTERS

Taking echinacea, vitamin C, and other supplements to boost the immune system and fight disease is part of many people's daily routine. Plant health boosters may be the equivalent treatment for plants, stimulating the plants to grow more vigorously and making them more disease resistant.

One of the best plant health promoters is simple compost. Added to the soil, compost feeds microorganisms, which in turn produce hormones that stimulate plant growth. Plus, compost hosts beneficial microorganisms that may prey on plant pathogens, produce antibiotics that kill pathogens, or compete with pathogens for nutri-

ents. Spraying plants with compost tea also can have a health-promoting effect. The teas work by stimulating growth but also by creating an environment on the leaf surface that is unfavorable to pathogens. To learn more, see Composting.

Some commercial plant health boosters contain a protein called harpin, which is very similar to a protein that occurs in many plant pathogens. Harpin is reported to stimulate plants to activate their natural defense systems and promote vigorous growth, but results from university trials are variable. Spraying plants with harpin may also help them better tolerate drought and heat stress. Harpin is nontoxic to humans and no phytotoxic or negative environmental effects are reported, but harpin is not listed by the Organic Materials Review Institute.

PREVENTION VERSUS CURE

Beyond the basics of healthy gardening, there are steps you can take to reduce disease problems. Most of these practices are preventive—there are few cures for plant diseases. So when a disease problem strikes, first identify the cause. Next, evaluate how serious the problem is. If the crop is at the seedling stage, or if only a few plants are affected, the easiest solution may be to pull out the seedlings and replant in a different part of the garden, implementing strategies such as choosing a resistant variety.

If disease symptoms appear on crops that are more mature, something as simple as switching from sprinkler irrigation to drip irrigation may slow the spread of the disease and the severity of the symptoms, although a complete cure is unlikely. Sprays containing *Bacillus subtilis*, bicarbonate, or

copper can help stop the spread of symptoms. You may be able to nurse the crop through to harvest.

With disorders, often there is no quick cure, either, but if you can figure out what caused the problem, you can try something different next time to solve the problem. For example, if your peppers suffered from sunscald because the plants toppled over, leaving the fruit exposed to the sun, plan to stake your peppers next year. The plants will stay upright and the foliage will shade the fruits.

At the end of the season, cleaning up the garden properly is an important part of breaking the cycle of disease. And for the next garden season, you can improve soil conditions and promote plant health to reduce the chance that disease will get a foothold in your crops.

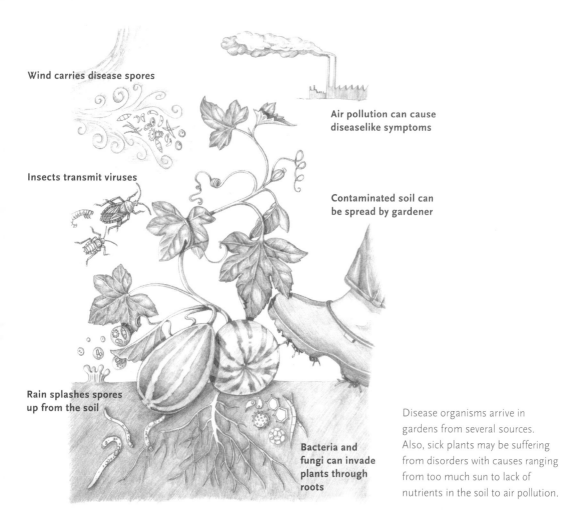

Wind carries disease spores

Air pollution can cause diseaselike symptoms

Insects transmit viruses

Contaminated soil can be spread by gardener

Rain splashes spores up from the soil

Bacteria and fungi can invade plants through roots

Disease organisms arrive in gardens from several sources. Also, sick plants may be suffering from disorders with causes ranging from too much sun to lack of nutrients in the soil to air pollution.

When to Use

Start applying plant health boosters to young plants early in the season, before disease symptoms appear. It's best to experiment by spraying some plants and comparing their growth to untreated plants. If you see no difference, it may indicate that the product didn't affect the plants, or it may show you that you've done such a good job of creating a healthy garden environment that your plants don't need any extra boost.

Plant health promoters may be especially good for heirloom varieties, that may be susceptible to a disease that's common in your area.

How to Use

For harpin products, mix the powder with water according to label directions and apply it with a pump sprayer, a misting bottle, or a watering can. For directions on making and using compost tea, see Composting. Begin applying a spray early in the growing season and repeat 2 to 3 weeks apart. If you live in an area where disease pressure generally slackens after midsummer, try stopping sprays then. In regions like the humid Southeast, you may want to continue using the sprays until heat and humidity ease up in fall.

Sometimes plants treated with plant health boosters still develop disease problems—the treatments are not protective enchantments that work equally well every time. If you decide to use a harpin product, be sure the powder is still viable. If you open a packet and use only part of the contents, keep the opened packet in a dry place (but not in the refrigerator). Use the remainder of the packet the next time you spray. Don't try to keep sealed packets of harpin powder longer than one growing season.

BACILLUS SUBTILIS

Bacillus subtilis is a bacterium that kills or outcompetes some types of plant pathogens. Particular strains (subtypes) of *B. subtilis* are used as a foliar spray to prevent powdery mildew and some other fungal diseases. The foliar spray contains bacterial spores and also a protein produced by the bacterium during fermentation. The protein is a natural fungicide, and it's the active component in the spray. The protein has no harmful effects on people, pets, or beneficial insects.

Commercial farmers also use a different strain of *B. subtilis* as a soil drench or seed treatment to prevent diseases such as damping-off that are caused by soilborne fungi.

When to Use

Try to time foliar applications of *B. subtilis* just before the time of year you would expect disease symptoms to appear. For example, if you live in the arid West and powdery mildew shows up on your cucumber plants each year in June, you should start spraying in early June. *B. subtilis* sprays may also help prevent anthracnose, downy mildew, early blight, late blight, and fungal leaf spot.

How to Use

Some *B. subtilis* products are available in ready-to-use form; others are concentrated powders. If you're using a concentrate, mix it with water according to package directions, and shake the sprayer occasionally while you're applying to keep the powder in suspension. Spray all plant surfaces thoroughly until liquid just begins to drip off the plants.

To give plants an extra boost of balanced nutrients and promote healthy growth, you can add fish emulsion to the spray mix as well (in the proper amount for a half-strength fish emulsion solution).

Don't store leftover spray because the protein in it will break down and lose effectiveness.

Repeat sprays every 7 to 10 days. If the weather is warm and moist, shorten the spray interval to every 5 days.

When Things Go Wrong

SYMPTOMS GET WORSE AFTER SPRAYING. If a disease problem worsens after you apply *B. subtilis*, it may be because you didn't start your spray program early enough. Although the natural protein in the spray is a fungicide, it acts only to kill spores on the leaf surface; the fungicide can't penetrate the leaf to stop an infection that's already moved into the leaf cells. Also, if you didn't follow up with weekly repeat sprays, spores that landed on the plants after you sprayed may have germinated and infected the plants.

It's also possible that *B. subtilis* isn't strongly effective against the disease that has appeared on your plant. The spray purportedly kills many types of fungi that attack plants, but results are not equally good for all kinds of pathogens or in all conditions.

Make notes on the crop you sprayed and the disease that developed. Try a different control measure for this crop next year and compare results.

TRICHODERMA HARZIANUM

Trichoderma harzianum is a beneficial fungus that helps prevent damping-off. In the soil, *Trichoderma* colonizes the root zone around young seedlings and outcompetes a variety of pathogenic fungi that cause seedlings to damp-off. The presence of *Trichoderma* also seems to spur root growth, resulting in larger root systems and more vigorous topgrowth.

When to Use

Apply *Trichoderma* to seed furrows or when watering in new transplants. It is a preventive only. If seedlings begin to show symptoms of damping-off, applying *Trichoderma* will not cure them.

How to Use

Trichoderma comes in powder form. Dilute it in water according to package directions. Reapply it every 3 months if you are replanting a bed with succession crops, because the fungus dies out in the soil over time.

When Things Go Wrong

If treated seedlings or transplants develop damping-off, consider two possibilities. First, temperatures may be too cold. *Trichoderma* is not effective when soil temperature is below 50°F. Another possibility is that the product is too old, and the fungus died out in storage. Don't try to keep *Trichoderma* products longer than one growing season. You may want to split the purchase of a container with a gardening friend. Then, if you find it useful, you can collaborate on the purchase of a new container each year.

BICARBONATE

Bicarbonate is a mild fungicide, although how it works is not precisely known. Commercial bicarbonate fungicides contain potassium bicarbonate; studies show that potassium bicarbonate is more effective for controlling powdery mildew than sodium bicarbonate (baking soda), but there are many reports of success with sodium bicarbonate sprays, too.

When to Use

Ideally, start bicarbonate sprays about 2 weeks before you expect to start seeing disease symptoms, and continue spraying every 7 days until conditions are no longer favorable for the disease you're trying to prevent. You can increase the frequency of sprays if it is rainy or there's a long period of humid weather. Bicarbonate will help control powdery mildew, downy mildew, Alternaria, leaf spot, and anthracnose. It's used mainly on squash-family crops and tomatoes.

How to Use

Mix bicarbonate fungicides according to label directions. Some products recommend adding a spreader-sticker, such as lecithin, for best results;

others say that none is needed. Repeat applications two or three times.

For a homemade baking soda spray to promote plant health, combine 1 teaspoon baking soda with 1 drop liquid soap and 2 quarts water and mix well. Use a pump spray bottle or backpack sprayer to apply. Another formula also includes lightweight mineral oil. For this spray, combine 1 tablespoon baking soda and 2 tablespoons of the oil in 1 gallon water.

When Things Go Wrong

Some types of bicarbonate sprays can burn foliage if they are too concentrated. If leaves on plants you've sprayed turn brown, pick them off but don't pull out the plants. New foliage should be undamaged and the plants will recover. Before you use a bicarbonate spray on this crop again, test-spray a small portion of foliage. If there is no damage, it's possible that you mixed the spray incorrectly the first time. If the leaves burn again, though, then don't use bicarbonate sprays on this crop anymore.

BREAKING THE DISEASE CYCLE

Once disease symptoms appear on a crop, your focus will change from prevention to stopping the spread of the disease and breaking the disease cycle. You'll do this by staying out of the garden when plants are wet and by spraying neem or copper for certain serious disease problems.

To break the chain and stop the disease from infecting your crops in the future, it's important to destroy crop debris and, for some diseases, to rotate crops. Biofumigation and solarizing soil, described on pages 185 and 186, are more advanced techniques for breaking the disease cycle.

DON'T SPREAD SPORES

If you handle plants when they're wet, spore-filled water droplets may travel on your hands, gloves, or tools from an infected plant to a healthy plant. Because of this, ideally you would never weed, stake, or harvest when plants are wet. Sometimes, though, this isn't realistic. For example, if your only free time for gardening is on the weekends, and the weekend is rainy, do you let the weeds keep growing or risk spreading disease by weeding in a wet garden?

Odds are your garden won't be ravaged by disease if you work among wet plants. Just do your best to touch as little foliage as possible. And don't try to remove diseased leaves from plants when they're wet. If you need to harvest, work carefully.

Keep in mind that spores can spread on the surfaces of wet tools and on soil clinging to tools also. Clean your tools after working in a wet garden, but don't wash or scrape the soil off them back onto garden beds.

Mulching garden beds also helps stop the spread of spores. When a hard rain strikes bare soil, it can splash soil particles up onto foliage. Spores in those soil particles then can germinate on the wet leaves. Organic mulch or plastic mulch provides a barrier that keeps the spores in the soil and away from foliage.

NEEM OIL

Neem oil is extracted from the tropical neem tree, and it's used both as an insecticide and a fungicide. Neem is effective for controlling powdery mildew, downy mildew, anthracnose, rust, and botrytis. Begin applying it at about the time you would expect to see the disease develop, or at the first sign of symptoms. Neem will not cure diseases, but it can prevent them by killing spores on leaf surfaces. For details on how to use neem and what to do if problems develop after spraying, see page 318.

COPPER

Copper is a natural mineral element, and applied as a dust or foliar spray, it is the strongest type of organic fungicide and bactericide available. It kills spores on leaves and also can penetrate into leaf tissue to some extent. Some copper fungicide products are combined with a soap to provide more even coverage. Copper sprays are not harmful to beneficial insects, but copper is toxic to people and animals and highly toxic to fish. Wear protective equipment when you apply it and don't let the spray come in contact with bodies of water or storm drains, and do not use it where runoff is a possibility. If these sprays are used heavily over a period of years, copper may build up to levels in the soil that are toxic to plants. Because of these risks, only use copper as a last resort when experience tells you that a serious disease problem may develop and ruin a crop. Whenever possible, find alternatives, such as planting disease-resistant varieties or stimulating the plants' natural defenses by applying plant health boosters.

When to Use

Apply copper sprays when there is a high risk of disease problems developing, or when you first see symptoms of a disease that could become a serious problem. Copper is effective against anthracnose, bacterial spot, bacterial blight, early and late blight, Septoria leaf spot, downy mildew, and other fungal leaf blights. Repeat every 7 to 10 days when disease pressure is high; respray if rain washes the fungicide off.

How to Use

Apply copper dust as directed on the label. Mix wettable powders and liquid concentrates according to label directions, using the most dilute strength recommended unless conditions are very wet and humid. For late blight, use the maximum recommended strength solution.

When Things Go Wrong

Copper can burn plant surfaces, causing brown spots to appear on leaves after spraying. If the copper does not stay well mixed in the sprayer, it

MORE PREVENTIVE POSSIBILITIES

Some of the products that repel or kill insects also act as mild fungicides. Garlic has antifungal properties, as do soap sprays and some oils. You'll find complete information about these sprays, including recipes for making your own, in Pest Control. Less is known about which diseases they will prevent and how effective they are.

Even ordinary milk has fungicidal properties. Research studies show that spraying a dilute solution of milk on zucchini plants and grapevines is effective at preventing powdery mildew and may even have a fungicidal effect. If you want to try this in your garden, add 1 to 2 tablespoons milk to 1 cup water and use a spray bottle to coat the foliage.

tends to come out of the spray nozzle in globs. The concentrated glob of copper will injure the area of the leaf it contacts. To prevent this damage, shake the sprayer occasionally as you work.

Copper can also damage leaf tissue if the wet solution stays on the leaf too long without drying. Try to time applications for the morning of a dry day.

While copper is usually quite effective, it may not prevent disease completely during long periods of wet weather. If symptoms develop after initial treatment, continue sprays at the recommended interval to prevent further spread of symptoms.

HANDLING CROP DEBRIS

Diseased crop debris is like a hot potato—hard to handle safely. The debris can shelter spores, allowing them to survive winter and infect next year's crop. If you add the debris to a compost pile, the spores may survive in the pile. But if you don't compost plant debris or return it to the soil, what can you do with it?

Commercial farmers sometimes burn diseased plant debris, but that's not practical (or legal) for most home gardeners. You could throw the debris away with your household trash. That removes the problem from your garden, but it contributes to the larger problem of overflowing landfills. Plus, if you throw away crop debris, you lose the benefit of the organic matter it could contribute to your garden soil. And even if you threw away every bit of crop debris, you wouldn't end the problem of disease in your garden. New disease organisms will still reach your garden each year, carried in on the wind, by insects, and on diseased seeds or transplants.

It's best to decide how to handle diseased crop debris on a case-by-case basis. One important factor is knowing which species of disease organism is in the debris. Some disease organisms—such as some viruses and the bacteria that cause bacterial wilt of squash-family crops—can't survive in crop debris. There's no risk in returning the debris to your soil or adding it to compost.

But for other disease problems, such as late blight and Verticillium wilt, removing diseased crop debris from your garden environment is almost always the best choice. Late blight can wipe out tomato and potato crops in a hurry, and the disease spores are persistent in the soil even after crop debris breaks down. It's important to remove and destroy plants that become infected with this disease, or you're likely to have serious problems with it again the following year. Verticillium wilt is another disease that requires cautious handling because it can infect such a wide range of crops. Controlling it by crop rotation isn't an option, so it's best to minimize the pathogen's opportunity to take up residence in your garden soil.

Another factor to consider is how you manage your compost and garden soil. If you compost actively, turning the compost frequently and monitoring moisture levels, your compost pile will be teeming with microbial activity. Some of those microbes will be the type that prey on other microbes—during the composting process, those microbes will attack and reduce the disease organisms in the crop debris. Similarly, if you've had opportunity to build soil organic matter and improve soil tilth, microbial activity in the soil will work in your favor as well when you turn under crop debris.

BIOFUMIGATION

Commercial farmers use synthetic chemical fumigants to kill some types of soil-dwelling pathogens. You can experiment with an all-natural alternative called biofumigation. This technique involves growing a crop and then turning it into the soil, where it will release natural fungicidal substances as it decays.

Cabbage-family crops such as mustards and rapeseed are high in glucosinolates, natural chemicals that suppress nematodes, fungi, and weed seed germination. The biofumigation treatment may suppress root rot fungi, Verticillium, damping-off, and scab.

Try planting a cover crop of rapeseed in late summer or early fall in the area you want to fumigate (see Cover Crops for planting instructions). Let it overwinter and then mow or cut the crop close to soil level about 4 weeks before you plan to plant. Immediately after mowing, chop the plant material, then till or dig it into the top few inches of soil. Set up a sprinkler to water the bed until the soil is as moist as a wrung-out sponge.

Another option is to fumigate during the growing season. Plant the biofumigant crop in spring, then chop or till it under and cover the soil tightly with clear plastic as you would when solarizing, which is described on page 186.

Thus, if disease problems in your garden tend to be minor, it's not going to change the balance greatly if you turn under diseased crop debris or compost it. It's not necessary to throw out or burn every bit of crop debris that shows a few brown spots or has some yellowed leaves. Discard any noticeably diseased material. As much as possible, use healthy, pest-free organic materials for your compost piles and as soil amendments.

When a disease problem in your garden becomes severe (plants die or produce little or no crop), be sure you identify the nature of the problem so that you make the right choices in dealing with the debris and preventing future problems. Each entry in this book on a specific disease offers advice on how to handle infected crop residues.

If you're digging diseased plant material into a garden bed, it is important to avoid spreading the debris all over the garden as you work, especially if you're practicing crop rotation to break the disease cycle, too. Chop the material as best you can, and then dig it in. If you use your tiller, be sure to clean the tiller tines thoroughly as soon as you've incorporated the diseased material, before you move on to tilling another part of the garden.

When adding diseased plant material to a compost pile, chop it first to speed decay, and put it at the center of the pile, not on the edge. Ideally, you'll manage the pile so that it will heat up to kill the pathogens, but this isn't always possible (turn to page 122 to learn how to manage a hot compost pile).

CROP ROTATION

Most plant disease organisms are specialists: A particular species of fungus or bacterium can infect only crops that belong to one crop family. For example, Alternaria fungi cause leaf blight in several different crop families. But the *Alternaria* species that causes early blight of potatoes and tomatoes can't hurt carrots or celery at all.

Conversely, the species of *Alternaria* that blights carrots and celery can't invade a potato plant.

We can outmaneuver crop-specific diseases by changing the location of crops in our garden from year to year. Called crop rotation, this practice also helps make the best use of soil nutrients. To learn how to set up a rotation, see Crop Rotation.

Crop rotation won't help break the disease cycle for pathogens such as rust, which is blown into gardens on the wind, or for pathogens such as club root, which can survive in a dormant state in the soil for up to 20 years.

BEYOND THE BASICS

SOLARIZING THE SOIL

Heat can be very effective for killing disease spores in soil and compost. In a hot compost pile, natural microbial activity generates enough heat to kill many weed seeds and disease spores. (See Composting for directions for managing a hot compost pile.) Microbes in the soil don't produce that much heat, but you can use solar energy to increase soil temperatures through a process called soil solarization.

Solarization is a project that takes considerable preparation and some expense. Plus it involves working in the sun at the hottest time of year. Still, the results can be dramatic, and worth the effort to break the disease cycle of tough problems such as Verticillium, some types of Fusarium, club root, southern blight, and root rot. Solarization also kills nematodes and some types of weed seeds.

It takes about a month to solarize soil. Plan to solarize during the hottest time of year in your area—July and August in most regions, or June and July in Florida.

The minimum area is 6 feet by 9 feet (smaller plots won't heat up enough). To prepare a plot for solarizing, remove all plant debris and till or dig the soil to create a uniform texture with no large soil clods. Shape the bed so that it's slightly raised in the center. This will encourage water to run off the plastic—water pooled on the plastic interferes with heat absorption (if pools form, use a broom to gently sweep the water off). Wet the soil thoroughly—it's important for the top 2 *feet* of the soil to be moist.

Water again 2 days later. Next, spread plastic over the soil. Use clear plastic that is 1 to 6 mils thick. Thin plastic provides the best heat absorption, but it also punctures more easily. If you're covering a large area, you can link pieces of plastic together with duct tape or transparent packaging tape. Transparent tape also works for patching tears that occur as you work with the plastic. Pull the plastic tight over the soil and bury the edges to maintain a tight seal.

Leave the plastic in place for 4 to 6 weeks. Your goal is to raise the temperature of the covered soil above 100°F for at least 4 weeks. The combination of heat and natural toxins produced in the moist soil kill the pathogens and weed seeds.

After you remove the plastic, disturb the soil as little as possible. Solarization is most effective in the top 2 inches of soil. Below that level, there may be lots of viable weed seeds as well as some disease spores.

Solarizing may also kill beneficial organisms, but they seem less affected—or better able to rebound—than pathogens. Solarization also stimulates release of nutrients from organic matter, and the total effect is such that the first crop planted in a solarized bed often shows remarkable vigor and productivity.

After you've solarized a bed, be very careful about the tools you use in it. If you work in another part of the garden, then move to the solarized bed and work there with the contaminated tool, you may reinfect the bed.

DOWNY MILDEW

Downy mildew blooms on crops during long periods of wet, cool weather. Infection happens rapidly, and plants appear to have been frosted overnight (but temperatures will be above freezing). Symptoms of downy mildew are somewhat similar to powdery mildew (see Powdery Mildew). However, with downy mildew the furry-looking masses of spores form only on leaf undersides, and the spores disappear when it's dry. Infected plants usually don't produce a good harvest.

RANGE: United States and Canada

CROPS AT RISK: A wide range of vegetable crops; can be a serious problem in cucumbers, melons, onions, spinach, and lettuce

DESCRIPTION: On many crops, the first symptom is irregular pale green or yellowish brown spots on older leaves. A downy coating appears on leaf undersides at these spots if weather conditions are right. The coloring of the mildew varies from crop to crop. On squash-family crops, it may be pale gray to nearly black; on lettuce, yellow or brown; on spinach, it will be purple or blue; onions develop a blue-gray mold. Infected leaves will turn brown and die.

Some crops show other symptoms, too. On lima beans, downy mildew appears on the pods. Infected beet plants form a rosette. Downy mildew doesn't infect squash fruits, but infected plants often produce stunted or sunscalded fruits because of the loss of foliage. Secondary rots sometimes invade the sunscalded fruits.

FIGHTING INFECTION: Early on, pick off infected plant parts and remove them from the garden. Switch to drip irrigation or using a handheld hose. If the weather turns warmer and drier, these steps alone may nip the problem in the bud. Apply copper as a last resort to prevent the disease from ruining a crop.

GARDEN CLEANUP: For squash-family crops, dig in or compost crop remains. For other crops, remove and destroy noticeably diseased crop residues. Compost other residues in a hot compost pile. Remove weeds that are alternate hosts, including shepherd's purse and wild mustard.

NEXT TIME YOU PLANT: Choose resistant or tolerant varieties when available. Plant clean seed. Shift planting dates so that crops won't grow during cool, wet weather (early spring in some areas; late fall in others). If spring plantings showed downy mildew symptoms, avoid planting fall crops of spinach or cabbage-family crops in the same spot or nearby. Space plants widely, and stake and prune plants to allow better air circulation. Apply preventive sprays of bicarbonate fungicide, compost tea, or *Bacillus subtilis*. If past problems have been severe, consider spraying neem as a protective fungicide, beginning when symptoms first appear.

CROP ROTATION: Crop rotation won't help much with downy mildew of squash-family crops because the disease spreads primarily by windblown spores. A 2- to 3-year rotation by crop family is helpful for other crops.

EARLY BLIGHT

Tomatoes are the crop hit hardest by early blight, but it can also infect potatoes, peppers, and eggplant. The fungus that causes it is Alternaria solani; *other species of* Alternaria *cause Alternaria blight of carrots and squash-family crops (see Alternaria Blight). In spring, early blight spores blow in the wind or are spread by insects or rain splashing up from the soil. Spores enter plants through wounds or directly through leaf surfaces. The fungus produces spores in several rounds each growing season. It spreads most rapidly in the hot, wet weather of late summer.*

Potatoes may not show symptoms until the crop is almost ready to dig, but spores can spread from infected potato plants to potato tubers during harvesting and to nearby tomato plants.

RANGE: United States; Canada; tends to be severe in the Midwest and humid eastern states

CROPS AT RISK: Tomatoes, potatoes, peppers, eggplant

DESCRIPTION: Dark spots appear on older leaves. Viewed through a hand lens, the spots show concentric rings. The spots enlarge, become leathery, and merge; withered leaves hang on or fall off plants. As the disease progresses, black spots also appear on younger leaves and on stems. Symptoms usually don't appear until midsummer or later. However, sometimes tomato seedlings infected by early blight show dark, sunken areas near the soil; this is called collar rot.

Tomato fruits will rot inside starting from the stem end. Potato tubers develop small dark spots; the flesh underlying spots becomes corky. The rot can spread in stored tubers, and other types of rot may also invade. Fruits sometimes suffer from sunscald if infected plants lose too many leaves.

FIGHTING INFECTION: Cut off infected leaves as soon as you spot them. If you haven't already done so, apply mulch to prevent spores from splashing up off the soil surface. Spray compost tea or *Bacillus subtilis* to prevent the blight from spreading (apply them early in the day so foliage will dry rapidly). Water by hand or use drip irrigation rather than a sprinkler. Don't harvest potatoes when the soil is wet.

GARDEN CLEANUP: The fungus survives winters in potato tubers, crop debris, and tomato seeds in the soil. Remove noticeably diseased plant material and discard or destroy it or incorporate it into a well-managed hot compost pile. Do your best not to let any diseased tomato fruits drop off plants and decay in the garden.

NEXT TIME YOU PLANT: Choose resistant varieties. With potatoes, long-season varieties generally are more resistant. A different strategy is to choose short-season potatoes and plant them early in order to harvest before early blight invades the crop. Separate your potato patch from your tomato plants to prevent cross-infection. Start your own tomato seedlings from clean seed, or buy transplants only from a reliable source. Drenching the soil with *Tricho-*

derma harzianum at planting time may help prevent infection (see Disease Control).

Space potatoes more widely than standard spacing. Stake eggplants, peppers, and tomatoes, and prune tomatoes to encourage good air circulation and keep fruit off the soil surface. Monitor soil moisture to prevent water stress. If you're concerned that your soil is low in nutrients, side-dress plants with a balanced fertilizer when fruit set begins. Control weeds that belong to the tomato family, such as black nightshade and bittersweet. If your season is long enough, make staggered plantings of tomatoes, and remove the early planting as soon as production slows or plants begin to show significant infection. Younger plants are better at resisting infection than older plants. If early blight has been a problem on tomatoes in your garden in the past, start potassium bicarbonate sprays about 2 weeks before you would expect to see symptoms.

CROP ROTATION: Start a 3-year rotation for tomato-family crops to reduce the reservoir of fungal spores in the soil.

EARWIGS

Earwigs look evil, but they aren't all bad. Their main role in the garden is to eat decaying plant matter, insects (including aphids), and insect eggs. The species that sometimes becomes a garden pest is an import, the European earwig. These earwigs also will feast on newly emerged seedlings, tender leafy crops, and ripening fruit.

Earwigs overwinter as adults in the soil or remain active year-round in mild-winter areas. Population explosions are most likely to happen in damp or coastal regions, especially when a wet spring follows a mild winter.

If you suspect that earwigs are damaging your seedlings, check after dark, as earwigs feed at night. They can cluster on plants by the hundreds. The best way to control excess earwigs is with simple homemade traps. Unless earwigs are threatening your crops, leave them be, because they could be hard at work eating insect pests.

PEST PROFILE

Forficula auricularia

RANGE: Throughout the United States and Canada

HOW TO SPOT THEM: Check your garden at night with a flashlight when earwigs are active or look underneath mulch and rocks during the day.

ADULTS: Slender reddish brown insects, up to 1 inch long, with leathery forewings and a pair of pincers at the tail end. They have wings but rarely fly.

Earwig

LARVAE: When they first hatch, the small white larvae remain in the soil, and their mother feeds them there. As they grow, they turn darker, resembling adults, and emerge to feed on their own. Adult females lay two separate batches of eggs per year.

EGGS: White eggs laid in clusters in the soil

CROPS AT RISK: Seedlings of most crops, tender leafy crops, and corn

DAMAGE: Earwigs can eat seedlings down to bare stems and also feed on tender crops such as lettuce, celery stems, and ripening tomatoes (especially those already injured). They also chew on corn silks, which interferes with pollination, resulting in poor kernel formation.

CONTROL METHODS

GARDEN CLEANUP. Since earwigs like to feed on decaying organic matter and hide in moist dark places, keep your garden clear of crop debris, pulled weeds, buckets, containers, and boards (except those you're using as traps).

REFUGE TRAPS. Use old flowerpots as traps by stuffing them loosely with crumpled paper.

Set the pots upside down around the garden, with the rim propped up by a stone or stick. Or loosely roll sections of damp newspaper or pieces of corrugated cardboard, secure them with a rubber band, and lay them around the garden. Still another option is to cut short pieces of garden hose and place them around the garden. Earwigs will crawl inside these refuges to hide during the day. In the morning, pick up the traps and shake them over a bucket of soapy water to dislodge the earwigs and drown them.

A reusable trap made of T1-11 plywood (a type of plywood with surface grooves) also works well. Earwigs will crawl into the interior openings, and you can dunk the entire trap into a bucket of soapy water to kill the earwigs, and then set out the trap again.

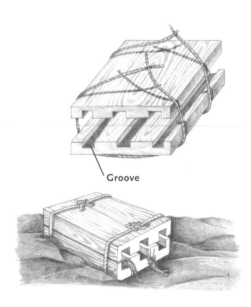

Groove

Cut two pieces of T1-11 plywood about 12 inches square and tie them together with the grooves facing in. Earwigs crawl inside the trap at night.

MONITOR MULCH. Earwigs like to hide under mulch during the day. If you're using an organic mulch such as straw, you may need to remove it from areas of the garden where you've planted seedlings, until the seedlings are well established and no longer susceptible to earwig damage.

REPELLENTS. A homemade hot pepper and garlic spray may help deter earwigs from feeding on garden plants. Apply the spray in the morning so the plants will be dry before nightfall.

CONTROL CALENDAR

BEFORE PLANTING

- Clear weeds, crop debris, and stray objects out of the garden.

AT PLANTING TIME

- Put out homemade traps that earwigs will seek out as hiding places. Each morning, empty the traps into a compost pile (not one located in your garden) or into a bucket of soapy water.

WHILE CROP DEVELOPS

- Spray plants with hot pepper and garlic repellent in the morning to deter earwigs. Reapply as needed.
- Continue to put out traps and empty them daily as needed when earwig populations are too high.
- Stop using traps as soon as earwigs cease doing significant damage (to allow them to carry out their beneficial functions).

EGGPLANT

Attention to details pays off when you grow eggplant. This crop suffers a big growth setback anytime it's stressed, which can reduce harvest from several fruits per plant to only one or two. Leave eggplant off your crop list unless it's one of your favorites. Buy it at the farmers' market instead. However, if you're an eggplant lover, read on to learn how to nurture a beautiful, bountiful crop.

CROP BASICS

FAMILY: Tomato family; relatives are peppers, potatoes, tomatillos, and tomatoes.

SITE: Give eggplant full sun and protection from strong wind.

SOIL: Eggplant needs well-drained, moderately fertile soil. For average soil, spread 1 inch of compost and work it in before planting, or add one shovelful of compost to each planting hole.

TIMING OF PLANTING: Plant transplants in the garden when the weather is fully settled and soil is more than 60°F. In some areas, this will be 2 weeks or more after the last spring frost.

PLANTING METHOD: Sow seeds indoors 8 weeks before your planned transplanting date. See "Start Your Own Seedlings" below for details.

CRUCIAL CARE: With eggplant, everything is crucial. Any kind of stress—water, temperature, nutrients, insect damage, or disease—throws the crop into shock.

GROWING IN CONTAINERS: Containers can be perfect for compact varieties of eggplant that grow only 1 to 2 feet tall. Use a rich planting mix in a spacious container at least 12 inches deep (a bushel basket is a good choice). If possible, put the container on casters so that you can roll the plants to a protected location anytime cool weather is predicted.

HARVESTING CUES: Check the variety description to learn the mature size of fruit (it could be anywhere from 2 inches to 10 inches). When fruits reach one-half to one-third mature size, you can begin harvesting. It's wiser to pick fruits on the young side than to let them become too mature. Press the skin with a finger; if the skin springs back, the fruit is ready to pick. Old fruit loses its shine and may taste bitter.

SECRETS OF SUCCESS

START YOUR OWN SEEDLINGS. Most greenhouse-grown transplants suffer from water, heat, or cold stress at some point during production. Many types of vegetable transplants bounce back, but not eggplant. Unless you can buy eggplant transplants from a reliable local grower, it's best to start your own. Eggplant seedlings need a little special care. First, provide bottom heat to speed the germination process. At 80°F, eggplant seeds sprout in about 5 days; below 70°F, they may languish 2 weeks before germinating, or never come up at all. Once seedlings produce a set of true leaves, transplant them to individual 4- or 6-inch pots. Transplant only vigorous, sturdy seedlings. Place the seedlings under strong lights in a warm place (with overnight temperatures of at least 65°F).

NEVER RUSH TO PLANT. With tomatoes, it's a rewarding challenge to set out some plants early to try to speed the harvest. Not so with eggplant; wait until the soil is warm and daytime temperatures reliably exceed 60°F before planting eggplant outdoors. That's why you plant the seedlings in big pots—they'll have room to continue growing if the weather refuses to warm up on schedule. If possible, prewarm the soil by covering it with black plastic. Once plants are in the ground, stand ready to cover them with baskets, cardboard boxes, or row covers overnight if predicted low temperature will be 55°F or less.

FEED AND WATER ON SCHEDULE. One round of water stress can sharply reduce later yields of eggplant. Thus, eggplant is a prime candidate for drip irrigation or watering reservoirs (see page 430 for installation instructions). Provide 2 to 3 gallons of water per plant per week. If you haven't used black plastic, wait about 1 month from planting and then apply an organic mulch to conserve moisture and keep down weeds. Plants need a steady supply of nutrients, too. Once the first fruits set, water plants every 2 weeks with half-strength fish emulsion.

REGIONAL NOTES

SOUTHEAST

Gardeners in the Southeast enjoy the benefit of a long warm season, which allows them to grow eggplant without much worry about cold stress. However, the humid, hot weather of the Southeast makes all disease problems more of a concern. Southeast gardeners should be on the alert for Phomopsis blight of eggplant, a fungal disease that spreads rapidly in hot, humid conditions and can't survive over winter except in mild-winter areas. Infected seedlings develop girdled stems and die. On larger plants, brown spots develop on leaves and stems. These spots enlarge and small black dots appear at the center. Sunken, dry cankers form on plant stems and may girdle and kill the plants (this is called collar rot). The worst symptom of the blight appears on fruits, which develop sunken circular or oval spots. These spots turn brown and enlarge, sometimes covering the whole fruit, which becomes soft and may rot. If the weather turns dry, infected fruits shrivel up.

Remove all eggplant crop debris and destroy it, or compost it and use that compost somewhere other than your vegetable garden. Build soil organic matter to encourage microorganisms that compete with or attack the Phomopsis fungus. Once the fungus is present in your garden, choose resistant varieties and start a 3-year rotation of eggplant (the fungus doesn't infect any other crops). The disease is carried on seed, so be sure to buy clean seed.

Support your eggplants to preserve the fruits' beauty and to reduce the risk of disease. A sturdy 4-foot stake ensures that the fruits will hang free.

Eggplants that produce rounded fruits will do well inside standard tomato rings or in a simple stakes-and-string corral.

PREVENTING PROBLEMS

BEFORE PLANTING

* Spread black plastic to prewarm the soil.

AT PLANTING TIME

* Encircle each transplant with a cutworm collar.
* Cover plants with row cover and seal the edges to keep out insects (see Row Covers).
* If not using row covers, spray plants with kaolin clay to discourage flea beetles.

WHILE CROP DEVELOPS

* If not using row covers, cover individual plants with cloches or other protection on chilly nights.
* Remove row covers 3 to 4 weeks after planting.
* Weed carefully around young plants to avoid damaging crop roots. When soil is thoroughly warm, apply an organic mulch.

AT HARVEST TIME

* Use a sharp knife or pruners to cut fruits off plants without breaking stems.

AFTER HARVEST

* If Verticillium wilt is a problem in your garden, end the harvest a couple of weeks early. Remove unripe fruits, plant tops, and roots from the garden and compost them. This helps to break the disease cycle by preventing the fungus from forming resting spores that will persist in the soil.

TROUBLESHOOTING PROBLEMS

NEW SEEDLINGS EATEN. Cutworms or slugs and snails are responsible. Cutworms often cut transplants off cleanly near soil level; see Cutworms for preventive measures. Slugs and snails leave behind a slime trail; see Slugs and Snails for controls.

PURPLE LEAVES ON YOUNG PLANTS. Phosphorus deficiency induced by cold soil is to blame. If you have extra transplants, pull out the stressed plants. Wait a week or so for the weather and soil to become warmer and then plant your reserve transplants. Protect them from late cold spells and they should grow fine.

SMALL HOLES IN LEAVES. Flea beetles love eggplant. When the beetles feed on young plants, the stress may set back the plants for the whole season, significantly reducing yield. Always think ahead about how you'll minimize flea beetle damage to your eggplant crop. See Flea Beetles for prevention and control options.

LARGE HOLES IN LEAVES. Tomato hornworms devour large sections of leaves, sometimes right to the midrib. They are easy to handpick; for other controls, see Tomato Hornworms.

SKELETONIZED LEAVES. Colorado potato beetles and their larvae like eggplant as much as—or more than—potato plants. Handpick them daily; see Colorado Potato Beetles for more controls.

YELLOW OR STIPPLED LEAVES. Spider mites are a problem on eggplant, especially during hot, dry conditions. Look closely at leaves for small dark specks and webbing on leaf undersides. To control mites, see Mites.

SPOTS ON LEAVES. Early blight begins as light-colored spots on leaves; severely infected leaves die and fall off. Leathery spots that show concentric rings develop on fruits, too. See Early Blight for controls.

SCORCHED LEAVES. Flea beetle damage, Verticillium wilt, and heat or cold injury are all possibilities. Check plants for small black beetles and holes in leaves. If you find them, see Flea Beetles for controls.

If Verticillium wilt is to blame, plants probably will be stunted as well, and part or all of the plant may wilt. Uproot infected plants. See Verticillium Wilt for preventive techniques.

If you find no sign of pests or disease, temperature extremes are the likely cause. The new growth should be normal; however, the stress the plant suffered may set it back for the rest of its life, reducing yield.

MOTTLED LEAVES. Mosaic viruses cause mottling as well as leaf crinkling and abnormal growth. For preventive measures, see Mosaic.

LOWER LEAVES TURN YELLOW. Deterioration of older leaves can be a sign of lack of nutrients or Verticillium wilt. If plants also begin to wilt, suspect Verticillium. See Verticillium Wilt for controls.

STUNTED PLANTS. Root-knot nematodes invade plant roots, causing yellowing leaves and poor growth. Uproot a plant and check the roots. Swelling and distortion of roots is a confirming sign. (Other types of nematodes also attack eggplant roots in some parts of the country but do not produce root swelling.) For controls, see Nematodes.

PLANTS WILT. Bacterial wilt, southern blight, and Verticillium wilt are possible causes. Bacterial wilt of eggplant can develop suddenly, and the wilted plants sometimes die quickly. Confirm the problem by cutting into the main stem at the soil line—if wilt is present, a brown slime will ooze from the stem. The bacteria can survive in soil for several years. To reduce future problems, enrich the soil with compost, and start a rotation of tomato-family crops. Be careful not to move infected soil from one bed to another in your garden.

Southern blight is a fungus that spreads in hot, moist weather. Leaves will turn pale and fall off. A white mold may appear around the base of the plants. Carefully uproot plants without knocking loose any of the mold, and discard or destroy them. To prevent future problems, see Southern Blight.

Verticillium wilt often causes wilting that begins on only one side of the plant. See Verticillium Wilt for controls.

BLOSSOMS DROP OR NO FRUIT FORMS. Because eggplant is so sensitive to stress, many factors may cause lack of fruit formation: Cold temperatures, high heat, and water stress are possibilities. Pamper the plants and fruit set may improve.

SMALL, MISSHAPEN FRUIT. If fruits have an odd shape and a large scar at the blossom end, it's a sign of poor pollination, which can be due to temperature extremes, lack of insects, or stress. If weather is to blame, new fruits should be normal. If stress was the problem, the plants may never bounce back completely. For the future, change your management practices to help prevent stress.

BITTER FRUIT. Stress while the fruit was developing is probably to blame. To prevent bitterness in the future, water consistently. Another possibility is that the fruit is too mature; refer to page 192 for tips on when to harvest eggplant for best quality.

ODDLY COLORED FRUIT. When you leave eggplants on the vine too long, they become dull and look bronzed. To avoid this problem, start picking fruit when it's only half the mature size (for large-fruited types); the quality will be much better.

DRY BROWN SPOT ON BLOSSOM END OF FRUIT. Blossom-end rot is the result of a lack of calcium usually due to moisture stress while fruit is forming. Fruits may be usable if you cut out the brown area. See Blossom-End Rot.

WATER-SOAKED SPOTS ON FRUITS. Anthracnose and Phytophthora blight are fungal diseases that can infect eggplant. Spots caused by anthracnose become leathery; infected fruits may be usable if you cut out the bad spots. See Anthracnose for prevention tips.

Phytophthora is a fungus, and white mold often appears on infected fruits. The flesh turns brown and the fruits drop off the plant. Controls are similar to those for other fungal blights; see Early Blight.

WHITE PATCHES ON FRUITS. If there is no other sign of disease except brownish or pale patches on the skin of the fruit, the problem is sunscald caused by overexposure to hot sun. Sunscald can occur if plants have lost leaves due to disease, or if branches sprawl under the weight of the fruit. To prevent it in the future, support the plants and take steps to reduce disease.

FRUITS ROT. Rot fungi can invade eggplant fruits through small wounds in the surface. The fruits soften, and sometimes white mold covers the fruit. Discard or destroy rotten fruits, taking care not to touch healthy fruits as you work. For preventive measures, see Rots.

ENDIVE

If you enjoy a sharp, slightly bitter taste in a tossed salad, add endive to your seed list. You can grow curly endive (or frisee), with its decorative frizzy leaves, or smooth-leaved escarole. Both are easy to grow in cool weather, with the best harvest available in fall.

CROP BASICS

FAMILY: Lettuce family; relatives are artichokes, Jerusalem artichokes, and lettuce.

SITE: Plant endive in full sun or light shade.

SOIL: Endive grows well in average garden soil; sandy loam enriched with organic matter is ideal. For direct-seeding, prepare a fine seedbed, raking away any large clumps.

TIMING OF PLANTING: Sow seeds outdoors 2 to 4 weeks before last spring frost for spring harvest, and about 15 weeks before first fall frost for fall harvest.

PLANTING METHOD: Sow seeds directly in the garden and/or start them in flats or pots.

CRUCIAL CARE: Moisture stress causes endive to turn especially bitter. Check the soil often and water whenever the top couple of inches are dry.

GROWING IN CONTAINERS: Endive grows well in containers. Potted plants will need more watering than those in a garden bed.

HARVESTING CUES: Harvest individual outer leaves or cut whole heads as desired. If you're blanching, harvest within 10 days or the plants may start to rot at the center.

SECRETS OF SUCCESS

BLANCH TO REDUCE BITTERNESS. Blanching endive for a week to 10 days before harvest causes the inner leaves to lose chlorophyll. These yellow or white leaves will be more ten-der, with a mellower flavor than green leaves. Blanch endive by using string or a rubber band to gather up the outer leaves around the center of the plant. Or cut the top and bottom out of a milk carton and slip it over a plant. For curly endive, you can use a small dish or plant sau-cer as shown below. If it rains, remove the cover or tie to let the foliage dry out, and then put it back in place.

TROUBLESHOOTING PROBLEMS

Endive has few problems when grown in rich soil with plenty of water. The following are the most common problems of endive. If you see problems on your endive plants that aren't men-tioned here, refer to Lettuce for a full list of symptoms and causes.

STUNTED PLANTS. Mosaic virus may cause stunted growth without causing the leaf mot-tling that's a common symptom of mosaic. The virus is spread by aphids, and there's no cure. To prevent it in the future, use row covers or other

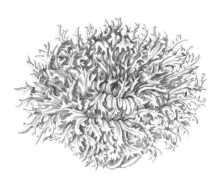 

A ceramic plate or clay plant saucer is a light-excluding cap that sits neatly on the center of a curly endive plant to blanch the center leaves. Leave the cap in place up to 10 days.

Not the same as regular endive, Belgian endive is a delicacy that's grown in stages, first by planting witloof chicory in the garden, then transferring the roots to pots in a dark place to produce tender blanched shoots called chicons. Be sure you buy the right kind of seeds for Belgian endive.

Witloof chicory needs the same conditions in the garden as regular endive. Sow seeds 2 to 3 months before your first fall frost. Let the plants produce lots of lush growth—you won't be eating these leaves. After some frosts but before the ground freezes, carefully dig out the plants, roots and all. Keep the straight roots that are 1 inch diameter or larger; compost small or forked roots. Cut off the tops 1 inch above the crown, and also cut off the bottom inch or so of the roots. Store the trimmed roots in a box of damp sand in a root cellar until you're ready to force them. If you don't have a root cellar, you can store them in a perforated plastic bag in your refrigerator.

Force the roots by planting them in a large pot or plastic bucket. Fill the pot halfway with sand or potting mix, then stand the roots in the pot and add more sand up to the top of the roots. Water the pot well and put it in a totally dark location, or cover it with another large pot (tape over the drainage hole) or with a black plastic garbage bag. Temperatures between 55° and 60°F are ideal to encourage sprouting. Check the pot every few days and water as needed to keep the medium evenly moist. If you let it dry out, the shoots may die back. It takes about 2 weeks for the sprouts to reach harvest size. Cut them at the base. Strong roots will provide a second or even third round of chicons.

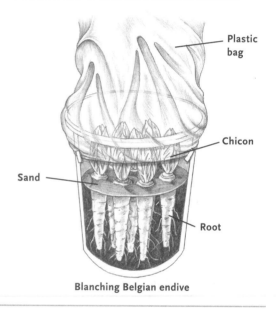

Blanching Belgian endive

measures to prevent aphids from feeding on the plants. See Mosaic for more information.

OUTER LEAVES TURN BROWN. This is a sign of basal rot or of cold damage. Uproot one plant. If the crown or roots are dark and soft, the problem is rot. Remove the damaged plants, salvage any healthy leaves, and compost the rest. See Rots for preventive measures. If the roots and crown look healthy, the outer leaves probably were injured by cold temperatures. Remove the damaged leaves, but allow the plants to continue growing. If temperatures lower than the high 20s are predicted, protect the plants with row covers or cloches.

FERTILIZERS

Fertilizer fades in importance as you gain experience with gardening organically. Eventually, you may not use fertilizer at all, or only for a few heavy-feeding crops such as broccoli. But when you first start an organic garden, and while you're building the soil, you'll probably need to rely on fertilizers to ensure steady, healthy crop growth throughout your garden. When you prepare the soil for planting, you'll add products such as alfalfa meal, rock phosphate, or an organic fertilizer blend to supplement the supply of nutrients in your soil. In a conventional garden, applying chemical fertilizer is a regular ritual. In an organic garden, building the soil becomes the standard practice instead. You'll add organic matter to the soil every year (see Soil for a full discussion of building soil organic matter). But how can you tell whether and when to add organic fertilizer, too?

One simple way to decide is to watch how your garden grows overall. If your plants are vigorous and growing steadily, then they're getting all the nutrients they need the natural way. If growth seems slow or weak even though the soil is moist and temperatures are optimal for growth, your plants are probably suffering nutrient stress.

TESTING YOUR SOIL

Another approach to fertilizer decisions is to test your soil and follow the recommendations of the laboratory. Keep in mind, though, that many soil test labs analyze soil from a chemical gardening perspective and make recommendations for building soil by chemical means. It's more difficult to analyze soil nutrient content from an organic perspective and recommend additions of organic fertilizers to adjust nutrient levels (for more information about the difference between organic

soil management and conventional soil management, see Soils). Soil testing can be expensive, too. In most cases, testing soil (except for testing soil pH) isn't necessary for home gardens.

WATCH FOR DEFICIENCIES

Examine individual plants for symptoms of nutrient deficiencies. Nutrient deficiencies are unusual in gardens where soil organic matter content is moderate to high. But in a new garden with low soil organic matter content, deficiency symptoms may appear. Nutrient deficiency symptoms and disease symptoms have a lot in common. Leaves turning yellow, for example, is an early symptom of many diseases, but it's also a possible sign of nitrogen deficiency.

To figure out whether a symptom is due to nutrient stress or disease, examine the pattern of symptoms. If a whole row is turning yellow at once, nutrient deficiency is one likely cause.

However, when symptoms show up first on a few plants and then spread, or when they first appear during wet weather, the cause is probably disease-related.

The chart on the opposite page lists some of the most common nutrient deficiency symptoms.

If you find a match to symptoms described in the chart, don't panic. There's probably a relatively simple solution. Here's what to do:

- Check soil moisture. It could be that the soil has dried out well below the surface and needs a deep, restorative watering (plants can't absorb nutrients from dry soil). Or your soil may be waterlogged. Roots can't absorb nutrients properly from soaking wet soil, either.
- To help your plants bounce back, apply a foliar fertilizer, such as compost tea or a fish/kelp product. These organic liquid feeds supply a broad range of nutrients, and plants can absorb many of them directly through their leaves.
- Before you replant in the area where the problem occurred, improve soil drainage if needed. Also check soil pH.
- Add organic matter by working in compost, planting cover crops, and covering the soil with organic mulch. Overall, it's better for the soil to build the nutrient supply by adding organic matter than by dosing it with fertilizer.

CHOOSING FERTILIZERS

Fertilizer labels offer lots of information, but it's not always easy to interpret. One critical thing to know before you buy: Some fertilizers that are labeled "organic" *do* contain synthetic chemical ingredients. How can that be?

FERTILIZER PHILOSOPHY

Throughout this book, you'll find detailed entries on specific crops, and these entries often include recommendations for applying fertilizer. Think about the big picture of your garden management before you follow these recommendations. If you feed your soil regularly and your plants seem to be growing well, with few pest and disease problems, perhaps you don't need to fertilize at all. Sure, if you apply fertilizer, it will boost yield, but at what cost?

Purchased fertilizer, even organic fertilizer, is an "outside input." It takes resources and energy to produce that fertilizer and to package and ship it. It costs you money to buy it, but there's an environmental cost as well—the pollution or habitat disturbance that results from mining or processing the fertilizer, shipping it, and disposing of the packaging.

Analyze your goal before you apply fertilizer. If you garden in sandy soil, sustaining and increasing soil organic matter content can be a serious challenge. Annual fertilizing may be essential to reap even a moderate harvest. On the other hand, if you're striving for broccoli heads as big as those in the grocery store, or to grow the largest pumpkin on the block, consider adjusting your attitude and cutting back on fertilizer. Garden for the wonderful flavor and rich food value of the vegetables you grow, not for appearance and impressive size.

NUTRIENT DEFICIENCY SYMPTOMS

Nitrogen, potassium, and phosphorus are the three primary mineral elements that plants need for healthy growth. Calcium, magnesium, and other minerals are also essential, but in much smaller amounts. If you see any of the symptoms described below in your garden and suspect that nutrient stress is the cause, take a sample of the plants to your local Cooperative Extension office for confirmation. If your Extension agent suggests fertilizers to apply, make sure he or she knows that you're an organic gardener.

NUTRIENT LACKING	COMMON SYMPTOMS
Nitrogen	Spindly or stunted growth; lower leaves pale green, then turn yellow
Phosphorus	Lower leaves and stems look reddish or purplish; young leaves pale; plants don't flower or form fruits
Potassium	Lower leaves mottled or curled; edges of leaves turn brown and dry; weak stems; small root systems
Calcium	Growing tips die back; tips of new leaves appear scorched; leaves are abnormally dark green; weak stems; blossom-end rot of fruits
Magnesium	Lower leaves mottled yellow or show white patches; leaves turn reddish purple
Iron	Young leaves turn yellow but veins remain green
Zinc	Young leaves mottled yellow; plant tips stop growing; plants wilt easily
Boron	Leaves curl under

ANALYZING A FERTILIZER LABEL

One way to analyze the "organicness" of a bag or bottle of fertilizer is to look at the NPK ratio. Many synthetic chemical fertilizers have fairly high numbers in the ratio, such as 10-10-10 or 20-20-20. Fertilizers that contain only organic materials have a much lower analysis, such as 2-5-1. That's because the NPK ratio measures the nutrient content in terms of nutrients that are readily available to plants (i.e., in a form that roots can absorb). However, in bona fide organic fertilizers, most of the nutrients are contained in complex compounds that aren't soluble in water and can't be absorbed by plants. Soil organisms must break down these compounds into simpler forms first. (This characteristic of organic fertilizers is a good thing, not a bad one. For one thing, it means that the nutrients in organic fertilizers are less likely than those of chemical fertilizers to leach out of the soil and cause water pollution.)

Reading ingredient lists is also helpful. Learn to recognize common terms used in chemical fertilizers, such as ammonium, urea, nitrate, and superphosphate. If you spot those on a fertilizer label, then it's not a fertilizer you'll want to use in your garden.

Another option is to look for the three-leaf or OMRI logo on a product (see page 6 for more information). Or seek out a supplier with a reputation for integrity and ask for help in finding fertilizer products that meet your personal standards for your organic garden.

TYPES OF ORGANIC FERTILIZERS

Organic fertilizers are derived from plants, from animal by-products, and from minerals. For example, kelp meal and alfalfa meal are plant-based fertilizers. Bloodmeal and bonemeal are two of the best known animal by-product fertilizers, and rock phosphate and greensand are mineral-based. You'll also find a wide variety of premixed fertilizer blends, which may combine plant, animal, and mineral products.

Deciding which specific fertilizer to use in your garden can be challenging. "Organic Fertilizer Choices" below lists some of the most common organic fertilizers and suggests how to use them for vegetable gardening. Some gardeners choose not to use particular fertilizers because of how they're made. For example, people who are vegetarians may decide not to use bloodmeal or bonemeal because these fertilizers are by-products of the slaughter of animals.

MAKING FERTILIZER TEAS

Fertilizing gardens with manure tea was a popular recommendation 20 years ago. However,

ORGANIC FERTILIZER CHOICES

Use these listings to decide which kind of fertilizers to apply to your crops. Keep in mind that fertilizers with a high nitrogen content (the first number in the ratio listed under the column labeled Average Analysis) will promote leafy growth but may suppress formation of fruits (such as tomatoes and peppers).

FERTILIZER NAME	TYPE	HOW IT HELPS	AVERAGE ANALYSIS	COMMENTS
Alfalfa meal	Plant derived	Supplies nitrogen	2.5-1-1	Generally low-priced; add to soil before planting; good for side-dressing
Blended fertilizer	May include plant, animal, and mineral components	Boosts nutrients overall when soil fertility is low	Analysis varies widely	Commercial blends are convenient, but mixing your own is less expensive; add to soil or use as side-dressing
Bloodmeal	Animal derived	Supplies nitrogen	11-1-0	Fast-acting and easy to apply; use as side-dressing
Bonemeal	Animal derived	Supplies phosphorus	3-11-0	Place in planting holes or use as side-dressing; use caution or avoid if you are concerned about mad cow disease
Cottonseed meal	Plant derived	Supplies nitrogen	6-2-1	Also supplies phosphorus and potassium; may contain pesticide residues; lowers soil pH—good for acid-loving plants
Epsom salts	Mineral derived	Supplies magnesium	10% magnesium, 13% sulfur	Dissolve in water for a soil drench or work into soil when soil test indicates serious magnesium deficiency
Feather meal	Animal derived	Supplies nitrogen	13-0-0	Strong, fast-acting source of nitrogen; work into soil or use as side-dressing

times have changed. These days, it's not easy for most home gardeners to find or buy fresh manure. Even if you can procure manure, it's probably not a good idea to use fresh manure in your vegetable garden. First, if the manure is from a nonorganic farm, it may contain chemical or antibiotic residues. Also, there's a concern about contamination of manure with bacteria that can cause disease in humans. Under NOP rules (see page 5), farmers must compost manure before they can apply it to food crops, and that's a good guideline for home gardeners as well.

Two nutrient-rich alternatives to manure tea are compost tea and comfrey tea. Commercially produced compost tea is available from some garden suppliers, or it's easy to make it yourself. For complete information on compost tea, refer to Composting.

Comfrey tea is brewed from Russian comfrey (*Symphytum × uplandicum*), a vigorous perennial

FERTILIZER NAME	TYPE	HOW IT HELPS	AVERAGE ANALYSIS	COMMENTS
Fish products	Animal derived	Supplies nitrogen	5-1-1 (fish emulsion); 5-3-3 (fish meal)	Fast-acting; use fish emulsion as a foliar feed or soil drench; use fish meal as side-dressing; some have a fishy odor
Greensand	Mineral derived	Supplies potassium	0-0-7	Also supplies a variety of trace minerals; may lower soil pH
Guano	Animal derived	Supplies nitrogen, phosphorus, and potassium	8-4-2 (bat guano); 13-12-2 (bird guano)	Also supplies trace minerals; has some odor
Gypsum	Mineral derived	Supplies calcium	23% calcium, 19% sulfur	Sulfur content lowers soil pH—don't use in acid soils; helps loosen compacted soil
Kelp (seaweed)	Plant derived	Supplies potassium	1-0.5-2.5	Also supplies trace minerals; excellent soil additive; use in liquid form as a foliar feed
Rock phosphate	Mineral derived	Supplies phosphorus	0-3-0	Slow-acting—apply a season ahead
Soybean meal	Plant derived	Supplies nitrogen and potassium	7-0.5-2.3	Buy soybean meal specifically labeled for organic gardens; most soybean meal sold as animal feed is derived from genetically modified soybeans; use as side-dressing
Wood ashes	Plant derived	Supplies potassium	0-1.2-8	Highly alkaline and will raise soil pH significantly with repeated use; use with caution

MIXING AN ORGANIC BLENDED FERTILIZER

Save money by mixing your own blended organic fertilizer. For a general balanced fertilizer to enrich poor to average soil, combine 3 parts cottonseed meal, 2 parts bloodmeal, or 3 parts fish meal; 3 parts rock phosphate; 1 part kelp meal; and $1\frac{1}{2}$ parts greensand. Spread 8 to 9 pounds per 100 square feet.

If your soil is rich, and you want a supplement for boosting crops, mix 2 parts cottonseed, bloodmeal, or fish meal with 2 parts rock phosphate and 1 part kelp. Sprinkle one trowelful around the base of each plant of heavy-feeding crops such as cabbage.

Use an old measuring cup or a plastic container or a large scoop as your measuring tool. Sweep an area of your garage floor clean (but not an area where motor oil has dripped) and mix the materials right there, using a rake or other long-handled tool. Always wear a dust mask when mixing fertilizers. Store any leftover fertilizer in a clean, covered container and label it clearly, including all ingredients.

herb that's hardy from Zone 3 to 9. The tea supplies a range of nutrients, notably nitrogen, potassium, and calcium. If you have an unused sunny spot away from your other garden beds (comfrey can spread if roots are disturbed), plant several comfrey plants in spring. Let them grow for a year, and then start harvesting leaves once the flowerstalks appear during the second season. Wear gloves, long sleeves, long pants, and shoes when working with comfrey because some people suffer from skin irritation after touching comfrey leaves.

Fill a bucket or other container half to three-quarters full of comfrey leaves and top off the container with water. Let it sit for 2 to 3 weeks (cover the container during this time). The water will turn into a foul-smelling dark tea, which you can apply full-strength or diluted up to 50 percent with water.

APPLYING FERTILIZER

Applying fertilizer is an easy task, especially if you use a commercial product. Follow the instructions on the package when applying commercial fertilizers. Keep these tips in mind for best results.

APPLY WHEN THE SOIL IS MOIST. Biological activity is higher in moist soil than dry soil. Thus, the fertilizer you supply will become part of the food web faster if you apply it when the soil is moist.

SIDE-DRESS FOR A LONG-LASTING BOOST. Side-dressing is simply applying a fertilizer to the soil surface near the plants you want to feed. If the soil surface is mulched, pull back the mulch first so the fertilizer is in direct contact with the soil. You can use a hand tool to lightly scratch the fertilizer into the top inch or two of the soil. Side-dressing won't cause a dramatic growth burst, but it will fuel steady growth as the fertilizer releases its nutrients to the soil over time.

USE LIQUIDS AND TEAS FOR FAST EFFECT. Plants can absorb some of the nutrients in liquid fertilizers and teas directly through their leaves, making foliar feeding the fastest way to deliver a boost to your plants. You can apply liquids (diluted according to package instructions) and teas with a watering can, a hose-end sprayer, a small pump-type sprayer, or through an irrigation system. Always strain teas and liquids through a coffee filter or several thicknesses of cheesecloth before using them in a sprayer or irrigation setup, or the sprayer tip and emitters may clog. Note that plants that have waxy leaves, such as cabbage and broccoli, may not respond as well to foliar feeding as other crops do.

FEED AT CRITICAL GROWTH TIMES. Supplying fertilizer can help plants through key transitions, such as germination or transplanting in the garden, setting flower buds and young fruits, and surviving dry weather or other stress.

FLEA BEETLES

When a pleasant morning early in the growing season draws you outside to sow seeds, flea beetles are lying in wait, snacking on weeds as a warm-up to attacking your precious seedlings. These tiny, shiny beetles hide in weedy areas near your garden, in crop debris, and under clumps of soil in garden beds. They emerge in hungry hordes, initially feeding on weeds and other plants around the yard. Their eating style is to chew lots of small, round holes, usually called "shotholes," in plant leaves.

Flea beetles are particularly attracted to seedlings, and vegetable gardens are a powerful draw. If a crowd descends on seedlings or small transplants, the plants may not survive. (Older plants usually outgrow the damage.) Flea beetles seem especially attracted to stressed plants, too, so keeping plants healthy is an important part of fighting these widespread pests.

Flea beetles emerge over a period of several weeks in spring and feed for several weeks. Female beetles then lay eggs at the base of host plants. Fortunately, the adults die off soon after egg laying. Eggs hatch about 1 week after they are laid; the tiny white wormlike larvae feed on roots for 2 to 3 weeks and then pupate. The pupae rest for about 2 weeks and then the new adults emerge. Thus, there's a window between generations when few adult flea beetles are around to pester your garden. In the North, there may be only two generations per year, but in the South, gardeners have to contend with several rounds of flea beetles.

Unfortunately, some of the crops that flea beetles like best are those where even minor leaf damage is annoying, because it's the tender young leaves that you want to eat; arugula, mustard, and spinach are a few flea beetle favorites. Larvae can be serious pests of potatoes and sweet potatoes. Flea beetle feeding can transmit early blight of potatoes and Stewart's wilt of corn as well.

PEST PROFILE

Many genera and species

RANGE: Throughout North America (flea beetles are found worldwide)

HOW TO SPOT THEM: Scout young plants frequently for small black beetles that jump rapidly and widely when disturbed. If you miss early arrivals, the first sign you may notice is leaves riddled with small holes or plants that look like they're covered with black dots (the beetles).

ADULTS: Long back legs allow these small beetles to jump or hop like fleas when disturbed. Flea beetles are shiny black, reddish brown, yellowish brown, or bronzy, from $\frac{1}{16}$ to $\frac{1}{2}$ inch long. Some have stripes on their wing covers.

LARVAE: Soil-dwelling, slender white grubs with brown heads; up to $\frac{3}{4}$ inch long

EGGS: Laid at base of host plants

CROPS AT RISK: Cabbage-family crops, sweet potatoes, leafy greens, spinach, beets, eggplant, tomatoes, potatoes, peppers, corn, basil

DAMAGE: Adults chew many small holes in leaves. Feeding weakens or kills seedlings; disfigures salad crops; larvae feed on plant roots but cause little damage to them. Feeding allows diseases to gain a foothold.

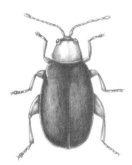

Flea beetle

Flea beetle feeding damage

CONTROL METHODS

TIMED PLANTING. Check with your local Cooperative Extension office to find out when flea beetles emerge in your area and the approximate date when the first generation of beetles dies off. Calculate sowing dates so that seedlings emerge when few beetles will be present. By the time the second generation of beetles emerges, your plants should be large enough to withstand some damage.

INTERPLANTING. Underplant flea beetles' favorite crops with a living mulch, such as clover, to make it harder for flea beetles to find and attack the susceptible plants.

ROW COVERS. One of the best ways to prevent flea beetle damage is to cover a seeded area immediately after sowing. Leave the cover in place, removing it only as needed to allow for pollination of fruiting crops and for harvest. Do check under the cover occasionally to be sure that no flea beetles emerged from the soil under the cover or otherwise invaded the refuge. To learn more about techniques for handling row covers, consult Row Covers.

STICKY TRAPS. Hold a homemade sticky trap close to infested plants and brush the foliage. Beetles will jump and some will end up stuck to the trap. This method for using sticky traps is preferable to mounting them permanently in your garden, because the traps also attract and capture beneficial insects. See page 301 for directions for making sticky traps.

REPELLENTS. Kaolin clay and garlic sprays may deter flea beetles from feeding. See page 306 for more information on repellents.

TRAP CROPS. Flea beetles are drawn to many cabbage-family crops, especially napa cabbages and daikon radishes. Try planting one of these trap crops a week or two before planting other crops attacked by flea beetles. See page 307 to learn how to manage trap crops. Note: If you *don't* like arugula, it's another good choice for a sacrifice crop because flea beetles love it!

VACUUMING. Use a handheld vacuum to suck flea beetles off plants (test to be sure plants are strong enough to withstand the suction). You'll have to do this frequently (perhaps daily) to make an impact on the resident flea beetle population (see "Vacuum Doom for Pests" in Pest Control).

USE BIOCONTROL SPRAYS. Spraying *Beauveria bassiana* or spinosad may kill off flea beetles. Repeat sprays will probably be needed. See page 312 for directions for using these biocontrol sprays.

CONTROL CALENDAR

BEFORE PLANTING

- Pull out all cabbage-family weeds in and around your yard to starve out emerging beetles.
- Sow a trap crop about 1 week before you plant susceptible crops.

AT PLANTING TIME

- Time planting to occur after the first generation of beetles has finished feeding.
- Interplant crops or sow a living mulch among transplants.
- Cover seeded areas with row covers immediately and leave them in place as long as possible.

WHILE CROP DEVELOPS

- Keep seedlings and transplants well watered to avoid moisture stress.
- Use sticky traps to catch flea beetles around your garden by holding traps near infested plants and brushing the foliage.
- Apply a repellent to deter flea beetles from feeding.
- Vacuum plants to remove beetles.
- Apply spinosad or *Beauveria bassiana* to kill flea beetles.

AFTER HARVEST

- Remove or turn under weeds and crop debris that may harbor the pests between growing seasons.

FLEAHOPPERS

You may never notice a fleahopper in your garden until you search for it. These tiny black bugs suck sap from leaves, as do their nymphs, but the damage they cause is fairly subtle unless their population is very high. At a glance, it's easy to mistake fleahoppers for aphids or flea beetles. One test is to shake the plants lightly to see how the insects respond. If they're aphids, they won't move much at all. On the other hand, flea beetles are champion long-jumpers that will scatter to other plants when disturbed. Fleahoppers content themselves with short hops of a few inches.

Beans are a favorite choice of fleahoppers, but they will feed on a wide range of crops. Fleahopper eggs overwinter in crop debris and weeds except in the Far South, where adults and nymphs can survive year-round on crop plants and in weedy areas. The number of generations per year varies with latitude from only a few to as many as ten. The extent of fleahopper damage fluctuates from year to year, but they're generally not among the most serious garden pests, especially if you keep your garden clear of weeds that host fleahoppers. Fleahoppers can be serious pests of alfalfa and clover. If you live near fields of these crops, be alert for the migration of fleahoppers from the field to your garden when the alfalfa or clover is cut.

PEST PROFILE

Halticus bractatus

RANGE: Eastern United States and eastern Canada; rare in the Plains states

HOW TO SPOT THEM: Brush your hands over plants and watch for tiny black insects hopping on leaves. Nymphs feed on leaf undersides, so examine those surfaces with a hand lens. The feeding damage resembles that of spider mites.

ADULTS: Tiny, shiny black bugs with long antennae; females may be oval-bodied with short wings or slender with long wings; all males are slender and long-winged.

NYMPHS: Tiny, wingless, pale green insects that darken as they develop

EGGS: Laid in plant tissues

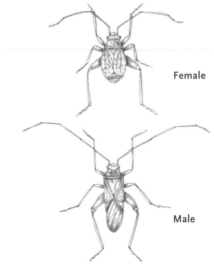

Female

Male

Fleahoppers

Fleahopper feeding damage

CROPS AT RISK: Beans and many other vegetable crops

DAMAGE: Yellow or white spotting or stippling of leaves; leaves turn pale and dry and drop off; plants may be stunted; young plants may die; fruits show catfacing.

CONTROL METHODS

REMOVING WEEDS. Fleahoppers feed on and take refuge in weedy areas. Controlling weeds around your yard and garden can be very helpful in reducing fleahopper problems. Host weeds include bindweed, mallow, pigweed, plantain, prickly lettuce, ragweed, smartweed, and wood sorrel.

ATTRACTING BENEFICIALS. Plant small-flowered annuals and perennials around your yard and garden to attract parasitic wasps that can significantly reduce fleahopper populations.

See the Beneficial Insects for flowers that attract beneficials.

ROW COVERS. Cover small plants with row covers and leave the covers in place as long as possible. However, be sure that the areas you cover are clear of crop debris and weeds that may harbor fleahoppers or their eggs. Seal row cover edges as described in Row Covers.

GARDEN CLEANUP. After harvest, remove all crop debris from the garden in case nymphs or adults are hiding there.

KAOLIN. Spray vulnerable plants with kaolin clay before fleahoppers become active, and reapply as needed. See Pest Control for more information on kaolin.

GARLIC SPRAYS. Spray plants before fleahoppers appear in your garden to deter feeding and egg laying. See Pest Control for information on garlic repellents.

BIOCONTROL. Spraying nymphs with *Beauveria bassiana* may provide control; see page 312 for details.

SPRAYS. As a last resort, spray infested plants with insecticidal soap or pyrethrins. You may need to spray a second time as new nymphs hatch from eggs protected in plant tissues.

CONTROL CALENDAR

BEFORE PLANTING

- Plan to include small-flowered annuals and perennials in or near your garden to attract and shelter parasitic wasps that help control fleahoppers.
- Pull weeds that can host fleahoppers as they appear.

- Cover seedbeds and new transplants with row covers.

- Continue weed control throughout the growing season.
- Spray plants with kaolin clay or garlic repellent before fleahoppers become active.

- Spray nymphs with *Beauveria bassiana*.
- For a severe infestation, spray fleahoppers with insecticidal soap or pyrethrins.

- Remove crop debris and weeds that can shelter fleahoppers or their eggs through the winter.

FUSARIUM WILT

Fusarium fungi are common in the vegetable garden, causing wilts, yellows, and root rot on a wide range of crops. Many species of Fusarium cause disease in vegetable crops, but each species attacks a particular crop. The fungi attack and destroy plant roots and then spread into the plant's vascular system near the crown, blocking the flow of water from roots to stems and leaves. The first symptom of Fusarium often is wilting of the foliage on one side of the plant.

Cabbage yellows is another Fusarium disease. It infects only plants in the cabbage family, causing a unique set of symptoms. Fusarium root rots are part of the complex of rot-causing bacteria and fungi (for controls for rot problems, see Rots).

Fusarium wilt occurs in hot weather, and plants often show no symptoms until they are close to maturity. Verticillium wilt, which causes similar symptoms, tends to occur in cool, wet weather, such as in late spring (see Verticillium Wilt).

RANGE: United States; southern Canada

CROPS AT RISK: Most vegetable crops

DESCRIPTION: On tomatoes and other upright plants, lower leaves wilt. Frequently, all leaves on one side of the plant will wilt. Yellow patches may show up on leaves; leaves may die. Beet and Swiss chard leaves will curl inward; potato and tomato leaves curl under. Dark lesions appear on lower stems and on roots near the soil surface. Cut stems in half lengthwise to reveal reddish brown or black streaks inside.

On vining crops such as sweet potatoes and squash, the first symptom often is yellowing and wilting of the older leaves. Other times, vines wilt without showing any other symptoms first. Vines may be stunted or die completely.

Cabbage-family crops infected with cabbage yellows become dull green and look lifeless. Stems and leaves warp or curl. Lower leaves turn yellow and may die and drop off the plants. Cut stems show a brown discoloration.

FIGHTING INFECTION: If plants are near maturity, harvest what you can before uprooting the infected crop debris. If the plants are young and symptoms are mild, you can try to nurse the crop through by watering regularly. In the case of cabbage yellows, if the weather cools off and soil temperatures drop, the plants may rebound and produce a decent harvest.

GARDEN CLEANUP: Remove infested crop residues from the garden, taking care not to spread the debris while you work. Discard or destroy this debris; if you compost it, be sure not to apply that compost in your vegetable garden or you'll end up spreading the fungi throughout your garden. It's okay to apply the compost to ornamentals because Fusarium fungi are so host-specific. Solarizing infected beds during the hottest part of the summer will help to reduce the levels of Fusarium fungi.

NEXT TIME YOU PLANT: Test soil pH several weeks before planting, and if needed adjust it to between 6.5 and 7.0. Planting resistant and tolerant varieties is the best defense when you know that your soil contains Fusarium fungi. Be sure to buy certified disease-free seed potatoes. Always handle plants carefully to avoid creating wounds. Mulch to keep soil cool and moist, and add compost to boost beneficial soil microorganisms. In the South, grow cabbage-family crops only during the winter, when the soil is cool and less favorable for Fusarium development. In the North, plant broccoli or other crops on the early side in spring with protection to produce an early harvest, before Fusarium becomes serious. If nematodes are a problem in your garden, take steps to control them as well, because nematode injury leaves crop roots even more susceptible to invasion by Fusarium. See Nematodes for control measures.

CROP ROTATION: Fusarium spores can persist so long in soil—up to 15 years—that you can't rid your garden of Fusarium by rotating crops. However, since each crop family or individual crop is attacked by a different species of Fusarium, practicing crop rotation will help to prevent a continual buildup of Fusarium fungi in your garden. Plan a 4-year rotation (or longer) if possible.

GARLIC

A favorite of vegetable gardeners who've tried it, garlic is easy to grow and perpetuates its own supply of planting stock year after year. Grow it for cooking, for making a pest-repellent spray, or both. Fall is the time to tuck garlic cloves into a garden bed, where it will take root and perhaps initiate some foliage before going dormant for winter. In spring and early summer, garlic grows vigorously, and each clove planted turns into a full-size bulb.

The quality of the garlic you plant has a major influence on how well your harvest will turn out. Take time to find a local grower of high-quality garlic for planting or ask fellow gardeners to recommend mail-order suppliers they trust.

CROP BASICS

FAMILY: Onion family; relatives are asparagus, chives, leeks, and onions.

SITE: Garlic needs full sun for vigorous growth and bulb production.

SOIL: Average well-drained soil will suffice for garlic, but for best results, work in 1 to 2 inches of screened, mature compost before planting. If you want to plant garlic in a bed where a cover crop has grown, be sure the crop debris has broken down thoroughly.

TIMING OF PLANTING: Plant garlic in fall— as early as September in the North and as late as December in the South.

PLANTING METHOD: Plant individual garlic cloves directly in garden beds.

CRUCIAL CARE: Control spring weeds early on because garlic can't compete well against weeds. But remove weeds carefully because garlic has shallow roots that are easily disturbed. For the best bulbs, feed garlic once or twice after it resumes growth in spring. Alfalfa meal or liquid fish fertilizers are good choices. Stop fertilizing at least 1 month before the expected harvest date. Also cut back on water once the bulbs start to mature (when foliage first begins to turn brown).

GROWING IN CONTAINERS: Tuck a few garlic cloves into containers where you've planted other sun-loving vegetables. These garlic plants

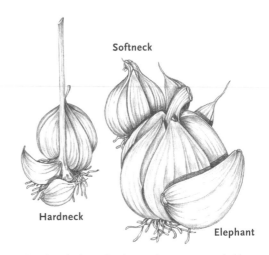

Softneck

Hardneck

Elephant

Hardneck garlic has a hard central core surrounded by one or two layers of cloves. Softneck garlic has soft neck tissue that works well for braiding. Elephant garlic produces bulbs as large as a softball.

In the South, fall-planted garlic will produce leaves shortly after planting and may continue growing slowly throughout winter. Simply plant cloves 1 to 2 inches deep and 4 to 8 inches apart in a prepared bed. Garlic can tolerate light frost, but if hard frost is predicted, cover the plants with a medium-weight row cover overnight and remove it when temperatures reach a safe range.

In areas with very mild winters, garlic may not grow well if it doesn't receive enough exposure to cold (below about 40°F). You can try a preplanting cold treatment by putting the separated cloves in some potting mix in a resealable plastic bag and storing the bag in the refrigerator. When the cloves begin to sprout, plant them outdoors in the garden. Another option is to grow elephant garlic (which is more closely related to leeks). Elephant garlic produces big, mild-flavored bulbs (so space cloves about 10 inches apart); however, the bulbs last only a few months in storage.

Northern gardeners plant garlic in fall, too, but they need to protect the cloves from frost heaving and severe cold. Plant cloves 2 to 4

inches below the soil surface. Once the soil freezes, insulate the bed with up to 6 inches of dry (not matted) grass clippings, straw, or shredded leaves. In spring, foliage grows quickly and then bulbs develop. Harvest starts as early as May in the South and West; July or August is harvesttime in the North.

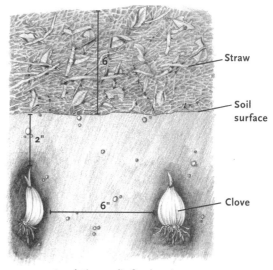

Straw

Soil surface

Clove

Insulating garlic for the winter

won't produce bulbs of any great size, but you can snip the leaves periodically to add to salads or stir-fries.

HARVESTING CUES: Garlic leaves turn brown as the bulbs mature. When about half the foliage has changed color, the bulbs are ready to harvest as described on page 214.

SECRETS OF SUCCESS

START WITH SMALL QUANTITIES. One bulb of garlic contains 6 to 15 or more cloves, depend-

ing on the variety. Each clove you plant yields one bulb of garlic. Thus, if you want to, you can increase the amount of garlic you plant each year just by setting aside a few of the bulbs you harvest in summer to replant in fall. Rather than buying several bulbs of one variety, try to find a local supplier and buy one bulb each of a few varieties (or find friends with whom to split an order from a mail-order supplier).

In general, hardneck garlic grows better in cold-winter areas and softneck garlic does well in hot areas and areas with mild winters. Elephant garlic grows well in all regions. If you'd like garlic to braid, there's no harm in experimenting with planting a small amount of softneck garlic in the North. Don't just plant a bulb from the grocery store, though. Order softneck garlic grown especially for planting.

With both hardneck and softneck garlic, pick out the largest cloves for planting because there's a direct correlation between clove size and the size of the bulb that results. Plant the small cloves in a garden bed or container for growing fresh garlic greens (you can take several cuttings of greens from plants when you're not worried about harvesting the bulb).

HARVEST RIGHT FOR LONG STORAGE. Yanking garlic out of the ground by the tops isn't a good idea. Use a spade to cut the roots and lift the bulbs from the ground instead. Insert the blade straight into the soil about 6 inches deep at a spot several inches away from the row of bulbs. Then press the tool handle toward the ground, which will cause the blade to pivot under the bulbs and loosen them. Gently shake or brush off excess soil. Handle the bulbs carefully, as any injury to the wrapper leaves will shorten the storage life.

If you live in an area with consistently dry, warm summer weather, you can move the bulbs to an airy, covered location, such as an open shed or covered patio, to dry. Garlic bulbs sunburn easily; drying them directly in the garden isn't recommended. In areas with humid summers, you may need to dry the bulbs in a spot where you can run a fan to circulate air. Brush the bulbs again after they're dry to remove any remaining soil or loose outer skin. Trim the roots and store your harvest in a dry spot at room temperature.

REGIONAL NOTES
WESTERN STATES

White rot is a soilborne fungal disease that can infect garlic and other onion-family crops anywhere, but the highest risk is in areas such as California, where large commercial crops are grown. The disease starts with yellowing and wilting of leaves, and then roots rot and plants die. Fluffy white mold appears at the base of the bulb and then spreads to destroy the whole bulb. Small, dark, resting bodies (sclerotia) may form in the mold. These sclerotia can last in the soil 10 years, possibly even longer.

Since white rot fungus can travel in and on garlic bulbs, there's a risk that it will invade your garden if you plant infected bulbs. There is no certification program for garlic bulbs. The best way to avoid white rot is to buy your bulbs from a local grower with a good reputation. Reject any bulbs that show white mold or black dots on the bulbs.

If white rot shows up in your garden, destroy the infected plants, and avoid planting any other onion-family crops in that bed. Be careful not to move soil from the infected bed to other parts of the garden. Solarizing the soil will kill off some of the fungi but won't wipe them out completely.

PREVENTING PROBLEMS

BEFORE PLANTING

* Enrich the soil with compost and build up raised beds if needed to avoid drainage problems for garlic.

AT PLANTING TIME

* In the South, apply a surface mulch between crop rows to suppress weeds.

WHILE CROP DEVELOPS

* In cold-winter areas, cover the bed with organic mulch after the soil freezes to protect the sprouted cloves below the soil surface.
* Be on the alert for early signs of thrips damage, and spray plants with water to wash off the pests.

TROUBLESHOOTING PROBLEMS

Garlic is often nearly trouble-free if you plant it in well-drained soil. The most common problems are listed below. If your garlic is suffering symptoms not listed here, refer to Onions for a more complete list of symptoms and causes.

YOUNG LEAVES DIE IN SPRING. Exposure to hard frost can cause this, and the plants will usually regrow from the bulb. To prevent the damage, cover plants with medium-weight row cover overnight when temperatures will dip below the mid-20s.

SILVERY STREAKS ON LEAVES. Thrips damage plant cells as they feed, leading to the appearance of streaks. To confirm thrips, hold a piece of white paper near the leaves and tap the plants. The thrips will show up as small dots on the paper. Spray infested plants with water. For other controls, see Thrips.

STUNTED PLANTS. Nematode feeding results in poor growth; infected plants often rot. See Nematodes for controls.

BULB OR NECK OF PLANT ROTS. Fusarium or other fungi are to blame, but the underlying cause is usually poorly drained soil. To prevent future problems, see Rots.

GOURDS

Gourds are part of the squash family, and the vigorous annual vines are prone to the same pests and diseases that cause problems for winter squash and pumpkins. Gourds include luffas, bottle gourds, and a delightful and colorful array of decorative winged and warty gourds. Some Asian gourds are good for eating, too. For growing tips and problem-solving information, see Squash.

GRASSHOPPERS

Grasshoppers are pests of late summer and fall, sometimes arriving in a sweeping swarm capable of ravaging a garden. Grasshoppers inhabit all parts of the United States and Canada, but most species are fairly harmless; some even eat other grasshoppers. A handful of species, though, can reach damaging levels in the middle section of the continent. Populations rise and fall from year to year, influenced by weather conditions, populations of predators and parasites, and other factors.

While it's hard to predict precisely when a bad grasshopper season will occur, there are some clues you can watch for. For example, if summer has been long and hot, grasshoppers will have plenty of time to feed and lay eggs, which could mean that the following year will be a bad grasshopper year. On the other hand, if spring is damp and cool, grasshopper populations may be low that summer if disease organisms kill off a large proportion of the nymphs.

Grasshoppers lay eggs in the soil in concentrated breeding grounds, which usually aren't found in home gardens. Grasshoppers like to lay eggs in undisturbed areas with compacted soil, such as along roadsides or in empty lots. The eggs overwinter in the soil and hatch in late spring. Wingless nymphs feed in the vicinity of the breeding ground for about 2 months until they reach adulthood. Adult grasshoppers can fly some distance in search of food, and they will feed and lay eggs until a hard frost kills them.

There are no organic pest-control products that kill adult grasshoppers, and that means your strategy for fighting this pest has to be preventive. On a home garden scale, this isn't easy to do. Check below for some creative strategies to try, but no matter what, be sure you stock a supply of screening or row cover on hand to protect your crops for those bad years when a grasshopper plague occurs.

PEST PROFILE

Many genera and species; *Melanoplus* spp. are some of the common garden pest grasshoppers.

RANGE: Throughout the United States and Canada; problematic only in an area that roughly spans from Montana to Minnesota and from Texas to New Mexico; rarely, grasshopper populations spike in arid desert areas, particularly after a heavy summer rain.

HOW TO SPOT THEM: Adult grasshoppers are easy to see, especially because they will hop vigorously when you disturb them in your garden. If you find wingless nymphs in your garden, too, it means there must be a breeding ground not far from your vegetable garden.

ADULTS: Yellow, gray, or brown insects, up to 2 inches long, with long wings and large hind legs. Some pest species have black markings on the outer side of their hind legs.

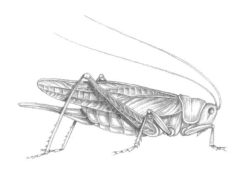

Grasshopper

Grasshopper nymph

LARVAE: Resemble adults but are smaller and wingless, with small wing pads on the back

EGGS: Laid in clusters in the soil

CROPS AT RISK: Grasshoppers love corn, carrots, lettuce, and onions, and seem to dislike tomato and squash vines. However, when a large swarm invades a garden, they will eat nearly any vegetable crop.

DAMAGE: Grasshoppers chew on foliage; large infestations can demolish a garden.

CONTROL METHODS

COVERING PLANTS. Cover as much of your garden as possible. Fine wire mesh is preferable to row covers because grasshoppers can chew through row cover fabric, but row cover is better than nothing. You don't need to seal the edges of row covers tightly, but do secure the corners with rocks, plastic bottles filled with water, or other weights.

HANDPICKING. Diligent handpicking is one reliable but time-consuming way to prevent drastic grasshopper damage. Collect grasshoppers daily and dump them into soapy water to kill them.

BIRDS. If you keep chickens, ducks, or turkeys, allow them to run in a confined area around the borders of your vegetable garden to eat the grasshoppers. Don't let the birds into your garden, though, or they may eat many of your plants, and their scratching can ruin a garden. Wild birds also eat grasshoppers, but on their own, they won't eat enough to protect your garden sufficiently.

MOWING GRASS SHORT. If your vegetable garden is surrounded by open lawn, keep the grass mowed short during late summer and fall to make it less attractive as a food source for grasshoppers. If they bypass your lawn, they may miss your vegetable garden, too. This practice will put stress on your lawn, so be prepared afterward to take steps to restore its condition.

FALL CULTIVATION. Till any areas that serve as grasshopper breeding grounds in fall to expose grasshopper eggs to predators and cold.

BIOCONTROLS. If you find the breeding grounds of grasshoppers near your yard, try applying *Nosema locustae*, a protozoan disease that infects the nymphs. *Nosema* is slow-acting, and it's most effective when applied on a broad scale as part of a long-term program for reducing grasshopper populations. It is not helpful for controlling an outbreak of adult grasshoppers in a garden. Another option is to treat the nymphs with *Beauveria bassiana*, again, as part of a large-scale program to control grasshoppers. If the breeding grounds are not on your property, be sure to ask the property owner's permission before you take any action.

CONTROL CALENDAR

BEFORE PLANTING

- Scout for grasshopper breeding grounds on and near your property, and if feasible (and allowed), treat young nymphs with biological controls.

WHILE CROP DEVELOPS

- In late summer, cover susceptible plants with fine wire mesh or row cover.
- Confine chickens, turkeys, or ducks in the area around (but not in) your vegetable garden to eat grasshoppers.
- Handpick grasshoppers, daily if possible.
- Keep lawn areas around vegetable gardens mowed short.

AFTER HARVEST

- Till grasshopper breeding grounds to expose eggs.

GRAY MOLD (Botrytis)

Gray mold is the fuzzy brownish or gray coating that sometimes appears on vegetable crop leaves, flowers, and fruits in cool to mild, wet weather. The fungi that cause gray mold are Botrytis fungi; Botrytis blight is another common name for gray mold.

Botrytis fungi aren't all bad. They're one of the common types of fungi that help break down dead plant material and release organic matter back into the soil. Botrytis becomes problematic when it attacks dead or dying plant parts—such as wilted flowers or old leaves—that are still attached to living crop plants, and then spreads to healthy leaves, stems, and fruits.

A wide variety of vegetables are susceptible to gray mold, and there's no way to eradicate Botrytis fungi from your garden (it can also be a problem in home greenhouses). The best approach is to avoid creating the wet, debris-rich environments in which the disease breaks out and spreads, and also to keep your crops' natural resistance high by avoiding stress such as lack of water and by using plant health boosters.

RANGE: United Stated and Canada; worst in the Northwest and Northeast

CROPS AT RISK: Asparagus, beans, lettuce, tomatoes, and many other crops

DESCRIPTION: The earliest symptom of gray mold is water-soaked or tan spots on leaves, stems, and flowers. In cool, moist weather, gray or brown mold grows quickly on these spots, and the mold may be the first symptom that you notice. The spots sometimes enlarge and merge to cover entire leaves. Lettuce heads may disappear under a moldy coating, and the interior becomes mushy. If the weather turns dry after infection, the heads may undergo a dry, dark decay and split off the main stem. Asparagus spears and artichoke flower buds are also potential victims of gray mold. Stems of tomato-family plants may develop elongated spots covered with brown mold. Infected tomato fruits turn brown and soft or develop small whitish areas surrounded by a yellow halo. Mold can also form on squashes, melons, and bean pods, and the fruits and pods may shrivel and rot.

Seedlings infected by Botrytis fungi show classic damping-off symptoms; see Damping-Off for details.

FIGHTING INFECTION: Pick off infected leaves, flowers, and fruits and remove them from the garden so they can't release spores and spread the disease. If you uproot whole plants, cultivate the soil lightly in that spot to encourage it to dry out. Switch from sprinkler irrigation to drip irrigation. Prune and stake plants to improve air circulation (but only when plants are dry). Prune off leaves that touch or are close to the soil. Rake up wet surface mulch (bury it in a compost pile because it may carry Botrytis), and let the soil surface dry out.

GARDEN CLEANUP: Since it's impossible to wipe out Botrytis, your cleanup efforts should be aimed at eliminating the sites from which Botrytis fungi spread to your crops. Don't dig in plant debris unless you're sure it will break down completely before you plant your next crop in that bed; instead, remove all crop debris and compost it. (If you suspect that the debris harbors Botrytis fungi, put it at the center of the pile.) Clean out weedy areas in or near your garden. If your vegetable garden includes flowers (such as sunflowers, marigolds, cosmos, or other plants to attract beneficials), deadhead frequently because Botrytis easily invades dying flower petals. Discard or destroy any fruits or other plant debris that are seriously infected with gray mold.

NEXT TIME YOU PLANT: Avoid planting susceptible crops in heavy soil or poorly drained spots; if possible, plant in raised beds. Add mature, screened compost to soil to build up populations of beneficial microorganisms that fight against Botrytis fungi (screening is important because partially decomposed chunks of organic material can be a breeding ground for Botrytis). Avoid overfertilizing with nitrogen because it leads to succulent growth that's susceptible to infection. Thin seedlings promptly to encourage good air circulation. Spray plants with compost tea every 2 weeks in spring or fall when conditions are favorable for gray mold to break out (rainy or high humidity and about 60°F). Prop up squashes and melons on a can or other support so they don't contact the soil surface. If Botrytis has been a problem in your garden in the past, apply a potassium bicarbonate spray when conditions are favorable for the disease.

CROP ROTATION: Not helpful

HARLEQUIN BUGS

Heat brings harlequin bugs to a gardener's attention. While these little bugs can be active in southern gardens year-round, they're usually most damaging when the weather turns warm. Adult harlequin bugs feed by sucking sap from plants, which causes plants to wilt and turn brown. Seriously infested plants may die.

Cabbage-family crops are the prime targets of harlequin bugs, although they sometimes feed on a variety of other vegetable crops. Cabbage-family plants are particularly prone to damage when hot weather stresses the plants. If you notice a spike in harlequin bug activity in your garden, it's a signal to pull your cool-season crops and focus on warm-season plants.

Adults also lay eggs on host plants, and the eggs hatch in anywhere from a few days to a month, depending on temperature. Nymphs feed for 1 to 2 months before reaching the adult stage. In the northern part of their range, harlequin bugs cycle through two to four generations before adults take shelter for winter in crop debris or weedy patches.

PEST PROFILE

Murgantia histrionica

RANGE: Most parts of the United States, but problematic only in the southern half of the country

HOW TO SPOT THEM: Harlequin bug eggs are distinctive. Check leaf undersides of cabbage-family crops to find these black-and-white "barrels." Splotchy areas on leaves or stems may be early signs of feeding damage.

ADULTS: Shield-shaped, shiny, flat, red-and-black or orange-and-black bugs, about ⅜ inch long

NYMPHS: Wingless oval insects that hatch nearly colorless; as they grow, they develop the characteristic red-and-black markings of adults.

EGGS: Barrel-shaped white eggs with black bands, laid in double rows on leaf undersides

Harlequin bug and eggs

Harlequin bug nymph

CROPS AT RISK: Arugula, broccoli, cauliflower, Brussels sprouts, cabbage, cauliflower, collards, horseradish, kale, kohlrabi, mustard, radishes, turnips, squash, corn, beans, peas, asparagus, okra, tomatoes, eggplant, beets

DAMAGE: Yellow and white splotches around feeding sites; wilting; foliage turns brown; plants may die.

CONTROL METHODS

CHOICE OF VARIETIES. Some varieties of cabbage, cauliflower, collards, and radishes are tolerant of harlequin bug damage. Early-maturing varieties are also a good choice; they are more likely to mature before serious damage occurs.

TIMED PLANTING. In the South, plant cabbage-family crops as soon as possible in late winter or early spring to allow for harvest before the weather turns hot. Or plant these crops for fall harvest, when harlequin bug damage usually isn't as severe.

HANDPICKING. In the morning, when bugs are still sluggish, knock them off plants into a can of soapy water. Check leaf undersides for eggs and crush them.

GARDEN CLEANUP. Cabbage-family crop residues, which can harbor harlequin bugs, often are large and woody. Break up the debris before incorporating it into the soil. If you want to compost it, try shredding it first. Uproot cabbage-family weeds around your yard and garden.

TRAP CROPS. Try planting a bed of mustard a week or two before the rest of your spring cabbage-family crops as a trap crop to attract harlequin bugs. Or plant mustard or turnips in late summer ahead of your fall crops. Watch the trap crop carefully and kill off the insects before they migrate to your other crops.

SPRAYS. As a last resort, spray nymphs with insecticidal soap or horticultural oil. See Pest Control for more information.

CONTROL CALENDAR

BEFORE PLANTING

- Select tolerant and early-maturing varieties.
- Plant a trap crop a week or two before planting your main crop.

AT PLANTING TIME

- Plant early in spring and/or plant cabbage-family crops in fall.

WHILE CROP DEVELOPS

- Handpick adults, nymphs, and eggs.
- Destroy harlequin bugs on trap crops before they migrate to desired crops.
- As a last resort, spray insecticidal soap or oil.

AFTER HARVEST

- Turn under or compost crop residues.

HORSERADISH

Planted in its own special space, horseradish thrives problem-free in vegetable gardens from Zone 3 through Zone 8. In Zones 9 and 10, horseradish may not succeed because the roots aren't good for harvesting while the tops are actively growing. Cold weather sends the plants into dormancy and improves root quality—the roots are best after frost.

If you're starting a horseradish patch for the first time, be smart and start small! Planting three or four roots is plenty. Horseradish is a vigorous perennial. If you have a passion for horseradish, it will be easy to increase the size of your patch over time. But if you decide you want to get rid of the horseradish you've planted, it takes persistence and lots of digging.

Although horseradish is a perennial, you'll dig up the roots each year (possibly every other year) to harvest large roots and replant small ones. It's wise to replant in the same spot each time. If you move horseradish from place to place in the garden, you'll probably end up with horseradish everywhere because even small pieces of root left behind will sprout and grow.

TIMING OF PLANTING: In the North, plant in early spring or in fall. In the South, plant any time in spring or in fall.

PLANTING METHOD: Plant root cuttings or divisions directly in the garden.

CRUCIAL CARE: Weed diligently until the plants become established. After that, horseradish grows so vigorously that it outcompetes weeds. If growth seems slow, feed plants with a liquid fish fertilizer and liquid kelp to boost growth.

GROWING IN CONTAINERS: It's unusual to grow horseradish in a container, except for a bottomless container sunk into a garden bed.

HARVESTING CUES: The only time to avoid harvesting horseradish is when the tops are actively growing.

CROP BASICS

FAMILY: Cabbage family; relatives are broccoli, Brussels sprouts, cabbage, kale, kohlrabi, radishes, and turnips.

SITE: Plant horseradish in full sun to light shade.

SOIL: Horseradish will grow well in average garden soil as long as it's well drained and loose at least 12 inches deep. In rich soil, horseradish will grow abundantly.

SECRETS OF SUCCESS

TRIM TWICE FOR NICER ROOTS. Horseradish roots tend to be rough and "hairy," with lots of side roots. You can improve root quality and encourage smooth roots by trimming side roots twice during the growing season. When horseradish leaves reach 8 to 12 inches tall, scoop away the soil around the crown of each horseradish root to expose some of the root. Use a sharp knife to slice off side roots all around the root, and then push the soil back into place. A few weeks later, repeat the process.

HARVEST TO SUIT YOUR SCHEDULE. You can dig horseradish roots at any time when the leaves are not actively growing (or are just barely sprouting) or the topgrowth has died back. That means gardeners in most areas can harvest in spring or fall, or even through winter from a well-mulched bed.

Fall is the best time to dig horseradish roots if your soil tends to be wet in early spring. One easy approach is to dig the roots, cut off pencil-size side roots, and replant the side roots the same day.

You may prefer to wait until spring to replant, either because you want to concentrate on processing the roots you've harvested or because you want to add compost or amendments to the planting site. In that case, cut side roots, making a slanted cut at the "root" end of the side root (so you'll know which end is which). Store the roots in a pot of sand in a root cellar over winter. In spring, dig a trench in the soil about 5 inches deep and place the roots in the trench on an angle, with the root end lower than the stem end. Cover the roots so the stem end is about 2 inches below the soil surface.

It's said that horseradish will be hottest if you wait to harvest it in spring, just as the topgrowth starts to sprout. If that's your choice, you'll need to replant the side roots the same day (or within a few days).

PREVENTING PROBLEMS

BEFORE PLANTING

- Loosen the soil at least 12 inches deep.

WHILE CROP DEVELOPS

- Mulch or cultivate to prevent weeds from becoming established in the horseradish bed.
- Water during droughty periods to prevent water stress.

AT HARVEST TIME

- Dig a trench 1 foot deep alongside the row. Working from there, it will be easier to pull and lever the roots free without damaging or breaking them.

AFTER HARVEST

- As needed, refresh the soil with compost and organic matter before replanting.

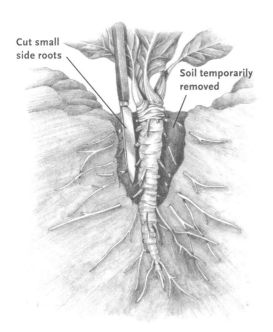

Cut small side roots

Soil temporarily removed

For long, smooth horseradish roots, cut off some of the side roots during the growing season.

TROUBLESHOOTING PROBLEMS

As a member of the cabbage family, horseradish is vulnerable to many of the same pests and diseases that trouble cabbage. In general, horseradish suffers few problems; the most common are described here. If you see symptoms that aren't listed below, refer to Cabbage for more symptom descriptions and causes.

SMALL HOLES IN LEAVES. Flea beetles will feed on horseradish foliage but rarely cause significant damage. See Flea Beetles for controls if needed.

UNWANTED SPROUTS APPEAR. If you decide to stop growing horseradish or move it to a different part of the garden, it's important to dig out the roots as thoroughly as possible. Even small pieces of root can resprout. If horseradish stubbornly hangs on, chop back the tops or mow them down repeatedly (don't try to uproot any of it). Eventually you'll starve out the roots. Or use a smothering technique for a full season, as described under "Bindweeds" in the Weeds, on page 441, to starve out the horseradish roots.

WOODY OR HOLLOW ROOTS. Age or excess nitrogen are possible causes. It's natural for horseradish roots to become woody or hollow over time. Prevent this from happening by digging up main roots at least every other year and replanting sections of pencil-size side roots. If you don't need all the roots you've harvested, give them to friends. If you must dispose of old roots, never put them in a compost pile. They may resprout, leading to a horseradish weed problem in your composting area and beyond.

If young main roots are woody, then the problem could be too much nitrogen. Next year, don't fertilize your crop or use fresh grass clippings as a surface mulch. Over time, the soil nitrogen level should return to a better level for horseradish.

ROTTED AREAS ON ROOTS. Fungal diseases can lead to root rot, but it's usually a problem only in poorly drained soil. Dig up and discard the rotted roots. Improve the drainage by building up a raised bed (at least 8 inches high) before replanting, or start a new horseradish bed in an area with well-drained soil.

IMPORTED CABBAGEWORMS

When white butterflies dance through your garden, take a minute to admire their beauty, but then get out your butterfly net. These cabbage butterflies, or cabbage whites, are the adult stage of the imported cabbageworm, a potentially serious pest of cabbage-family crops.

In most areas, the butterflies emerge in mid-spring and search for egg-laying sites on cabbage-family crops and weeds. The eggs hatch 3 to 5 days later, and the small green caterpillars begin chewing on outer leaves. Their feeding can ruin young transplants. Older larvae tunnel into cabbage and lettuce heads. The caterpillars camouflage themselves when resting by aligning their bodies with leaf veins on the underside of leaves.

After feeding for 2 to 3 weeks, the larvae pupate, making a green chrysalis on or near plants. The chrysalis turns brown as it matures, and about 2 weeks later, the cycle starts again. In some southern states, the cycle continues all or nearly all year, with up to eight generations. In the North, two or three overlapping generations occur each growing season.

This pest is similar to the cabbage looper (see Cabbage Loopers), and controls are similar as well. It's best to be on the lookout for both, as one may be a problem in your garden this year, but the other may be the star pest of the future.

PEST PROFILE

Pieris rapae
RANGE: United States and Canada

HOW TO SPOT THEM: The white butterflies are easy to see. To find larvae, look carefully on leaf undersides, especially toward the base of wrapper leaves. Feeding along leaf margins and green frass on leaves are signs that cabbageworms are hiding somewhere in the plants.

ADULTS: White butterflies with 1½-inch wingspan; females have two black spots on the

Cabbage butterfly

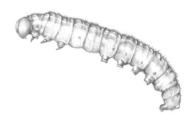

Imported cabbageworm

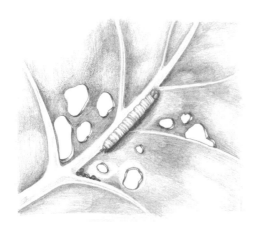

Imported cabbageworm feeding damage

forewing, males have one spot; the underside of the wings is yellowish.

LARVAE: Green caterpillars with a velvety coating and a narrow yellow stripe down the back; three pairs of slim front legs and five pairs of stumpy legs behind; up to 1¼ inches long

EGGS: White, bullet-shaped, turning yellow or yellow-orange as they mature; usually laid singly on the underside of leaves

CROPS AT RISK: Asian cabbages, arugula, broccoli, Brussels sprouts, cabbage, cauliflower, collards, horseradish, kale, kohlrabi, mustard, radishes, rutabagas, turnips, lettuce

DAMAGE: Feeding on leaves; green frass contaminates and stains plants; larvae tunnel into cabbage and lettuce heads.

CONTROL METHODS

ENCOURAGING BENEFICIALS. Many kinds of beneficial insects, including parasitic wasps, help provide natural control of imported cabbageworms. Plant small-flowered annuals in or near your vegetable garden to attract them (see Beneficial Insects for suggestions).

CHOICE OF VARIETIES. If cabbageworms are a regular problem in your garden, try growing red or purple cabbage so you can spot the worms more easily. Avoid savoy-type cabbages because the crinkled leaves make worms even more difficult to spot.

TIMED PLANTING. In the North, planting as early as possible allows crops the maximum time to grow before larvae hatch and begin feeding.

ROW COVERS. Cover plants at planting time and leave the covers in place until harvest if possible to prevent butterflies from laying eggs on the plants (seal row cover edges as described in Row Covers).

GARDEN CLEANUP. In spring, scout for cabbage-family weeds and uproot them to eliminate possible egg-laying sites. As you harvest, remove all wrapper leaves or side leaves from the garden and put them at the center of a compost pile. In fall, dig or till in crop residues to bury the pupae. (This will kill the pupae or butterflies as they try to emerge.)

HANDPICKING. Crush any eggs you spot on leaves. Pick off cabbageworms and dump them in soapy water. Catch cabbage butterflies in a butterfly net.

GARLIC OR HOT PEPPER SPRAY. At about the time when cabbage butterflies appear in your area, spray plants with a garlic or hot pepper spray. Repeat sprays weekly until plants are close to harvest (see page 308 for directions).

BIOCONTROLS. Apply spinosad or *Bacillus thuringiensis* var. *kurstaki* when small cabbageworms are present. (See Pest Control for information on these sprays.) Biological control sprays are less effective against large larvae because they tend to hide and feed near the center of plants. If you're applying Bt in fall, wait for a warm day, when larvae are more likely to be feeding.

SOAP AND NEEM. Insecticidal soap sprays and neem will kill imported cabbageworms. In most cases, though, biocontrol sprays will do a better job of bringing an imported cabbageworm problem under control.

CONTROL CALENDAR

BEFORE PLANTING

- Uproot cabbage-family weeds that can serve as egg-laying sites.

AT PLANTING TIME

- In the North, plant as early as possible so crop is well established before larvae hatch.
- Cover plants with row cover.

WHILE CROP DEVELOPS

- Handpick cabbageworms and crush eggs.
- Spray garlic and hot pepper sprays to repel butterflies and larvae.
- Spray *Bt* var. *kurstaki* or spinosad to kill young larvae.
- If you're caught off guard and don't have biocontrol sprays at hand, use insecticidal soap or neem to stop imported cabbageworms from getting out of control.

AT HARVEST TIME

- If cabbageworms have been active, cut cabbage-family crops and head lettuce on the young side to minimize amount of damage.
- Remove all wrapper leaves and cuttings from the garden.

AFTER HARVEST

- Dig or till in crop residues to bury pupae.

JAPANESE BEETLES

Rose growers curse Japanese beetles, but in vegetable gardens they're usually an annoyance rather than a crisis. These familiar metallic beetles show up in early summer, congregating in groups at the tops of plants to chew the tender tissue between leaf veins. The first beetles to arrive on a host plant release a pheromone that attracts more beetles to the site. Damaged plants also release natural chemicals that draw even more beetles. When a large group attacks a plant, they strip foliage quickly, leaving behind leaf skeletons.

Over a period of several weeks, female beetles fly to the soil to lay eggs, a few at a time, then return to host plants to feed and mate again. Each female will mate and lay eggs many times before the beetles die out in late summer and early fall. Meanwhile, the larvae hatch as white grubs in the soil, where they feed on plant roots. They can cause considerable damage to lawn grasses, a preferred food, but generally aren't significant in vegetable gardens. The grubs overwinter below the frost line in the soil and then move up again in spring to feed awhile and pupate. There is one generation per year, or in some northern areas, the beetle may take 2 years to complete its life cycle.

Naturally occurring predators and parasites have some effect on Japanese beetles but not enough to save a crop if the beetles show up in large numbers. The beetles are easy to spot and to handpick, and that's probably your best bet for controlling them. There are biological controls for the grubs, which can reduce problems with Japanese beetles over the long term, but only when they're applied on a community-wide basis.

PEST PROFILE

Popillia japonica

RANGE: Eastern United States and Canada; pockets of beetles occur as far west as Utah; pockets also show up along the West Coast.

HOW TO SPOT THEM: Start scouting in early summer, especially on beans and corn. The beetles are distinctive looking and feed on the topside of leaves—they're easily seen.

ADULTS: ⅜-inch, metallic green beetles with copper-brown wing covers, five tufts of white hair along each side of the body, and a pair of tufts at the tip of the abdomen

LARVAE: Soil-dwelling white grubs up to 1¼ inches long, with dark heads; the grubs curl in a C shape when at rest.

Japanese beetle

Japanese beetle larva

EGGS: Laid in the soil

CROPS AT RISK: Beans, corn, rhubarb, and sometimes other vegetable crops

DAMAGE: Larvae feed on plant roots but prefer grasses to other types of plants. An infestation of larvae may be a problem for sweet corn, but generally it's the beetles that cause damage in vegetable gardens. Beetles chew leaf tissue between veins, reducing leaves to "skeletons" that shrivel and die. Beetles also feed on corn silks, which can interfere with pollination.

CONTROL METHODS

ROW COVERS. Cover susceptible crops with row cover early in the growing season, before the beetles become active in your area. Seal edges of covers as described in Row Covers. Remove covers in late summer.

KAOLIN. To discourage Japanese beetles from feeding, spray susceptible crops just before beetles become active in your area; repeat sprays as needed. See Pest Control for more information on kaolin.

HANDPICKING. The best time to handpick Japanese beetles is early in the morning, when they are sluggish and won't try to fly away. Pick them off plants individually, or shake the foliage to prompt the beetles to drop to the ground (where you've spread a sheet to catch them). Dump the beetles in soapy water.

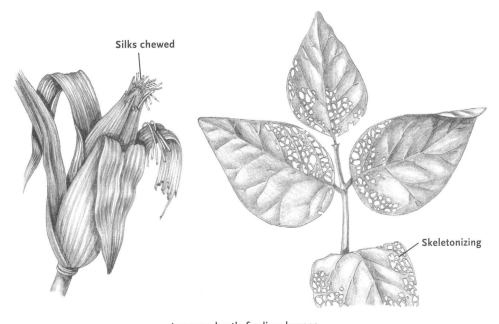

Silks chewed

Skeletonizing

Japanese beetle feeding damage

BIOCONTROLS. Milky spore is a bacterial disease that infects and kills Japanese beetle grubs. The disease is available as a powder that contains the bacterial spores. There are different formulations, and it's important to follow application instructions carefully. Some products must be applied multiple times. Using milky spore is a long-term strategy to reduce grub damage to your lawn by establishing a permanent population of milky spore bacteria in your soil. If milky spore is being applied on a community-wide basis, you should notice a drop in the number of adult beetles. Milky spore disease has been most effective in the southern half of the United States, where it persists longer in the soil.

Heterorhabditis nematodes are also effective for controlling white grubs; see Nematodes for information on this biocontrol. Again, applying nematodes may not help with protecting a vegetable garden against adult beetles, but it is a positive step you can take to fight against the overall problem these pests pose.

INSECT TRAPS. Commercial Japanese beetle traps that lure the beetles using pheromone and floral lures are very popular, but their effectiveness is disputed. Studies show that the traps attract more beetles than they catch, thus potentially worsening a Japanese beetle problem. The traps may be useful for diverting the beetles from a particular part of your yard, such as your vegetable garden. To do this, set up traps upwind of the garden on two sides, at least 30 feet from the perimeter. Follow the directions that come with the trap regarding specific placement. The traps will lure beetles to that part of your yard, where some will die in the traps and others will feed—but with luck, they won't find your garden. Keep in mind that overall you may be increasing the number of eggs laid in your lawn by using these traps. Some experts say that the best strategy is to hope that your neighbors will put up Japanese beetle traps and draw the beetles away from your yard.

SPRAYS. Apply neem to deter beetles from feeding; neem will probably not be sufficient to remedy a heavy infestation. As a last resort, insecticidal soap or pyrethrins will kill Japanese beetles, but only if the spray contacts the beetles directly. See Pest Control for more information.

CONTROL CALENDAR

AT PLANTING TIME

* Cover crops with row cover just before adult beetles become active in your area.

WHILE CROP DEVELOPS

* Coat susceptible crops with kaolin clay just before adult beetles become active in your area.
* Handpick beetles as often as possible.
* Apply milky spore disease at the appropriate times (which depends on the formulation you choose to apply).
* Experiment with Japanese beetle traps as a method of luring beetles away from your vegetable garden.
* Apply neem to deter beetles from feeding.
* For a heavy infestation of beetles, apply insecticidal soap or pyrethrins.

AFTER HARVEST

* Apply parasitic nematodes in late summer to infect white grubs in soil.

KALE

For tasty cooking greens in every season, plant both kale and collards. These closely related crops are in the cabbage family, but they're easier to grow than cabbage, broccoli, and cauliflower. Kale is so cold-hardy that you can dig down through a covering of snow to harvest it. Southern gardeners appreciate collards because they withstand heat yet can also tolerate light to moderate freezes. In most areas, you can plant a spring crop and follow up with a fall crop that will continue producing over a long period, especially if you provide protection from the coldest weather.

CROP BASICS

FAMILY: Cabbage family; relatives are arugula, broccoli, broccoli raab, Brussels sprouts, cabbage, cauliflower, horseradish, kohlrabi, mustard, radishes, turnips.

SITE: Kale and collards need full sun. For fall and winter crops, choose a spot that's protected from wind if possible. Try to plant fall kale where tall crops will shade it until summer heat abates.

SOIL: Kale and collards grow best in soil that's been enriched with organic matter over time. If soil quality is poor, spread 1 to 2 inches of compost over the bed and work it in before planting.

TIMING OF PLANTING: Start seeds indoors 10 to 12 weeks before the last spring frost. Move transplants to the garden or sow seeds outdoors 4 weeks before the last frost. For a fall crop, sow seeds 10 to 12 weeks before the first fall frost.

PLANTING METHOD: Start seedlings indoors or sow seed directly in garden beds.

CRUCIAL CARE: Keep the soil consistently moist but not too wet. Apply a thick layer of organic mulch to cool the soil during hot summer conditions. To extend the harvest as long as possible, cover collards with row covers when temperatures drop below 20°F.

GROWING IN CONTAINERS: Kale and collards will grow well in compost-enriched potting mix in containers at least 8 inches deep. Keep the plants on the small side by harvesting from the growing tip rather than side leaves.

HARVESTING CUES: You can begin harvesting collard and kale leaves at any size. On large plants, don't bother with lower leaves that have started to turn yellow—they are past prime. Cut them off and compost them. Cool weather and frost improve the flavor of both collards and kale.

SECRETS OF SUCCESS

HOT-WEATHER STRATEGY. If you're eager to start your fall crop of kale but summer heat is dragging on, sow kale seeds in containers and place them in a partially shaded spot outdoors, such as on a patio with a lattice roof or under a shade tree. Keep the soil moist but not saturated, and check the plants frequently for aphids and leaf-eating caterpillars. The seedlings will be ready for the garden in about 6 weeks, by which time the worst heat should be over.

With collards, you can follow the same strategy or simply leave your spring plants in the garden throughout summer. Keep them watered and mulch the soil to cool it. Once the weather cools in late summer or early fall, start harvesting leaves again as desired.

SUCCESSION THINNING FOR BIGGER HARVEST. The standard recommendation for kale and collards is to sow seeds 6 inches apart and thin the plants to 12 inches apart; large varieties need even wider spacing (mature collard plants can grow to 3 feet across). To get the most from your planting along the way, start by thinning plants to 6 inches

apart. Let these seedlings grow until the leaves of adjacent plants start to touch. Then, for kale, cut every other plant in the row at ground level, and enjoy your early harvest. The remaining plants will continue to grow, and you can harvest individual leaves from them as needed.

With collards, thin two out of every three plants at this stage, leaving plants 18 inches apart. That allows room for the collards to expand to their impressive mature height and spread.

PREVENTING PROBLEMS

BEFORE PLANTING

- In spring, plant small-flowered annuals and perennials to attract aphid predators to your garden. Plan for staggered bloom times throughout the growing season.
- Set up commercial insect traps to lure and trap cabbage looper moths.

AT PLANTING TIME

- Put a cutworm collar around each transplant in spring.
- Cover spring transplants with row covers and seal the edges to keep out pests (see Row Covers).
- If you're not using row covers, put a paper or foam barrier around the stem or cover each plant with a screen cone to prevent cabbage maggot flies from laying eggs at the base of the plant.
- For summer plantings, water well and mulch the soil immediately to prevent overheating.

Enjoy the full variety of colors and textures that kale can add to the vegetable garden. Collards look more like oversize cabbage transplants, but they're as tasty as kale and more heat resistant, too.

WHILE CROP DEVELOPS

- Check fall crops frequently for aphids, and spray infested plants with a strong stream of water.
- Handpick eggs and larvae of leaf-eating caterpillars frequently.
- For plants that aren't covered, spray a garlic or hot pepper spray to deter aphids and other pests (but not on leafy crops that are close to harvest).

AFTER HARVEST

- Clear out crop debris, including roots. Compost healthy residues. Destroy or discard noticeably diseased residues.

TROUBLESHOOTING PROBLEMS

Kale and collards suffer from few pest problems, especially when planted in a garden where other cabbage-family crops are growing. Pests seem to prefer broccoli and cabbage, with the exception of the diamondback moth, which seems to like collards more than related crops. If you see symptoms or pests on your plants that aren't described below, refer to Cabbage for a complete listing of potential problems.

NEW SEEDLINGS OR TRANSPLANTS EATEN. Cutworms or slugs and snails can damage spring plantings of kale and collards. Cutworms often cut transplants off cleanly near soil level; see Cutworms for preventive measures. Slugs and snails leave behind a slime trail; see Slugs and Snails for controls.

MANY HOLES IN LEAVES. Diamondback moth larvae chew on leaves and can damage growing tips of young plants. Look for very small (about ¼ inch) greenish yellow caterpillars on foliage that wiggle rapidly when disturbed. See page 84 for control information.

LARGE HOLES IN LEAVES. Cabbage loopers, imported cabbageworms, or slugs and snails will chew on kale and collard leaves. Green caterpillars with white stripes on plants are cabbage loopers. Imported cabbageworms are velvety green caterpillars. Handpick them. For more controls, see Cabbage Loopers and Imported Cabbageworms.

Since slugs and snails feed at night, you may not find them on plants, but look for silvery slime trails on the plants or on the ground nearby. See Slugs and Snails for controls.

CURLED OR CUPPED LEAVES. Cabbage aphids suck plant sap. They can build up quickly on kale and collards in late summer, resulting in stunted growth. Check leaf undersides for colonies of pale green aphids (see Aphids for controls).

LEAVES TURNING YELLOW. Downy mildew is a fungal disease that can infect kale or collards in cool, wet weather. Look for a white coating on leaf undersides to confirm the diagnosis (see Downy Mildew).

V-SHAPED YELLOW AREAS ON LEAVES. Black rot is a bacterial disease that spreads easily in hot weather. Uproot and destroy infected plants. See Rots for controls.

PLANTS FORM A FLOWERSTALK. This is a natural reaction of kale and collards after overwintering in the garden. Once a plant starts to flower, it's done producing leaves, so uproot it and add it to your compost pile.

PLANTS COLLAPSE IN COLD CONDITIONS. Kale is extremely hardy, and collards can withstand moderate freezes. However, a sudden hard cold snap, when temperatures drop overnight into the teens, can injure or even kill plants because they haven't had time to toughen up. In the future, provide overnight protection in fall or winter, especially for collards, if the weather forecast predicts a sudden temperature switch.

↬ KOHLRABI

Kohlrabi means "cabbage turnip" in German, but the main part you eat is a bulbous stem, not a root. Kohlrabi has a mild, sweet flavor when it's grown at the right time of year and harvested promptly once the stem enlarges. Kohlrabi leaves are good for eating, too—cook them as you would collards or kale.

Kohlrabi is relatively easy to grow, especially if you pass over the old standard varieties ('Early White Vienna' and 'Early Purple Vienna') and choose newer, fast-maturing varieties that are ready to harvest 6 to 7 weeks after sowing.

CROP BASICS

FAMILY: Cabbage family; relatives are arugula, broccoli, broccoli raab, Brussels sprouts, cabbage, cauliflower, collards, horseradish, kale, mustard, radishes, turnips.

SITE: Kohlrabi needs full sun.

SOIL: Kohlrabi will produce succulent, better-tasting stems when it grows in soil that's been enriched with organic matter over time. If soil quality is poor, spread 1 to 2 inches of compost over the bed and work it in before planting.

TIMING OF PLANTING: Sow seeds (indoors and out) 4 to 6 weeks before last expected spring frost. Transplant started seedlings to the garden when they reach about 4 inches tall. Sow seeds again 8 to 10 weeks before first fall frost. In areas with mild summers, make succession plantings for harvest through summer. In Zones 9 and 10, repeat-sow through fall for winter and early spring harvest.

PLANTING METHOD: Start seedlings indoors or sow seed directly in garden beds.

CRUCIAL CARE: Watering as needed from planting to harvest is the most important task. Check soil moisture twice a week and water whenever the top few inches of soil are dry. In spring, after the soil warms, apply at least 1 inch of organic mulch to maintain even soil moisture. For fall crops, mulch as soon as you transplant or seedlings have emerged. In poor to average soil, fertilize by watering with a nutrient-rich tea when plants are about 1 month old (see "Making Fertilizer Teas" on page 202).

GROWING IN CONTAINERS: Kohlrabi isn't a prolific container crop, but if you want to try it, plant transplants in wide containers at least 8 inches deep. Enrich the potting mix with compost and set plants 4 to 6 inches apart. Water frequently and apply a nutrient-rich tea as needed to maintain steady growth. Harvest bulbs when they reach 2 inches in diameter.

HARVESTING CUES: The most common mistake that gardeners make with kohlrabi is waiting too long to harvest. On the day you plant kohlrabi, count out the expected days to maturity on a calendar and mark the date. Start checking your kohlrabi plants several days before that.

SECRETS OF SUCCESS

TREAT IT LIKE BROCCOLI. Kohlrabi stems need careful tending, just as broccoli flowerheads do. If you leave the harvest too late, the flavor and quality of the stems decline. And kohlrabi is best for eating within a day or two of picking, although it will keep well in the refrigerator. When you sow seeds, stick with small quantities and make succession plantings, or plant small amounts of a few varieties with different maturity dates.

As the crop approaches maturity, check the bulbs frequently and begin harvesting when the bulbs are about the size of a plum. Some varieties remain sweet until they reach the size of a large orange, but until you gain experience, err on the side of early harvesting. In the South, summer heat may turn the leaves bitter.

BEFORE PLANTING

- Set up commercial insect traps to lure and trap cabbage looper moths.

AT PLANTING TIME

- Put a cutworm collar around each transplant in spring (see page 162).
- Cover spring seedbeds and transplants with row covers and seal the edges to keep out pests (see Row Covers).
- If you're not using row covers for transplants, put a paper or foam barrier around the stem or cover each transplant with a screen cone to prevent cabbage maggot flies from laying eggs. Also spray plants with kaolin clay to deter flea beetles (see Pest Control for more information).
- For summer-planted transplants, water well and mulch the soil immediately to prevent overheating.

WHILE CROP DEVELOPS

- Handpick eggs and larvae of leaf-eating caterpillars frequently.
- For plants that aren't covered, spray a garlic or hot pepper spray to deter aphids and other pests.

AT HARVEST TIME

- Remove plants entirely, including the roots. Compost healthy roots and leaves (if you don't want to eat them). Discard or destroy any noticeably diseased roots or leaves.

TROUBLESHOOTING PROBLEMS

Serious pest and disease problems are rare with kohlrabi. The most common problems are listed below, but kohlrabi is potentially susceptible to the other problems that plague the cabbage family. If you see symptoms or pests on your plants that aren't described below, refer to Cabbage for a complete listing of symptoms and causes.

NEW SEEDLINGS OR TRANSPLANTS EATEN. Cutworms or slugs and snails can wipe out spring-planted kohlrabi. Cutworms often cut off transplants cleanly near soil level; see Cutworms for preventive measures. Slugs and snails leave behind a slime trail; see Slugs and Snails for controls.

KOHLRABI PLANTING SCHEMES

To maximize your kohlrabi harvest, sow seeds in short rows 6 to 8 inches apart and thin the seedlings to stand 6 to 8 inches apart in staggered formation. As the plants grow, the leaves will overlap and shade the bulbs, which will promote tenderness, especially for a spring crop that matures as temperatures are on the increase.

Another option is to tuck kohlrabi transplants between widely spaced fall broccoli plants. You'll harvest the kohlrabi before the broccoli plants reach full size.

6"

6"

MANY SMALL HOLES IN LEAVES. Flea beetles chew small holes in leaves, creating a shot-hole appearance. Look closely to spot the tiny black beetles on the foliage. See Flea Beetles for controls.

LARGE HOLES IN LEAVES. Cabbage loopers, imported cabbageworms, or slugs and snails will chew on kohlrabi leaves, which can slow the growth of young plants. Handpick the caterpillars and dump them in soapy water. For more controls, see Cabbage Loopers and Imported Cabbageworms.

Since slugs and snails feed at night, you may not find them on plants. Check for silvery slime trails on the plants or on the ground nearby. See Slugs and Snails for controls.

V-SHAPED YELLOW AREAS ON LEAVES. Black rot causes this symptom. It is a bacterial disease of cabbage-family crops; it spreads easily in hot weather. Uproot and destroy infected plants. See Rots for controls.

BULBS ARE WOODY IN TEXTURE. Kohlrabi doesn't like heat, and stems that mature in hot weather often are woody. If the texture is too unpalatable, compost the bulbs. To avoid this problem in the future, change the timing of planting, starting earlier in spring or later in summer. Light frost is said to improve the flavor.

BULBS HAVE UNPLEASANTLY STRONG FLAVOR. Stress can slow the growth and ruin the flavor of kohlrabi. Water stress is the most common, but nutrient stress is a factor, too. To ensure better flavor next time you grow kohlrabi, water the plants regularly and feed them at least once during the growing season with a liquid fertilizer or nutrient-rich tea.

LATE BLIGHT

The appearance of new strains of late blight in the United States in the past two decades has boosted it to the top of the "worst diseases" list for tomatoes and potatoes. Late blight can ruin a crop in only a few days when nights are cool and moist and days are warm and humid. Unfortunately, once symptoms appear in your garden, there's no reliable way to stop the disease. Because the spores can be spread for miles by moist winds, it's critical to destroy infected plants as soon as possible to prevent this disease from spreading.

A rainy period followed by heat and humidity can also lead to a sudden onset of late blight. If the weather turns dry after symptoms appear, they may stop spreading. But it's a temporary reprieve, and the disease will intensify again as soon as rain returns.

The fungus survives winter in potato tubers, living crop debris, and sometimes as resting spores in the soil. Late blight shows up in a spotty fashion. One year it may devastate your garden but next year not appear at all. Or it may infect tomato plants in your garden but not in your neighbor's garden.

RANGE: United States; southern Canada (humid regions only)

CROPS AT RISK: Tomatoes and potatoes

DESCRIPTION: Leaves develop greasy-looking or water-soaked spots. In moist conditions the spots enlarge rapidly, turning brown or purplish black. Shoots blacken and plants sometimes collapse quickly. White moldy growth sometimes appears. Rotting plant parts give off an unpleasant odor. Potato tubers show purplish or brown corky spots. Tomato fruits show grayish green water-soaked areas. These spots enlarge and turn dark. Often, secondary rot sets in on both potato tubers and tomato fruits, and they turn mushy and foul-smelling.

FIGHTING INFECTION: Check for symptoms at least every other day when the weather is wet. Remove infected plants from the garden immediately. Even if symptoms show only on a few leaves, uproot the whole plant or plants, place the plant material in a plastic bag, and seal it. Remove the bag from the garden and put it with household trash for disposal (or bury the plants deeply in soil well away from your vegetable garden). Afterward, change your clothing and wash your hands before you return to the garden—spores on your clothes or hands could infect other plants. Hill potato plants to stop spores from washing down to the tubers.

If potato crops show symptoms as harvest time approaches, remove all potato foliage from your garden and then wait at least 2 weeks before you dig the tubers. Wait for dry conditions to dig tubers, and be sure to remove all tubers from the soil. Check all harvested tubers for spots and don't put spotty ones in storage. Spray foliage with compost tea.

GARDEN CLEANUP: If late blight appeared at all in your garden, clear out all tomato and potato crop debris at the end of the season and burn it or bag it for disposal.

NEXT TIME YOU PLANT: For future crops, improve soil drainage. Choose certified disease-free seed potato varieties that are resistant or tolerant. For tomatoes, choose resistant varieties if you can—not many are available. Keep your tomato plants dry at all times by sheltering them under a clear-plastic tent (open at the sides). Water only at soil level. Clean your tools every time before working around susceptible plants. If you can't shelter plants from rain, use preventive sprays of compost tea or *Bacillus subtilis*, but note that these sprays are not always effective against late blight. Copper sprays also protect against infection, but repeated and thorough sprays are required.

CROP ROTATION: In general, rotating tomato-family plants is a good idea to prevent the buildup of diseases in the soil and produce healthy, vigorous plants. However, for late blight, crop rotation isn't a guarantee against infection because winds can transport the spores for miles.

BEYOND THE BASICS

A PROTECTIVE PLASTIC TENT

There are two ways to make a protective plastic tent to keep rain off leaves of plants susceptible to late blight. One method is to stretch clear plastic over the framework of an old, discarded canvas gazebo or shade tent (look for one at yard sales).

Another option is to sink tall wooden stakes or poles 1 foot into the ground around the perimeter of the tomato (or potato) bed. Set the stakes no more than 4 feet apart. Use 8-foot stakes along the back of the bed and 7-foot stakes along the front of the bed. Hammer or screw wooden crosspieces across the tops of the back row of stakes. Staple one end of the plastic to these crosspieces. Drape the plastic over the plants (it should be well above the tops of the plants) and staple it to the front stakes. You will end up with a sloping plastic "roof" that will keep plant foliage dry and shed rain. Just remember that air circulation under the cover is important so that condensation doesn't wet the foliage.

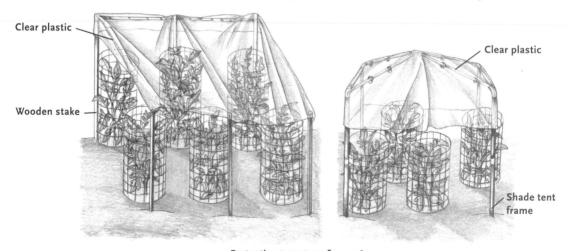

Clear plastic

Wooden stake

Clear plastic

Shade tent frame

Protecting tomatoes from rain

LEAFHOPPERS

Tiny but potentially deadly, three species of leafhoppers can damage a range of garden crops. Aster leafhoppers may transmit a mycoplasma as they feed, and the organism causes aster yellows disease, which cannot be treated. Beet leafhoppers can transmit viral curly top disease to beets and yellows virus to tomatoes. Potato leafhoppers don't transmit disease, but their feeding results in damage called "hopperburn."

Some species of leafhoppers overwinter as adults in weeds, others as eggs. Potato leafhoppers can overwinter only in the Gulf States, but they fly and are blown long distances north each year. Adults feed by sucking on plant tissue. They lay eggs in leaves and stems. Eggs hatch as tiny nymphs about 1 week later, and nymphs feed for several weeks. Adult leafhoppers can be relatively long-lived; aster leafhoppers sometimes survive for months. There are one to several generations per year.

Leafhopper activity varies from year to year, depending on their migration patterns and on the weather. Rainy weather and drought conditions seem unfavorable for leafhoppers, but when spring and early summer are moderate and pleasant, be on the watch for heavy leafhopper activity.

PEST PROFILE

Various genera and species

RANGE: Aster leafhoppers throughout the United States and Canada; beet leafhoppers in the western United States and Canada; potato leafhoppers in the eastern United States and some parts of southern Canada

HOW TO SPOT THEM: Leafhoppers hop or fly away fast when they're disturbed, so you may never get a good look at them on your plants. If you suspect leafhoppers in a crop, attach a sticky card to a stick, hold the card near the plants, and shake the foliage. If leafhoppers are present, they'll fly up and some will stick to the card. Use a hand lens to identify them.

ADULTS: Wedge-shaped, ⅛-inch-long, light green or olive green (aster leafhopper) insects with transparent wings. Aster leafhoppers have three pairs of spots on their heads.

NYMPHS: Pale green, wingless insects that resemble adult leafhoppers

EGGS: Laid in plants

CROPS AT RISK: Aster leafhoppers feed on lettuce, celery, carrots, endive, and parsnips. Beet leafhoppers feed on beets, potatoes, and tomatoes.

DAMAGE: Potato leafhopper feeding causes "hopperburn"—curling and browning of leaf tips and margins and abnormal vein development. Feeding damage by aster leafhoppers is minor, but they can transmit aster yellows, which causes a variety of symptoms and can ruin a crop. Beet leafhopper feeding transmits beet curly top, which causes deformed growth and can ruin a beet crop (see Beet Curly Top).

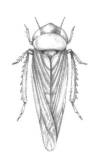

Leafhopper

Healthy foliage   Hopperburned foliage

Carrot infected with aster yellows

CONTROL METHODS

TIMED PLANTING. In the North, plant early to avoid the first wave of leafhopper activity.

ROW COVERS. Cover seedbeds and new transplants with row covers and seal the edges. Leave the covers in place as long as possible, removing the covers from long-season crops about 1 month before harvest is okay. Seal cover edges as described in Row Covers.

GARDEN CLEANUP. If plants develop symptoms of beet curly top or aster yellows, uproot them and remove them from the garden so that leafhoppers can't feed on them and ingest the disease organisms. For symptom information see the entries on Beet Curly Top and Aster Yellows. If beet leafhoppers are a serious problem in your area, remove host weeds such as wild mustard from your yard and garden.

REFLECTIVE MULCH. Spread light-colored straw mulch, commercial reflective mulch, or aluminum foil on the ground under susceptible plants to confuse leafhoppers. For more on reflective mulch, see Mulch. If you use straw mulch, renew it whenever it begins to turn dark.

KAOLIN CLAY. Coating plants with kaolin clay before leafhoppers find them may discourage feeding and egg laying. See Pest Control for information on kaolin.

WATER SPRAYS. Train a strong spray of water on infested plants, and try to cover all leaf surfaces. This will knock leafhoppers off plants and drown some of them.

GARLIC SPRAYS. Apply garlic spray just before leafhoppers become active in your area to deter feeding and egg laying.

BIOCONTROL. Apply *Beauveria bassiana* as soon as you detect leafhopper activity in a crop.

SPRAYS. Insecticidal soap, lightweight oil sprays, and pyrethrins will kill leafhoppers, but assess carefully whether spraying is worthwhile. Because leafhoppers are very fast-moving, usually only the first few blasts of spray make contact with the pests—the rest will already have fled the plants you're spraying. Plus, if the insects have already infected plants with disease, then killing the insects is of little help. Use sprays only if all other control avenues have failed.

CONTROL CALENDAR

AT PLANTING TIME

* Plant early in the North to avoid leafhopper damage.
* Cover seedbeds and plants with row covers.
* Use reflective mulch or light-colored straw mulch around susceptible plants.

WHILE CROP DEVELOPS

* Before leafhoppers are active, coat plants with kaolin clay.
* Spray plants with a strong stream of water to kill leafhoppers.
* Spray plants with garlic repellent to protect them from damage.
* Remove from the garden any plants that show symptoms of aster yellows or beet curly top virus.
* Apply *B. bassiana* as soon as leafhoppers are present.
* As a last resort, spray insecticidal soap, horticultural oil, or pyrethrins.

AFTER HARVEST

* Clear host weeds of beet leafhopper out of your garden.
* Turn under any crop residues that may shelter leafhoppers.

LEAFMINERS

Spring spinach and summer tomatoes are both prime targets of leafminers. Leafminer flies emerge from the soil to lay eggs on leaf undersides or directly inside leaf tissue. The small maggots that emerge create long irregular tunnels or mines through the leaves as they feed.

Leafminer damage rarely causes crop failure, but it detracts from the appearance of leafy crops such as spinach and chard. Leafminer feeding can cause tomato and pepper plants to lose some of their foliage, exposing the ripening fruits to sunscald.

When the miners finish feeding, they emerge from leaves and drop to the soil to pupate. There are several generations per year; in favorable conditions, the full life cycle may require only 3 weeks to complete. With spinach and beet leafminers, the first generation of miners is the most troublesome because they are active when spring crops are relatively small. Damage from later generations is less noticeable because plants are larger. By midsummer, this species generally fades out of the picture. In contrast, vegetable leafminer, a southern species, is present but not damaging during the cold weather, but can build to high populations by midsummer.

If leafminers have invaded your crop of spinach, beet greens, or chard, don't let it ruin your enjoyment of the harvest. Cut away any badly damaged portions of leaves—the undamaged areas will be just fine to eat.

PEST PROFILE

Liriomyza spp., *Pergomya* spp.

RANGE: Various species range throughout the United States; also some areas of Canada. Vegetable leafminers are a garden problem only in the southern United States.

HOW TO SPOT THEM: Inspect leaves of young spinach and beet plants in spring for clusters of leafminer eggs or for telltale squiggly lines. Hold a damaged leaf up to the light, and you may be able to see an outline of the larva inside the mined area.

ADULTS: ¼-inch-long gray flies with black bristles

LARVAE: Pale wormlike maggots found in tunnels inside leaf tissue

EGGS: Small groups of white oval eggs laid on undersides of leaves; vegetable leafminer eggs laid under leaf surface

CROPS AT RISK: Spinach and beet leafminers feed on spinach, beets, and Swiss chard, and occasionally on cabbage, radishes, and turnips. Vegetable leafminers feed on tomatoes, peppers, potatoes, eggplant, cucumbers, melons, watermelon, pumpkins, squash, beans, peas, onions, okra, and sometimes other crops.

DAMAGE: Larvae feed on leaf tissue between upper and lower leaf surfaces, creating long, winding mines or irregular white or gray blotches. Damaged areas may blister. Small plants that suffer a serious infestation sometimes become stunted. If leafminer feeding causes tomato or pepper plants to lose foliage, fruit may suffer from sunscald.

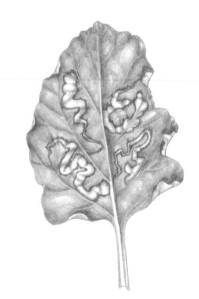

Leafminer damage

CONTROL METHODS

TIMED PLANTING. Plant cold-hardy leafy crops such as spinach in fall, and mulch well to overwinter. The plants will begin leafing out again in early spring, providing a nice harvest before the first generation of leafminers causes damage.

ATTRACTING BENEFICIALS. Several species of parasitic wasps attack leafminers and can keep levels low enough that no other controls will be needed. Create plantings to attract and shelter these wasps (see Beneficial Insects).

ROW COVERS. Cover seedbeds and young transplants with row covers to prevent the first generation of beet or spinach leafminer flies from laying eggs on the plants in spring. Do not use covers on beds where leafminer pupae may be present in the soil. Seal cover edges as described in Row Covers.

HANDPICKING. Early on, remove damaged leaves to prevent the miners from moving on to undamaged ones. Destroy the leaves or bury them deeply in a compost pile. After the first generation of miners, leave damaged leaves on the plant so that they can continue to photosynthesize; removing too many leaves can reduce yield.

GARDEN CLEANUP. Throughout the season, remove pigweed, plantain, chickweed, and lamb's-quarters from the garden, as they are also host plants for leafminers.

SPRING TILLING. Before planting, work the soil to bury beet and spinach leafminer pupae, thus preventing leafminer flies from reaching the soil surface.

OIL SPRAYS. If you find many clusters of eggs around your garden, spray leaf undersides of susceptible crops with a lightweight oil to smother the eggs.

NEEM. As a last resort, spray neem to kill larvae in the mines (this will not undo any of the damage already present).

CONTROL CALENDAR

BEFORE PLANTING

- Plant an insectary garden area to shelter parasitic wasps.
- Work the soil before spring planting to bury beet and spinach leafminer pupae.
- Pull weeds that can serve as alternate hosts.

AT PLANTING TIME

- Cover seedbeds and young plants with row covers and seal the edges.
- Plant leafy crops in fall to overwinter and bear an early spring crop.

WHILE CROP DEVELOPS

- Handpick damaged leaves early in the season.
- Pull weeds that can serve as alternate hosts.
- Spray leaf undersides with lightweight oil to destroy leafminer eggs.
- As a last resort, spray neem.

AFTER HARVEST

- Remove crop debris and weeds that serve as alternate hosts.

LEEKS

In the early stages, leeks require some special attention and care. But once leeks settle into place, they grow steadily with few problems. Leeks are related to onions and have a mild onionlike flavor, but they don't produce bulbs. What you'll harvest is the blanched stalk that consists of many tight layers of leaves.

Choose varieties depending on where you live and how you want to harvest. Some leek varieties (generally those that mature early) aren't hardy enough to overwinter outdoors. Long-season varieties can withstand temperatures down to 20°F; harvest them as needed from the garden throughout winter (or until the soil freezes in the North). There are also varieties especially suited for harvesting early as baby leeks.

CROP BASICS

FAMILY: Onion family; relatives are asparagus, garlic, onions, and shallots.

SITE: Plant leeks in full sun.

SOIL: Leeks need rich soil that's loosened at least 8 inches deep. Enrich a bed for planting leeks by spreading a 2- to 3-inch layer of compost and working it in. If your soil is stony, screen out the rocks and other debris that could interfere with the leeks' growth.

TIMING OF PLANTING: Sow seeds indoors 8 to 10 weeks before last spring frost. Plant seedlings in the garden in late spring. For a planting that will overwinter, start seeds 8 to 10 weeks before first fall frost.

PLANTING METHOD: Sow seeds indoors or buy started seedlings. It's possible, but not easy, to start leeks from seed directly in garden beds. Starting leek seeds indoors is very similar to starting onions from seed, as described in the Onions entry. Sow seeds less thickly than you would onions (about 1 inch apart).

CRUCIAL CARE: Leeks need lots of water and a steady supply of nitrogen as they grow to reach full size. Begin feeding leeks as early as 4 weeks after sowing seeds, and continue on a monthly basis. Fish emulsion, cottonseed meal, and other nitrogen fertilizers are good choices.

GROWING IN CONTAINERS: You can grow leeks in a deep container, but they probably won't reach the same large size as in a garden bed. Plant seedlings in deep holes (as described on the opposite page), and gradually fill in the holes with soil.

HARVESTING CUES: You can harvest leeks before they reach full size. Harvest short-season leeks before frost hits in fall. Long-season leeks store well in the garden, and they will continue to grow until weather turns too cold. See the opposite page for tips on extending the leek harvest. Always use a digging fork to loosen the extensive root system before you try to pull up a leek plant.

SECRETS OF SUCCESS

PLANT DEEPLY FOR BLANCHED STALKS. Hilling is a traditional way to blanch leeks, but in the home garden, planting leeks in individual deep holes does the trick with less work. Water your seedlings the day before you plan to transplant them. The following day, use the rounded handle of a hoe or other tool (about 1½ inches diameter) to poke holes in your prepared planting area. The holes should be about 6 inches deep and 8 to 10 inches apart (or 4 to 6 inches apart in rows 12 inches apart). Leek seedlings look quite spindly, but they're surprisingly strong. Simply place one plant in each hole so that the youngest leaf sticks up above the soil surface. Don't try to fill in the holes with soil. Instead, set up a sprinkler to gently irrigate the newly planted seedlings. The water will wash just enough soil into the holes to cover the roots. After that, rainfall and watering will take care of filling the holes, leaving you with nicely blanched leeks. If you're concerned that not enough stem is covered by soil, it's fine to hill up more soil around the plants. Avoid hilling any higher that the point where the leaves separate, or the soil will work its way down between layers of leaves (making the leeks hard to clean after harvest).

EXTEND THE HARVEST. Southern gardeners can enjoy leeks throughout winter, simply pulling them as needed. In areas where temperatures drop below 20°F, mulch leeks with straw to extend the harvest. Snow can also serve as mulch, as long as it's reliable. Continue digging mulched leeks until the ground freezes; then leave the crop alone until late winter or early

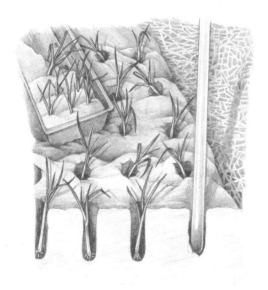

Use a tool handle to poke deep holes for planting leek seedlings.

Rain and irrigation water will move soil into the planting hole, and the mature leek will have a nicely blanched stalk.

spring. When temperatures rise back into the 20s, uncover any remaining plants. They'll start growing again in spring but may bolt quickly. Harvest before the flowerstalk starts to form, or the leaves will become stringy and tough.

REGIONAL NOTES

CANADA

The leek moth, a European pest, was accidentally introduced in Ontario in the early 1990s. Since then, it's become a problem in organic leek, garlic, and onion crops in Ontario and could show up in home gardens, too. The moths lay eggs on leek stalks and the yellowish green or bright green larvae bore through the outer layers of leaves to feed on the young leaves at the center. Growth may become distorted, but if it doesn't, you may not realize that plants are infested until you harvest and cut up the stalk.

In garlic and onions, larvae mine in the leaves, and their feeding creates a translucent "windowpane" effect. The larvae delve into the bulbs and cloves, too, opening pathways for rot organisms.

There may be two generations of this pest per year, and it could spread to other parts of southern Canada and into the United States. If you think you've found leek moth larvae in your garden, take samples to offices of your provincial ministry of agriculture or to local Cooperative Extension offices for identification.

PREVENTING PROBLEMS

BEFORE PLANTING

- Enrich the soil with compost and build up raised beds if needed to avoid drainage problems.

AT PLANTING TIME

- Plant late to avoid onion maggot flies.
- Cover planted areas with row cover and seal the edges (see Row Covers).

WHILE CROP DEVELOPS

- Be on the alert for early signs of thrips damage and spray plants with water to wash off the pests.

TROUBLESHOOTING PROBLEMS

The most common problems in growing leeks are listed below—in general, pests and diseases don't cause much trouble in leeks. If your leeks develop a problem that isn't described here, refer to Onions for a complete list of symptoms and causes.

SILVERY OR RUSTY STREAKS ON LEAVES. Onion thrips cause this damage. Spray plants with water to wash off thrips. See Thrips for other controls.

SMALL WORMS IN LEEKS. Onion maggots are small white larvae that feed near the base of the plant, tunneling into the stalk. Salvage what you can from infested leeks and discard the rest. To prevent problems in the future, cover young plants with row covers to prevent flies from laying eggs on the plants. See Onions for more information.

STALKS DON'T THICKEN. Lack of nitrogen or planting too late in short-season areas can leave you with skinny leeks at the end of the growing season. Go ahead and harvest the plants—they'll be fine to eat. To enjoy a more satisfying crop next time, start it earlier in the season, and side-dress or foliar-feed plants monthly with a nitrogen fertilizer.

LETTUCE

Lettuce belongs in every vegetable garden. You can grow a combination of looseleaf, butterhead, romaine, and crisphead lettuces. Plus lettuce is easy to grow in average soil. With experience, you can enjoy fresh-picked lettuce for most of the year, no matter where you live. In the Deep South, you'll probably choose to take a summer hiatus from lettuce. And in the North, you'll be without lettuce for winter unless you have a greenhouse or hoop house.

Overplanting lettuce is a common problem. You can't freeze lettuce, dry it, or can it. Lettuce, especially leaf lettuce, easily loses its delightful fresh-picked crispness. It can be difficult to save lettuce for donation to a food pantry or soup kitchen unless you plan carefully. Make it your first goal to learn how to grow a moderate, steady supply of lettuce over as long a season as possible.

CROP BASICS

FAMILY: Lettuce family; relatives are chicory and endive.

SITE: Lettuce grows well in full sun in cool conditions, but in hot weather it does better in partial shade. Choose a site that receives about 4 hours of direct sun daily.

SOIL: Average garden soil is fine for lettuce, and sandy loam soil is ideal. For direct-seeding, prepare a fine seedbed, raking away any large clumps of soil or organic matter.

TIMING OF PLANTING: Start seeds in pots 3 to 4 weeks before you want to transplant seedlings into the garden. Begin planting seeds and seedlings outdoors about 4 weeks before the last expected spring frost. Continue sowing or planting until 1 month before average daytime temperatures will exceed 80°F (or longer if you can shade the plantings and choose heat-resistant varieties). In late summer, begin sowing about 8 weeks before first expected frost.

PLANTING METHOD: Sowing lettuce seeds directly in the garden and planting lettuce transplants are both effective practices.

CRUCIAL CARE: Proper watering is the most important aspect of caring for lettuce. Lettuce has shallow roots, and it suffers easily when the soil dries out. However, a constantly moist soil surface can lead to disease problems. If possible, water lettuce deeply once a week. However, in hot conditions, lettuce may need watering every other day or even daily in dry climates. If you water daily, water first thing in the morning so the soil dries quickly, or at midday, which also helps to cool the soil at the hottest time of day.

Early weeding is important, too, especially for direct-seeded plantings. Cultivate shallowly to avoid injuring the crop roots.

GROWING IN CONTAINERS: Lettuce is an easy crop for containers. Plant a container or two of mixed lettuces to grow just outside your kitchen door, and add leaf lettuce as an edging for any type of container garden. In hot weather, lettuce in containers will probably need daily watering and will fare better in partial shade than full sun.

HARVESTING CUES: Begin picking outer leaves of leaf lettuce as soon as they are big enough to use. For romaine, butterhead, and crisphead types, harvest shortly after the heads have formed. Romaine and crisphead types should feel solid but probably not as firm as the heads of lettuce sold in grocery stores.

SECRETS OF SUCCESS

SET A PLANTING PACE. In general, lettuce heads mature uniformly. If you plant 10 plants in 1 day, about 1 month later, you'll have 8 to 10 heads of lettuce ready for picking. You can spread the harvest over several days, but if you wait longer than that, those beautiful heads will probably bolt, which changes their flavor from fine to bitter.

For most gardeners, it works well to plant lettuce once a week. Ten plants (or fewer) is a good amount to start with. Sowing seeds in individual pots or cell packs will help you maintain a disciplined approach. Planting seeds in pots also ensures good results whatever the weather—lettuce seed doesn't germinate well in hot or dry soil.

Sow two or three seeds in each small pot or cell (lettuce transplants don't need much rooting space). Set the pots under lights, on a sunny windowsill, or in a sheltered spot outdoors. When plants are 1 inch tall, snip off the weaker seedlings, leaving one plant per pot. If the seedlings are growing indoors, harden them off for several days outdoors. Your goal is to transplant them when they are 3 to 4 weeks old. You may think the transplants look awfully small, but don't delay. Lettuce seedlings more than 4 weeks old transition poorly to the garden.

HELP LETTUCE COPE WITH HEAT. Options abound for extending the lettuce growing season into the heat of summer. First, switch to planting heat-resistant varieties about 1 month before the time when average daytime highs will exceed 80°F in your area. Once those hot summer days arrive, it's important to provide some shade. Drape 40 to 50 percent shade cloth over your lettuce bed. Or situate summer lettuce on the north side of a trellis, and plant fast-growing pole beans or Malabar spinach to cover the trellis.

Placing shade cloth over lettuce is a daily ritual for summer crops in most areas. Remove the cloth at night to allow air to circulate.

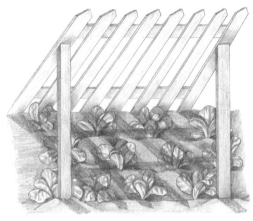

Prop a section of picket fence or lattice over summer lettuce to provide light shade without restricting air flow.

Plant lettuce in containers and position the containers where they'll receive morning sun and afternoon shade.

Mulch the soil with a light-colored organic mulch and plan your watering to help cool the plants. Daily watering is often necessary. If you're home at midday, water then, to help lower soil temperature when the sun is at its peak. Otherwise, water in the morning.

PLAN FOR FALL HARVEST. As days grow shorter and cooler in fall, lettuce growth slows. This means that the time from planting to harvest will increase to double or perhaps triple that listed on the seed packet. If you live in an area where autumn temperatures drop below the mid-20s, you'll have to protect your lettuce under row covers or a plastic tunnel. This should keep the plants safe from frost burn even if temperatures drop below 20°F. Since the plants will virtually stop growing at this point, it's as if you're storing your lettuce supply right in the garden. You can continue harvesting until late November or into December. This is the one time when you will want to plant a large amount of lettuce all at once. To time the planting right, count back from the average date of your first fall frost. Add 2 weeks (because of the slow fall growth), and that's the date to plant your undercover crop. Plant enough to supply yourself with lettuce for about 6 weeks.

All of the lettuce will be close to mature size in mid-October, and from there on it will grow very slowly, if at all. Monitor the weather forecast. Once nighttime temperatures in the mid-20s are predicted, it's time to cover lettuce with row cover or vented clear plastic covers supported by hoops. Uncover the plants as needed to harvest. Harvest whole heads or plants rather than picking individual leaves because plants are more prone to disease in the moist, cool environment under the cover.

If you're transplanting on a hot late-summer day (high temperatures more than 80°F), shade the plants for at least the first 5 days (see page 328 for examples of shading setups). After that,

you can remove the shade device as soon as the weather turns cooler.

You can also plant a small final planting in a coldframe or garden bed about mid-September to overwinter. These plants won't reach harvestable size in fall, but instead they will go dormant. Cover garden beds planted with lettuce with a heavyweight row cover. Switch to a lightweight row cover in early spring; feed them once with a diluted fish fertilizer and they'll be ready to pick about 3 weeks before your first spring-planted lettuce reaches harvest size.

BEYOND THE BASICS

THE CHALLENGE OF CRISPHEAD LETTUCE

Crisphead lettuce is the generic term for what most people call iceberg lettuce. 'Iceberg' is actually a specific variety of crisphead lettuce, well-known for its large heads and succulent, densely packed inner leaves. Growing crisphead lettuce is challenging because it takes longer to mature than other lettuces, and it's susceptible to bolting unless conditions are just right throughout its growth.

Crisphead varieties for home gardeners are different than those developed for commercial gardeners. Avoid varieties described as "supermarket" types. Batavian or summer crisp varieties are a good first-time choice. These lettuces don't form as tight a head, but they're more heat tolerant and faster growing than standard crisphead types. It takes about 2 months for crisphead lettuces to produce a full-size head, and the plants need temperatures averaging in the 60s throughout development. Ideally, maximum daytime high won't exceed 75°F; overnight lows won't drop below 55°F. Few places in the United States can match that ideal, but with

excellent care, crisphead will still grow well even if temperatures occasionally rise as high as 80°F or dip down to 50°F.

If your climate never offers a 2-month stretch of such moderate weather, then opt for starting early in spring. Start seedlings indoors under lights (keep them on 12 hours per day). Begin hardening off seedlings when they're 2 weeks old. Set 3-week-old transplants outside under row covers even when overnight lows still dip into the 30s. Plant the seedlings on slightly raised hills—this helps prevent bottom rot, even in a raised bed.

As your crop nears maturity, monitor the weather forecast carefully. Wet weather can lead to rot just as the heads are perfect for picking. Heat also causes lettuce heads to suddenly elongate and send up a flowerstalk. If you're an optimist (or a gambler), pick a few heads early and leave the rest, hoping they'll fare okay. If you're more cautious, you may want to harvest most of the heads and leave only a few to face the unfavorable weather.

REGIONAL NOTES
SOUTHWEST

The saltmarsh caterpillar is a woollybear-type caterpillar that sometimes feeds on lettuce and other vegetable crops. It's widespread in the United States and Canada, but it's most likely to be a problem in the Southwest in fall. The caterpillars chew on the leaves, skeletonizing them or leaving large, ragged holes. They tend to migrate out of farm fields in fall, and nearby home gardens with lush plantings of tender, leafy crops such as lettuce make an attractive destination. Handpick the hairy, buff-colored caterpillars, or apply *Bacillus thuringiensis* var. *kurstaki* when the caterpillars are small. Caterpillars are the migratory form of this pest, and you can stop them by digging a steep-sided trench alongside your garden nearest to the farm field (the direction from which the caterpillars will come).

PREVENTING PROBLEMS
BEFORE PLANTING

- Plant annuals and perennials with small flowers to attract beneficial insects that prey on aphids and cabbage loopers.
- Set up commercial insect traps to catch armyworm moths.

AT PLANTING TIME

- Cover seedbeds and transplants with row cover and seal the edges to keep out pests.
- Surround young plants with reflective mulch to deter thrips.

WHILE CROP DEVELOPS

- Spray young plants with compost tea or other plant health booster to help prevent disease problems.
- Handpick armyworm and looper egg masses.
- To forestall bolting, water regularly, especially during hot weather.

AT HARVEST TIME

- Remove all stray lettuce leaves and crop debris and compost it. Discard or destroy noticeably diseased plant material.

TROUBLESHOOTING PROBLEMS

SEEDLINGS DAMAGED OR DEAD. Cutworms, corn earworms, and armyworms are possible culprits. See Armyworms, Cutworms, and Corn Earworms to learn about controls.

CURLED LEAVES AND STUNTED PLANTS. Aphids and aster yellows are potential causes. Aphids suck sap from leaves; look for clusters on leaf undersides or at the growing tip. Wash aphids off plants with a strong stream of water. See Aphids for more controls.

If you don't find aphids, the problem is probably aster yellows, which is spread by leafhoppers. Pull out infected plants and compost them. When you replant, protect plants from leafhoppers (see Leafhoppers for methods).

MOTTLED LEAVES. Mottling is a symptom of mosaic virus. Uproot and compost infected plants. If mosaic is a serious problem in your garden, choose resistant cultivars for future plantings. See Mosaic.

LARGE HOLES IN LEAVES. Cabbage loopers, armyworms, slugs, snails, and birds will feed on lettuce plants. Check plants for caterpillars or slime trails (a sign of slugs and snails). Watch for bird activity around lettuce plants. Handpick caterpillars. Put out slug traps. Use row covers or screened cages to protect lettuce from birds. See Armyworms, Cabbage Loopers, Birds, and Slugs and Snails for more information.

PALE OR YELLOW SPOTS ON LEAVES. These spots are an early sign of downy mildew. Confirm the diagnosis by looking for white mold on leaf undersides. See Downy Mildew for controls.

RUSTY SPOTS ON LEAVES. If the spots are sunken and oozing, the problem is bottom rot, a fungal disease. Infected heads quickly turn slimy and rot. To prevent bottom rot on future plantings, see Rots.

SILVERY STREAKS ON LEAVES. Thrips are tiny sucking insects that feed on lettuce leaves. To spot thrips, hold a piece of white paper near the damaged leaves and shake the plants. The thrips will appear as specks on the paper. Spraying plants with a strong spray of water will dislodge the thrips. See Thrips for more controls.

SILVERY LEAVES. If leaves turn silver after a cold night, it's due to frost damage. The upper leaf surface may appear to have separated from the rest of the leaf. Harvest the head and remove damaged leaves. Undamaged leaves are fine to eat. To prevent future damage, cover plants with straw, row cover, or plastic when overnight temperatures are expected to sink to the mid-20s.

LEAF VEINS TURN CLEAR AND ENLARGE. Big vein is a viruslike problem that's transmitted by a soil fungus. Leaves also appear puckered and ruffled. Poorly drained soil favors the development of this disease, so avoid overwatering. Some varieties are more susceptible than others. Remove and destroy infected plants, including the soil surrounding the roots. Add compost to the infected bed to improve drainage and introduce beneficial organisms that will compete with the fungus. Replant in a different spot.

DARK EDGES ON INTERIOR LEAVES. Tipburn of lettuce leaves happens when plants can't absorb enough calcium, usually due to moisture stress or sudden temperature changes. It's similar in action to blossom-end rot. See Blossom-End Rot for tips on avoiding this problem.

CENTER OF PLANT ELONGATES. This is a sign that a flowerstalk is forming at the growing point. Bolting can happen quickly when temperatures exceed 80°F. Pull out the plants and compost them.

STUNTED PLANTS. Feeding by root-knot nematodes causes stunting. Uproot a plant and check for galls and misshapen roots. See Nematodes for control measures.

STEM AND LOWER LEAVES ROT. Check for a gray or brown fuzzy coating on the rotting plant parts; this is a sign of gray mold, which is caused by Botrytis fungi. Remove and compost infected plants. Replant in areas with better drainage. See Gray Mold for other preventive measures.

LOWER LEAVES OR WHOLE PLANTS WILT. White mold, a fungal disease, causes lettuce to wilt. To confirm that white mold is the cause, look for a white cottony growth under the leaves and around the crown (see White Mold).

MELONS

Melons need heat to be sweet, and some also need as much as 3 months of warm weather to reach maturity. Gardeners in Zones 8 through 10 have the most flexibility in choosing melons because there's less worry about whether the melons will reach maturity before the growing season ends. However, more and more fast-maturing melon varieties are becoming available, and northern gardeners can now enjoy Crenshaw, honeydew, Charentais, Galia, Asian, and Spanish melons as well as the usual muskmelons and small watermelons.

You can switch types from year to year until you find your favorite—flavor preference among melons is a very individual choice.

CROP BASICS

FAMILY: Squash family; relatives are cucumbers and squash.

SITE: Plant melons in full sun in the warmest part of your garden.

SOIL: Melons do best in loose, rich soil. If your soil isn't rich in organic matter already, spread 2 to 4 inches of compost and work it into the top foot of soil. In arid climates, loosen the soil even more deeply if you can to promote root penetration (deep rooting is important to survive high summer heat), and work in up to 6 inches of compost if possible.

TIMING OF PLANTING: Sow seeds or set out transplants after danger of frost is past and the soil is warm. In the Deep South and parts of the Southwest, sowing multiple crops is possible, with the final planting about 100 days before first expected fall frost.

PLANTING METHOD: Sow seeds directly in the garden or start seedlings indoors in large peat pots.

CRUCIAL CARE: Plant early enough to allow the fruits to ripen before temperatures drop. Melons need 75 to 100 days of warm weather from seeding to harvest. Check maturity dates before you buy seeds.

GROWING IN CONTAINERS: Choose a dwarf variety. Plant in a half-barrel using a mixture of equal parts potting soil and compost. Train the vines on a sturdy trellis and use cloth slings to support the fruits.

HARVESTING CUES: When ripe, muskmelons develop a crack at the spot where the melon attaches to the stem. They will slip off the vine at a very light touch. Muskmelons also give off a strong fragrance when ripe, and you'll be able to smell it even without cutting into the melon.

Watermelons and honeydew and Crenshaw melons do not slip, even when ripe. With watermelons, check the ground spot; it should change from white to yellow. The tendril closest to the fruit will turn dark, and when you thump the fruit, you'll hear a dull rather than a ringing sound. Watch for honeydew and

Crenshaw melons to become white or yellow. Also, press on the blossom end, which will become a little soft when ripe. These melons continue to ripen after picking if you keep them at room temperature.

SECRETS OF SUCCESS

TRIM EXCESS FLOWERS AND FRUIT. As growth of melon vines picks up in midsummer, pinch the growing tips of all shoots to concentrate the plant's energy on the developing fruits. Prune off baby melons after midsummer in the northern half of the country—they probably won't reach maturity. Six melons per vine is probably the maximum you can expect (two or three for large watermelons). In hot-summer areas, pinch the main growing tips of the vines but allow side shoots to grow. This promotes denser foliage

Can pushed into the soil about 1 inch

Set a young melon on a pair of bricks, a coffee can, or a flat stone to prevent disease problems and speed ripening.

cover, which will help prevent the sun from burning the rind of the developing melons.

SET FRUITS ON A THRONE. Prop up melons off the soil to keep them clean, to discourage rot, and to encourage faster ripening. As each melon forms, place a large metal can (such as a coffee can), a large flat rock, or two bricks beside the fruit. When the melon is the size of a baseball, set it on its "throne." Rocks and bricks absorb heat during the day and release it to the melon at night, which will help speed ripening.

MANAGE WATERING FOR MAXIMUM SWEETNESS. Melons need a steady but not overwhelming supply of water for good growth. Monitor soil moisture and water whenever the top 4 inches dries out, or if the plants begin to wilt before noon (afternoon wilting on hot days is not a cause for alarm). Water deeply. Setting up a watering reservoir (like those shown on page 430) at planting time helps with this. For an extra boost, water plants with a solution of fish emulsion or compost tea as the first fruits form. Maintain steady soil moisture until 2 weeks before you expect to harvest your melons, and then stop watering. The reduction in water supply will boost sugar content in the melons as they ripen.

REGIONAL NOTES

FAR NORTH AND HIGH ALTITUDES

It's a challenge to produce melons in areas with a short frost-free season. To succeed, pre-heat the soil by covering it with an IRT plastic mulch (see Mulch for more information). Choose early-maturing cultivars, and start seedlings indoors about 4 weeks before the last expected frost date. Challenge the season by setting out the transplants 1 week before the last expected spring frost, inserting the rootballs through slits in the row cover. Cover the new plants with medium or even heavyweight row covers to protect against chilly nights. Switch to a lightweight row cover when temperatures rise. Leave the covers in place as long as possible, even after flowers form. You'll need to hand-pollinate to assure that fruits will develop (see the illustration on page 390). Trim off excess flowers that won't have time to form mature fruits.

EASTERN STATES

Bacterial fruit blotch is a problem mainly in commercial watermelon crops, but the bacterium could infect other kinds of melons as well as squash, cucumbers, and tomatoes.

Symptoms on seedlings are small water-soaked spots on the cotyledons or leaves. The infected spots turn dark brown and may be surrounded by a yellow halo. Infected seedlings may develop as normal-looking plants. It may take close inspection to spot the small dark lesions on the leaves. However, bacteria from the leaf spots can infect the fruit, and that's where the dramatic symptoms show up. Fruits show a water-soaked blotch or stain on the upper surface that starts out small but grows to cover most of the fruit in as little as 1 week. The rind may split. Secondary bacteria can invade and cause rotting of the flesh.

Watermelon varieties with dark green rinds are less vulnerable to infection than those with light green rinds, but no varieties are resistant.

The bacteria can overwinter on seeds, and hot-water treatment of seeds does not seem effective for killing the bacteria. Thus, one way the disease could reach your garden is on infected transplants raised from infected seed.

If melon plants in your garden develop suspicious symptoms, be sure to take a sample to your local Cooperative Extension office for a diagnosis. If bacterial fruit blotch is confirmed, remove and destroy the infected plants.

PREVENTING PROBLEMS

BEFORE PLANTING

- Sow a trap crop to attract cucumber beetles.

AT PLANTING

- Cover seeded rows or seedlings with row cover and seal the edges as described in Row Covers to keep out insect pests.
- In hot-summer areas, sow seeds or set out transplants through a reflective mulch to deter aphids and other insect pests and to keep soil from overheating.

WHILE CROP DEVELOPS

- Spray exposed plants with kaolin clay to deter cucumber beetles (see Pest Control for information).
- Remove row covers when plants start to flower, or open covers regularly to hand-pollinate.
- Spray vines with plant health boosters or compost tea to help prevent disease problems (see Disease Control for information on plant health boosters).
- Weed out squash-family weeds and volunteer seedlings.

AFTER HARVEST

- Remove and destroy any noticeably diseased crop residues, including damaged fruits.
- Turn under or compost all other crop debris.

TROUBLESHOOTING PROBLEMS

SEEDLINGS CHEWED. Cucumber beetles can vigorously attack melon seedlings, sometimes eating the plants right down to the ground. Don't try to save young seedlings; replant instead and cover the seeded areas with row cover. See Cucumber Beetles for more controls.

PALE AREAS OR YELLOW FLECKS ON LEAVES. Squash bugs or spider mites may be the culprits. Look on leaf undersides and on young shoots for the small greenish squash bug nymphs or shield-shaped gray adults. Spider mites look like tiny specks, and there will be fine webs on the leaf undersides.

Handpick squash bugs. For serious infestations, spray pyrethrins. See Squash Bugs for more controls.

If a spider mite problem is severe, spray insecticidal soap. For other controls, see Mites.

YELLOW AND/OR WATER-SOAKED SPOTS ON LEAVES. Angular leaf spot, anthracnose, downy mildew, gummy stem blight, and scab are possible causes. As angular leaf spot develops, the spots become geometric and turn brown, and the diseased tissue sometimes drops out, leaving holes in the leaves (see Angular Leaf Spot).

Leaf spots that drop out are also a symptom of anthracnose, and it's more common on honeydew melons and watermelons than on musk-melons. Check plant stems, too; if the problem is anthracnose, you will find dark streaks and a pink jellylike substance (during wet weather). If the disease has progressed this far, you'll probably have to uproot the vines (see Anthracnose).

To confirm that the problem is downy mildew, check leaf undersides, where purplish mold will grow on the spots. Remove infected leaves. Applying a copper spray may stop the problem from spreading. See Downy Mildew to avoid future problems.

Gummy stem blight fungus can be very damaging to melons and watermelon, especially in the Southeast. Symptoms appear first at the center of the plants and spread outward. Look for water-soaked streaks on leaves and stems. Drops of dark, gummy ooze will appear along the streaks. Leaves may turn yellow and die. Fruits will be infected, too, usually leading to development of a black rot. Remove and discard infected vines and fruits. Feeding by cucumber beetles and aphids leaves plants more susceptible to gummy stem blight, as does infection by powdery mildew. Thus, the best way to prevent the disease is to grow powdery-mildew–resistant cultivars and to cover young plants with row covers to prevent insect damage. The fungus persists in the soil, so start a 2-year or longer rotation of squash-family vegetables.

Muskmelon is more prone to scab than other melons. In wet weather, spots due to scab become covered with a velvety mold. Water-soaked spots will form on fruits too, leaving the fruits prone to rot. If young vines are already infected, uproot them and replant, using resistant varieties. See Scab for other controls.

SPOTS ON OLDER LEAVES. Yellow or brown spots, sometimes with a pattern of concentric rings, are typical of Alternaria blight. Remove infected leaves and take steps to improve air circulation and decrease moisture around the plants. See Alternaria Blight for more controls.

EDGES OF LEAVES DIE. Fusarium wilt causes older leaves to turn yellow; leaf margins dry up and turn brown. One or all runners may wilt. At times, the vines wilt first without showing other symptoms. Check plant stems. You may see a reddish ooze similar to that of gummy stem blight. Cut open the stems, too. Infected plants will show a darkening of the stem interior. Remove and destroy infected plants, and see Fusarium Wilt for more information.

POWDERY WHITE COATING ON LEAVES. Powdery mildew is a widespread fungus. The coating may cover stems and fruits, too, if not checked. Handpick infected leaves and see Powdery Mildew for more controls.

MOTTLED LEAVES. Cucumber mosaic virus causes mottling. Leaves may turn under. There is no treatment, and the virus will infect and ruin the fruits, too. Uproot and destroy the infected vines. See Mosaic for prevention tips.

CURLED AND TWISTED LEAVES. Melon aphids suck sap, causing leaves to curl and twist. Look for colonies of yellow to dark green aphids on leaf undersides. Spray leaves with a strong stream of water to dislodge aphids; if the problem persists, see Aphids for more controls.

VINES WILT. Squash vine borer injury or bacterial wilt can cause vines to wilt suddenly. Check the base of the plant for borer holes with yellowish debris around them. If the problem is borers, there's some hope of reviving the vine (see Squash Vine Borers for details).

If you don't see evidence of borers, try cutting through a stem. If a thick ooze is present, the problem is bacterial wilt, and there is no cure. Watermelons are not susceptible to bacterial wilt but may suffer from Fusarium wilt. Uproot and compost the diseased vines. Cucumber beetles spread bacterial wilt as they feed; see Cucumber Beetles for prevention tips.

FLOWERS DON'T FORM FRUITS. The first flowers that form on melon vines are male flowers, which produce pollen but no fruit. Inspect the vines; you may just need to wait until female flowers form. However, if you find female flowers, and they aren't forming fruit, the problem is a lack of pollination. This can happen during cloudy, cool weather when insects aren't active, because melon flowers are insect-pollinated. To ensure better pollination, try hand-pollinating, as shown on page 390.

MOLD ON BLOSSOMS OR SMALL FRUITS. Some fungal organisms can invade melon blossoms and then grow into the fruits; the disease is commonly called blossom blight or wet-rot. Remove and destroy the infected blossoms and fruits. See Rots for more information.

HOLES IN BLOSSOMS AND FRUITS. Pickleworms feed on blossoms and fruits of summer squash (rarely on winter squash and pumpkins). Their tunneling in fruits allows rot to set in, so fruits usually aren't salvageable. See Pickleworms.

MOLDY OR DRY, ROTTEN SPOTS ON FRUITS. Many of the disease organisms that cause symptoms on leaves and stems of melon plants can also infect the fruits. These disease organisms can cause rotting or allow infection by secondary bacteria that produce wet or dry rots. The fruits probably won't be usable. Removed diseased fruits to a hot compost pile or bury them deeply so that the seeds inside won't sprout in the future as volunteer seedlings (the seedlings could be infected with disease organisms). To prevent fruit rots in the future, improve air circulation around vines, use a support structure to keep vines and fruit off the ground, or place supports under individual fruits.

CRACKS ON FRUIT SURFACE. Cracks and dead-looking areas on melons are the result of sunscald, which is most common in arid, hot-summer areas. The portion of the flesh below the sunburned skin won't ripen properly. To prevent sunscald, encourage lush foliage by feeding and watering melon vines. If necessary, shield exposed fruits with a brown paper bag.

DRY ROTTED AREA AT BLOSSOM END OF FRUIT. Blossom-end rot is the result of a lack of calcium while fruits are forming, usually due to water stress. Water uptake can be limited if the soil is too dry or too wet. Affected fruits may still ripen, or they may crack or succumb to secondary rots. See Blossom-End Rot to prevent future cases.

MEXICAN BEAN BEETLES

Mexican bean beetles awake from winter dormancy about the time that young bean plants are unfurling leaves. Although these beetles look like lady beetles (and are related), they're no friend to gardeners. The beetles lay eggs on undersides of bean leaves, and they also feed on leaves, chewing areas between veins to skeletonize the foliage. Fuzzy-looking yellow larvae hatch a week or two after eggs are laid, and they join the feast. After feeding for 3 to 5 weeks, the larvae attach themselves to the leaf underside and pupate. Pupae rest briefly, and then a new generation of adults emerges. At the end of the growing season, adult beetles seek shelter in crop debris or in weedy areas to wait out winter.

Mexican bean beetles will feed on all kinds of beans, and they seem to especially favor wax beans. Bush varieties of snap beans tend to suffer more damage than pole varieties do. Northern gardeners have to contend with only one generation of bean beetles, but in the Southeast the beetles can produce four generations per year.

One biological control measure that commercial growers rely on to help reduce bean beetle damage is releasing a parasitoid wasp called *Pediobius foveolatus*. Native to India, this wasp cannot overwinter in the United States, but it's quite effective at curbing populations of the beetle. The wasp lays eggs in bean beetle grubs, and the wasp larvae feed inside the grubs. Eventually, the grubs die, becoming mummified. Inside the mummy, the wasp larvae pupate. The next generation of wasps escapes by pushing a hole through the mummy skin. As long as they continue to find hosts, the wasps will continue their cycle throughout the growing season and then will die out when the weather turns cold.

Pediobius is released on a broad scale in some eastern states. You can buy a small quantity of mummies and release them in your garden, but they aren't cheap. If you want to try this biocontrol tactic, be prepared to spend $30 or more, and keep in mind that there's no guarantee that the wasps will lay eggs on your bean plants. They may just fly off in search of host plants in nearby yards or farms.

Mexican bean beetle

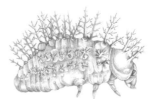

Mexican bean beetle larva

Mexican bean beetle feeding damage

PEST PROFILE

Epilachna varivestis

RANGE: Throughout the United States; rare in the Pacific Northwest and in Canada

HOW TO SPOT THEM: Check the underside of bean leaves to spot egg masses, beetles, and larvae. If you miss their arrival, the first cue you'll see will be areas of leaves that look as if they've been scorched, killing the leaf tissue between the veins.

ADULTS: Oval, ¼-inch beetles that are light yellow when young, becoming coppery or bronze when older, with 8 black spots on each wing cover.

LARVAE: Humpbacked yellow grubs with branching spines on their backs, up to ⅓ inch long

EGGS: Clusters of yellow or orange-yellow eggs laid on leaf undersides

CROPS AT RISK: All types of beans

DAMAGE: Larvae and beetles chew on leaves, feeding from under the leaf. They consume leaf tissue between the veins, leaving behind leaf skeletons. Adults also sometimes feed on young pods.

CONTROL METHODS

PLANTING TO ENCOURAGE BENEFICIALS. To help keep *Pediobius* wasps that you released (or that came from nearby farms) to stay in your yard, plant small-flowered annual and perennial flowers for the wasps to feed on (see Beneficial Insects).

TIMED PLANTING. In the North, plant beans late to avoid the first round of egg laying. In the South, plant early so that plants can grow as large as possible before beetles start to feed.

ROW COVERS. At planting, cover seedbeds with row covers and seal the edges (see Row Covers for tips).

HANDPICKING. Crush egg masses on bean plants (these eggs look identical to the eggs of beneficial lady beetles) and handpick larvae and adult beetles. Wear gloves when handpicking because the beetles may secrete a smelly yellow liquid when handled.

SPINED SOLDIER BUGS. Make a trap for spined soldier bugs and release them on your bean plants or put out commercial lures to entice the bugs to come to your yard.

RELEASING PREDATORY WASPS. If you plan to release *Pediobius* wasps, scout bean plants regularly and place the order as soon as you spot egg clusters on leaf undersides. Follow release instructions that come with the wasp mummies. If your introduction is successful, you should find dark-skinned grub mummies on your plants 2 to 4 weeks later.

GARDEN CLEANUP. After harvest, remove or dig under all bean vines and debris to reduce overwintering sites for beetles.

KAOLIN CLAY. Thoroughly coat plants with kaolin clay to discourage feeding and egg-laying. See Pest Control for details on kaolin.

BIOCONTROL SPRAYS. Apply *Beauveria bassiana* to kill beetle larvae.

INSECTICIDAL SPRAYS. To control a serious infestation of larvae, spray insecticidal soap or oil or a neem product, especially on leaf undersides. For more, see Pest Control.

CONTROL CALENDAR

BEFORE PLANTING

- Plant small-flowered annuals and perennials with staggered bloom time to supply food sources for *Pediobius* wasps and other beneficial insects.

AT PLANTING TIME

- Plant early or late to minimize feeding damage.
- Cover seedbeds with row covers.

WHILE CROP DEVELOPS

- Crush egg masses and handpick grubs and adults.
- Coat plants with kaolin clay before adults start laying eggs.
- Apply *Beauveria bassiana* to kill grubs.
- Release *Pediobius* wasps as soon as grubs begin feeding in your garden.
- If an infestation is threatening to ruin your crop, spray insecticidal soap, oil, or neem.

AFTER HARVEST

- Remove or turn under all bean vines and debris.

MITES

Be suspicious when speckles pop up on crop leaves, especially during hot, dry weather. Speckling and stippling of leaves is a sign that tiny spider mites are feeding on the protected surfaces of leaf undersides. Mites are eight-legged creatures related to spiders and ticks, and they're so small that you can't see them without using a hand lens.

Your garden is home to both beneficial and harmful mites, an important point to remember when you take steps to control crop-eating mites. Many measures that kill harmful mites kill the beneficials, too. In conventional gardens and farm fields, gardeners and farmers sometimes create new pest problems when they spray synthetic pesticides because the sprays also kill off beneficial mites that were providing excellent natural control of other types of pests.

BENEFICIAL MITES

Beneficial mites eat spider mites, small insects (such as thrips), and insect eggs. They are widespread in yards and gardens, but since they're just as small as spider mites (less than 1/50 of an inch long), you may never have noticed them. Through a hand lens you can view the subtle differences between the predator mites and spider mites. Predatory mites have longer legs and move much more quickly and often than spider mites; the only time they seem to stop moving is to feed. Also, the mouthparts of predatory mites stretch out in front of their bodies, while spider mite mouthparts project downward for feeding

on leaf surfaces. Predatory mites are available for sale and release. Releasing beneficial mites to control pest outbreaks is effective in farm fields, greenhouses, or indoors among a collection of houseplants, but releasing them in a home garden probably isn't. Instead, the best way to encourage predatory mites in your garden is to avoid insecticides, including biological insecticides, insecticidal soap, and especially sulfur sprays.

PEST MITES

The twospotted spider mite (*Tetranychus urticae*) is the most widespread garden pest mite, but from the home gardener's perspective, identifying the particular species of spider mite in your garden isn't important. Control measures are the same for any spider mite you'll find in your garden.

Because spider mites are so tiny, you usually won't spot them until the damage they cause motivates you to search using a hand lens. You'll find them on leaf undersides, feeding in groups, often under the cover of light webbing. The mites look like specks of dust traveling in the web.

As temperatures rise, spider mites speed up their growth and rate of egg-laying, which is why spider mite infestations seem to manifest out of nowhere when the weather is hot and dry. Natural enemies of the mites don't fare well in hot, dry weather, which may also be a factor in spider mite population explosions.

Adult mites wait out winter in a dormant state in crop debris, but in mild-winter areas, they may be active year-round on evergreen plants.

CROPS AT RISK

Spider mites will feed on almost any crop. They can be a serious problem on beans, squash, melons, eggplant, and tomatoes in hot, dry weather.

SYMPTOMS

Spider mite feeding causes stippling, bronzing, silvering, or scorching of leaves. Rub a leaf between your fingers and you'll notice a gritti-

When crop leaves look stippled or bronzed, check for spider mite webbing.

ness if spider mites have been feeding. As feeding intensifies, leaves turn yellow and will drop off the plants. If this happens to squash and melon plants, the fruits often end up sunscalded. Mites also feed directly on pea and bean pods, causing stippling and bronzing.

Mites will leave damaged plants in search of a better meal elsewhere. The tiny mites can't move very far on their own, so instead they'll ride wind currents to reach new host plants. Always check damaged plants for mites before you use a control measure—you may discover that the mites have already abandoned the plant in search of fresher food.

REDUCING MITE PROBLEMS

AVOIDING PLANT STRESS. Healthy plants are less likely to become targets of spider mites. As the weather turns hot and dry, make sure you water often and apply mulch to preserve soil moisture. This is one situation where using a sprinkler to irrigate is a good idea. If plants seem to be growing slowly, use a balanced fertilizer or compost tea to boost growth. Avoid fertilizers that are rich in nitrogen, because the lush leafy growth that nitrogen stimulates is attractive to spider mites.

WATER SPRAYS. Aim a strong stream of water from your garden hose at plants infested with spider mites. Turn the stream at many angles to wash all plant surfaces clean. The mites may reinfest the plants, but their growth will be slower because they must rebuild their webs before they can start to lay eggs again. Also, the water spray will remove dust and increase humidity around the plants, which favors the beneficial insects that prey on mites.

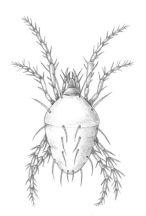

In this magnified view, it's easy to see that spider mites are related to spiders.

REPELLENT SPRAYS. Hot pepper and garlic sprays may repel spider mites. Spray plants in the morning. See page 308 for a recipe for a homemade spray.

INSECTICIDAL SPRAYS. Soap sprays, neem, and sulfur dust or spray will kill mites, but the results are variable. Use these sprays only as a last resort because they will kill beneficial mites, too. Spray in the morning, and be sure to spray leaf undersides thoroughly. Repeat sprays may be needed. Note that sulfur is only moderately toxic to mammals, but it is very irritating to skin and eyes. Wear complete protective gear when you apply it.

Sulfur can damage foliage when applied in hot, dry conditions. Don't spray sulfur if temperatures are more than 80°F. Also, sulfur and oil sprays together are even more likely to cause foliar damage. Wait at least 30 days between spraying sulfur and spraying any oil-based product (such as an oil-soap spray).

MOSAIC

Mosaic shows up in most vegetable gardens sooner or later. Several viruses can cause this disease. Some of the common forms are tobacco mosaic virus, cucumber mosaic virus, and turnip mosaic virus. Symptoms of specific mosaic diseases are subtly different, but all include mottling or streaking of foliage, blistering, and a condition called "shoestringing"—leaves are skinny and elongated, looking somewhat like fern leaves.

Mosaic can infect vegetables, flowers, and weeds. Cucumber mosaic virus, for example, has more than 700 different host plant species. Diagnosing which specific form of mosaic has infected a crop isn't easy, but prevention and control is similar no matter which particular virus is to blame. One key step is to remove weeds that can host mosaic early in the season. Aphids feed on these weeds and pick up the virus, and winged forms of the aphids then spread the virus to crop plants. Some common flowers, such as marigolds and zinnias, are subject to mosaic, too.

Tobacco mosaic virus is spread from plant to plant on tools or people's hands. Smokers may pick up the virus by handling infected tobacco products, so they should be very careful to clean their hands before working with tomato-family crops.

Mosaic viruses survive winter in crop debris or in living weeds; some species can persist outside a host plant for several years.

RANGE: United States; southern Canada

CROPS AT RISK: Tomatoes, peppers, potatoes, eggplant, squash, cucumbers, melons, lettuce, turnips, spinach, and other crops

DESCRIPTION: Infected plants develop leaves that are mottled yellow and green. Leaves sometimes develop puckers and blisters or exhibit "shoestringing"—elongated, fernlike growth. Leaf veins may turn yellow or brown. Leaf edges may turn or cup downward.

As the infection grows worse, plants become stunted and yield poorly, but they usually don't die. If squash-family crops produce fruits, the fruits usually are mottled, with unpalatable or bitter taste.

FIGHTING INFECTION: Pull up any infected plants as soon as you spot them and destroy them. Wash your hands and tools after you touch infected plants. Note: Wash tools with a strong soap solution, and put them in boiling water for 5 minutes if possible. A chlorine solution does not kill mosaic virus.

GARDEN CLEANUP: At the end of the season, uproot and compost all crop residues. At the beginning of the following growing season, pull out all weeds that may host mosaic, including shepherd's purse, chickweed, plantain, and purslane. As you weed, keep an eye out for infected annuals or perennials in your flowerbeds and remove them, too.

NEXT TIME YOU PLANT: For future crops, choose resistant varieties when possible, as well as certified virus-free seed potatoes and disease-free crop seed. Inspect tomato-family transplants before buying, and reject any that show mottling or other symptoms of mosaic. Cover young plants with row covers to prevent aphids from feeding. For tomato-family crops, if you are a smoker, change your clothing and wash your hands thoroughly before you touch plants.

CROP ROTATION: Crop rotation is not an important form of control for mosaic.

MULCH

Mulch is a marvelous problem preventer. Mulch suppresses weeds, prevents erosion, conserves soil moisture, adds organic matter to the soil, moderates soil temperatures, and more. Using mulch isn't entirely problem-free, however. The art of mulching involves figuring out exactly how and when to use mulch to maximize the benefits and minimize the problems.

DECIDING WHAT TO USE

The easy way to choose mulch is to use what you have on hand. For most gardeners, lawn clippings and leaves are easily available. If you have a chipper/shredder, you can shred the leaves easily and make wood chips, too. If you don't have enough mulch, you can scrounge mulch materials from neighbors or local businesses that generate organic wastes such as sawmills, landscapers, and nut growers or processors (for nut shells). Or you can buy mulch in bags or in bulk. Many municipalities also offer mulch for free or for a small fee. Here are some pros and cons of common mulch materials and tips on how to get the best results from them in your vegetable garden.

COMPOST

If you have a large supply of compost that is nearly free of weed seeds, it's fine to use it as a weed-suppressing mulch. Apply it 1 inch thick. By the end of the growing season, soil organisms will have done the work of mixing it into the soil for you. If you garden in an area with dry summers, you may want to top the compost with a thin layer of shredded leaves or straw to protect the microorganisms and other living things in the compost from "cooking" in the heat of intense summer sun.

GRASS CLIPPINGS

Grass clippings are easily available every time you mow your lawn. Just hook up your bagging attachment and you're in business.

Fresh grass clippings are nice for tucking all around crops (avoid the area right around the main stem of plants though, or the clippings may promote disease problems). Limit layers of fresh clippings to about ½ inch thick, or the clippings may begin to decay anaerobically from the center of the layer, leading to bad smells and too much heat. A moderate layer breaks down fairly quickly, adding some nitrogen to the soil as well as organic matter.

Another option is to let the clippings age and dry out before spreading them. You'll lose some of the nitrogen benefit but won't have the worry about odors and overheating.

The big risk in using grass clippings is introducing weed problems—especially crabgrass problems—into your garden. If you get grass clippings from a neighbor, you can assume that they contain weed seeds. A hot compost will kill most. Also, ask first whether the lawn has been treated with herbicides. If it has, don't use the clippings. Some herbicides may persist through the composting process and injure your plants.

LEAVES

Leaves are easy to find in most areas of the country. Shredded (chopped) leaves are an excellent mulch for summer use because they buffer the soil so well against the heat of summer sun. An area mulched with 4 inches of leaves can be more than 10°F cooler than bare soil would be. With very sensitive crops such as eggplant, let the soil warm up before you apply a mulch of shredded leaves, and use a soil thermometer to check the temperature.

To chop leaves with a lawn mower, spread the leaves over a patch of your lawn. Set the mower height at 3 inches (and set it for mulching). Mow over the leaves on a circular track, working from the outside edge of the leaves toward the center. If you want them more finely chopped, repeat the process, but use a back-and-forth pattern. Then put on your mower's bag attachment and use the mower to suck up the leaves.

To use a string trimmer, put the leaves in an old, clean plastic trash can—about three-fourths full—and put the string trimmer down into the leaves to chop them up. It's like running an oversize kitchen blender. (Always use eye protection and wear long sleeves and gloves for this job.)

LIVING MULCH

A cover crop seeded around vegetable crops or in garden pathways provides a living mulch. This mulch suppresses weeds, moderates soil temperatures, and when turned under, adds organic matter to the soil. It's important to carefully time the planting of a living mulch in garden beds, or the mulch plants may outcompete the crop for water and nutrients. (See the Cover Crops entry for details on using white clover as a living mulch.)

Some gardeners choose lawn grass as a living mulch for garden pathways. Sod paths are attractive and nice for kneeling on. The drawback of

Use your lawnmower on the mulching setting to chop leaves. Then use the bagging attachment to collect them. Store the leaves in plastic bags or use them right away as mulch.

sod paths is that grasses, which spread by lateral shoots, can invade adjacent garden beds. To prevent this, use a well-sharpened edging tool to edge pathways a few times per year, and always pull out grass that infiltrates crop beds as soon as you spot it.

NEWSPAPER

Newspaper is the ideal mulch extender. Start by covering open spaces between plants or crop rows with 4 to 6 sheets of damp newspaper (let the paper soak in a tub of water briefly to wet it). This base layer very effectively suppresses weeds. For a better appearance, top the newspaper with an inch of shredded leaves, grass clippings, or straw. The loose top layer also shields the newspaper from hot sun, which can "fry" the paper to a hard surface that won't rewet well.

A base layer of newspaper around young plants allows you to extend your supply of organic mulches such as leaves and straw.

In an arid climate, the newspaper may dry out too much even when topped with a surface mulch. If you try this technique in a dry climate, recheck the newspaper periodically. If it turns into a hard shield, or if it seems to absorb all of the rain or irrigation water, leaving none for your plants, then switch to a porous mulch, such as a loose covering of straw, that water can infiltrate easily.

PINE NEEDLES

Pine needles work well as a mulch for perennial garden crops such as asparagus. They may break down too slowly to use as a weed-suppressing mulch. Sprinkled lightly over newly seeded rows, though, they will prevent seeds from washing out during a heavy rain. As the seedlings germinate, you can gently lift away the pine needles (if a few cling on, it's fine to leave those in place).

PLASTIC

Black plastic is not the ideal mulch for an organic garden, both in practical terms and for reasons of principle. On a practical level, you have to buy black plastic, and yet it doesn't add any organic matter to the soil. In terms of principles, using black plastic isn't a sustainable method. It's a petroleum-based product, and it doesn't last for the long term. After a few seasons of use (at best), you'll have to throw it away, and it ends up in a landfill.

One occasion when black plastic can be useful is when you need to eradicate a troublesome perennial weed. Covering a bed with black plastic can kill almost any kind of weed, so you won't have to resort to herbicides (you're choosing the lesser evil).

Black plastic also makes it possible for northern gardeners to grow heat-loving crops such as sweet potatoes and melons because it warms the soil.

If you use black plastic, follow the tips under "Better than Basic Black," below, for optimum results.

STRAW

Depending on where you live, straw may cost as little as $3 per bale or as much as $10. When straw is cheap, it's great to use both on beds and paths, but if it's expensive, you may want to reserve it for garden beds. In pathways, straw may wear down and break down before the garden season ends.

 BEYOND THE BASICS

BETTER THAN BASIC BLACK

New types of plastic mulch, such as reflective silver mulch, infrared-transmitting (IRT) mulch (which is green or brown in color), and red mulch boost yields better than black plastic mulch. These mulches are more expensive than black plastic, but if you like to experiment with new products, you can test them in your own garden. Rolls that are 30 to 48 inches wide are available for about $15 (for 50 feet) to $20 (for 100 feet).

These mulches suppress weeds, but not as well as black plastic; don't choose them for smothering an invasive weed such as bindweed or thistle.

Red mulch can heat soil more effectively than black plastic does. Because it works so well, be sure to cover the mulch with straw as summer heat arrives, unless the crop you're growing (such as melons) spreads to cover the mulch as it grows.

Silver mulch doesn't heat the soil as well as black plastic, but it confuses or deters aphids, whiteflies, and thrips, which is important for tomatoes and other crops where these insects can infect plants with serious viruses.

Keep in mind that, like black plastic, these mulches are petroleum-based products and eventually have to be disposed of in landfills. It's up to you to decide whether plastic mulch offers benefits that outweigh these drawbacks. If so, follow these tips for effective use:

• Use them on raised beds (6 inches high or more). Water will run off the mulch surface and it will stay cleaner (dirty mulch cannot absorb or reflect light as well).

• Water the soil before applying the plastic if it is not already moist. Heat transfer is better into moist soil. Also, plan how you'll supply water during the growing season—the best choice is to lay a soaker hose or irrigation tape along the beds before you spread the plastic.

• Wait to spread IRT mulch until a week or so before you plan to plant. IRT mulch does allow some light to pass through, and weeds may sprout under the mulch. If the weeds receive a sizable head start, they may compete too much with your crop for water and nutrients.

Choose straw when you want to buffer the soil against summer heat—soil covered with a 4-inch layer of straw stays as much as 8°F cooler than unmulched soil.

WOOD CHIPS

Use wood chips for vegetable garden paths. They're cheap, easy to work with, and long last-ing. Use newspaper or cardboard as a base layer for even better weed suppression.

WHEN THINGS GO WRONG

In almost all gardens, the benefits of using mulch outweigh the drawbacks. However, mulching a garden occasionally does lead to

• Tightly stretch the mulch over the bed and bury the edges. This installation improves soil-mulch contact, which leads to more efficient warming. With a tight fit, weeds can't grow as easily under the plastic.

• At the end of the season, remove the plastic from the beds on a day when the plastic is dry. Try to keep as much plastic as possible for reuse the following season. Store the plastic (and any unused plastic) in a dry place out of direct sunlight (sunlight contributes to its breakdown).

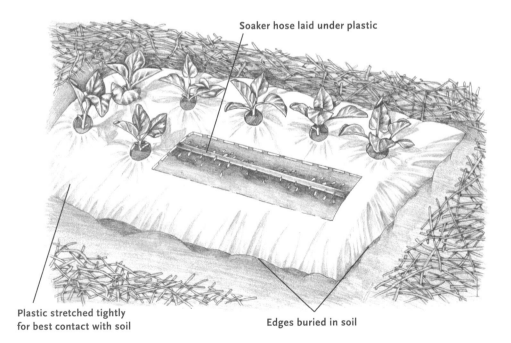

Soaker hose laid under plastic

Plastic stretched tightly for best contact with soil

Edges buried in soil

problems with soil temperature, pests (on your plants), or allergies (for the gardener).

SOIL BECOMES TOO WET OR COLD

Chances are you'll need to remove mulch at some point during the year to let the soil dry out and/or warm up. For example, in the Northeast, if you've mulched beds for winter, you'll want to pull back the mulch in spring a few weeks before planting. Don't turn under the mulch because the soil is probably too wet to work.

In the Desert Southwest, if you maintain a garden through winter, you may want to remove mulch as the weather cools to minimize the drop in soil temperature. In this case, turning in the mulch to add organic matter is fine because the soil probably won't be too wet to work.

SLUGS, SNAILS, AND OTHER PESTS

Some gardeners find that mulching seems to increase the intensity of slug and snail damage in their garden. Other gardeners will swear that slug and snail damage is the same whether they mulch their beds or not. Who's right? Probably all of them. It's possible that in one garden, the mulch becomes a spot where slugs and snails congregate, while in another garden, the slugs and snails may have a perfect hideout adjacent to the vegetable garden in a dense groundcover. If you're using mulch and slugs and snails are damaging your plants, try putting a different kind of mulch on one bed and compare damage—perhaps the particular type of mulch you're using is very attractive to slugs and snails. In another bed, pull back the mulch from individual plants or crop rows to leave islands of bare soil around the vegetable plants.

In yet another part of the garden, try leaving mulch in place and using an iron phosphate slug bait. Among these different strategies, you'll probably find at least one that reduces your slug and snail problem.

ALLERGIES

If you're allergic to grass or straw, using it as mulch could cause an allergy flare-up. Try wearing a dust mask when you spread it. If that doesn't help, enlist the help of a friend or family member to spread mulch for you (but first demonstrate specifically how you want it done).

Even if you're not prone to allergies, working with mulches can make you feel sick. Many kinds of organic mulch—including wood chips, straw, hay, grass clippings, and shredded leaves—can contain large quantities of mold spores. This is true even if the materials don't look or smell moldy. If you're exposed to the mold spores, you could end up with a flulike reaction or shortness of breath. It doesn't last long, usually only a day.

To avoid an adverse reaction, avoid spreading mulch on windy days. Try to avoid bending close over the mulch, and move the mulch gently to avoid stirring up dust and particles. Wearing a dust mask may help.

If you have a mulch pile in your yard, it may become noticeably moldy after a period of wet weather. Breaking the pile apart to dry will help reduce the problem. Use a long-handled shovel or other tool rather than using your hands for this task. Wear a dust mask or respirator to avoid inhaling the mold spores.

NEMATODES

Although you can't see nematodes, they're present in every garden. These microscopic wormlike animals live in the film of water that coats soil particles, and they're helpful to gardeners in two important ways. First, nematodes are an essential part of the soil food web that transforms organic matter into the nutrients that crops need, especially nitrogen. In this way, nematodes are nearly as important as earthworms in maintaining a healthy soil.

Nematodes also help gardeners by preying on soil-dwelling plant pests such as white grubs and root maggots. Scientists have discovered ways to propagate and package these nematodes, called insect-parasitic nematodes, for sale to both farmers and home gardeners.

A few notorious types of nematodes are garden pests, including root-knot nematodes, lesion nematodes, ring nematodes, and sting nematodes. They feed in the roots of a wide range of crop plants, resulting in stunted growth and other symptoms, and sometimes killing the plants. There are no organic methods that will permanently clear garden soil of these nematodes, but there are techniques that will reduce nematode populations enough to allow a good harvest.

Two species most widely used in home gardens and landscapes are *Steinernema carpocapsae* and *Heterorhabditis bacteriophora*. These nematodes invade soil-dwelling insects through their natural body openings. Once inside, the nematodes release a bacterium that paralyzes and kills the insect. The nematodes feed on the tissue of the insect carcass and also eat up the bacteria. They reproduce inside the carcass, moving through up to three generations before they leave to find a new host.

In fall, when soil temperature drops, the nematodes move deeper to survive. They can return when the soil warms up again in spring, but they won't move quickly enough to be "on the job" when soil pests become active. If you want to rely on nematodes to control soil-dwelling pests, you'll need to reapply the nematodes each spring.

BENEFICIAL NEMATODES

Insect-parasitic nematodes are harmless to humans, wildlife, bees, and earthworms. It takes care to apply these nematodes properly, but when conditions are right, they are effective for controlling pests ranging from cutworms to squash vine borers.

WHEN TO USE

Apply *Steinernema carpocapsae* to combat cutworms, armyworms, corn rootworms, and fire ants. Apply *Heterorhabditis bacteriophora* to reduce problems with cabbage root maggots and Colorado potato beetle larvae, white grubs, and root weevils. Using *S. fetiae* can suppress plant-pathogenic nematodes including root-knot nematode, ring nematodes, and sting nematodes.

Some companies offer formulations for different regions. These formulations contain more than one type of nematode; ask your supplier which pests the formulation will control.

HOW TO USE

You'll receive the nematodes packaged in a gel, in a powder on a small sponge, or perhaps mixed with peat and vermiculite. Read package directions carefully and follow them closely. Generally, you'll need to mix the nematodes with water and apply the suspension with a sprayer or watering can. Also, most formulations should be refrigerated until you use them.

For best results, apply the nematodes to moist soil in the late afternoon or early evening, or during light rain. If you're using them at planting time, sprinkle the nematode suspension directly into seed furrows or transplant holes to concentrate the nematodes in the root zone of your seedlings and transplants. For an established bed, water the entire bed, applying the suspension directly to the soil rather than sprinkling plant foliage. It won't hurt to overapply the nematodes, but if you apply them at less than the recommended rate, they may not control the pests.

One specialized technique is to use a pump spray bottle to apply a nematode suspension to squash or cucumber blossoms to kill pickleworms and to corn silks to kill corn earworms or fall armyworms. You can use a garden syringe to inject the nematode solution into squash stems to kill squash vine borers.

WHEN THINGS GO WRONG

PEST PROBLEM DOESN'T GO AWAY. As long as the entomopathogenic nematodes come in contact with soil pests, they should attack and kill the pests. If the pest problem is unchanged after you've applied nematodes, it means that something happened to the nematodes.

Sometimes weather is the problem. A heavy downpour can wash nematodes deep into the soil, out of range of the pest insects. At the opposite extreme, hot and dry conditions can kill nematodes. Application instructions for nematodes specify how often and how much to water after you've applied nematodes—be sure to follow these instructions closely.

Another possibility is that the soil was too cold for the beneficial nematodes. *Steinernema carpocapsae* is most effective between 72° and 82°F; the nematodes become inactive below 60°F. *Heterorhabditis* nematodes are even more sensitive and stop being active when soil temperature is below 68°F.

If you've had the nematodes in storage, they may have died before you applied them. Although some formulations can live for 1 month or longer when refrigerated, it's best to use nematodes as soon as possible after you receive them. Once you mix them with water, their life span is very short.

PEST NEMATODES

Root-knot nematodes are a common problem in vegetable gardens, especially in the South, where more species can survive overwinter. They tend to be more damaging in sandy soils than in clay soils. These nematodes can move in soil water to find roots of susceptible crops. They enter the roots and settle in to feed, secreting a substance that causes knots of enlarged cells to form. After feeding, nematodes lay masses of eggs, from which new larvae hatch. Eventually, the infected roots split open, releasing the new population of nematodes into the soil.

CROPS AT RISK

Root-knot nematodes can infect lettuce, tomatoes, peppers, eggplant, potatoes, squash, carrots, and other crops. Some species are host-specific; others infect many kinds of plants.

SYMPTOMS

The first symptom of root-knot nematode infection often is stunted growth. Leaves turn yellow and plants wilt and sometimes die, especially

Normal roots are white and uniform in appearance. Roots infested with root-knot nematodes develop galls, swellings, and scabby patches.

under dry conditions. The infection is easy to see if you uproot a plant. Roots will show galls, swelling, and/or scabby patches. The roots may be rotting, too, as fungi or bacteria invade cracks where roots burst open.

Often, some plants show severe symptoms and others show very little.

REDUCING NEMATODE PROBLEMS

Managing a garden that's infested with nematodes takes trial and error. In some parts of your garden, the population may be low enough to allow a reasonable crop, as long as you keep the plants well watered and fed. Other parts of the garden may need treatment every other growing season to reduce the nematode population enough to allow good crop growth.

RESISTANT VARIETIES. Nematode-resistant varieties of susceptible crops such as tomatoes may still develop symptoms, but the symptoms will be less severe or show up later in the season than they would on susceptible varieties. Use resistant varieties as part of your overall strategy in managing nematodes (an "N" in the catalog listing for a resistant variety indicates nematode resistance). If you can plant a bed with a nematode-resistant variety 2 years in a row, it should reduce the population enough that a susceptible variety will make good growth in the third year.

You may find that broccoli, Brussels sprouts, mustard, onions, leeks, garlic, rutabagas, and globe artichokes can produce reasonably well in a nematode-infested bed. However, growing these crops won't reduce the population of nematodes in the same way as growing a resistant variety will.

SANITATION. Nematodes are slow travelers. In the course of a growing season, they can move perhaps 3 feet through the soil. Don't help them spread any faster. Tiller tines and hand tools coated with soil can transport nematodes across a distance it would take them a decade to accomplish on their own. Identify the problem nematodes spots in your garden. After you dig or cultivate in that area, wash your tools thoroughly (either directly over the problem spot or in an area far away from your vegetable garden).

BUILDING UP BENEFICIALS. Naturally occurring fungi, bacteria, and other organisms will prey on and compete with root-knot nematodes. Add a wide variety of organic material to your soil by working in compost, growing cover crops, and using organic mulch to support a diverse, strong community of beneficial organisms.

SOLARIZING SOIL. Covering soil with clear plastic during the hottest part of summer kills both beneficial organisms and pest nematodes, but beneficial organisms tend to rebound more quickly after the treatment. Solarizing isn't a permanent cure for nematode problems, but it will buy you a year or two without nematode problems. See Disease Control for directions on solarizing soil.

COMMERCIAL NEMATICIDE PRODUCTS. Chitin is a natural component of the body of nematodes. Many fungi that attack nematodes do so by breaking down the chitin. Adding extra chitin to the soil stimulates the fungi and seems to speed their action against nematodes, too. Commercial chitin products currently available also contain urea, however, which is a synthetic fertilizer not used by organic farmers or gardeners. If you see chitin products at garden centers, be sure to ask whether they meet organic standards.

Another product, which does meet organic standards, is one made from ground sesame stalks. It suppresses many species of harmful nematodes. A 20-pound bag costs about $60, and the application rate is ½ pound per 100 square feet.

It is also possible that essential oils such as mint and oregano oils can be used to control harmful nematodes.

CROP ROTATION. Standard crop rotation doesn't help with controlling root-knot nematodes because many vegetable crops are nematode hosts. If nematodes are so serious in your garden that you're thinking of giving up, first try planting cover crops of marigolds or rye every other year, as described in "Knock Back Nematodes with Cover Crops" on the opposite page.

FALLOWING. If a particular section of your garden is heavily infested with nematodes, plant nothing there for a full year. Keep the soil scrupulously weeded. If denied all food for a year, the nematode population should decline, perhaps enough to allow reasonably good crop growth the following year.

WINTER TILLING. In the South, if you can work your soil in winter, till it lightly two or three times to expose nematodes to killing sunlight and dryness. The downside of this approach is that it's detrimental to soil structure and organic matter content. You'll need to rebuild the soil afterward by adding organic matter and mulching.

KNOCK BACK NEMATODES WITH COVER CROPS

Planting marigolds for 2 months and working them into the soil will significantly reduce populations of harmful nematodes—but only if you grow the right type of marigolds. Choose French marigolds (*Tagetes patula*) or African marigolds (*T. erecta*). Be sure the marigolds are not hybrids—root-knot nematodes can infest hybrid marigolds. Effective varieties include 'Nema-gone', which grows 4 feet tall and has golden orange flowers, and 'Golden Guardian', which grows only 2 feet tall.

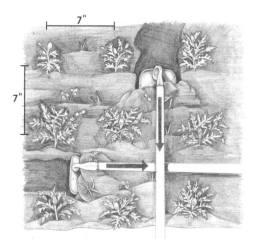

Cultivating to keep marigold cover crop weed-free

It's essential to create a continuous cover of the marigolds and to weed out any other plants that might host nematodes. You can establish a stand by scattering seed, but a more reliable method is to start young plants in a flat and transplant them to the infested bed. Set the plants 7 inches apart in all directions (for a 4-foot × 10-foot bed, you'll need about 125 plants). If you sow seeds, follow up with thinning and hand weeding until the stand is established; if you use plants, you can easily cultivate the bed weekly by pulling a hoe between rows of plants as shown below left.

After the plants have grown for 2 months, use a lawn mower or string trimmer to cut down the plant tops. Let them dry for a few days, and then turn the crop under.

Some cereal crops help to reduce nematodes. Sudan grass is especially effective if allowed to grow for one month and then is turned into the soil as a biofumigant. In the South, winter cover crops of oats and Elbon rye are helpful.

Cover crops are not a permanent reprieve from nematodes. You'll be able to grow vegetables in a marigold-treated bed for one season (possibly two), and then you'll need to repeat the marigold treatment again.

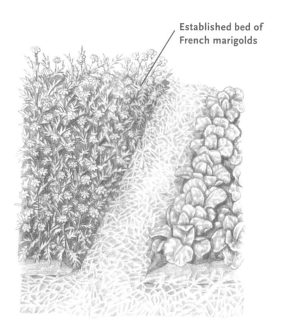

Established bed of French marigolds

OKRA

Okra is a dramatic feature of southern vegetable gardens, where it may grow 8 feet tall with beautiful red and yellow flowers similar to hibiscus. The heat-loving plants are sturdy and don't need support. Northern gardeners will find okra more of a challenge but can reap a good crop if they choose the right varieties and keep plants warm.

CROP BASICS

FAMILY: Mallow family

SITE: Plant okra in full sun.

SOIL: Okra tolerates many kinds of soil but grows best in light, deeply worked soil that's been steadily enriched with organic matter. If you're concerned about soil fertility, spread compost in the planting furrow (if direct seeding), or add compost to each planting hole for transplants.

TIMING OF PLANTING: Sow seeds or set transplants in the garden after all danger of frost is past, when soil is at least 60°F.

PLANTING METHOD: Sow seeds directly in the garden or start seeds indoors. Soaking or presprouting seeds is recommended.

CRUCIAL CARE: Control weeds until plants are about 6 inches tall, and then mulch to suppress weeds. Gently hand-pull weeds near the okra plants to avoid disturbing the okra roots. Fertilize with a balanced organic fertilizer immediately after thinning direct-sown seedlings and again when plants start to set pods. Okra is fairly drought tolerant but will need watering once a week if the weather is dry.

GROWING IN CONTAINERS: Okra will grow well in a container in full sun if you give it enough space. Use at least a 5-gallon container; a half-barrel is preferable. For a 5-gallon container, sow three seeds and then thin to the strongest seedling. Water several times weekly.

HARVESTING CUES: It's important to harvest okra pods when they're young and tender. See below for details on harvesting.

Treat okra plants with respect because the leaves, stems, and pods of many varieties are covered with short spines that can irritate skin.

SECRETS OF SUCCESS

TENDING OKRA SOUTH AND NORTH. Okra is well adapted hot Southern gardens, but plants that bear heavily in the heat of summer may wear themselves out. If you notice that the plants are dropping their lower leaves, try a trim-and-fertilize regimen to rejuvenate them. In late summer, use pruners to cut stalks back to 2 inches above secondary buds (wear gloves and long sleeves while you do this). After pruning, water the plants well with half-strength fish emulsion. The plants should respond by sending out a flush of new growth and flowers for fall.

In the North you need to coax okra to produce well. Choose dwarf and/or early-yielding varieties. Start seeds indoors 2 to 3 weeks before last spring frost—use newspaper or peat pots to minimize root disturbance and avoid transplant shock. Use black plastic to preheat the soil and cover plants with row covers if nighttime temperatures threaten to dip below 55°F.

HARVEST EARLY AND OFTEN. Okra flowers bloom for 1 day only, and pods are ready to harvest several days later. Cut tender pods when they are 2 to 4 inches long; larger pods will be tough. Harvest at least every other day to catch all pods in their prime. Use pods immediately after harvesting because they begin to lose quality fast after picking. Refrigerating pods can cause them to turn black (chilling injury). If some pods become oversize, cut them off the plants and compost them to encourage formation of more young pods.

PREVENTING PROBLEMS

BEFORE PLANTING

- Plant flowering plants and maintain permanent mulched or sheltered areas to provide habitat for beneficial insects that help control whiteflies (see Beneficial Insects).

AT PLANTING TIME

- Space plants widely to avoid disease problems.

WHILE CROP DEVELOPS

- Inspect plants frequently and handpick pests.
- Spray plants with a strong spray of water to remove aphids and thrips.

AFTER HARVEST

- Chop the woody stems before turning them under or adding them to a compost pile.

TROUBLESHOOTING PROBLEMS

HOLES IN LEAVES. Corn earworms, armyworms, and cabbage loopers sometimes feed on okra. Handpicking may be sufficient. Consult

Use scissors or pruners to cut young okra pods, which form at the juncture between the main stem and leaf stalk.

Corn Earworms, Armyworms, or Cabbage Loopers for other controls.

CURLED LEAVES. Aphids suck sap from leaves. Look for clusters of small insects at the growing tip and on the undersides of leaves. Spray plants with a strong spray of water to wash off the aphids. See Aphids for other controls.

WHITE COATING ON LEAVES. Powdery mildew is a common fungal disease in hot and humid conditions. (See Powdery Mildew.)

LEAVES TURN YELLOW. Yellowing may be due to feeding by silverleaf whitefly, Fusarium wilt, or root-knot nematodes. Brush plants with your hand. If whiteflies are present, they will fly up in response. See Whiteflies for controls.

Plants infected by Fusarium wilt usually grow poorly or start to wilt. Cut into the stem of one plant at the base. Dark coloration confirms Fusarium wilt. Discard infected plants. See Fusarium Wilt for controls.

If no whiteflies are present, uproot one plant and check the roots for knotty areas or swelling, which is a sign of root-knot nematodes. Infected plants probably won't produce a good crop. To reduce future problems, consult Nematodes.

FLOWERS APPEAR BUT NO PODS FORM. High heat or cold both interfere with okra pollination. Temperatures above 90°F or below 55°F can cause poor pollination. Lack of light, water stress, or excess nitrogen also can affect okra pod formation. Wait for temperatures to change. If pods still don't develop, try watering plants thoroughly. If that doesn't help, plant okra elsewhere in your garden next time, in a spot with full sun and enriched, but balanced, soil.

DEFORMED PODS. Southern green stink bugs suck sap from pods, causing them to become twisted and distorted. Spray plants with insecticidal soap to kill the pests; chop and compost the plant debris (see Stink Bugs).

ONIONS

Beginning gardeners can grow onions from sets or purchased seedlings without shedding tears, and experienced gardeners will enjoy growing onions directly from seed or raising their own seedlings in order to sample a wider choice of varieties. Whether you're a beginner or longtime gardener, though, it's important to understand what triggers onion plants to form fleshy bulbs.

When you plant an onion seed, seedling, or set, it first grows roots and upright green leaves. At a certain point in time, the plant switches from producing more leaves to developing a fleshy bulb. The amount of daylight is the signal that flips the plant's internal switch.

There are three categories of onions: short-day, long-day, and intermediate-day. Short-day onions form bulbs in spring as soon as days reach 10 to 12 hours long. Long-day onions need days that are 14 hours long or longer to initiate bulbing, so they don't start to bulb until summertime (and won't form sizable bulbs at all in southern areas where maximum daylength barely reaches 14 hours per day). Intermediate-day onions form bulbs when days are 12 to 14 hours long.

Some suppliers use the term "day-neutral" for widely adapted intermediate-day onion varieties such as 'Super Star' because gardeners in all parts of the United States and Canada can grow them successfully. These varieties allow northern gardeners to grow sweet onions that can come close to the sweetness of classic southern onions such as Vidalia types. Always check the variety description when choosing onions and select the type that's appropriate (this is explained in more detail below).

Green onions are easy to grow. You can simply harvest regular onions young, before they form large bulbs, or grow true scallions (also called bunching onions), an onion variety that's a close cousin to standard bulbing onions. Chives, another onion relative, are a favorite culinary herb grown for their onion-flavored leaves and flowers. Shallots are another type of perennial onion; to learn how to grow them, see Shallots.

CROP BASICS

FAMILY: Onion family; relatives are asparagus, chives, garlic, leeks, and shallots.

SITE: Plant onions in full sun.

SOIL: Onions thrive in light loam or sandy soil well enriched with organic matter. They will not grow as happily in heavy clay soil (build up raised beds with lots of organic matter added if your soil is heavy). See "Prepare an Ideal Planting Site" on page 280.

TIMING OF PLANTING: Planting time for onions depends on where you live and what type you're growing. See "Timing It Right North and South" on page 280 for details.

PLANTING METHOD: Many gardeners plant sets—small bulbs that grow quickly to produce full-size onions. Starting onions from transplants also works well and allows greater choice of varieties. Starting seeds directly in the garden is possible but requires careful attention weeding while the small, slow-growing seedlings become established.

CRUCIAL CARE: Weed early and often. Onions don't compete well with weeds because of their narrow foliage and shallow roots. Check soil moisture frequently, and if the soil is drying out, apply water to wet the top 6 inches of the soil. Leave off watering when the tips of the foliage start to turn yellow, which is a sign that the bulbs are maturing. Fertilizing young onion plants will encourage strong growth, but plan to stop fertilizing about 7 weeks before the expected harvest date. Late applications of fertilizer will prevent bulbs from maturing properly.

GROWING IN CONTAINERS: Chives and scallions grow well in containers.

HARVESTING CUES: Pull young green onions as desired once they are big enough to suit you. For your main crop of bulbs, watch for the plant tops to start to die back—a sign that the bulbs are enlarging. It's best not to bend over or break onion tops because the bruising and damage to the necks leaves the bulbs vulnerable to disease. If the foliage of some plants refuses to die back, harvest those bulbs and use them first. Onions for storage need to cure in a warm place for 1 to 2 weeks. After that, store them in a cold (close to 32°F) dry place.

SECRETS OF SUCCESS

PREPARE AN IDEAL PLANTING SITE. A traditional companion planting technique is to tuck a few onions in the corners of garden beds or between rows of other crops. That's fine for a casual crop, but the bulbs will probably be small because the plants are shaded by the neighboring crops or because soil at the edges of beds tends to be dry and less fertile. To excel at onion growing, prepare part of a garden bed especially for onions. Choose a site in full sun, and enrich the soil with 1 to 2 inches of screened compost. Avoid sites where you recently turned in a cover crop unless you're sure the plant material has broken down thoroughly. If you prepare the soil early, allow a flush of weed seeds to germinate and kill off the weed plants so your onions will have minimal competition.

If you plan to plant onions in rows, create 4-inch-deep furrows midway between the rows and spread a band of a balanced organic fertilizer in the bottom of each furrow—about 1 cup per 10 feet of row—and cover it with soil. That puts the fertilizer in easy reach of the roots as the plants grow. If you plan to plant onions in a block (as shown below), open furrows every 6 inches across the block, spread bands of fertilizer at a rate of ½ cup per 10 feet of furrow, and then fill in the furrows and plant.

TIMING IT RIGHT NORTH AND SOUTH. In general, onions are planted in fall in the South and spring planted in the North. Southern gardeners grow short-day onions, which start to form bulbs in late spring. The best way to ensure that short-day onions produce lots of healthy foliage before bulbing is to plant seedlings in fall. Seedlings will grow readily in warm autumn soil, and the plants will produce roots and some foliage before going dormant as temperatures drop below freezing. Mulch the overwintering plants with straw or leaves, but remember to remove the mulch in early spring. You can plant in early spring, too, especially if you choose a widely adapted intermediate-day onion.

In the North, wait until early spring to plant because onions can't survive the intensity of winter cold north of about Zone 7. Sow seeds indoors about 8 weeks before the planting-out date. Plant sets or transplants in the garden 4 to 6 weeks before the last expected frost date.

PLANTING POINTERS FOR ONION SETS. Sets are easy to plant and don't require the special attention that onion seedlings do. Simply poke the sets into well-worked soil so that the top of the set is about level with the soil surface. The sets should sprout and grow readily in moist soil.

One drawback of planting sets is limited choice of varieties, and a common problem with sets is that plants form seed stalks prematurely before they produce a bulb. To avoid bolting, sort your sets before planting. Discard any sets

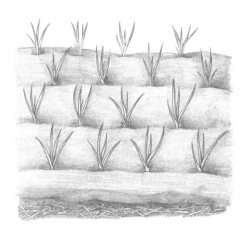

Use garden space efficiently by planting onions on 4-inch centers across the width of a garden bed.

that are soft or already sprouting. Separate remaining sets by size. Small sets are less likely to bolt. Plant sets smaller than a dime in one block, spacing them 4 inches apart—these plants should grow well and form sizable bulbs. Plant large sets in a separate area close together, so the sets are nearly touching their neighbors. You'll harvest these young as green onions, thus avoiding the risk that they'll bolt.

Exposure to cold—either a period of several days below 50°F, or a day or two below 30°F—tends to trigger sets of any size to bolt, so don't rush to plant spring sets too early in the season.

SOW A PERFECT POT OF SEEDLINGS. A shallow 6-inch plastic pot filled with moist seed-starting mix is ideal for growing onion seedlings. Sow the seeds thickly over the surface of the pot (use up to 100 seeds) and add a sprinkling of the mix over the seeds. Top the pot with plastic wrap and then a board or some newspaper. Set the pot in a protected spot where the temperature will remain between 70° and 75°F until the seeds sprout.

Immediately after sprouting, remove the cover and plastic wrap. If possible, move the pot to a cooler location, ideally to a coldframe. If you don't have a coldframe, figure out another place with bright light where the temperature will hover between 40° and 50°F. Keep the seedling mix moist but not too wet, and use a liquid fertilizer every 3 weeks. If the leaves grow more than 5 inches tall, use scissors to trim them back to 3 inches.

Transplants grown in a coldframe won't need hardening off, but if you grew transplants indoors, harden them off for at least a week before planting outside.

Onion transplants will be very skinny, but don't worry; they will settle in and grow well. To plant them, hold one seedling against your finger and push your finger into the soil about 2 inches deep. Then, still holding the seedling, begin to pull your finger out of the hole. After withdrawing your finger about ½ inch (which helps straighten the roots), release the seedling. Firm seedlings gently in place and water well.

Trim onion seedlings to prevent them from tangling and to encourage stocky topgrowth and vigorous roots.

RAISING SCALLIONS. If you'd like to grow varieties especially adapted for use as green onions, choose varieties described as scallions or bunching onions. These varieties have a reputation for milder flavor than regular onions do when harvested as green onions. Many varieties are ready to harvest about 60 days after seeding. Sow seeds or plant seedlings in spring or fall. These onions are hardy enough to overwinter even in the North. Start fall plantings about 4 weeks before your first expected frost. In the North, mulch the young plants with 3 to 4 inches of loose straw or leaves. Remove the mulch in early spring.

PREVENTING PROBLEMS

BEFORE PLANTING

- Enrich the soil with compost and build up raised beds to avoid drainage problems.

AT PLANTING TIME

- Plant late to avoid onion maggot flies.
- Cover planted areas with row cover and seal the edges (see Row Covers).

WHILE CROP DEVELOPS

- Be on the alert for early signs of thrips damage, and spray plants with water to wash off the pests.

AT HARVEST TIME

- Handle bulbs carefully to avoid injuring them as you pull them out of the soil. Cure your crop by laying the plants on the soil surface with the tops covering the

CARE-FREE CHIVES

Make room for a few clumps of chives in your vegetable garden, flower garden, or in a flowerpot. This easy-to-grow perennial supplies onion-flavored foliage that's always at the ready to liven up a cooked dish or salad. Chives suffer few pest or disease problems, and they're hardy as far north as Zone 3. You can start chives from seed or buy started plants at a local nursery.

Snip chives leaves as needed for seasoning, garnish, and salads. The flowers are edible, too. Be sure to cut off all flower buds as they fade, because chives self-sows prolifically. Volunteer seedlings can become a weed problem.

You'll only need to plant chives once. In mild-winter areas, they'll produce new leaves year-round. In cold-winter areas, the tops die back but resprout in spring. Dig and divide the clumps every 3 years, replanting divisions to increase your supply or giving them away to friends. Mix two trowels of compost into the soil with each division you replant.

bulbs, or by putting them in a warm, dry place, such as a garage or shed. If conditions are humid, run a fan to keep air circulating.

TROUBLESHOOTING PROBLEMS

SILVERY OR RUSTY STREAKS ON LEAVES. Onion thrips can be a serious pest of onions, especially in the South. Spray plants with water to wash off thrips. Spraying plants with kaolin clay can help also; thorough coverage is important. See Thrips for other controls.

DARK STREAKS AND BLISTERS ON LEAVES. Onion smut is a fungal disease that can infect onion plants at any stage of development. The fungus produces galls under the skin of the leaves, and the dark coloring is due to the spores developing in the galls. Pull and destroy infected plants to prevent the disease from spreading.

YELLOW LEAVES AND/OR WILTING. Nitrogen deficiency, onion maggots, or rot diseases may be to blame. If the weather has been cold and wet, the plants may not have been able to absorb many nutrients from the soil. Try sidedressing the plants with bloodmeal or another nitrogen source and the problem should clear up quickly when the weather warms up.

Onion maggots are small white larvae that bore narrow tunnels into the bulbs. The maggots are the larvae of small gray flies that lay eggs in soil cracks or at the soil line near onion plants. The larvae feed for a few weeks before pupating in the soil. Remove and destroy infested plants. The best way to prevent maggot infestations on future plantings is to cover plant with row cover at planting time, sealing the edges of the row covers at least 6 inches away from the plants. Another alternative, but probably less effective, is to sprinkle hot pepper around the young plants and reapply it after rains. Cabbage maggots tend to infest spring crops more than fall-planted onions.

A variety of fungal diseases can cause onions to rot, especially in cool or wet soil. The bulbs may develop a white, green, or black mold or the neck area may turn mushy. The best way to prevent rots is to plant in well-drained soil that's been enriched with compost. Some cultivars are resistant to smudge (the green mold).

SLOW GROWTH AND THICK NECKS. These symptoms are a sign that plants aren't getting enough phosphorus and/or potassium. If the plants are young, spray them with compost tea or kelp to boost their growth.

STUNTED PLANTS. Stunting can be a sign that root-knot nematodes have attacked the roots. Uproot one plant and check for galls on the roots. Onions infested by nematodes often have thick necks and never produce a sizable bulb. The bulbs are edible but won't keep in storage. To prevent future problems, take steps to reduce nematode populations in the soil (see Nematodes).

ROTTED, PINK ROOTS. Pink root is a soilborne fungus that thrives in wet conditions. In the future, select resistant varieties and plant in a different part of your garden.

SPLIT BULBS. Onion bulbs divide into two or three separate sections in response to water stress or late application of nitrogen fertilizer. The bulbs are still edible, but they may not store well. In future, check soil moisture frequently and stop fertilizing at least 7 weeks before the expected harvest date for your crop.

STRONGER FLAVOR THAN EXPECTED. If your sweet onions turn out to be pungent, weather is probably to blame. Heat stress or moisture stress can cause onions to develop more pungent flavor. The composition of your soil also plays a role. If you want sweet onions, try to grow them during cool weather, and never allow them to suffer water stress (but be careful not to overwater while the bulbs are enlarging).

PARSNIPS

For a steady fall and winter supply of roots that are sweeter than carrots, grow parsnips. Parsnip roots store well in the soil, and cold weather brings out the best of their sweet, nutty flavor. Parsnip seed is slow to germinate and the seedlings are delicate. Once established, though, parsnips are generally undemanding and suffer from few if any pest or disease problems.

CROP BASICS

FAMILY: Carrot family; relatives are carrots, celery, cilantro, dill, and parsley.

SITE: Plant parsnips in full sun.

SOIL: Rich soil that's loosened as much as 2 feet deep and free of rocks and debris are the ideal for parsnips, but they'll tolerate.a range of soil conditions.

TIMING OF PLANTING: Sow seeds in spring or early summer so that the crop reaches maturity at about the time of first expected fall frost. In mild-winter areas, plant seeds in fall after soil temperatures drop to about 75°F.

PLANTING METHOD: Sow seeds directly in the garden.

CRUCIAL CARE: Planting seeds with special care to help ensure good germination is the most critical aspect of growing parsnips, as described below. Once plants are established, mulch them to prevent weed competition. About 1 month after seedlings germinate, side-dress with compost or a balanced organic fertilizer.

GROWING IN CONTAINERS: It's possible to grow parsnips in deep containers (at least 5 gallons in volume). Use a rich, light potting mix. Insulate the containers in fall to prevent the soil and roots from freezing too quickly.

HARVESTING CUES: There's no urgency to harvest parsnips unless you live in an area with lots of winter rain. Parsnips left in cold, soggy soil may rot, so harvest your full crop in late fall and store it in a root cellar or refrigerator (parsnips keep for several weeks in a crisper bin). Elsewhere, harvest parsnips as needed. Frosts improve their flavor.

Parsnips send roots deep and wide. Always use a digging fork to loosen parsnip roots carefully before you try to uproot them.

SECRETS OF SUCCESS

PAMPER PARSNIPS AT THE START. Fresh seed, well-worked soil, and careful planting technique are the needed combination for establishing a strong stand of parsnip seedlings. Soak parsnip seed overnight before planting. In the planting area, loosen the soil at least 12 inches deep, and if it's rocky or contains lots of debris, screen the soil or remove as many stones and clods as you can by hand. Make trenches ½ inch deep and 12 inches apart. Sow seeds 1 inch apart, cover them with a noncrusting material such as vermiculite or sifted compost, and top the planted rows with a board or other items to hold in moisture (for tips on materials to use, see Planting and Transplanting). Parsnip seeds take as long as 1 month to germinate. Check under the board regularly, and water gently if the seedbed is drying out. As soon as seedlings appear, remove the board. If carrot rust flies are a problem in your area, cover the seedlings with row cover and seal the edges (see Row Covers). Check the area weekly for weed seedlings and carefully cultivate or hand-pull them. When the seedlings reach 6 inches tall, thin them to about 4 inches apart and mulch between the rows with 2 to 3 inches of organic mulch to suppress weeds. From then on, your parsnips should grow slowly and steadily until harvest.

WINTER PARSNIP HARVEST. In the North, begin harvesting parsnips after the first frost sweetens the roots. Continued exposure to cold will only make them sweeter. The plant tops will die back naturally. If you want to continue digging parsnips into winter, mulch them with up to 1 foot of straw before the ground freezes. Pull back the straw as needed to harvest. Or you can allow the parsnips to freeze in place in the soil, and then dig them in early spring when the ground thaws, but before the roots sprout new leaves.

PREVENTING PROBLEMS

AT PLANTING TIME

- Cover seeded areas with row cover to keep out carrot rust flies.

WHILE CROP DEVELOPS

- Weed carefully when crop is young to prevent weed competition.
- Hill up soil around exposed shoulders of roots.
- Set out wireworm traps and check them frequently for wireworms (see Wireworms).

AFTER HARVEST

- Remove any remaining crop debris and compost it. Discard or destroy any noticeably diseased debris.

TROUBLESHOOTING PROBLEMS

Parsnips often are pest- and disease-free. The list below describes the most common problems that may occur in parsnips. If your crop shows symptoms that aren't listed here, consult the Carrots entry for more descriptions of symptoms and causes.

BROWN SPOTS ON LEAVES. A fungal canker disease can cause leaf spots and reddish brown cankers at the crown or on the shoulders of parsnip roots. Uproot and destroy infected plants to stop the disease from spreading. The disease tends to develop in cool, wet weather in fall. For future crops, choose canker-resistant varieties and hill up soil around exposed shoulders of parsnip roots.

SUNKEN BROWN SPOTS ON ROOTS. A disease called scab causes these symptoms. Cut away the damaged areas and eat the rest of the root. For your next crop, plant in a different part of the garden. See Scab for controls.

TOUGH, BITTER ROOTS. Overmature parsnips sometimes become woody and bitter. This can happen if you plant parsnips so early in spring that they mature while the weather is still warm enough for active growth. Avoid this in the future by carefully calculating the spring planting date. Check the days to maturity of the variety you plan to grow, and count back that number of days from your first fall frost date. Parsnips also may turn bitter during a particularly mild winter in the South if the plants don't go dormant. In mild-winter areas, it's safer to harvest parsnips on the young side rather than waiting for a spell of frosts—the flavor may not be as sweet, but at least the roots won't be bitter.

PEAS

Peas are a classic spring crop, but you may find that you can grow a fine crop of peas during summer or in fall as well. The trick is to understand how pea plant responses to weather change as the plants grow. Young pea plants are quite frost tolerant, but pea blossoms and pods are frost sensitive. In spring and summer, you can plan for extended harvest until heat overcomes the vines. The fall harvest will be short because frost will end it.

Having too many peas to eat is rarely a problem because they're so tasty and sweet. If you grow more than you can eat yourself, you'll find it easy to give them away. However, keep in mind that picking peas takes time, and they need to be picked within a narrow window of time. If you plant a large quantity, recruit family and friends to help with the picking to enjoy your pea harvest at its best.

CROP BASICS

FAMILY: Legume family; relatives are beans.

SITE: Full sun

SOIL: Average garden soil is fine for peas as long as it's well drained. Avoid adding high-nitrogen materials just before planting. Raised beds are a good choice because soil will warm more quickly for the spring crop.

TIMING OF PLANTING: For a spring crop, sow seeds as soon as you can work the soil and at least 3 weeks before the last expected frost. In the South, plant peas 8 to 10 weeks before the first fall frost, and throughout winter in mild-winter areas (frost will damage blossoms).

PLANTING METHOD: Sow seeds directly in garden beds. Apply garden pea inoculant (available from seed suppliers) to the seeds.

CRUCIAL CARE: Don't overwater while young; seeds and seedlings tend to rot in wet soil. Increase water when the vines bloom and form pods.

GROWING IN CONTAINERS: Peas are a productive choice for containers. You can plant extra early because you don't have to wait for garden beds to dry out. See "Buckets of Peas" below for details.

HARVESTING CUES: Check the base of the vines first because that's where the youngest flowers are and therefore probably the most mature peas. Snow pea pods should be tender, easy to bend without breaking, with very small peas forming inside. Snap pea pods should be crisp, with peas swelling inside the pod. If too flat, they won't have reached prime sweetness. Shelling peas should be bright green and pods should not yet be waxy, the peas inside plump but sweet. Harvest frequently to stimulate the vines to continue producing.

SECRETS OF SUCCESS

PLANT TO SUIT THE SEASON. The most effective way to extend your spring pea harvest is to plant as early as possible, sowing three different varieties on the same date: one early, one mid-season, and one late (soil temperature should be at least 40°F on planting day). An early variety

Warm season crops such as these tomatoes will get a good start at the base of a pea trellis. The young transplants will grow up to fill the space after the spring pea harvest ends.

BUCKETS OF PEAS

Container gardening is a great idea for peas. Use 5-gallon buckets or other deep containers, and position them against a wall or fence or at a spot where you can run strings up to porch eaves for the vines to climb. Plant peas in the back portion of the bucket, with salad greens in front. Water the buckets frequently to avoid moisture stress.

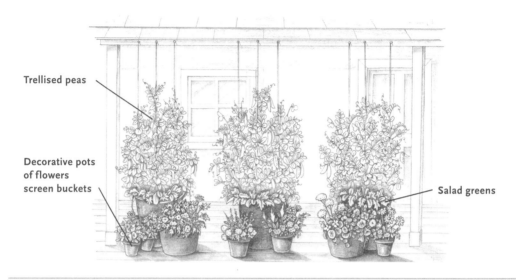

Trellised peas

Decorative pots of flowers screen buckets

Salad greens

will mature in about 8 weeks, while a late variety will take 10 to 11 weeks, so you can harvest peas for a month or longer. This strategy gives all of the pea crop maximum growing time during the cooler part of spring, which is ideal for pea vine growth.

If you want to try a summer crop of peas, sow seeds again 3 weeks later, choosing a heat-tolerant variety. In mild-summer areas, continue sowing at 3-week intervals as desired.

For fall crops, you'll want the peas to mature all at once, a week or two before the first fall frost, to avoid losing the crop to frost damage. That means you'll have to stagger planting times based on a variety's days to maturity. To deter-

mine the planting date for a particular variety, add 14 days to the days to maturity and then count back that total number of days from your first frost date. You'll probably be planting while the weather is hot, so shade the young plants and mulch them well to cool the soil.

ALWAYS SUPPORT YOUR PEAS. Peas grow best when their tendrils can cling to some kind of support. Tall varieties need a trellis. Shorter varieties do well clinging to pea brush (twiggy branches stuck in the soil) or to each other if planted in a block. Avoid sowing spring peas against a metal fence, such as a chain-link fence, because the excess heat radiated by the metal could stress the pea vines.

PLANT A WARM-WEATHER CROP AMONG THE PEAS. Some pea-loving gardeners devote up to a quarter of their garden space to spring peas, knowing they can plant a variety of warm-season crops right on schedule at the base of the vines. It's easy to slip tomato, pepper, eggplant, cucumber, melon, and other transplants into the soil at the foot of a pea trellis. When the pea harvest ends, cut the vines at ground level and pull them free of the trellis. The warm-season crops will happily stretch up in place of the peas for summer.

PREVENTING PROBLEMS

BEFORE PLANTING

- Cultivate soil to expose cutworms to predators.

AT PLANTING TIME

- Set up a trellis or pea brush to support pea vines.
- Apply parasitic nematodes to the planting area to reduce cutworm populations (see Nematodes).

WHILE CROP DEVELOPS

- If past crops suffered disease problems, spray young plants with a plant health booster (see Disease Control for information).
- Check soil moisture frequently once flowers start to form; don't allow the soil to dry out.
- Inspect plants frequently for pests. Use a strong water spray to remove aphids and spider mites; handpick larger pests. Take other action as needed if pest populations climb.

- In hot weather, spray plants with milk, potassium bicarbonate, *Bacillus subtilis,* or compost tea to promote health and prevent powdery mildew.

AT HARVEST TIME

- To avoid damaging the vines, use one hand to hold the vine while you pull the pod with the other, or use a pair of scissors to neatly clip off the pods.

AFTER HARVEST

- Clear out all residues that suffered pest or disease problems and put them at the center of a compost pile. Chop and turn under healthy plant residues to boost soil organic matter.

TROUBLESHOOTING PROBLEMS

SEEDLINGS NEVER EMERGE OR NEW SEEDLINGS DIE. Damping-off disease or seedcorn maggots can ruin a new planting, especially during cool, wet conditions. If seedlings have dark, water-soaked stems, damping-off is to blame. Try replanting in a spot with better drainage. See Damping-Off for more information.

If seedlings emerge but are deformed or have no leaves, seedcorn maggots are the cause. These pests are the larvae of small gray flies (similar to cabbage maggot flies). Dig around in the soil to find the small, yellowish white maggots. Remove the seedlings and treat the soil with parasitic nematodes as described in Nematodes. Replant when the soil is a little drier and warmer.

NEW SEEDLINGS EATEN. A single cutworm can destroy several plants in one night. These pests hide just below soil level (see Cutworms).

HOLES IN LEAVES. Cucumber beetles and bean leaf beetles chew holes in pea leaves. Handpick these pests. See Cucumber Beetles for other controls; see page 44 for more information on bean leaf beetles.

TUNNELS IN LEAVES. The larvae of leafminer flies feed inside leaves, leaving tunnels as they go. Handpicking damaged leaves is usually enough to limit the problem. See Leafminers for other controls.

YOUNG LEAVES AND SHOOTS CURLED. Aphids suck sap from leaves, but they rarely cause enough direct damage to hurt yields. However, because aphids transmit several serious viruses to peas, it's important to prevent aphid activity as much as possible (see Aphids).

CURLED, MOTTLED LEAVES. Mosaic viruses cause mottling and curling of leaves. Uproot and compost infected plants. Aphids spread these viruses, so the only feasible control measure is to prevent aphids from feeding.

WATER-SOAKED SPOTS ON LEAVES. These spots are a sign of bacterial blight, which can spread to the pods as well. It tends to become a problem in cool, overcast weather. Remove and destroy infected plants. To avoid problems next time you plant, choose a resistant cultivar and plant in a different part of your garden.

LOWER LEAVES TURN YELLOW. Root rot is a common problem in cool, wet weather, especially in the Midwest and Northeast. Uproot a stunted plant to check for water-soaked areas on the lower stems and roots. Remove infected plants and replant in an area with better drainage. See Rots for more information.

YELLOW SPOTS ON LEAVES. Downy mildew is a problem on peas, especially when conditions are cool and moist. To confirm the diagnosis, check leaf undersides for fluffy gray fungal growth (see Downy Mildew).

WHITE STIPPLING ON LEAVES. Spider mites will feed on peas during hot and dusty conditions. Look for webs on leaf undersides; the mites are tiny dots moving in the web. Spray plants with water to remove mites and webbing. See Mites for more controls.

WHITE COATING ON LEAVES. Powdery mildew is a common problem on peas in dry regions and on fall crops. Misting plants with water can help prevent the disease from worsening. See Powdery Mildew for more controls.

STREAKS ON STEMS. Light brown or purple streaks are a sign of pea streak, which is caused by a virus or combination of viruses. Pods may show brown to purple areas and be malformed. Aphids spread pea viruses, so protect future plantings from aphids (see Aphids).

STUNTED PLANTS. Root rot or viruses are potential causes. Root rot is a common problem in cool, wet weather, especially in the Midwest and Northeast. Uproot a stunted plant to check for water-soaked areas on the lower stems and roots. Branch roots may have died back. Remove infected plants and replant in an area with better drainage. See Rots for additional information.

Peas are vulnerable to several kinds of viruses, which cause a range of symptoms. The pea stunt virus also causes leaves to curl and twist. The growing tips may be deformed and plants may die. Aphids spread this disease, so preventing aphids from feeding is the only way to avoid the virus (see Aphids).

DEFORMED PLANTS. Pea enation virus causes such pronounced abnormalities that infected plants hardly look like pea plants. Leaves develop blisterlike growths and show translucent, windowlike areas. Pods will be highly deformed, too, with few peas inside. This virus tends to be most troublesome in the Northeast and Northwest. Some varieties are resistant. Aphids spread the virus (see Aphids).

WATER-SOAKED SPOTS ON PODS. Bacterial blight or gray mold can infect pea pods. If bacterial blight is the problem, the spots will develop purplish margins and run together, leaving the pods looking dark and blighted. Harvest all the healthy pods you can, then remove and destroy the vines. Choose a resistant cultivar for your next planting and plant it in a different part of your garden.

If gray mold has infected the pods, a gray coating will form and spread. Gray mold is more prevalent in humid areas (see Gray Mold).

SCARRED, DEFORMED PODS. Thrips feed on the surface of pea pods, resulting in scarring and puckering, especially on snap peas. The pods are edible—the damage is only "skin deep." To control thrips on future plantings, see Thrips.

MISSHAPEN PODS. Viral diseases affect pods as well as foliage. The pods probably won't produce normal peas. Uproot and compost the infected plants. Aphids spread pea viruses, so protect future plantings from aphids (see Aphids).

PEPPERS

Not too hot, not too cold, peppers like the temperature just right. In fact, protecting peppers from unfavorable weather conditions tends to be the prime challenge in growing them. Peppers sometimes suffer from diseases or infestations of insects, but in general, they tolerate pest problems better than their eggplant and tomato cousins.

Cold conditions slow growth and fruiting of all kinds of peppers, and bell peppers also stop producing fruit when temperatures rise too high. Hot peppers such as habaneros, cayennes, and jalapenos are a bit more heat tolerant than bell peppers. If you learn the right techniques for helping peppers cope with heat and cold, though, you can enjoy a good crop in almost any region. And since pepper plants are perennials by nature, gardeners in the Deep South can enjoy tending pepper plants year-round. Plants set in the garden in July offer a fall harvest, take a midwinter break, and then start producing again in spring until summer temperatures rise too high. But even in warm climates it's best to treat peppers as annuals and replant each year with vigorous new plants.

CROP BASICS

FAMILY: Tomato family; relatives are eggplant, potatoes, tomatoes, and tomatillos.

SITE: Peppers need full sun to produce a bountiful crop of large, firm fruit.

SOIL: Plant peppers in well-drained, fertile soil enriched with lots of organic matter. If you have clay soil or poor drainage, make raised beds for planting. Beds 6 inches high will suffice because peppers tend to be shallow rooted.

Prepare the soil by growing a cover crop and turning it under before planting if possible. Add one generous shovelful of compost to each planting hole. To boost phosphorus levels, add bonemeal to each planting hole, too.

TIMING OF PLANTING: In spring, wait to set transplants in the garden until the soil is at least 65°F. Gardeners in Zones 8 and warmer can also plant bell peppers in late June or July for a fall harvest.

PLANTING METHOD: Sow seeds indoors 8 to 10 weeks before planting-out date.

CRUCIAL CARE: Peppers need steady moisture as fruits develop, but the plants can't toler-ate being waterlogged. As the peppers ripen, cut back on watering for best flavor. Unless your soil is perfect, peppers will need fertilizing to fuel both foliage cover (to prevent sunscald) and bountiful fruiting. Options include adding alfalfa, cottonseed, or fish meal to the soil at planting; side-dressing with compost or blood-meal (2 tablespoons per plant) when the first fruits start to form; or drenching soil with fish emulsion/kelp solution every 3 to 4 weeks (2 cups per plant). It's also critical to support pepper plants. Branches laden with fruit can snap off, or the whole plant may sprawl to one side, leaving the fruits vulnerable to sunscald and disease problems. In the North, it may be enough to plant a bed of peppers inside a stakes-and-string corral (as shown on page 193). In the South, where pepper plants can reach eye level, use sturdy stakes or cages.

GROWING IN CONTAINERS: Containers are a good choice for peppers, especially in areas where low nighttime temperatures are a danger. Combine equal parts potting mix and compost, and add fish meal or alfalfa meal to the mix, too. Be sure to provide a cage or other support.

HARVESTING CUES: Harvest bell peppers when just starting to turn color and let them ripen indoors at room temperature. If you leave some fruit on the plants to reach full ripeness, pick them as soon as they are fully colored to encourage more fruit set. Refrigerating them slows ripening but also causes chilling injury. Harvest chili peppers as you need them. If you've planted enough for drying, harvest them when they are fully mature but before frost harms them.

SECRETS OF SUCCESS

TREAT TRANSPLANTS WITH LOVING CARE.
Pepper seedlings need special treatment, but it's worth growing your own, both to enjoy your choice of varieties and to be confident your transplants are sturdy and properly hardened. Start pepper seeds indoors 8 to 10 weeks before the planned outdoor planting date (add 2 more weeks if your seed-starting area is on the cool side).

Sow four seeds per cell or container; you'll thin to the strongest seedling. Peat pots are a good choice to minimize root disturbance. Set the seed flat in a warm place to germinate (optimum temperature for germination is 80° to 90°F). Add bottom heat or heat from above using an incandescent light (for details on adding heat, see Seed Starting). Seedlings like warm conditions, too: 75° to 80°F daytime and 60°F at night. Water seedlings with lukewarm water.

Pepper transplants seem more susceptible to transplant shock than other vegetables. If the transplants go into shock after being planted in the garden, they may never recover completely. To reduce the risk of transplant shock, harden off seedlings for at least 1 week before the transplanting date. Set the plants outside during the day to harden, but bring them in each evening (don't risk exposure to shocking cold).

Flower buds sometimes form on transplants, and that's okay. However, if any small fruits have begun to set, pinch them off. Set the plants with the top of their soil ball level with the garden soil surface—peppers can't form roots along the stem they way tomatoes can. (If

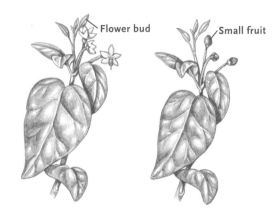

Don't mistake unopened flower buds for small pepper fruits. Leave buds in place, but if fruits have formed on transplants, pinch them off before planting outside.

your plants are in peat pots, rip away the top edge of the pot so it won't protrude from the soil and wick away moisture.) Transplant your peppers in the late afternoon or on a cloudy day to avoid shock from heat or sunlight in the first day of growth. Water them in with compost tea or alfalfa tea.

PROTECT PLANTS FROM THE COLD. Northern gardeners use tricks both early and late in the season to buffer peppers from temperatures that slow growth (in spring) or stop ripening (in fall). First, be sure to choose bell pepper varieties that ripen quickly, and varieties of chilies and other hot peppers that are adapted to cooler conditions. Preheat the soil by covering it with black plastic. Don't rush to set the plants outside; wait until nighttime lows stay above 50°F. Set transplants through slits in the plastic. If your peppers are caged, wrap the cages with row cover or plastic to create a mini-greenhouse. As daytime temperatures rise close to 80°F, open up the enclosure during the day. For pepper plants that aren't caged, use a medium-weight row cover.

Late in the season when nighttime temperatures are starting to decline, use row cover to cloak your peppers in pleasantly warm air overnight. If temperatures below 55°F are predicted, double up the row cover, which will provide about 8°F of insulation. Let the first layer rest directly on the plants, and if possible, support the second layer with arches (see Row Covers for support ideas).

In some areas, such as high altitudes in the Rocky Mountains, covering peppers with row cover may be a nightly ritual throughout the gardening season.

AVOID OVERHEATING. Temperatures higher than 90°F are hard on peppers, causing blossom drop and sunscald (if sunlight is intense). Plan ahead to shelter peppers from high heat by planting them between rows of tall sunflowers or trellised tomatoes that will provide afternoon shade when temperatures are peaking in summer. If that's not possible, plant peppers in pairs: Set two plants 6 inches apart, then plant the next pair about 2 feet away, and so on. This creates a better cloak of foliage to shield fruits from sunscald. If necessary, sprinkle peppers lightly with a cooling spray of water in the middle of the day (not in humid areas, though). Or you can cover your peppers with shade cloth, using the same kind of supports you would for row covers.

To prevent soil overheating, remove black plastic mulch if possible and replace it with several inches of a cooling organic mulch. If you can't remove the plastic, cover the plastic with organic mulch to prevent it from heating up. Just be sure to monitor soil moisture under the plastic and mulch.

Keep in mind that in some arid hot-summer areas, temperatures drop dramatically at night. Plants that needed protection from heat during the day will benefit from some nighttime insulation from cold air. Put a single layer of row cover over peppers if temperatures below 60°F are expected, but be sure to remove the cover promptly in the morning.

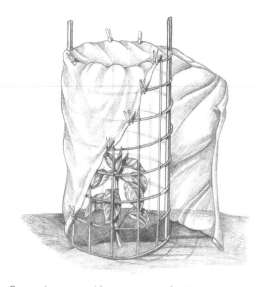

Cover wire cages with row cover or plastic to create a warm, sheltered environment for young pepper plants.

REGIONAL NOTES

FAR SOUTH

Pepper weevils are tiny black beetles with a long beak. These weevils remain active year-round in the southernmost parts of the United States, from Florida to southern California. The first sign you may notice when weevils invade is misshapen flowers and small, yellowing fruits. Weevil larvae bore into pepper fruits to eat the seeds, causing internal browning and deterioration of the fruit. Fruits may fall off the plant. The larvae are white, less than ¼ inch long. Destroy the damaged buds and peppers and clean out crop debris so the weevils can't overwinter. Weed out nightshade around your yard; pepper weevils also feed on nightshade and overwinter in the weedy debris.

FAR WEST

Powdery mildew doesn't produce quite the same symptoms on peppers that it does on other crops. The white powdery growth characteristic of this fungus shows up only on undersides of leaves. After that, the coating turns brown, and yellow spots appear on the upper side of the leaves. Leaves tend to fall off the plant. Powdery mildew of peppers occasionally also appears in Florida and other states. For controls, see Powdery Mildew.

PREVENTING PROBLEMS

BEFORE PLANTING

- Spread black plastic to prewarm the soil.
- Clear out nightshade and other weeds that can host tarnished plant bugs, pepper maggots, or pepper weevils (see Tarnished Plant Bugs for a list).
- Plant insectary plants to attract and shelter beneficials that prey on aphids (see Beneficial Insects).

AT PLANTING TIME

- Encircle each transplant with a cutworm collar.
- Cover plants with row cover to protect them from cold and insect pests.
- If not using row covers, cover individual plants with cloches or other protection on chilly nights.

WHILE CROP DEVELOPS

- Remove row covers when plants outgrow them or temperatures rise.
- Spray plant health boosters to promote natural disease resistance (see Disease Control for information).
- Pick off individual leaves that show symptoms of disease and destroy them.
- Check leaf undersides for white masses of European corn borer eggs and crush them.

AT HARVEST

- Inspect fruit for holes or signs of disease and separate it from healthy fruit.

AFTER HARVEST

- Remove noticeably diseased crop residues and discard or destroy them.

TROUBLESHOOTING PROBLEMS

NEW SEEDLINGS EATEN. Cutworms or slugs and snails are responsible. Cutworms often cut transplants off cleanly near soil level; see Cutworms for preventive measures. Slugs and snails leave behind a slime trail; see Slugs and Snails for controls.

SPOTS ON LEAVES. Early blight begins as light-colored spots on leaves; severely infected leaves die and fall off. Early blight is rarely a serious problem in peppers, but if it appears on your peppers, take steps to control it so that it doesn't spread to tomatoes or potatoes (which can be seriously damaged). See Early Blight.

WARTLIKE SPOTS ON LEAVES. Bacterial spot causes these raised brown spots; sometimes, the spots appear water-soaked. To control bacterial diseases, pick off infected leaves. Switch to drip irrigation or soaker hoses for watering. Don't touch plants when wet. If the problem has been severe in the past, use copper sprays to prevent the spread of the disease. For the future, choose resistant pepper varieties. Also, start a 3-year rotation of tomato-family crops, be sure to buy clean seed or healthy transplants, and disinfect all stakes and cages if you plan to reuse them (use a solution of 1 part household bleach to 12 parts water).

WATER-SOAKED SPOTS ON LEAVES. Cercospora leaf spot is a fungal disease that infects many vegetable crops. Spots tend to have dark edges; as symptoms progress, centers may drop out of spots, leaving a shothole effect. See Cercospora Leaf Spot for other controls.

BLACK COATING ON LOWER LEAVES. When aphids feed, they produce sticky honeydew, and mold sometimes grows on the honeydew. Spray plants with water to dislodge aphids. For the future, encourage beneficials; they usually keep aphids under control.

SKELETONIZED OR CHEWED LEAVES. Colorado potato beetles or blister beetles are possible culprits. If you find them on your plants, handpick them daily (wearing rubber gloves to remove the blister beetles). See Colorado Potato Beetles and Blister Beetles for more controls.

SMALL HOLES IN LEAVES. Flea beetles or European corn borers are possible culprits. Flea beetles aren't a serious pest of peppers except on very young plants (see Flea Beetles). If the damaged foliage is close to fruits, check the fruits for holes, a sign that the borers have tunneled into the peppers. You'll find the yellowish brown larvae inside the peppers. For the future, scout leaf undersides for masses of white eggs (see illustration on page 129). If you notice the holes in leaves and fruits are still less than 1 inch in diameter, spraying *Bacillus thuringiensis* or spinosad may be effective to kill the borers. If fruits are larger, though, the borers will already be inside and sprays won't affect them.

LEAVES ROLL UP. Leaf rolling sometimes happens in response to heavy rain or very strong sun (you may find rolled leaves on the sunny side of the plant). Even if the leaves don't recover, the plants should still produce fruit.

YELLOW MOTTLING OF LEAVES. Mosaic virus is the cause, and infected plants may be stunted or have twisted growth. For the future, choose resistant pepper varieties. See Mosaic for more information.

YELLOW STIPPLING ON LEAVES. Spider mites cause this damage. Look closely at leaves for small dark specks and webbing on leaf undersides. See Mites.

GROWING TIPS DIE OUT. This can be due to spotted wilt virus, which is most common in the Southeast. See page 416 for information.

BLOSSOMS DROP OR NO FRUIT FORMS. Pepper plants drop their blossoms when temperatures are too high or too low, when they're water-stressed, or when tarnished plant bugs are feeding in the blossoms. See "Secrets of Success" on page 293 for methods to protect pepper plants from temperature extremes. Check soil moisture regularly and water as needed whenever the top inch of soil is dry.

If temperatures have been moderate and you've watered your peppers regularly, use a hand lens to check the buds carefully for tarnished plant bugs. These pests can be difficult to control; see Tarnished Plant Bugs.

FEW FRUITS FORM. If no tarnished plant bugs are present, and the weather has been moderate, the problem is probably excess nitrogen. Too much nitrogen causes lots of lush leaf growth and poor fruit set. For the future, switch to a fertilizer that's lower in nitrogen content.

PLANTS WILT. Aphids, bacterial wilt, southern blight, Phytophthora blight, Verticillium wilt, root rot, and white mold are possible causes.

Aphids suck sap from leaves, and when the weather is hot and dry, they may stress the plants enough to cause wilting. Look for aphids on leaf undersides and growing tips. Wash them off with water, and plants should recover.

Bacterial wilt can develop suddenly, and the wilted plants sometimes die quickly. Confirm the problem by cutting into the main stem at the soil line—if wilt is present, a brown slime will ooze from the stem. The bacteria can survive in soil for several years. To reduce future problems, enrich the soil with compost, and start a rotation of tomato-family crops. Be careful not to move infected soil from one bed to another in your garden.

Southern blight is a fungus that spreads in hot, moist weather. Leaves will turn pale and fall off. Look for cottony white mold on the soil at the base of the stem. Carefully uproot plants without knocking loose any of the mold and discard or destroy them. To prevent future problems, see Southern Blight.

Phytophthora blight is also a fungus; it affects foliage and fruits. Fruits that come in contact with the soil containing Phytophthora fungi often rot. Controls are similar to those for other fungal blights; see Early Blight.

Verticillium wilt often causes wilting that begins on only one side of the plant. Internal brown steaks in the stem near the soil and in the roots confirm the diagnosis. See Verticillium Wilt for controls.

If roots are wet and brown, fungal root rot is to blame. To prevent this in the future, see Rots.

White mold is a fungal disease that starts out as water-soaked spots on stems. White moldy growth appears on the stems and then the branches wilt. To prevent future problems, see White Mold.

PLANTS TOPPLE OVER. Strong wind or the weight of developing fruit can cause pepper plants to keel over. Don't try to lift the plants back up because branches break easily. Instead, put straw mulch or newspaper under the plants to prevent fruit rots. Pick fruits as early as possible because they may be more susceptible to sunscald.

STUNTED PLANTS. Peppers don't grow well in acid soil; other symptoms include yellow spots on leaves and puckering. Confirm that soil acidity is the problem by testing pH. If needed, correct the problem by adding lime in fall or winter (in the South). Be sure to include the area where you plan to grow peppers next year.

SMALL, FLATTENED FRUITS. When pepper fruits are flat and undersized, it's a sign of poor pollination, usually due to low light or low temperature. Pick the small fruits; they won't get much bigger, and picking them will encourage more fruit set. Try tapping open flowers during midday (as long as the weather is dry) to improve pollination.

WATER-SOAKED SPOTS ON FRUITS. Anthracnose and Phytophthora blight are fungal diseases that can ruin pepper fruits. Spots caused by anthracnose become leathery; infected fruits may be usable if you cut out the bad spots. See Anthracnose for prevention tips.

When Phytophthora is the cause, white mold often appears on infected fruits. See Early Blight for controls.

DRY TAN SPOT ON OR NEAR BLOSSOM END OF FRUIT. Blossom-end rot is the result of a lack of calcium while fruit is forming, usually due to moisture stress. On peppers, spots sometimes show up on the sides of the fruit instead of at the end. Fruits may be usable if you cut out the damaged area. See Blossom-End Rot.

YELLOW RINGS ON FRUITS. This is a sign of spotted wilt virus, which is most common in the Southeast. See page 416 for information.

WHITE PATCHES ON FRUITS. If there is no other sign of disease, the problem is sunscald, which happens when pepper fruits are exposed to strong sun. The areas may become water-soaked or dry and papery. This can happen if plants drop leaves due to disease or if unsupported plants topple over. In future, side-dress plants to boost foliage production, take steps to prevent disease, and support plants with stakes or cages.

HOLES IN FRUITS. Corn earworms, European corn borers, pepper maggots, and pepper weevils are possible culprits. Pepper maggots (fly larvae) and pepper weevil larvae are small white worms, corn earworms have dark heads, and European corn borers are yellowish brown. See Corn Earworm for earworm controls. To prevent future problems with corn borers in peppers, check leaf undersides frequently for masses of white eggs and crush them (see page 129). To prevent future problems with maggots and weevil larvae, clear all pepper plant debris out of the garden at the end of the season, and also weed out all nightshade. In general, damaged fruits are usable if you cut away the damaged portions. Injured fruits won't store well, so use them as soon as possible.

FRUIT DROPS EARLY. This is another sign of pest insects feeding inside the fruits. See "Holes in Fruits" above for control measures.

PEST CONTROL

Tiny insects can leave you hopping mad when a sneak attack catches you unprepared. Even if the damage isn't serious, it's hard to stay rational when broccoli or bean plants that were perfect just yesterday are now marred by holes in the leaves. But it's important to stay calm when you assess pest damage. Most of the time, you'll spot pest problems early, when they're not serious enough to threaten the harvest. Simple, safe techniques such as handpicking or applying a biocontrol product can prevent a pest explosion that might require more invasive measures to control.

EVALUATE BEFORE YOU ACT

The sooner you spot invaders in your garden, the more easily you'll be able to deal with them. But whether an insect pest problem is serious or minor, start by identifying the cause. Keying out insects isn't too difficult, and it can be fun. You'll learn how on page 301.

Once you've identified the pest, evaluate the damage. How many plants are infested? How much foliage (or root) injury is there? How close to maturity is the crop? Are any predators or parasites present?

For example, if you find pests on only a few plants, you could simply pull up the plants and follow up with a daily pest check on the rest of your crop. That simple action may solve the problem. If the crop is near maturity, though, don't pull out any plants! Chances are the pests won't have time to significantly harm the crop. Get rid of as many pests as you can by handpicking or spraying with water, and harvest the crop as soon as it's ready to pick. (Then do your best to destroy the pests so they don't overwinter in your garden.)

When pests move into a large stand of young plants, weigh your options before you take action. If you decide that the damage is going to kill the plants or injure them severely enough to ruin the harvest, you may need to use a product such as insecticidal soap or neem to kill most of the pests quickly and save your crop.

FROM PROBLEM SOLVING TO PREVENTION

Be an optimist. Even if you're a busy gardener, you *will* find time to prevent pest problems rather than solve them. Prevention is actually a time-saver. Experience will teach you which pests will haunt your garden season after season. For those foes, plan to prevent problems using a range of techniques:

- Planting strategies such as crop rotation and timed planting.
- Barriers, repellents, and traps that prevent pests from reaching target crops.
- Trap crops planted to entice pests to ignore crops that you want to protect.
- Encouraging beneficial insects or using biological controls that prey on or parasitize plant pests.

HOW TO BE A
GOOD PEST SCOUT

Throughout the season, whenever crops are actively growing—even if it's only one bed of early spinach and peas—go pest hunting in your garden once a week (twice is better).

Gather some equipment before embarking on a pest-scouting foray. You'll need a magnifying glass and a heavy piece of dull white paper, as well as a small plastic container with a lid (such as a deli container) or resealable plastic bags to collect insects for identification. Hand sanitizer gel works well as an insect preservative. Take along a small bottle or just squirt some gel into the plastic container before you head outside. If you don't have any hand sanitizer, you can stow the container or bags in your freezer until you have time to identify the insects.

THE TELLTALE SIGNS

What does a pest scout look for? Insects, of course, but that's not all. Look for signs of insect activity such as chewed leaves or tiny droppings.

DAMAGE PATTERNS. The pattern of damage can offer important clues to the identity of a pest. For example, mottling or stippling on leaves is a sign of damage caused by sucking pests such as thrips. Holes in leaves usually indicate pests that feed by chewing, such as caterpillars or slugs.

Keep in mind that damage symptoms may look different depending on whether the insects are young or nearly mature. Tiny imported cabbageworms, for instance, may just scrape at the leaf surface, while bigger cabbageworms can devour large sections of cabbage or broccoli leaves.

HIDING PLACES. Turn leaves over and look at the underside; here you might find aphids, spider mites, and whiteflies, as well as eggs of armyworms, Colorado potato beetles, and squash bugs. Peer down into the center of bushy plants. Check the growing tips and leaf axils carefully for small pests such as aphids, which are especially attracted to tender new growth. Hit foliage against the white paper and check the paper for spider mites and thrips, which might not be visible on the foliage itself. If the plant has flowers on it, check those, too, for holes or insects hiding inside.

UNDERGROUND PROBLEMS. Check below soil level, too, especially if you find that plants

NEW PESTS
ON THE HORIZON

After gardening for several years, you'll recognize all the pests that trouble your vegetable garden, right? Perhaps not. Once every 3 to 4 years, a plant pest is imported into North America from another continent and becomes a serious problem in some region. Forty to 60 percent of the agricultural pests in the United States are exotics, including famous pests such as the imported cabbageworm, European corn borer, and Japanese beetle. One example of a new pest that's a problem in Canada and is venturing southward into the Northeast is the swede midge, a sucking insect native to Europe that can devastate cabbage-family crops (see page 70 for more information).

are languishing but no pests are present. Pull up a single bean seedling or one carrot or beet plant. Are the roots distorted or rotting? Are larvae feeding on or in the roots? Check soil moisture, too, by sticking your finger down into the top few inches of soil. Make note of any areas in need of watering.

While you're scouting for insect damage, it also makes sense to look for other types of problems, such as discolored leaves, mold growing on a leaf underside, or distorted growth. These could be signs of disease problems or nutrient deficiencies.

USING TRAPS TO MONITOR PESTS

Sticky traps add a level of sophistication to your scouting. Bright yellow surfaces attract many kinds of insects. You can take advantage of this attraction and make simple traps for monitoring pest populations. Cut bright yellow cardstock into pieces about 3 inches by 8 inches, or paint similar-size pieces of cardboard or plywood bright yellow. Coat the traps with a sticky substance such as Tanglefoot. Insects attracted to the yellow color land on the trap, get stuck, and die. If you prefer not to take time to make the traps from scratch, you can buy commercial traps from garden suppliers. Because sticky traps attract (and kill) beneficials, don't mount them permanently in your garden. Instead, take them along on scouting trips. Hold a card near crop plants and shake the plants to disturbs pests. As they fly up, some will be attracted to the sticky traps.

IDENTIFYING INSECTS

Suppose you've finished scouting and you've collected a few insects, which may be pests, benefi-cials, or neutral. How do you identify them? You can do it yourself by consulting references, or you can take the samples to your Cooperative Extension office and ask for expert help.

Want to try it on your own? Then it's helpful to know the basic terms used to describe insects. Also, it's important to understand how insects grow and develop, because some insects go through dramatic changes of size and appearance during their lives.

Adult insects have these basic characteristics in common:

- All adult insects have bodies divided into three parts: head, thorax, and abdomen.
- All adult insects have three pairs of jointed legs. Some immature insects, such as caterpillars, may have only rudimentary legs or no legs.
- Adult insects have one pair of antennae.
- Adult insects usually have 2 pairs of wings.

Larvae (immature insects) don't necessarily share these characteristics; they may look nothing like the mature form of their species. The change that occurs as an insect grows is called

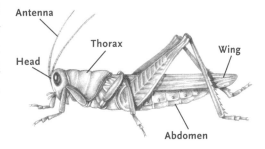

Listings of insects in identification guides often describe the appearance of the basic body parts shown here.

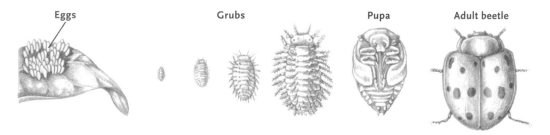

Eggs | Grubs | Pupa | Adult beetle

Mexican bean beetles undergo complete metamorphosis. Bright yellow eggs hatch as spiny grubs. The grubs feed, grow, and then pupate. Adults are yellowish orange beetles with black spots.

metamorphosis. The classic metamorphosis is the transformation of a caterpillar (larva) into a butterfly or moth.

Every insect begins life as an egg. From there, metamorphosis can be "complete" or "incomplete." Certain categories of insects, such as beetles and moths, undergo complete metamorphosis. Other types, such as aphids, develop via incomplete metamorphosis.

COMPLETE METAMORPHOSIS. In this type of life cycle, larvae often have a strikingly different form from the adult they will become. Caterpillars and grubs are the most common types of larvae you'll find in your garden. As these wingless eating machines feed, they grow, shedding their skin several times. After several weeks of eating, a larva senses that it's time to transform. It shifts to a resting phase called the pupal phase. A pupa may be exposed, or the larva may coat itself in a cocoon or pupa case as part of the process. When the transformation is complete, the adult insect breaks out of the pupa. Male and female adults mate, females lay eggs, and more larvae hatch, continuing the cycle.

INCOMPLETE METAMORPHOSIS. In this type of metamorphosis, the immature creatures that emerge from eggs are called nymphs. Nymphs are wingless but sometimes look similar to the adults they will become. Squash bugs are an example. Nymphs spend their time relentlessly eating and growing. They molt as they grow, finally becoming full adults without going through a resting phase. Adults mate; females lay eggs, and the life cycle is complete.

As a first step in identifying a specimen, turn to the illustrated key of common vegetable garden pests and beneficial insects on page 452; perhaps you can figure out the identity simply by comparing the specimen to the drawings in the key. Keep in mind that even if you found a critter on a damaged leaf, that doesn't mean it's a pest. (Most of the living things in your garden are beneficial or neutral.) If you see an insect in the key that resembles the one you found, turn to the entry for that pest or to the Beneficial Insects entry for a detailed description of all life stages. This should confirm the identification. If you're really stumped, take the specimen to your local Cooperative Extension office, where a knowledgeable staff member will help you.

If you confirm that the insect is a vegetable pest, next learn about its life cycle and the tactics you can use to keep it from ruining your crop.

PLANTING STRATEGIES THAT PREVENT PROBLEMS

Pest prevention can start even before you sow a seed. Sometimes you can avoid a potential pest problem simply by the choices you make about what, where, or when to plant.

TOLERANT VARIETIES

Insect-tolerant plants suffer less of a setback than other varieties when pests feed on them. Here's one example: Some varieties of sweet corn are less vulnerable to corn earworms because the wrapper leaves on the ears are very tight. The corn earworms literally have a hard time tunneling their way into the ear, so they don't do as much damage.

For other crops, the reasons why one variety tolerates pest damage better than another are not always known. It may be because they have hairier leaves or a thicker cuticle (waxy layer on plant surfaces). Your observations over time will help you discover whether one variety you're growing is more pest-tolerant than another. Sharing experiences with fellow gardeners in your area can help you, too. Eventually you will start to identify varieties that produce a great yield even when they've been chomped on by caterpillars, beetles, or other kinds of pests. Let the search for tough varieties motivate you to keep good garden records from year to year!

CROP ROTATION

On organic farms, crop rotation is an essential part of controlling pest and disease problems. From year to year, farmers change the locations where they plant crops from different plant

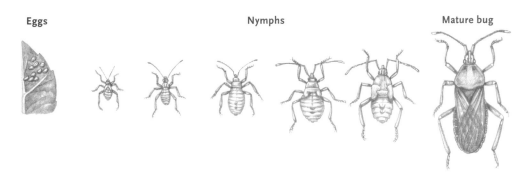

Eggs **Nymphs** **Mature bug**

The squash bug's life cycle is an example of incomplete metamorphosis. Nymphs hatch from eggs laid in spring. The nymphs feed and enlarge, molting five times before reaching the mature bug stage in late summer.

families to avoid insect pests and to prevent disease problems from escalating. On the home garden scale, though, rotating crops probably won't help in controlling insect pests. Unless you have a very large vegetable garden, the pests that overwinter in one bed will be able to fly or crawl to the crops they want to feed on, even if you've planted them at the far end of your garden. One case where rotating crops will play a role in controlling an insect pest is the Colorado potato beetle (see Colorado Potato Beetles).

TIMED PLANTING

Planting a crop earlier or later than usual can lessen pest problems by avoiding peak pest populations. For example, gardeners in the Northeast can avoid the main flush of Colorado potato beetle larvae by planting a month late and using a fast-maturing variety. In all regions, bean leaf beetle populations build up throughout summer. In this case an early bean planting is less likely to suffer than a late planting. Conversely, flea beetle damage often is worst in spring. Fall plantings of arugula, broccoli, and other susceptible crops may suffer little or no flea beetle damage.

Contact your Cooperative Extension office or talk with knowledgeable local gardeners to find out approximately when key pests emerge in your area and whether you can plant early or late to avoid them.

HANDS-ON CONTROL

You can reduce pest populations every time you visit your garden, and you don't need com- mercial pest control products or special equipment to do it. Handpicking insects, spraying infested plants with water, and removing crop debris are simple hands-on pest control techniques that are surprisingly effective for combating leaf-eating caterpillars, aphids, and other common pests.

HANDPICKING

Slugs, snails, caterpillars, and some beetles and true bugs are relatively large, slow-moving pests that you can pick off plants one by one. Handpicking these pests is effective, involves no costs, and has no negative effect on your garden. The only negative impact may be on you, if you're squeamish about handling insects. Avoid getting the willies by wearing latex gloves when you handpick, or use an old pair of chopsticks or tweezers to grasp the insects.

After you've collected the insects, you have to dispose of them. Squishing them works, but it can be messy and unappealing (children may do it happily, though). Instead, put some water and a few drops of dish soap in a wide-mouthed container or a small plastic bucket. Drop the insects into the soapy solution to kill them. When you're done, dump the water in an out-of-the-way corner of your yard.

If you have an insect or butterfly net, take it along when you handpick. It's surprising how many cabbage butterflies and squash vine borer moths you can snag while patrolling your garden. Check the contents of your net carefully. Put the pests into soapy water to die; release beneficials and other insects you can't identify.

WATER SPRAYS

Handpicking aphids and other very small insects isn't practical. Instead, remove aphids, thrips, spider mites, and other tiny, soft-bodied pests from plants by having a "plant wash." Turn a strong spray of water from your garden hose on infested plants. Direct the water at as many angles as possible to reach undersides of leaves and into the center of rosette-type plants such as spinach. A blast of water also can dislodge young caterpillars from cabbage-family plants and wash away their droppings. Spray plants in the morning so that they'll dry quickly, lessening the chance of disease problems. Recheck plants a day or two after spraying, and spray again if you find more pests. Check soil moisture around plants if you spray repeatedly, and stop if the soil is waterlogged. Soggy soil can lead to root diseases, which could be worse for your plants than the pests.

CLEANING UP AFTER HARVEST

In ornamental gardens it's fine to leave plants standing through winter to provide shelter for wildlife and winter interest in the landscape. But in your vegetable garden, pull out plants or turn them under after harvest because crop debris is a favorable overwintering site for insect pests. If you plan to turn plants under to build up soil organic matter, break up the plants as best you can. Here's one situation where a tiller can be put to good use. If you're going to dig plants under by hand, first chop them into pieces with a machete or the sharp blade of a spade. Your goal is to make the debris break down fast so that pests have nowhere to hide.

If you pull up plants to add to a compost pile, be sure to pull out the roots, too. Chop up the debris before adding it to the pile, and try to put it right at the center of a pile, where the biological activity is highest. Keep in mind that pests may survive in the compost pile if it is slow-acting; on the other hand, beneficial organisms in the compost pile may prey on the pests.

After pulling a crop, plant another vegetable or a cover crop, or mulch the bed. If the crop had problems with soil-dwelling pests such as wireworms, it's better to leave the soil exposed for several weeks during fall, cultivating weekly. This turns up pests for birds to eat and may expose some pests to killing frost. After this

VACUUM DOOM FOR PESTS

Suck pests off plants with a handheld portable vacuum cleaner. This technique can work well for gathering whiteflies, tarnished plant bugs, leafhoppers, and slow-to-fly pests such as Japanese beetles. Practice first to develop a technique that avoids damaging plants. For whiteflies, try brushing your hand over plants and following after with the vacuum (a hose attachment works well if you have one). For other pests, such as Japanese beetles, put a piece of cardboard on the ground under plants, shake the plants to dislodge the pests, then vacuum them up.

When you finish, dump the insects into a plastic bag and put them in the freezer overnight. Next day, empty the bag outdoors—the pests will be dead.

treatment, cover the soil for winter to prevent erosion.

As you clean up your vegetable garden in fall, uproot grassy weeds and weeds that belong to the cabbage family and tomato family, such as wild mustard and black nightshade, because they also harbor crop pests.

KEEPING PESTS AT A DISTANCE

Save crops from egg-laying moths and flies and a variety of chewing and sucking pests by putting a barrier between the pest and the crop. Another option is to apply a substance that will confuse or repel the pest and cause it to look elsewhere for food. Barriers and repellents are well proven as pest protectors; insect traps are a newer type of contraption for keeping pests away from plants, and you'll have to experiment for yourself to see how well they work.

PEST-SPECIFIC BARRIERS

For certain species of pests that need to gain entry to a plant at a specific site, simple homemade barriers will stop the pest in its tracks. For example, you can make cardboard collars to place around plant stems to stop cutworms, as shown in Cutworms. Block egg laying by cabbage maggot flies by protecting each transplant with a collar placed flat on the soil surface, as shown in Cabbage Maggots. Copper barriers are more expensive, but they can help keep voracious slugs and snails out of garden beds; see Slugs and Snails for details.

ROW COVERS

The most foolproof way to block leaf-feeding pests such as flea beetles and cucumber beetles is to cover plants completely before the pests show up. Covers made of lightweight synthetic fabric can very effectively protect crops from these damaging and widespread pests. Consult Row Covers for information on their use.

KAOLIN CLAY

Your garden may look like a frost hit when you use kaolin clay, but this frost helps your plants rather than hurting them. Kaolin helps prevent damage by some tough pests, including flea beetles, striped cucumber beetles, and grasshoppers. The clay coating seems to disorient or repel the pests so that they don't feed, but it doesn't kill them. Some disease organisms can't penetrate the coating, either, making this a doubly useful problem preventer.

The coating doesn't interfere with photosynthesis; in fact, in high heat and intense sunlight, kaolin relieves heat stress and helps prevent sunburn on foliage and fruit because it reflects harmful rays. Kaolin does not harm beneficials.

When to Use

It's important to apply kaolin before pests arrive. It's useful on a wide variety of crops to suppress bean leaf beetles, cucumber beetles, flea beetles, Japanese beetles, leafhoppers, Mexican bean beetles, and thrips. Results aren't equal for all crops or all conditions. Trial and error will show you when kaolin will work well in your garden.

How to Use

Kaolin clay is sold as a dry powder. Always wear a protective mask or respirator when working with kaolin powder because inhaling the dust can damage your lungs. It's also an eye irritant, so wearing goggles is a good idea, too.

You must mix the powdered clay into suspension in water and apply it using a sprayer. The clay sometimes resists being mixed with water. It helps to premix the clay and water in a container that is easy to shake, such as a large plastic jug with a wide mouth. Start by putting 1 quart of water in the container and adding ½ pound (about 3 cups) of kaolin slowly. Wait until the powder sinks into the water, then shake the mixture for 30 seconds. Add the rest of the water gently (don't use a hose) and shake the mixture well for half a minute.

Another way to mix kaolin and water is in a plastic bucket, either stirring by hand or using a paint mixer powered by an electric drill. Once the clay is in suspension (it doesn't dissolve in water), pour the mixture into your sprayer.

Spray all plant surfaces, including the underside of leaves, coating them to the point where spray almost drips off. It helps to make two passes over a crop to ensure full coverage. Wash out your sprayer thoroughly when you're done. It's okay to store leftover mix in a container to use the next time you spray, but don't keep it longer than 2 weeks. Shake it thoroughly before you pour it back into the sprayer.

When Things Go Wrong

SPRAY WON'T STICK TO PLANT SURFACES. Although you apply kaolin as a liquid suspension,

---

BEYOND THE BASICS

TRAP CROPPING

Trap cropping is a daring method of pest control. The idea is to attract pests to a decoy crop (one you don't plan to harvest) and then hope that they stay there, allowing your desired crop to grow damage-free. Choose a trap crop—also called a sacrifice crop—that will be very attractive to the pests. For example, mustard or arugula may lure flea beetles away from broccoli. With luck, predators or parasites of the pest will arrive and feast on the insects devouring the trap crop, which will keep the pest population under control. If they don't, you may have to kill off the pests before they move elsewhere by spraying them or using a handheld vacuum to suck them off the trap crop foliage.

If you want to try trap cropping, plant a row of a trap crop about 10 feet away from your desired crop. Timing is important; plant the trap crop 1 to 2 weeks before the other crop. You can also try spraying your desired crop with a repellent such as hot pepper spray or kaolin clay to discourage the pests and drive them toward the trap crop.

Monitor the trap crop to be sure the pests aren't devastating it too quickly (because then they'll look elsewhere for food). If you think you need to kill the pests to prevent them from migrating to another crop, use a fast-acting product such as insecticidal soap or pyrethrin. Pull out the trap crop immediately when its purpose has been fulfilled so that you'll remove any larvae before they pupate.

it doesn't stick well on wet surfaces. Spray on a dry day after the morning dew has evaporated. Also, kaolin may not adhere well to waxy plants such as cabbage or broccoli. Add a spreader-sticker, such as lecithin, to the suspension when you spray plants that have a waxy coating. Let the sun warm the plant surfaces first, then spray.

SPRAY COVERAGE IS UNEVEN. Kaolin clay tends to settle to the bottom of a spray tank. If you notice that the spray is becoming thin or colorless, stop spraying and shake the sprayer vigorously for half a minute to remix the clay and water.

PEST DAMAGE APPEARS ON SPRAYED PLANTS. Kaolin discourages pests, but it doesn't kill them. If pest pressure is high, the pests may cause some damage despite the deterrent of the clay coating. Also, thorough coverage is important because insects may feed on uncoated areas. Respray the plants, making sure you cover all surfaces, including leaf undersides.

Pests may also start to feed when the kaolin coating wears off. If there has been heavy rain or the crop no longer appears white, it may need a fresh coat. To double-check this, wipe a sprayed plant surface with a piece of dark cloth. If you can't see a white residue on the cloth, you should reapply kaolin. Note: Since water can wash kaolin off plant surfaces, don't use sprinklers to water coated plants. Water by hand at soil level or use a soaker hose or drip tape.

To supplement the kaolin barrier the next time you plant, plant a trap crop as well. See "Trap Cropping" on page 307 for an explanation.

FRUITS ARE COATED WITH CLAY AT HARVEST TIME. Spraying kaolin clay up to the day before harvest is allowable, but it's unlikely you'll need to do that in a home garden setting. Usually, the clay will have washed away before you harvest. However, if you find the clay coating on harvestable leaves or fruits, just use a soft cloth to rub it off.

FOIL MULCHES

Shiny silver surfaces reflect sunlight, and the glare can repel aphids, leafhoppers, and some other insects. You can use silver mulch to cover the soil around plants, especially those susceptible to viruses spread by aphids and leafhoppers. Rolls of metallized plastic mulch are available from some garden suppliers, or you can improvise by spray-painting clear plastic mulch with silver paint. For a small planting, tucking pieces of aluminum foil around plants may work. You'll have to experiment to see whether this repellent protects a crop thoroughly enough to be worth the time and expense. Try using silver mulch or foil on one bed and a conventional organic mulch (or black plastic) on another bed. Plant the same crop, same variety in each bed and see what happens.

Reflective mulch offers other benefits for plants as well. See Mulch to learn more.

HOT PEPPER AND GARLIC SPRAYS

The heat and bite that hot pepper and garlic lend to soups, stews, and other cooked dishes are culinary pleasures for us, but the strong flavors are reputed to repel a wide variety of garden pests. You can buy commercial products containing garlic or capsaicin (the active ingredient in hot peppers) or both, or you can make your own. Don't worry, the crops sprayed will not absorb the flavor of garlic or hot peppers.

When to Use

Claims abound for garlic's repellent effect on a wide variety of pests, including aphids, caterpillars, flea beetles, leafhoppers, leafminer flies, maggot flies, spider mites, thrips, and whiteflies. Garlic extracts also kill mosquito larvae (see Mosquitoes). Hot pepper sprays are recommended for repelling aphids, leafhoppers, mites, thrips, and whiteflies.

Manufacturers of commercial hot pepper and garlic sprays recommend applying the sprays preventively—before pests show up on plants. Applying these sprays seems harmless, but there's evidence that garlic sprays kill some types of beneficial insects, including beneficials that prey on aphids. (Hot pepper sprays apparently have no effect on beneficials once the spray dries.) Also, since the sprays should be applied preventively, you might take action against a pest that never would become a problem (and thus waste time and money). Don't take a *carte blanche* approach with these sprays. The time to try these repellents is when past experience tells you that a crop is vulnerable to a particular pest. These repellents are probably less expensive to use than row covers, but they are less reliable.

How to Use

Follow label instructions for diluting and applying commercial products.

You can make your own hot pepper/garlic spray in a blender. Recipes abound, and there's no magic formula that's perfect for all situations. In general, recipes call for combining two to six cloves of garlic with one or two hot peppers (or 1 to 2 teaspoons ground cayenne pepper or pepper flakes) and 1 quart water. Some recipes call for adding a small onion; also, some recommend letting the mixture steep overnight before using. Do strain the mixture through cheesecloth before putting it in a pump spray bottle. You can also add a few drops of liquid soap, not detergent (but if you do, keep in mind that spray will be even more likely to kill beneficial insects). Shake the bottle to combine the soap with the mixture.

Whether you make your own or use a commercial product, spray plants thoroughly, including the undersides of leaves. Capsaicin and garlic juice are very irritating to human skin and eyes, so wear the proper protective gear when you work with these sprays.

Hot pepper dusts also work as a repellent. If you grow hot peppers, dry them and grind them, and store them in a glass jar. Use a mortar and pestle to grind the peppers, wearing a mask, goggles, and gloves as you work. Dust plants when they are wet with dew. To discourage carrot, cabbage, and onion maggot flies from laying eggs, sprinkle the dust on the soil surface at the base of the plants.

Spray on a dry day when no rain is expected. If a garlic spray seems to be keeping away pests, reapply it about every 2 weeks. Reapply hot pepper sprays every 3 to 7 days.

When Things Go Wrong

Although garlic and hot pepper have a long-standing reputation for repelling pests, results will vary from application to application and place to place. Perhaps your crops are especially vulnerable to pests because of stress or nutrient deficiency, and the garlic and/or capsaicin didn't provide a strong enough deterrent. Once you find insect pests on crops, give up on repellent sprays. If the pest problem is becoming severe, identify the pests and treat them appropriately.

COMMERCIAL INSECT TRAPS

Homemade sticky traps help to monitor insect populations, but commercial insect traps are designed using pheromone lures or floral lures to entice insects to enter a prison from which they can't escape. Most gardeners are familiar with Japanese beetle traps, which are baited with a floral lure and a pheromone—a chemical cue that insects emit to attract other insects of their species. These pheromones are powerful lures; however, many experts caution that the lures may draw beetles to your yard but not necessarily into the trap itself. For more details on Japanese beetle traps, see the Japanese Beetle entry.

Another type of trap baited with a floral or molasses lure attracts and catches night-flying moths such as armyworm moths, cabbage looper moths, corn earworm moths, and cutworm moths. These traps attract both male and female moths, which potentially can significantly reduce the number of eggs these pests lay in your garden—and that means many fewer caterpillars chewing on your plants.

When to Use

Set up floral lure traps in spring, before caterpillars begin damaging crops. Replace the lures as needed. Data on the effectiveness of these traps is limited. But if you've had serious trouble with loopers or armyworms in your garden in the past, they may be worth the investment.

How to Use

Follow the manufacturer's instructions on the packaging to set up traps using floral lures around your garden.

When Things Go Wrong

OTHER TYPES OF INSECTS TRAPPED. Some types of beetles are attracted to the floral lure as well, so you may find beetles among the moths, too. Ants may enter the traps in search of a meal, and spiders may also come looking for a captive meal, so to speak.

CATERPILLARS STILL DAMAGE PLANTS. It can take more than one season for floral lure traps to reduce populations. Also, moths continue to migrate into the area throughout the growing season. Be sure you've replaced the lures on schedule, or the traps may become ineffective. Also keep in mind that the traps won't lure every moth to its doom. The goal is to reduce caterpillar feeding to an acceptable level, not to eliminate caterpillars from your garden.

BIOLOGICAL CONTROLS FOR VEGETABLE PESTS

Beneficial insects and microorganisms that attack crop pests are a great help to gardeners, whether we see them in our gardens or not. Planting garden areas that attract beneficial insects will repay your investment of time and money many times over. See Beneficial Insects for information.

Research on microorganisms that attack crop pests is blossoming, but it's a long process to identify a potential bacterium or fungus, test it, and then formulate an effective product. For home vegetable gardeners, the primary microbial biocontrols are *Bacillus thuringiensis; Beauveria bassiana*; spinosad; beneficial nematodes, which control a range of soil-dwelling pests; and milky spore disease, which kills Japanese beetle larvae. To learn about beneficial nematodes, see Nematodes. See Japanese Beetles for information on milky spore.

Biological control products do not infect people, pets, or wildlife, but there are some precautions to observe when you apply them. A very small percentage of people have an allergic reaction to Bt, for example. And any spray material or dust can be irritating if it gets into your eyes or your lungs. Use common sense to protect yourself: Don't apply products during windy conditions, protect your eyes, avoid inhaling sprays and mists, and use a dust mask when spreading dusts or granules. Wear gloves, shoes, socks, long sleeves, and long pants, and wash the clothing afterward.

BT (BACILLUS THURINGIENSIS)

First released in the 1960s, Bt was a breakthrough in organic pest control: a product containing a toxin produced by a bacterium. Insect pests sicken and die when they eat leaves sprayed with products made from Bt. There are several subspecies of Bt; each one is effective against a different class of insects. *Bt* var. *kurstaki* and var.

aizawai (Btk) produce toxins that kill many types of leaf-chewing caterpillars. *Bt* var. *san diego* and var. *tenebrionis* (Btsd) toxins kill beetle larvae. *Bt* var. *israelensis* (Bti) affects mosquito larvae, but it is most effective only when applied over a large area, not in a single yard (for better control options, see Mosquitoes).

Bt products contain the toxin and bacterial spores. Inside a caterpillar's gut, the toxin attaches to the gut wall, where it destroys a portion of the wall. Gut contents then spill into the insect's body cavity and poison it. The spores also contribute to the poisoning. The sick insects stop feeding and die within a few days.

Some commercial Bt products are manufactured using genetic engineering technology, which is not approved as organic. Check labels; if you're unsure, ask your supplier to recommend a Bt product that meets current organic standards.

When to Use

Young larvae are most susceptible, so apply Bt products when pest caterpillars or beetle larvae first appear. Bt won't be effective against adult moths or beetles. Different species vary in their susceptibility. You should get good results when you use Btk against armyworms, cabbage loopers, imported cabbageworms, diamondback moths, and tomato fruitworm. It's more difficult, but not impossible, to control boring insects, including corn earworms or squash vine borers, with Bt. Timing is important, and the Bt has to be placed in the parts of the plants where borers are active rather than simply sprayed onto leaves or stems.

In vegetable gardens, Btsd is useful to combat Colorado potato beetle larvae.

How to Use

Bt is available as a liquid concentrate or wettable powder that is mixed with water and sprayed on plants. Or you can buy it as a ready-to-apply dust or granules. Apply Btk to all parts of the plant where insects are likely to feed, including the undersides of leaves or the growing tips.

Btk kills not only pest caterpillars but also benign caterpillars such as butterfly larvae. When you apply Btk, keep the spray or dust away from plants that host butterfly caterpillars (such as milkweed); spray only plants that you're sure are infested with pest insects. Since insect predators and parasites don't eat plants, they shouldn't ingest or be affected by Btk.

When Things Go Wrong

PESTS DON'T DIE. You may see larvae on plants for 2 or 3 days after you spray because Bt doesn't kill pests on contact. The pests should have stopped feeding within a few hours of ingesting the Bt, though. If the pests you see are still eating, one of three things went wrong. First, the pest may not be susceptible to the type of Bt you applied. Reconfirm the identity of the pest and check the product label. If your diagnosis was accurate, a second possibility is that you didn't apply enough Bt. The pests are continuing to feed without ingesting any Bt. A third possibility is that the Bt became inactive very soon after spraying, so it had no effect on the pests when they ate it. Exposure to sunlight and heat can degrade Bt. When you reapply, do so in the late afternoon or on a cloudy day when no rain is expected (rain can wash Bt off plants).

MORE PESTS SHOW UP A FEW DAYS LATER. Bt doesn't spread naturally from infected insects to healthy insects. The only way to continue killing newly arrived or hatched pests is to repeat applications about 7 days apart (it's fine to hand-pick the pests between applications). Keep in mind that insects can become resistant to Bt. This is usually a problem only in large-scale agricultural settings, but if you live near commercial vegetable farms, it's possible that Bt-resistant pests will show up in your garden. (Colorado potato beetles and diamondback moths have developed resistance in some areas.) You may need to try a different control measure.

BEAUVERIA BASSIANA

In the 1800s silk makers wished they could wipe out the fungus *Beauveria bassiana*, which was infecting and killing precious silkworms. But you can be glad this fungus still exists. It's available as a spray that kills a variety of pests, including aphids, whiteflies, and Colorado potato beetles. (It is also naturally occurring—especially in soil.)

The spray contains fungal spores, which germinate and send hyphae into the bodies of susceptible insects. The fungus feeds on the blood and body contents, causing the insects to die of weakness and dehydration. The fungus also kills insect eggs. When conditions are right (cool, high humidity), the fungus produces white mold that covers the insect cadavers. New spores form in the mold and disperse, potentially providing another round of control.

Insects don't have to ingest *Beauveria*; they only have to come in contact with the sticky spores, either when spray drops hit them or when they crawl across sprayed plant surfaces. Susceptible insect species have not developed resistance to *Beauveria*.

When to Use

Wait until you see pests on the plants, then spray as soon as you can. Research on susceptible pests is ongoing; good results have been shown for whiteflies, aphids, thrips, diamondback moths, imported cabbageworms, grasshoppers, ants, and Colorado potato beetles. The fungus can colonize corn plants (without harming the plants) and then attack European corn borers that have tunneled into the stalks. The fungus may also control armyworms, ants, flea beetles, leafhoppers, mole crickets, Mexican bean beetles, weevils, and white grubs; research involving these pests has so far been inconclusive. You will need to experiment in your garden to see if you get good results.

Beauveria is most effective at killing young larvae. If a pest population has blossomed before you first spray, it will take several sprays, 3 to 5 days apart, to achieve full control. For example, if clouds of whiteflies spiral up from your plants when you enter the garden, it may be too late to tackle the problem with *Beauveria*.

B. bassiana infects insects best in cool, moist conditions. It may not do a good job of controlling pests in summer in arid climates.

How to Use

Mix the spray according to label directions, and use it immediately (or the spores may die). Spray all parts of plants where insects may be hiding, including flowers, growing tips, and undersides of leaves.

B. bassiana can infect and kill beneficial insects. To minimize harm to beneficials, spray it only on crop plants that have pest insects on them. Don't apply it to bodies of water because it could be toxic to fish.

To prolong the life of the spores, store the product container out of direct sunlight and below 80°F.

When Things Go Wrong

PESTS DON'T DIE. Although the fungus will infect pests quickly, it may take 3 to 5 days for the pests to die. Keep watching the pests to see what happens. If they're still alive and feeding 5 days after you sprayed, it's possible the fungus was inactivated by heat or dry conditions. Spray again in the early morning or in the evening on a cool day. If conditions have been dry, water your garden before you spray.

Another possibility is that the *Beauveria* was inactivated by a fungicide. If you applied a fungicide (such as a bicarbonate spray) within 48 hours before or after spraying *Beauveria*, the fungicide may have destroyed the *Beauveria* spores.

MORE PESTS APPEAR. *B. bassiana* won't kill 100 percent of the pests present at the time you spray. Plus, more pests may hatch or fly into the area after you've sprayed. Try a second application 5 to 7 days after the first one. You may need to apply a third spray as well. After that, you may still see a few pests on the plants; your goal is to reduce damage to a tolerable level.

SPINOSAD

Leaf-chewing caterpillars and fire ants suffer loss of muscle control after ingesting spinosad. Spinosad contains toxins called spinosyns produced by the actinomycete (a type of bacterium) *Saccaropolyspora spinosa*. Pests have to ingest the toxin in order for it to take effect; within as little as 2 hours, they die of exhaustion. The spray can persist on leaves for up to a week. There's also evidence that plant tissues can absorb spinosad.

Predatory insects don't seem to be affected by spinosad, but laboratory studies show that the fresh spray is toxic to honeybees and some other types of beneficials.

When to Use

Spinosad is a relatively new pest control product, and researchers are still studying its effectiveness against various pest species. Spinosad is a good control option when you spy young caterpillars such as diamondback moths, cabbage loopers, and tomato fruit borers on crops. It will kill asparagus beetle larvae on asparagus fronds. Spinosad also is somewhat effective for controlling leafminers and thrips. Spinosad is not effective against aphids, whiteflies, plant bugs, or most types of adult beetles. It may control flea beetles and shows some effect on small beetle grubs.

How to Use

Spinosad is available as a liquid or a wettable powder. Mix it according to label directions. Apply the spray in the morning or late afternoon. Repeat sprays every 4 to 7 days. Don't apply spinosad to bodies of water, as the toxins won't break down in water unless exposed to direct sunlight; also, it may be toxic to fish.

When Things Go Wrong

If you live near commercial vegetable farms, it's possible that some of the pests in your garden are resistant to spinosad. For example, some populations of diamondback moth have already become resistant to it. Like any other insect-killing spray, spinosad is not a magic bullet. Look on it as a last resort or as one possible control method for a problematic pest, not the only control.

HOW'S THE WEATHER?

Smart gardeners watch the forecast carefully. Not only can weather itself cause problems (such as strong wind toppling a bean trellis), but weather can also play a role in pest problems. For example, if slugs are attacking your plants like never before, check the rainfall pattern for the last few weeks. Was there more rain than normal? More cloudy, humid days? These are the conditions that slugs love, and their populations may be larger than usual. Conversely, if thrips have suddenly appeared on some of your crops, the infestation may be related to droughty conditions. If you keep track of weather conditions during the gardening season and note when you have insect problems, you can learn to anticipate which pests may rise to prominence and take some preventive action.

Sometimes, a change in the weather can help relieve a pest problem too. For instance, a drenching rain may wash away a burgeoning aphid colony on your pea vines, as well as many other pests.

HOMEMADE BIOCONTROL

When you handpick insect pests, you'll probably find some that appear diseased or are already dead. They may be discolored or mummified, or perhaps they are decaying into mushy goop. Often when a pest population swells, a disease epidemic builds up, too; it's part of the cycle of life in natural systems. Some gardeners take advantage of such a disease outbreak by using the bodies of the diseased dead insects to make a biological insecticide.

When to Use

Look for diseased insect pests in your garden. Identify the pests carefully before you collect them. If you find diseased cabbage loopers, for example, you can prepare them as a dust or spray and apply the material to plants that loopers might attack, such as broccoli and cauliflower. If the diseased insects are beneficials, however, it makes no sense to use them—the resulting spray may kill more beneficials.

How to Use

Crush dry dead insects and sprinkle the dust directly on foliage. For soft-bodied insects, mix the crushed dead insects with water, strain the water through cheesecloth, and use a pump spray bottle to apply the "bug juice."

Keep in mind there are no scientific studies to show whether "bug juice" works or whether it carries any health risk to humans or pets. If you try this technique, approach it as an experiment and be cautious. Wear a filter mask, long sleeves, long pants, and gloves when you apply the bug powder or spray, and don't apply it to plant parts that you plan to harvest later (such as lettuce leaves).

When Things Go Wrong

There are no guarantees with a home-prepared concoction. If it doesn't work, try a different method or product instead.

THE FINAL LINE OF DEFENSE

Concern about the environmental and health effects of pesticides motivates many gardeners to switch to organic methods. Our goal is to tend our gardens without ever using substances that have any harmful side effects on people, animals, earthworms, or beneficial insects. However, many of us—and many organic farmers—don't always reach that ideal goal. The National Organic Program allows farmers to use a range of insect-killing substances such as biocontrol products, insecticidal soap, and botanical poisons, but only as part of an overall farm plan that requires the use of preventive measures such as crop rotation, encouraging beneficials, row covers, and trap cropping.

We can be even more selective in our approach. For example, rotenone, ryania, and sabadilla are botanical poisons that farmers can use as an emergency last resort, but these substances have a variety of toxic effects on people, animals, and beneficial insects. Because of their toxicity, *Organic Gardening* magazine advises home gardeners *never* to use rotenone, ryania, or sabadilla. There are plenty of other options—including preventive techniques, handpicking, barriers, and biological controls—for controlling home garden pests without resorting to these three harsh botanical poisons.

On rare occasions, you may decide that you have to use one of the insect-killing sprays described below to stop pests from wrecking one of your crops. If you decide to spray, keep your choice in perspective. The benefits that come from your organic vegetable garden outweigh the drawbacks of limited judicious use of these organically acceptable insecticides. As your garden system matures and your gardening experience grows, there will be fewer and fewer occasions when spraying is required.

For your own safety when you spray, always wear appropriate protective clothing, goggles, and a dust mask or respirator as recommended below.

INSECTICIDAL SOAP

Spraying a solution of soapy water to control pests is an old-time method for controlling pests. You can make your own or use commercial insecticidal soap products, which contain salts of fatty acids. Insecticidal soaps kill many kinds of soft-bodied pests; soaps also kill soft-bodied beneficial insects. Soaps may have less effect on adult insects such as pest beetles and lady beetles, but they do kill the eggs and larvae.

When to Use

Use insecticidal soap to control a serious outbreak of aphids, spider mites, thrips, or whiteflies. For caterpillars, Bt or spinosad would be more effective and less harmful to beneficials.

How to Use

For commercial insecticidal soap products, mix the concentrate with water in a sprayer according to label directions. Spray all surfaces of infested plants, especially leaf undersides and the central growing points. Insecticidal soap kills insects by damaging the cuticle (outer coating) so that the pests dehydrate. Insects must come in direct contact with the spray when it is wet. Soap sprays aren't toxic to people or pets, but you should still take precautions to avoid inhaling the spray or getting it in your eyes.

Insecticidal soap can kill lady beetle larvae and other soft-bodied beneficial insects and mites, so apply it only to crops that you know are infested with pests.

To make your own soap spray, mix 1 tablespoon nondetergent dishwashing liquid with 6 cups water. Use a product with few or no additives such as perfumes and whiteners because these may harm the plants.

When Things Go Wrong

LEAVES TURN BROWN AFTER SPRAYING. Applying soap sprays can damage foliage, flowers, or fruit. Soap damage is more likely to occur during hot and/or humid weather. To avoid damage, don't spray when temperatures are above 80°F; if possible, spray on an overcast day or in the late afternoon. Squash and some cabbage-family crops tend to be vulnerable to leaf damage. Plants may or may not recover from this injury.

If you want the option of spraying soap to control pests in the future, be proactive. Test spraying will reveal which crops are prone to soap damage. Mix a small batch of soap spray and apply a bit here and there on as many different kinds of plants as you can. Mark sprayed areas with colorful pieces of ribbon. Check the sprayed leaves 2 days later and record the severity of damage. Store your notes right with the soap container or tag them in your garden journal to refer back to. Then when a crisis comes up, you'll know whether it's safe to spray a particular crop with soap.

If a crop does show damage, it's still possible to apply soap spray safely. The trick is to hose down the plants with water about 3 hours after you apply the soap spray. This allows enough time for the spray to kill the pests, but not enough contact to produce leaf injury.

Using hard water (water with high mineral content) in a soap spray also increases the risk of foliage burn. If your tap water is hard, try collecting rainwater for making soap spray.

PESTS REAPPEAR. Insecticidal soap kills insects on contact, but it has no residual effect. It tends to dry up quickly. If you miss any insects when you spray, they can continue feeding and reproducing. Also, new pests may arrive after you spray and attack the crop. Keep in mind that the plants were probably stressed to begin with, and that's why pests had taken hold. Be sure the plants are well watered. They may need side-dressing with a balanced or low-nitrogen fertilizer (too much nitrogen can actually worsen aphid problems).

It's okay to repeat soap sprays weekly. Even better is to employ other control techniques for the pest as well, rather than relying on soap sprays alone. For example, 2 days after you spray insecticidal soap, spray plants with a strong blast of water to dislodge pests tucked in hidden areas that you didn't reach with the soap spray.

OIL SPRAYS

Mineral oil, vegetable oil, and certain essential oils have insect-killing properties. Mineral oil and vegetable oil kill insects and insect eggs by smothering them or by penetrating the cuticle. Essential oils such as those of cloves, thyme, and peppermint kill insects by affecting a neurotransmitter inside them (a type of neurotransmitter not found in humans or pets). These oils also tend to repel pests or discourage them from laying eggs. Often insecticidal oils are combined with insecticidal soap because there is a synergy among the ingredients. The combination product is more effective than soap alone or oils alone.

COMBINATION SPRAYS

Soap spray isn't just soap spray anymore. Garden centers and garden supply catalogs stock several kinds of sprays combining soap and other substances. One combination product is soap spray with seaweed extract added; the seaweed is reputed to protect foliage from soap burn. Other formulations combine soap with pest-control substances such as pyrethrins, sulfur, or oil. The combination may have a synergistic effect, or it may increase the range of pests that the spray will treat. This is a convenience, but it can also lead to overuse of pesticides. The best course of action is to identify the specific pest that's attacking your crop and choose a product that will control it.

If you choose a mineral oil product, be sure it is labeled as lightweight oil that is safe for actively growing crops. Some oil products that are less refined are safe only for spraying on dormant woody plants.

When to Use

Oil sprays will kill soft-bodied insects such as aphids, small caterpillars, leafhoppers, mites, thrips, and whiteflies. In general though, Bt or spinosad are a better choice for controlling caterpillars. Some essential oil products also claim to be effective for controlling beetles, crickets, grasshoppers, and millipedes. Oil sprays have only minor effects on beneficial insects, and they do not persist in the environment.

How to Use

Choose a ready-to-use formulation or mix the oil with water according to label directions. Apply the spray in the morning or late afternoon rather than in the heat of midday. Spray all leaf surfaces until spray drips from the leaves; the spray must come in direct contact with the insects to kill them.

Repeat sprays every 5 to 7 days until the problem is under control.

For a combination soap-oil spray, combine 1 tablespoon dishwashing liquid with 1 cup oil (corn, peanut, safflower, soybean, or sunflower). Put 2½ teaspoons of the soap-oil mixture in a pump spray bottle and add 1 cup water. Or, if you're spraying a large area, put ¾ cup plus 1 teaspoon soap-oil mix in a backpack sprayer and add 1 gallon water.

When Things Go Wrong

Like soap sprays, oil sprays can damage plant cells, resulting in burned-looking foliage. See page 316 for precautions to use to prevent damage in the future and a technique for testing plants in advance of a crisis to be sure they can withstand being sprayed with oils.

NEEM

Neem oil has a safety record (for humans) that stretches back thousands of years. Neem trees are native to southern Asia, and neem oil is commonly used in India for soap, toothpaste, and even medical products. However, neem didn't draw attention as a botanical poison until the 20th century. It's now widely used as a pesticide and fungicide in organic agriculture and horticulture.

Neem products may contain neem oil, or they may be formulated with azadirachtin, a pesticidal substance extracted from neem oil. Some products contain both azadirachtin and neem oil (these may be more effective than products that contain only one or the other). Some insecticidal soaps are derived from neem oil, and their action is similar to that of other insecticidal soaps.

When to Use

Neem can cut back an infestation of leaf-eating caterpillars, leafminers, and young beetle larvae, such as Colorado potato beetle larvae. Neem also is somewhat effective in killing aphids, thrips, spider mites, squash bugs, weevils, and whiteflies.

Neem has multiple effects on insects. When insect larvae eat neem, it affects hormone balance, making it impossible for the larvae to molt to the next stage of development; they become deformed and die. Insects may lose the ability to swallow or digest food after eating neem-treated foliage. And some insects just don't seem to like the taste of neem, so they won't feed on sprayed leaves. Also, neem seems to disrupt mating in some species and to repel adult insects searching for egg-laying sites.

Because neem affects insects in so many ways, there's less chance that insects will develop resistance to it. Neem is not a contact poison, so it has little harmful effect on beneficial insects.

How to Use

Mix a spray according to label directions, keep it shaken well, and use within 8 hours of mixing. Apply two or three sprays 7 to 10 days apart. For a heavy concentration of pests, spray every 4 days.

When Things Go Wrong

It can take several days for neem to kill insects, but they should stop feeding shortly after spraying. If you see pests on plants that you have treated with neem, observe them closely. You'll probably find that they're not causing any further harm to the plants.

If the pests are still feeding, it's possible that the insect-killing component in the spray was broken down by heat, moisture, or ultraviolet light. Apply in the early morning or late afternoon for best results. In general, neem is more effective in warm weather than in cool weather.

Studies of neem's effectiveness are ongoing, and results apparently vary with different pests and crops. For example, studies thus far show that neem seems to work better against onion thrips than other species of thrips that attack vegetables.

Also, some brands of neem products seem to show better results than others. If you have repeated failures with neem, switching to a different brand of neem product may help.

PYRETHRUM

Pyrethrum is a general term for compounds called pyrethrins that are extracted from the pyrethrum daisy (a species of chrysanthemum native to southwest Asia). Read product labels carefully when you're selecting a pyrethrum product. Avoid buying pyrethroids, which are synthetic chemical insecticides similar in form to pyrethrins. Pyrethrum shows moderate toxicity to humans, and restrictions are now being imposed for indoor uses.

When to Use

Spray pyrethrins if an infestation of pest insects threatens to ruin a crop. It will control true bugs, caterpillars, beetles, aphids, mites, whiteflies, thrips, and leafhoppers. Pyrethrum is not equally effective for all species, though. For example, it is not highly effective against flea beetles, imported cabbageworms, or diamondback moths.

Pyrethrum is toxic to beneficial insects, producing the same symptoms of paralysis. Some beneficials will survive; others won't. It's also highly toxic to bees. Because of these effects, recommended application rates for pyrethrum products have been lowered. Apply pyrethrins only as a control method of last resort.

How to Use

Some commercial products combine pyrethrins with other insecticides to increase the overall effectiveness. For example, you can find products that combine pyrethrins and insecticidal soap, or pyrethrins and diatomaceous earth. Pyrethrins stop the insects in their tracks, and the other component kills the insects or provides longer-term control, since pyrethrins break down rapidly in the environment.

Check product labels for the ingredient piperonyl butoxide (PBO). This is a synthetic synergist (a substance that increases effectiveness of other ingredients) that is not allowed for organic production. You'll want to avoid it also because there are concerns about its toxicity to humans.

Spray plants thoroughly. The insecticide is only effective if it contacts pests directly. The spray breaks down quickly in sunlight and has almost no residual effect. Pyrethrins are very toxic to fish, so don't spray open areas of water.

Pyrethrum has moderate toxicity to humans. It can produce unpleasant symptoms if you inhale large quantities of spray. A small percentage of people experience an allergic reaction to pyrethrum. Also, there is some concern that repeated used of pyrethrum increases the risk of cancer.

When Things Go Wrong

INSECTS DON'T DIE. Insects can rebound after you spray pyrethrins, and repeat applications may be needed. It's also possible that pyrethrins aren't particularly effective against the specific pest you've treated. If results aren't good, respray or try a different control method. When you reapply the pyrethrins, do so on a cool morning when flying pests are sluggish. That way you'll be more certain of good contact between spray and pests. Spraying early in the day also helps avoid any harm to bees.

LEAVES TURN BROWN AFTER SPRAYING. If you use a pyrethrum product that also contains insecticidal soap, the soap may cause leaf damage. Mix a small quantity of the product and test spray as described on page 316. That way you'll know which crops you can safely spray with the product.

PICKLEWORMS

You'll make no pickles if pickleworms invade your cucumber plants. These caterpillars like to feed inside young cucumber, summer squash, and muskmelon fruits, leaving tunnels filled with ugly frass and often souring the flesh.

Pickleworms are the larvae of night-flying moths. This pest overwinters only in southern Florida and Texas, where it may be active nearly year-round. In these semitropical areas, adult moths as well as pupae hidden in leaf litter or soil can survive through winter. When spring arrives, moths lay eggs on young squash plants and also begin migrating north. Their life cycle is a short one, sometimes taking as little as 3 weeks. The strong-flying moths arrive at the northern end of their range about midsummer.

On young plants, pickleworms seek out tender growing tips to feed on. As the vines develop, the caterpillars switch to feeding on flowers and then inside young fruits.

After feeding for about 2 weeks, the larva spins a cocoon in a rolled section of a leaf. The pupa rests for 7 to 10 days, and a new moth emerges to start another generation. Even in the northern parts of their range, pickleworms can complete as many as four generations per year, and late plantings may suffer damage even before plants bloom.

PEST PROFILE

Diaphania nitidalis

RANGE: Southeastern United States, ranging north to Connecticut and New York and west as far as Nebraska and Oklahoma

HOW TO SPOT THEM: Examine the growing tips of squash and cucumber plants to find young caterpillars. Also peer into squash flowers. Look for rolled leaves, and crush pupae if you find them hidden there. Check young fruits for entrance holes, from which brown frass may protrude.

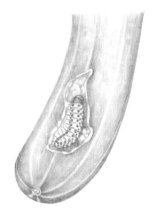

Young pickleworm larva and damage

Pickleworm moth

ADULTS: Yellowish brown moths with brown wings with central light yellow area; wingspan about 1 inch. The moths have a purple sheen and a tuft of hair at the tip of the abdomen.

LARVAE: Pale or yellowish white when small, these caterpillars turn yellow as they grow and develop many dark spots on their backs; spots disappear and the larvae become green or copper-colored when mature.

EGGS: White or yellowish, laid on leaves singly or in small clusters

CROPS AT RISK: Summer squash, cucumbers, muskmelons, and pumpkins; rarely watermelon and winter squash

DAMAGE: A single caterpillar can tunnel into and ruin several fruits.

CONTROL METHODS

CHOOSING RESISTANT VARIETIES. If pickleworms have been a serious problem, select only resistant varieties of susceptible crops. Ask your seed supplier for recommendations.

TIMED PLANTING. Plant fast-maturing varieties as early as possible in spring to promote strong growth and fruit set before pickleworms reach damaging levels.

ROW COVERS. Cover seedbeds and young plants, especially of later plantings, with row cover and seal the edges (as described in Row Covers) to prevent moths from laying eggs on plants. Keep plants covered as long as possible. If you leave covers on when plants flower, hand-pollinate the flowers.

TRAP CROPS. Summer squash flowers are a favorite feeding site for pickleworms. Plant summer squash a week or two ahead of cucumbers or melons; check growing tips and flowers of the trap crop daily for pickleworms. Kill any caterpillars you find. When blossoms become infested, destroy the trap crop. Plant a new hill of summer squash every few weeks as a continuing trap for pickleworms. (If you want to grow some summer squash for harvest, cover those plants with row cover.)

INSECT TRAPS. Early in the season, set up two insect traps that include a floral lure to attract pickleworm moths. Trapping female moths will reduce the number of eggs laid, and may reduce damage to a tolerable level. See Pest Control for details on these traps.

GARDEN CLEANUP. When you finish harvesting an early squash or cucumber crop, uproot and destroy all crop remains if pickleworms were present. Shredding is best, if possible, to destroy hidden larvae and prevent them from pupating. Turn the soil before replanting to bury pupae that have already formed. If you live in southern Florida or Texas, practice crop cleanup year-round.

BIOCONTROLS. Spraying open flowers with beneficial nematodes once a week is helpful because the nematodes can survive in the damp enclosed areas of the flowers and attack pickleworms feeding there.

Bacillus thuringiensis and spinosad sprays will kill pickleworms, but only if the pests consume treated foliage or flowers. The sprays won't kill worms burrowing in fruits or stems. See Pest Control for information on applying Bt.

NEEM. Neem is labeled for pickleworm control, but the worms must come in contact with the spray. It will not kill worms hidden in fruits or stems.

CONTROL CALENDAR

BEFORE PLANTING

- Plant a trap crop of summer squash a week or two before planting your desired crop.
- Select pickleworm-resistant varieties.
- Set up insect traps to catch pickleworm moths.

AT PLANTING TIME

- Plant as early as possible to avoid pickleworm damage.

- Cover seedbeds and young plants with row covers.

WHILE CROP DEVELOPS

- Renew insect trap lures as needed.
- Spray Bt or spinosad to kill caterpillars.
- As a last resort, spray neem.

AFTER HARVEST

- Shred all crop remains to kill hidden larvae. Work the soil to bury pupae.

PLANTING AND TRANSPLANTING

Planting is not without its challenges and frustrations. Fortunately, most planting problems are easy to solve. Read on to discover how to take the problems out of planting.

WHEN TO PLANT

Recommendations on when to plant vegetable crops is available from so many sources: seed packets, magazines, Web sites, books, Cooperative Extension offices, and more. Sometimes the wealth of information is overwhelming or contradictory. As you gain gardening experience, you'll find one or two sources of planting information that you like best, probably one that is tailored to your region. And if you keep garden records, they will eventually become your main guide in deciding when to plant.

PLANT WHEN YOU HAVE TIME

Guidelines or no guidelines, your life schedule is probably the critical factor in deciding when to plant. For example, let's say you have the whole afternoon free on a beautiful Saturday 2 weeks before your last average frost date. Why not spend it planting? There are plenty of crops that are perfect for planting at this time. But don't limit yourselves to those—try some tricks that push the limits. Use row covers, cloches, or raised soil ridges to create warmer conditions for cold-sensitive crops such as beans and tomatoes.

PLANT WHEN THE SOIL IS WARM

One extremely helpful tool in deciding when to plant is a soil thermometer. Most seeds have an optimum temperature range for germination, and that temperature refers to the soil, not the air. It's handy not only in spring but also in fall. For example, southern gardeners should check soil temperature before sowing parsnips in fall because the seed won't germinate well until soil temperature drops below 75°F.

If you don't own a soil thermometer, you can use cues from nature to figure out when to plant. For example, when lilacs bloom in your neighborhood, you can safely plant corn, tomatoes, and other warm-weather crops. It is also said that it's safe to plant tomatoes outdoors when you can comfortably walk barefoot in your garden without getting cold feet.

PLANT FOR LATE-SEASON HARVEST

For a fall harvest, calculate planting dates mathematically by counting backwards from your first expected frost date. For example, if you want to make a late planting of bush beans that mature in 60 days, you'd count back 60 days from your fall frost date. To be safe, though, add a buffer of a few weeks to allow for the lower temperature and shorter daylength of early fall. The bean plants will grow more slowly then than they do in midsummer; it may take them 75 to 80 days to mature.

HOW MUCH TO PLANT

With most crops, overplanting isn't a problem, it's an opportunity. If you end up with more produce than you and your family can eat, donate it to people in need. Early in the gardening season, find out which local food banks or soup kitchens in your area accept donations of fresh produce.

If you decide to plant extra, avoid crops that are labor-intensive to pick but won't keep on the vine. Two notorious examples are cherry tomatoes and bush beans. If you plant too much, it can take hours to harvest because the scores of fruits or pods need to be carefully removed individually without injuring the plants. Chances are the bounty will rot in the garden.

The best way to avoid an overplanting crisis is to start small with your planting efforts, especially when you try a new crop. Sow a 5-foot row or a few square feet, and sow a second row or block a week or two later. Staggered plantings often result in a nicely spaced-out harvest. If you end up with fewer beans, peas, or cherry tomatoes than you wished for, you can plant a bit more next year.

FITTING IT ALL IN

Another common planting problem for home gardeners is running out of space before you've planted all the different crops you'd like to grow. One solution is to increase the size of your garden, but that's not always practical. When it's not, try using double rows, interplanting, and succession planting to reap the maximum benefit from your garden space. Planting crops to climb trellises and planting vegetables in containers are two other ways to "create" more room for gardening.

PLANT IN DOUBLE ROWS

Standard spacing for crops such as tomatoes, peppers, and Brussels sprouts is to set plants at least 12 inches apart in rows 3 feet apart. This spacing works well on a commercial scale, but in a home garden, there's a more space-efficient approach. Plant these large crops in double rows that are only half the distance of the standard recommendation, and stagger the position of plants in the rows. This way, you can plant tomatoes 18 inches apart within each row in two rows 18 inches apart. Staking or trellising large plants in double rows is essential.

A garden bed planted with a double row of tomatoes becomes a jungle of foliage as the plants reach mature size. To encourage good air movement, plan your garden layout so that adjacent beds are planted to low-growing crops (such as salad greens or root crops).

INTERPLANTING AND SUCCESSION PLANTING

Interplanting is simply planting two or more different kinds of crops in one bed. The strategy is to fill spaces between long-season crops such as cabbage with crops that mature quickly. Succession planting is sowing or transplanting a second crop into a bed as soon as (or even before) the first crop is harvested.

Radishes are the prime example of a crop to interplant. When you plant transplants of nearly any crop, sow radishes in the open space between the young transplants. The radishes will be ready to pick before the transplants shade them out. Any crop that matures in 30 to 45 days is a good choice for interplanting. Try any of the following: baby beets, baby turnips, broccoli raab, cilantro, dill, kohlrabi, leaf lettuce, radishes,

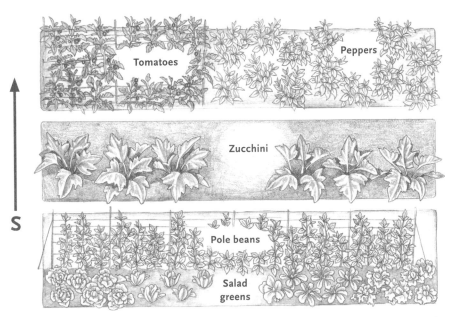

When planting in double rows, offset the plants in a staggered pattern. Also alternate beds of tall and short crops to encourage better air circulation.

salad greens (most types), scallions (from sets), and spinach.

Another way to interplant is to literally sow two kinds of seed in the same space. For example, mix together leaf lettuce and carrot seed for broadcasting (see directions below for how to broadcast small seed). Lettuce germinates fairly quickly, producing small but harvestable plants within 30 days. Carefully cut off the lettuce plants at soil level, and the carrot seedlings, which germinated slowly in the protection of the lettuce plants, will be ready to take off and fill the bed.

With succession planting, think cool-warm-cool. Start the season with a cool-season crop that matures quickly (60 days or less). After you harvest it, plant a warm-season crop. If it's a crop that takes a long time to reach maturity, leave enough space between the plants to set in the transplants of the next cool crop before you harvest the warm-season crop.

Some successions require that you start your own transplants because few garden centers and nurseries offer vegetable transplants for the later growing season. For example, garlic and Brussels sprouts can be good succession partners in areas with long growing seasons. The trick is to have Brussels sprouts transplants ready to place into the bed as soon as you harvest your garlic (plan to start your own from seed).

BALANCING YOUR GOALS

Succession planting helps you achieve the maximum yield per foot of garden space. Weed control is a secondary benefit—the crop cover will be nearly continuous, and weeds will have little opportunity to grow. However, if you're also trying to break the cycle of soilborne diseases in your garden by crop rotation, succession planting can get in the way. For example, the example shown below includes three major crop families: tomato family, cabbage family, and lettuce family. If you've set up a rotation that calls for cabbage-family crops to be planted the year after tomato-family crops, this type of succession will break the pattern. It's possible to do both, especially if you grow crops from a wide range of crop families, but it's challenging.

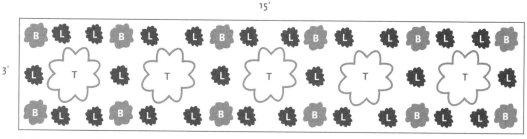

T=Tomato transplants   B=Broccoli transplants   L=Lettuce transplants

Some planting schemes combine succession cropping and interplanting. In this garden, tomato transplants are set out in early May. Lettuce and broccoli are added 4 to 6 weeks later. Lettuce is harvested by the end of July, leaving room for the other crops to continue growing. Tomato harvest proceeds through summer, and broccoli is ready for fall cutting.

SEED-SOWING SOLUTIONS

Sowing seeds in the garden is fun, but it's also painstaking work that seems to take way too much time. Plus, sometimes after our careful effort to sow the seeds, our reward (or lack of it) is a spotty stand of seedlings with lots of gaps, or a thick stand that suddenly dies back because of damping-off disease. Read on for tips on improving your seed-sowing technique and ensuring even germination. To learn how to prevent problems with damping-off, refer to Damping-Off.

TRICKS FOR HANDLING SEEDS

Handling seeds—especially small seeds—is tricky, but with the right tools and techniques, you can sow seeds of all sizes evenly and efficiently.

USE A FURROW MAKER. Seed furrows that are straight and uniformly deep help ensure a good stand of seedlings. A homemade furrow maker is a great tool for this purpose. Use a length of furring strip up to 5 feet long. Hold it on edge over your seedbed and gently rock it back and forth to open a straight furrow of the desired depth. Lay the strip beside the furrow and sow seeds at the proper spacing.

PLAN A PERFECT BROADCAST. With small seeds such as those of carrots and lettuce, you may find it easier to broadcast the seed than to plant it in rows. It pays to prepare the seed for sowing indoors before you head out to the garden. Although the preparation takes some time, it's worth it for the fine results. Plus, you can do the "advance" work on the seeds in the evening while watching TV, perhaps.

Start by calculating how many seeds you should sow in your planting area. For example, let's say you're planting a 1-foot-wide strip across a 3-foot-wide bed, and instructions are to sow seeds about 1 inch apart (later, you'll thin the seedlings). The area you're planting is 3 square feet, and there are 144 square inches per square foot. Multiply 3 times 144, and the result is 432. Round that up to 450 or down to 400. Pour seed from the packet onto a tray or plate, and use a small plastic knife to roughly count off seed by groups of 10—you don't need to be exact. Mix the seeds with fine sand (about ½ cup sand per 100 seeds) and pour the mixture into a shaker dispenser (such as a grated cheese dispenser). Now your seed is ready for sowing.

In the garden, smooth out the planting area and remove surface debris and stones. Shake the dispenser over the bed, emptying it completely. After you broadcast the seed, you must incorporate it into the soil. Use a rake to work it in by lightly chopping the soil surface with an up-and-down motion. Then use the flat of the rake to tamp the soil lightly. Finish by gently watering the bed.

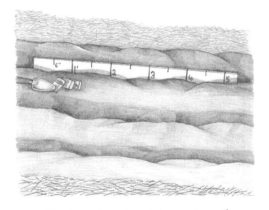

A piece of furring strip works well for opening uniform furrows. Use a permanent marker to draw lines at 1-inch intervals, and your furrow maker serves as a spacing guide, too.

IMPROVING GERMINATION

If the weather cooperates, your newly sown seeds will germinate right on schedule. However, it's rare that the weather supplies just the right conditions for seed germination. More often than not, the weather will be too dry or too wet. Covering your seedbeds provides a buffer against both hot sun and heavy rain.

- Presoaking is recommended for some seeds that are slow to germinate, such as parsley and parsnips.
- If you're sowing seeds in late summer, the soil may be dry several inches deep. If so, make your seed furrows a little deeper than the standard recommended depth, then gently fill the furrows with water and let it soak in. Then sow your seeds and cover them lightly. Apply mulch between the seed rows right away to conserve moisture and help cool the soil.
- Instead of covering seed furrows with soil, cover them with vermiculite or sifted compost, which will hold moisture and resist crusting.
- In areas where you broadcast seed, spread a sprinkling of grass clippings or straw over the area to hold in moisture or break the impact of a downpour.
- Cover seeded areas with burlap or a piece of row cover for the same kind of benefits. Use small stones to weight the edges of the cover.
- For seeds that are slow to germinate, such as carrots and parsnips, or to cool off hot soil,

cover seed rows with a board, as shown below.

THINNING SEEDLINGS WITH EASE

Once your seedlings come up, it's important to thin them to prevent damping-off and to allow the plants to grow without stress from too much competition. It's easier to thin seedlings when the soil is moist. If you can, wait to thin until just after a rain. If not, use a hose or watering can with a rose attachment to gently water the area, as shown on page 429.

After watering, allow the seedlings to dry off before you thin them. If the seedlings are growing close together, use nail scissors to cut them off at soil level rather than pulling them out by the roots.

A piece of plywood works well to shade a seedbed where you broadcast seeds, but the wood is too heavy to rest directly on the soil. Prop it up with bricks or stones as a "roof" for the bed. Remove the roof as soon as seedlings appear.

AVOIDING TRANSPLANT SHOCK

Transplanting is fast and easy compared to sowing seeds, and the head start that using transplants allows is often crucial for northern gardeners. You can grow your own (see the Seed-Starting entry) or buy them. Your goal is to help transplants adjust to their new home in a garden bed without suffering stress that sends them into shock. Transplant shock shuts down the plant's ability to produce new roots or take up water and nutrients properly. It sometimes has a lasting effect that slows growth and production throughout the plant's life.

The most important way to avoid transplant shock is to harden off transplants before you put them in place in your garden. If you grow your own transplants, you can control the hardening off process from start to finish. (To learn how, turn to page 362.) If you buy transplants, ask the salesperson whether they've been hardened off. If the transplants are in greenhouses and look lush and bright green, chances are they're not fully hardened off. If they're in outdoor display racks, then they've probably been hardened off (perhaps even stressed).

Even hardened off transplants appreciate a little pampering when you first set them in the garden. Try some of these tricks to help your plants avoid shock.

- Plant transplants in the early evening if possible. This gives them half a day to adjust before they are hit by strong sun and heat.
- If you're planting at a hot, dry time, stop every 5 minutes or so to water the transplants you've just set in place. Or dig all your transplant holes first. Then, as you set each plant in place, refill the hole to within 2 inches of the soil surface. Gently pour water into the planting hole and let it soak in. Then finish filling the hole and firm the soil in place around the stem.
- If sun will be very strong, shade transplants as described on the opposite page.
- Wait 1 week after planting before cultivating. This way the transplants will have anchored in with new roots and can tolerate a little disturbance.

For insurance, reserve any leftover transplants for at least another week. Then if some of the plants in the garden die, you have replacements.

PLANTING ROOTBOUND TRANSPLANTS

It's easy to recognize rootbound transplants. When you slip the transplant out of its container, the first thing you'll see is roots, not soil. The roots may be so densely packed that they perfectly holds the shape of the container.

The best thing to do about rootbound transplants is: nothing special. Don't invest time in teasing apart the roots or cutting through them to open up the rootball. The roots of annuals grow quickly and sprout new roots easily. As long as you're putting the rootballs into a favorable environment, new roots will grow. In fact, teasing or cutting roots may just send the transplant into shock.

POTATOES

Growing potatoes in your home garden can be an adventure. Choose unusual varieties you can't buy at the grocery store, or a favorite variety that you've discovered through your local farmer's market. A 20-foot row of potato plants can yield 25 to 30 pounds in a good year. You may want to place your seed potato order with a couple of gardening friends so that you can plant small quantities of several varieties.

The list of pests and diseases that potentially bother potatoes may seem daunting, but in practice it's usually not difficult to grow a good crop. It's critical to buy certified disease-free seed potatoes and to plant in well-drained soil.

CROP BASICS

FAMILY: Tomato family; relatives are eggplant, peppers, tomatillos, and tomatoes.

SITE: Plant potatoes in full sun. If the site isn't well drained, build a raised bed, or at least surface-plant your potatoes, as described below.

SOIL: If your soil is moderately fertile and enriched with organic matter, potatoes will grow fine. If your soil fertility is low, add 4 pounds of balanced blended organic fertilizer per 100 square feet and work it into the top several inches of soil before planting.

TIMING OF PLANTING: Timing varies considerably by region, as explained below. Check soil temperature before planting; potatoes grow best when soil is at least 50°F.

PLANTING METHOD: Plant seed potatoes—tubers raised especially for planting to produce a crop.

CRUCIAL CARE: Pay attention to watering your potatoes because they have shallow roots. Potatoes especially need lots of water once foliage is fully developed and tubers are enlarging. If your soil is sandy, you may need to water as often as every 3 days. Cut back on water when foliage starts to turn yellow; otherwise, the tubers will rot. Crop rotation is also very important for potatoes because many of the problems that affect the tubers are carried in the soil.

GROWING IN CONTAINERS: Potatoes grow well in large containers such as bushel baskets and even plastic trash cans. Always provide drainage holes. Cut seed potatoes into pieces as described in "Presprouting Potatoes" on page 332. Put a few inches of soil in the bottom of the container, lay the seed pieces on the soil, and then cover the pieces with a couple of inches of soil. Continue adding soil, compost, or mulch as the stems lengthen. To harvest, carefully tip the container on its side and let the contents spill out.

HARVESTING CUES: Blooming flowers are a cue that potatoes have reached "new potato" size. If you want to harvest new potatoes, dig out a full plant and its tubers, leaving neighboring plants undisturbed. This minimizes the risk of injury to the tubers of the neighbor plants. Vines may die back on their own as the potato crop matures, but if they don't show signs of doing so, cut them off at soil level 2 weeks before you want to dig your potatoes. This will trigger the potatoes to harden, which helps them last longer in storage.

SECRETS OF SUCCESS

TIMED PLANTING PREVENTS PROBLEMS. Spring is traditional potato planting time—about 3 weeks before the last expected spring frost. However, in many areas, you can choose a different planting time for some or all of your potatoes to minimize problems with diseases and certain pests and to avoid hot, dry conditions. If you love homegrown potatoes, try planting crops in two different seasons for extended harvest. Here's a tour of regions and potato planting times:

- In the Rocky Mountains, Northern Plains, Great Lakes, and upper reaches of the Northeast (both United States and Canada), plant potatoes in mid-spring for late summer harvest.
- In lower New England, the Mid-Atlantic, and Southern Plains, plant potatoes in late spring and/or in midsummer (early varieties only) for a fall harvest. The advantage of the summer planting is avoiding Colorado potato beetle damage. Ask your supplier to deliver your seed potatoes as late as possible, and refrigerate the seed pieces until you're ready to cut them for planting.
- In the Southeast, plant in February or March. You may need to plan ahead and

order these seed potatoes in fall. Refrigerate them or put them in cool storage (40° to 50°F), and bring them out in time to warm up before planting. Also plant a fast-maturing crop in early April. Plant again in late summer or early fall for a fall crop.
- In the Gulf Coast region, plant only in fall after high heat is over. The plants survive winter (they will go dormant) and resume growth in early spring.
- Pacific Northwest gardeners plant potatoes in both spring and fall. For a fall crop, cover the tubers well with mulch and then with a plastic tarp because cold, wet fall and early winter conditions can lead to rot. Remove the tarp when temperatures rise in spring; potatoes will be ready to dig in spring.
- In most parts of California, potatoes do best when planted throughout the fall months,

Use soil or mulch to protect tubers from light, injury, and temperature fluctuations. A thick layer of soil or mulch also protects the tubers from fungal spores and insects that sometimes reach tubers through soil cracks or by washing through a shallow mulch layer.

PRESPROUTING POTATOES

Presprouting potatoes is sort of like starting seedlings indoors. It gives you a jump on the season and reduces the length of time from planting to harvest, which also lowers the risk of pest and disease problems. Even if you live in an area where the length of the growing season isn't a concern, you can try this technique with a portion of your seed potatoes to extend the harvest. Presprouted potatoes also have the potential for higher yields because all the buds sprout uniformly.

Presprouting potatoes isn't hard, but it's important to monitor temperature and handle the sprouted seed potatoes extremely carefully (broken sprouts won't grow and will provide an entry point for disease).

When your seed potatoes arrive, immediately put them in a cool, dark storage place. Then, about 2 weeks before you plan to plant them in the garden, move the seed potatoes to a warm spot, about 70°F. Spread them on a sheet of newspaper on a counter, or lay them in a single layer in a shallow box or seed flat. They need light, but not direct sunlight. In the light, the sprouts stay short and stubby, so they're less in danger of being accidentally snapped off

or broken. Once the sprouts are $\frac{1}{2}$ inch long, it's time to cut the seed potatoes. (The potatoes will have turned green, and that's okay.)

If a seed potato is the size of an egg or smaller, don't cut it (you'll plant it whole). Use a sharp knife to cut larger potatoes into pieces so that each piece has at least two eyes and weighs no less than 1 ounce.

The flesh is the stored food that will fuel plant growth for up to 3 weeks after planting. If any pieces show dark rings or discoloration in the flesh, discard them—don't plant them.

Presprouted seed potatoes cut and ready for planting

You can plant the cut pieces immediately, but if you have clay soil, it's better to let the cut surfaces harden for a couple of days.

and early varieties are most successful. Some inland California gardeners also can plant early varieties in spring.

- Desert gardeners in southern Arizona and California plant potatoes from November through March; farther north, spring planting is best, but it's critical to choose fast-maturing varieties.

KEEP TUBERS UNDER COVER. Whether you plant potatoes in a trench or on the surface, it's

important to use soil or mulch to cover the base of the plants well and shield tubers from light, injury, and pests. Trench planting is the traditional method, but surface planting is a good choice to minimize digging and bending, both at planting and harvest.

If you plant potatoes in a 4-inch-deep trench, cover the seed potatoes with at least 2 inches of soil right away. Fill the trench as the foliage emerges, and then use a hoe to hill up loose soil around the plants at least once (twice is better) as they continue to grow. By the end of the season, the plants will be covered by a low mound of soil.

If you surface-plant, first loosen the top few inches of soil. Lay the seed pieces on the soil cut side down. Mulch with shredded leaves, leaf mold, or clean straw. Add mulch as the plants grow, maintaining a layer several inches thick over the tubers. In addition to protecting tubers, surface mulch minimizes weeding. The only disadvantage is that yields will probably be lower than they would if you hilled the plants with garden soil, because mulch doesn't supply as many nutrients as soil. Spraying foliage with a fish emulsion/seaweed spray, shortly after sprouts emerge and again just before plants flower, may help make up the difference.

REGIONAL NOTES

WESTERN STATES

Potato psyllids are jumping insects that suck sap from potato plants (and occasionally tomatoes, peppers, or eggplant). They look like miniature cicadas, but they're so small that you'll need to use a hand lens to see the resemblance. The insects also inject a toxin as they feed, which causes poor growth. Leaves curl and older leaves turn yellow. The edges of leaves may turn purple. Plants may end up brown and stiff. The condition is referred to as psyllid yellows. Tubers are small and sprout prematurely. Spraying sulfur can help control psyllids on small plants. If tubers have already formed when the psyllids invade, there's no need to spray. Just cut and

compost the foliage, and then harvest the crop. Potato psyllids survive over winter only in the Southwest, but they migrate north each year, reaching areas of the Rocky Mountains and Plains states.

SOUTHEAST

Hot weather can produce some unusual effects on potato tubers in the soil. One possible problem is heat necrosis, which appears as black streaks below the tuber skin. Another is that tubers may develop sprouts before they are harvested. To avoid these problems, harvest tubers as early as possible and keep them out of heat after harvest.

PREVENTING PROBLEMS

BEFORE PLANTING

- Find a source of certified disease-free seed potatoes.
- If early blight has been a problem in your garden, choose resistant varieties.
- Pull out tomato-family weeds such as nightshade that might host Colorado potato beetles.

AT PLANTING TIME

- Plant early or late to avoid damage by Colorado potato beetles. Plant early in the North to avoid leafhopper damage and early blight infection.
- Cover newly planted areas with row cover and seal the edges to protect young sprouts from aphids, Colorado potato beetles, and other pests (see Row Covers).

- Scout plants for foliage-feeding pests and handpick any that you find.
- Spray garlic or hot-pepper sprays to repel insect pests.
- Coat plants with kaolin clay before leafhoppers become active.
- Hill or mulch plants well to protect developing tubers.

- Dig plants carefully to avoid wounding tubers.

- Remove all crop debris from the garden. If it's healthy, compost it. If it's visibly diseased, dispose of it with household trash.

TROUBLESHOOTING PROBLEMS

NO SPROUTS EMERGE. Blank spaces along a row of potato plants are due to seed pieces rotting before sprouting. Several types of bacteria and fungi can infect seed potatoes. In future, plant certified disease-free seed, plant shallowly, and be sure the soil is warmer than 50°F before planting.

YOUNG SPROUTS DIE BACK. Blackleg, black scurf, or frost damage are possible causes. With blackleg, a bacterial disease, the sprouts will show a wet, black rot from the soil level up (the "blackleg"). Remove and destroy infected tubers. Replant, using certified disease-free seed, in warm, well-drained soil.

Black scurf is a fungal disease. Look for brown, sunken spots on the stems below soil level. Remove infected plants—including the seed potatoes—when conditions are dry. Prevent both problems in the future by planting certified disease-free seed and waiting for the soil to warm up before planting.

If it's early in the season and there are no signs of disease, the problem is probably cold temperatures. Frost can kill tender potato foliage. New healthy growth should resprout from the soil line.

LEAVES CURL UNDER. If leaves are also puckered, check leaf undersides for potato aphids, which are small pinkish or green insects. The damage they cause isn't serious, but they spread viral diseases that can ruin a crop. See Aphids for controls.

LEAVES ROLL UP. Potato leaf roll is a viral disease. Leaves also turn light green and feel leathery. Infected plants won't produce edible tubers. Aphids spread this virus. Remove infected plants and take steps to control aphids.

YELLOW LEAVES. This is an early sign of bacterial wilt or Verticillium wilt. With bacterial wilt, infected plants may wilt dramatically. Cut into stems and drops of gray or brown slime may ooze out. If the bacteria invade the tubers, they turn grayish brown and also ooze slime when cut. To prevent bacterial wilt, plant in well-drained soil. Also control root-knot nematodes because nematode injury leaves the plants more susceptible to wilting and tuber infection (see Nematodes).

With Verticillium wilt, the foliage will also droop. Cut open a stem and you'll see dark streaks inside. See Verticillium Wilt for controls.

MOTTLED LEAVES. Mottling is a sign of mosaic or other viruses. Confirm the diagnosis and remove infected plants. Many viruses are spread by aphids; controlling aphids will help prevent virus problems. See Mosaic.

GRAY BLOTCHES ON OLDER LEAVES. These patches are a sign of feeding by potato tuberworms, which tunnel through the interior of leaves. Handpick damaged leaves and destroy them. Hill plants well so the worms (which are actually small caterpillars) can't invade the tubers as well. If you've seen signs of tuberworm activity in your potato patch, harvest the crop as soon as it matures to minimize the chance that tuberworms will invade the tubers.

SMALL, DARK SPOTS ON OLDER LEAVES. Check these spots with a hand lens; concentric rings are a confirming sign of early blight. Symptoms will spread quickly. Remove infected leaves immediately. See Early Blight for controls.

WATER-SOAKED SPOTS ON LEAVES. Late blight is a potentially devastating disease of potatoes and tomatoes. In moist weather, the spots spread rapidly. If you suspect late blight in your garden, take samples for identification. If the problem is late blight, destroy all infected plants as soon as possible. See Late Blight for more information.

LEAF EDGES TURN BROWN. Windburn and leafhoppers are possible causes. With windburn, usually only the leaf tips turn brown, and no further symptoms develop. It should not affect yield, and no controls are necessary.

Leafhopper feeding causes more serious browning and scorching of foliage—the condition is called hopperburn. Brush the leaves and you'll see small white dots moving around—these are the leafhoppers. See Leafhoppers for controls.

NEW GROWTH TURNS PURPLISH. This is a sign of a disease called purple top in potatoes; in other crops it's called aster yellows. New leaves stay small and wilt. Small aboveground tubers form at points where leaves attach to stems. See Aster Yellows.

FOLIAGE CHEWED. Look for slender beetles on the leaves, but don't touch them. They secrete an oil that can cause skin to blister. Wear gloves and handpick the beetles. See Blister Beetles.

SMALL HOLES IN LEAVES. Check the leaves for small black beetles that jump when disturbed. These are flea beetles, and their damage to foliage usually isn't serious. The larvae injure tubers. See Flea Beetles for information.

LARGE HOLES IN LEAVES. Colorado potato beetles are the most notorious pest of potatoes. Loopers and climbing cutworms also feed on potato foliage. Check leaves for humpbacked yellow potato beetle grubs and orange beetles with black stripes on their backs. Handpick them; plants can withstand moderate damage. For other controls, see Colorado Potato Beetles.

Loopers and climbing cutworms are green caterpillars. If you find caterpillars on plants, see Cabbage Loopers and Cutworms.

ONE OR TWO STEMS WILT. This is a sign of bacterial ring rot. Leaves may roll up. Cut a stem, and a cheesy material will come out of the stem when you squeeze it. The bacteria also infect tubers. The tubers will be discolored and a cheesy bacterial material will form. To prevent ring rot in the future, buy certified disease-free seed, and clean your cutting knife between cuts when preparing seed potatoes for planting.

STUNTED OR WILTED PLANTS. Bacterial wilt, Verticillium wilt, Fusarium wilt, or root-knot nematodes can cause stunting and wilting. To figure out what is causing the problem, cut through a stem. If brown ooze drips out, bacterial wilt is the cause. If the stem shows brown streaks inside, the problem is Verticillium or Fusarium. Uproot a plant to check for bumps on tubers and small galls on other roots. These are signs of root-knot nematodes.

Whichever disease is the cause, stunted and wilted plants won't yield well. Remove the infected plants. To prevent bacterial wilt in the future, plant in well-drained soil. For other controls, see entries on Fusarium Wilt, Nematodes, or Verticillium Wilt.

GREEN AREAS ON TUBERS. Potato tubers will develop green spots if they're exposed to light because of the formation of chlorophyll. Light also stimulates the formation of poisonous alkaloids, so cut away the green areas before using the tubers in the kitchen. To prevent this problem, hill plants well with soil or mulch to shield tubers from light. After harvest, always store tubers in the dark.

KNOBBY TUBERS. Abrupt changes in soil moisture results in tubers with odd bumps or sometimes an hourglass shape. Did heavy rain follow a dry period while your potatoes were developing? The potatoes will be fine to eat. In future, monitor soil moisture closely while tubers are enlarging.

SMALL BLACK PATCHES ON TUBERS. These patches look like spots of dark soil, but they won't wash off. They're actually resting spores of the disease called black scurf. You can peel the potatoes and use them. To prevent future problems, plant certified disease-free seed and plant in well-drained soil. Planting a cover crop of oats and turning it under 3 weeks before you plant potatoes may also reduce black scurf problems. Applying *Trichoderma harzianum* to the soil may help prevent the disease (see Disease Control).

CORKY SCABS ON TUBERS. Scab is a common potato disease. You can peel and eat affected potatoes. To prevent scab in future crops, see Scab.

CORKY SPOTS IN TUBERS. These spots are due to early blight, a fungal disease. The tubers often develop secondary rots as well. See Early Blight for more information.

ROUGH SURFACE ON TUBERS. Flea beetle larvae may feed on tubers, causing rough areas covered with tiny cracks. Cut away damaged portions and use the rest. Don't place damaged tubers in storage. See Flea Beetles for controls.

PINK AREAS ON TUBERS AROUND THE EYES. This condition is called pinkeye, and it usually occurs on tubers in wet soil. The cause isn't known. To avoid it in the future, plant potatoes in well-drained soil.

BROWNISH PURPLE SPOTS ON TUBERS. This is a sign of late blight. Destroy all infected tubers. See Late Blight for more information on this serious disease.

TUBERS WITH HEALED-OVER CRACKS. These are growth cracks. They form when the tubers grow rapidly due to a sudden increase in soil moisture, such as when heavy rain follows a dry period. To prevent cracks, use mulch and monitor soil moisture to prevent soil from drying out when tubers are developing.

LARGE, SHALLOW HOLES IN TUBERS.
White grubs sometimes feed on potato tubers. Cut away damaged areas and use the rest of the tubers, but don't try to store damaged tubers. If this problem is serious in your garden, see Japanese Beetles for controls.

TUNNELS IN TUBERS. Look for the insect that made the tunnels. Wireworms are leathery, segmented worms; potato tuberworms are white or light green caterpillars with dark heads and upper bodies. Wireworms are most common in gardens that were recently converted from lawn; see Wireworms for controls.

Potato tuberworms are the larvae of a moth that lays eggs on potato foliage or on potato tubers at or near the soil surface. The larvae can also wend their way down soil cracks to reach shallow tubers. To prevent tuberworms from reaching your potato crop, hill plants with several inches of soil or mulch.

CAVITY INSIDE TUBER. Hollow heart and Verticillium wilt are possible causes.

Hollow heart occurs when potatoes grow too fast because of an oversupply of moisture or fertilizer. The cavity will be discolored, but you can eat the potatoes if you cut away the brown areas. To prevent hollow heart, fertilize plants early (when they are 4 to 6 inches tall) rather than when tubers are about to form. Avoid overwatering. When planting, don't space seed potatoes too far apart.

If the cavity is lined with powdery decay, the problem is probably Verticillium wilt, which can develop in wounded potatoes in storage. To prevent it in the future, inspect potatoes carefully and don't store any wounded tubers.

ROTTEN TUBERS. Bacterial soft rot can invade tubers that are wounded by tools or insects or weakened by another disease such as ring rot. The tubers will not be usable. To prevent future problems, take steps to protect tubers from injury, insects, and diseases; also see Rots.

POWDERY MILDEW

The white coating that covers leaves, stems, and flowers of plants infected by powdery mildew looks awful, but this disease can be fairly easy to manage. Powdery mildew spores germinate on dry leaf surfaces when humidity is high; the spores can't germinate on wet leaves, and they don't germinate very well on leaves in full sun. Thus, the disease tends to be a problem in dry climates or during dry spells, taking hold in areas of dense, shaded foliage where air circulation is poor.

Managing your garden to prevent pockets of sluggish air movement will help prevent powdery mildew. There are also several organic sprays that can help prevent or slow the spread of the fungus.

RANGE: United States and Canada

CROPS AT RISK: Beans, peas, lettuce, cucumbers, melons, pumpkins, squash, and other crops

DESCRIPTION: Round white or gray spots form on leaves. As the mildew spreads, plants look like they've been sprinkled with flour or talcum powder. Leaves may turn yellow or brown, dry up, and drop off; infected flower buds do not open.

FIGHTING INFECTION: Pick off affected leaves and compost them. To slow the spread of mildew, spray stems and leaf surfaces with 1 part milk mixed with 9 parts water. Potassium bicarbonate or homemade baking soda sprays also help prevent and reduce powdery mildew problems. Try sprays of *Bacillus subtilis*, compost tea, or comfrey tea.

GARDEN CLEANUP: The spores overwinter in plant debris and are spread by wind.

Remove all crop debris at the end of the season. Compost diseased plant material (in a hot pile if possible). If powdery mildew is a serious problem in your garden, destroy diseased plant material.

NEXT TIME YOU PLANT: For future crops, plant resistant or tolerant varieties. Space plants widely to ensure good air circulation. Trellis beans and squash if possible. Be cautious in applying nitrogen-rich fertilizers. Plant susceptible crops in full sun. If conditions are dry but humid, spray susceptible crops with water during the day to wet the foliage and prevent powdery mildew spores from germinating. If powdery mildew has been severe in the past, apply neem as a preventive measure (see Pest Control for information on neem).

CROP ROTATION: Rotating crops won't stop powdery mildew from developing in home gardens because the fungus is so widespread and spores travel by wind.

PUMPKINS

Jack-o-lanterns, pie pumpkins, and miniature pumpkins are part of the squash family. Pumpkins are planted and grown in the same way as winter squash—vine crops that produce fruits that mature fully on the vine before harvest. Pumpkins can take up a lot of space, but they're still a garden favorite. To learn how to tend your personal pumpkin patch, consult the Squash entry.

RADISHES

Watering, prompt thinning, and a little weeding are important for a fast-growing, successful crop of radishes. You'll encounter occasional pest problems, too, but with such a fast-growing crop, it's easy to start anew.

CROP BASICS

FAMILY: Cabbage family; relatives are arugula, broccoli, broccoli raab, cabbage, cauliflower, collards, horseradish, kale, kohlrabi, mustard, and turnips.

SITE: Radishes grow best in full sun. In the summer, planting them in the shade of neighboring crops helps prevent overheating, but too much shade results in leafy tops with little root growth.

SOIL: Radishes tolerate a range of soil conditions but do best in rich, loose, well-drained soil. Roots may be small in heavy clay soil. Long radishes require deeply worked soil so the roots can stretch to mature size.

TIMING OF PLANTING: In spring begin sowing early radish seeds about 5 weeks before the date of last frost. Make succession plantings weekly or as desired. Switch to heat-resistant varieties for summer. In areas with intensely hot summers, stop planting when temperatures reach the 80s. For spring crops of black and Chinese radishes, wait until the soil warms to about 60°F. For fall crops of any type of radish, begin sowing in late summer—count back the number of days to maturity from the first frost date. In mild-winter areas such as southern California and the Deep South, continue planting radishes throughout winter.

PLANTING METHOD: Direct-sow seeds in the garden.

CRUCIAL CARE: A steady supply of water is essential for high-quality radish roots. In hot conditions, cover the soil surface with a few inches of organic mulch to help retain moisture and also to cool the soil.

GROWING IN CONTAINERS: Tuck a few seeds of small radishes into any and all container gardens, including window boxes and hanging baskets. Deep containers such as half-barrels are ideal for daikons because the roots easily grow to full size in a loose, rich potting mix. Put the containers in full sun and water frequently to keep the soil evenly moist.

HARVESTING CUES: Check early radishes every few days, and harvest them before they exceed 1 inch in diameter. The quality of early radish roots declines quickly. For other types of radishes, mark the days to maturity on your calendar at the time of sowing, and start to monitor the crop at that time.

SECRETS OF SUCCESS

AIM FOR SPEEDY TURNAROUND IN SPRING.
Hot weather and hot soil are hard on most radishes. Roots turn pithy or woody and develop a stridently hot flavor. To avoid this, encourage

your spring radishes to grow as fast as possible. Plant them in full sun, sow seeds no more than ½ inch deep, thin seedlings promptly so they don't compete with one another for water and nutrients (use both the young leaves and immature roots in spring salads), and never let the plants suffer water stress. If you're concerned about the quality of your soil, feed the plants with compost tea to supply nutrients. Harvest the roots on the young side. Spring radishes taste best when 1 inch in diameter or smaller.

PLANT A STORAGE CROP FOR FALL. All radishes grow well in cool fall weather, but large daikons are especially well suited for a fall crop. As you prepare to plant them, check the seed packet to determine the expected mature length. Be sure to loosen the soil to at least that depth, and work in a couple of inches of organic matter (especially if you're planting in an area where you've just harvested an earlier crop). Let the soil settle for a couple of days, and then sow seeds. After the seedlings emerge, mulch the soil surface to help cool the soil and retain moisture. It's normal for the top third of these large radishes to jut up above the soil. Use a heavy mulch of straw to protect the radishes from hard frosts, and harvest the roots as needed. Or, at the end of the season, dig up the whole crop and store the roots in a box of sawdust or sand in a root cellar.

SAMPLING RADISHES OF EVERY KIND

Everyone's familiar with early radishes—small, round red or pink roots that are ready to pick a few weeks after planting. Make small sowings (about 1 foot square) of spring radishes about once a week, or sow mini-grids in the corners of garden beds and between young plants of long-season crops such as tomatoes and corn (see page 95 for this seeding technique).

Experiment with other types of radishes, too. Black radishes tend to be too hot to eat raw but are delicious in spicy soups and stews. Chinese radishes vary in color and have mild, tender flesh. Japanese radishes (daikons) are the largest type, with mild flavor as long as they grow in cool weather. Large radishes need a bed area of their own; they're a good choice for fall gardens. Like most cool-season crops, their flavor improves after frost. Rat-tail radishes produce edible pods with a mustardlike flavor.

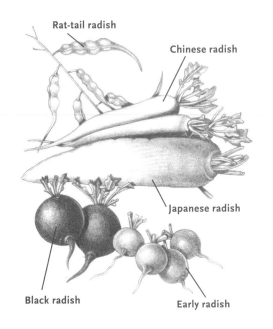

Rat-tail radish

Chinese radish

Japanese radish

Black radish

Early radish

PREVENTING PROBLEMS

BEFORE PLANTING

- Plant a trap crop of mustard 1 week before planting fall crops of radishes.

AT PLANTING TIME

- Plant early in spring to avoid harlequin bug infestations and poor root quality. However, if downy mildew has been a problem in your garden, avoid early spring plantings.
- Cover seedbeds with row cover and seal the edges to keep out flea beetles and cabbage root maggot flies (see Row Covers).

WHILE CROP DEVELOPS

- Destroy pests on trap crops before they migrate to your radishes.
- Uncover rat-tail radishes when they produce flowerstalks to allow for proper pollination.
- Spray plants with compost tea to help prevent disease and stimulate growth.

TROUBLESHOOTING PROBLEMS

Radishes rarely suffer from disease problems. A few common insect pests trouble radishes; those that feed directly on the roots are the most troublesome. The most common radish problems are listed below. If you find pest or disease symptoms on your radishes that aren't described here, refer to Cabbage for a full listing of potential problems.

SMALL HOLES IN LEAVES. Flea beetles like to feed on radish leaves, and a heavy infestation can ruin young plants. It's easier to prevent flea beetles from feeding on your radishes that to save a crop once flea beetles attack. Refer to Flea Beetles for preventive measures.

PLANTS WILT. If radish plants wilt mysteriously, chances are that root maggots have attacked the roots. Uproot a wilted plant and check the roots for scarring, distortion, or tunnels with maggots inside. Discard infested roots and replant in a different part of the garden. See Cabbage Maggots for controls.

If the roots appear normal, check the foliage for small orange-and-black bugs. These are harlequin bugs, which suck sap from leaves. They tend to be problematic during hot weather, so if harlequin bugs have invaded your radishes, quit planting them until the weather turns cooler (see Harlequin Bugs).

RADISHES HAVE INTENSELY HOT FLAVOR. Hot flavor is due to hot growing conditions and/or water stress. Before you give up on super-hot radishes, peel a couple and taste the flesh. Sometimes the worst of the "heat" is confined to the skin of the radish.

GRAYISH BROWN ROOTS. Discolored roots are a symptom of root rot. Infected roots develop soft sunken spots, too. Various fungi cause radishes to rot. Discard the rotted roots and replant in a different area of the garden. If the problem persists, look for resistant varieties. See Rots for other controls.

PITHY, WOODY, OR HOLLOW ROOTS. These problems result when radishes grow too slowly or when temperatures are high while roots are expanding. Hollow roots are usable if the flavor is palatable. To avoid this in the future, change the timing of planting and take steps to encourage fast growth.

CRACKS IN ROOTS. Early radish roots tend to split open if they're left in the ground too long. In the future harvest roots while they are 1 inch in diameter or smaller.

If the cracked roots have rough skin and dark flesh, the problem may be downy mildew, a fungal disease that can be a problem in humid regions (see Downy Mildew).

RHUBARB

Trouble-free rhubarb is easy to grow as a perennial crop in most areas. Healthy rhubarb plants in a sunny spot thrive for 20 years or longer with little care other than dividing the crowns when they become crowded. If you like desserts, sauces, and baked goods with a rhubarb tang, you will want to plant a crown or two. Don't worry if you can't find room in your vegetable garden for this large plant. Stately rhubarb blends beautifully into a sunny ornamental garden, too.

CROP BASICS

FAMILY: Buckwheat family

SITE: Plant rhubarb in full sun. In areas with very hot summers, rhubarb may grow best in a site that receives full sun in spring but filtered shade in summer.

SOIL: Fertile, well-drained soil is important for rhubarb's longevity and productivity. See the instructions below for preparing an enriched planting site.

TIMING OF PLANTING: Plant crowns or potted plants in spring or fall. Sow seeds (in mild-winter areas) in late summer.

PLANTING METHOD: Planting crowns or potted rhubarb plants is recommended unless you're growing rhubarb as an annual (see Regional Notes below).

CRUCIAL CARE: Pay special attention to rhubarb's watering needs when the temperature rises above 80°F. In prolonged heat, the plants often go dormant, but they'll resprout when temperatures drop. Divide plants when their growth becomes crowded and spindly.

GROWING IN CONTAINERS: Plant a single rhubarb crown in a rich soil mix (such as half compost and half good garden loam) in a container that's 2 feet wide and 2 to 3 feet deep.

HARVESTING CUES: In the first year of growth, harvest only a few stalks (or none at all) In the second year, harvest for several weeks, choosing only stalks that are as wide across as your thumb. After that, there's nearly no limit to the harvest, as explained below. Note: Rhubarb stems are the edible portion of the plant; the leaves contain toxic levels of oxalic acid.

SECRETS OF SUCCESS

DIG A BIG PLANTING HOLE. Rhubarb crowns for planting will measure only 3 or 4 inches across, but for best results, dig a hole about 2 feet wide and 1 foot deep for each crown. Space crowns 3 to 4 feet apart. Mix a large shovelful of compost and a large shovelful of aged manure into the soil you remove from the hole (add some additional compost if your soil is heavy clay). Refill the hole partway with this mixture until you can set the crown in place with the crown tip 1 inch below the soil surface. Then refill the hole the rest of the way.

If you garden in an area with wet winters, such as parts of the Pacific Northwest, try elevating your rhubarb planting. Dig the planting hole, but then refill it (mixing plenty of compost into the soil) and even build it up about 6 inches above the surrounding soil (firm the soil as you work). Then open a planting hole and set the crown in place, with the top of the crown about 1 inch below the surface of the mound.

Mulch the planting area lightly to control weeds. Water the young plants weekly unless there's sufficient rain to keep the soil moist. After the first year, the vigorous leaf canopy will shade out weeds.

HARVEST FROM SPRING TO FALL. You can cut rhubarb stems right through summer and fall

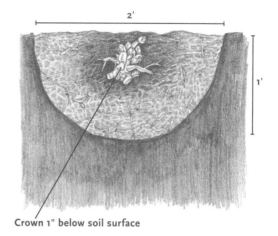

Crown 1" below soil surface

A king-size planting hole with enriched soil will result in a royal harvest of tasty rhubarb stems.

if weather conditions are right. The trick is never to remove all the green stalks at once—limit yourself only to those stalks as wide as your thumb.

In summer, if you don't harvest for a while, large stalks may turn corky and unpalatable. If that happens, cut out and compost all the large stalks, and water the plant well. New leaves and stems will sprout, and plants will be back in form within a few weeks.

Plants also produce flowerstalks, which are not good to eat. Remove the flowerstalks or allow the flowers to bloom—they are white and showy. Cut off flowers before they go to seed to avoid possibility of self-sowing and to allow the plant to expend energy on forming stalks rather than seeds.

South of Zone 6, rhubarb plants may go dormant in summer in response to high heat. Keep the plants watered (but don't overwater), and they will resprout again when the weather cools. It's fine to harvest more stalks then as long as they are broader than your thumb.

REGIONAL NOTES

DEEP SOUTH

In Florida, southern Texas, and similar mild-winter areas, rhubarb performs better as an annual crop planted from seed (or crowns, but it's expensive to buy new crowns each year). Start seeds indoors, generally in late August or early September. Soaking seeds enhances germination. Once the seeds sprout, you can move the containers outdoors to light shade and gradually expose them to more sun over time. Once the plants have three to five leaves, they're ready for planting in a garden bed in fertile, well-drained soil. Side-dress the plants monthly with balanced organic fertilizer. Their growth will slow during winter. If a hard freeze is predicted, protect the plants with cloches or a heavy row cover to prevent foliage injury. In late winter, growth will accelerate and you can start to harvest stalks. The harvest will end in May or June as heat sends the plants into decline. Pull up and compost all the residues. If possible, rotate your planting from bed to bed as you would other vegetable crops.

PREVENTING PROBLEMS

BEFORE PLANTING

- Prepare the soil well to ensure that plants establish easily and grow vigorously.

WHILE CROP DEVELOPS

- Harvest very lightly or not at all during the first year of growth.
- Remove and destroy diseased crowns to prevent problems from spreading to nearby healthy plants.
- If aphids show up on rhubarb, spray plants with water to dislodge the pests.

AFTER HARVEST

- Divide plants every 4 to 5 years to spur vigorous new growth.

TROUBLESHOOTING PROBLEMS

PLANTS FAIL TO RESPROUT IN SPRING. If you have an established plant that doesn't sprout in spring or sprouts but then dies back, it's probably suffering from crown rot. Check the crown for brown or rotted areas. Dig out and destroy infected crowns. Replant in a different spot in your yard, using disease-free crowns from a reliable supplier.

LARGE HOLES IN LEAVES. Slugs like to dine on rhubarb leaves, but their feeding probably won't reduce yields. Handpick the slugs and put out slug traps to control them.

LEAVES WILT. Check for brown, sunken spots on stems and brown areas on crowns. These are signs of fungal crown rot. The crown may be soft and rotted, too, if secondary rot has set in. Dig out and destroy infected plants. When you replant, buy crowns from a reliable source and plant in a well-drained spot.

STICKY DROPS ON STALKS. Sticky sap on rhubarb stalks is a sign that rhubarb curculios have been feeding and laying eggs on your rhubarb plants. The curculio is a grayish or yellow-gray insect with a long snout. The curculios suck plant juices and also pierce the stems to lay eggs.

Curiously enough, the eggs can't hatch in rhubarb stems, but they will hatch and complete their life cycle in other host plants—dock, thistle, and sunflowers. Injured rhubarb stems may start to decay, so remove them as soon as you notice them—use the uninjured parts and compost the rest. Control curculios by handpicking and by removing their other host plants around your yard.

NEW GROWTH IS SPINDLY AND CROWDED. Crowded, skinny stalks are a sign that your rhubarb is overcrowded and needs to be divided. Divide plants in fall after growth stops or in spring before new growth starts. Dig away soil from around the crown and use a sharp shovel or knife to cut cleanly through the crown. Slice off pieces about the size of a fist, and be sure each piece contains at least one active bud. Dig and replant the pieces. Leave about one-third of the original plant in place and fill in around it with a mix of soil and fresh compost.

CROWN ROTS. Fungal crown rot or red leaf, a bacterial disease, may be the cause. Fungal rots are firm but may turn soft and mushy if a secondary bacterial rot sets in. Red leaf is a problem on rhubarb in Canada and some northern states in the United States. The rot begins at the center of the plant, and leaves of infected plants turn red. Dig and destroy crowns infected by either crown rot or red leaf. Buy new crowns from a reliable source and plant them in a different part of your garden. Aphids may transmit red leaf as they feed, so prevent aphids from building up on your rhubarb plants by spraying the foliage with a strong spray of water.

ROTS

Your healthy garden soil is home to a prodigious collection of fungi and bacteria that cause plant material to rot. Pythium, Rhizoctonia, Phytophthora, and Fusarium are some of the common fungi that cause rot diseases, but many other fungi may be to blame as well.

Decomposition is an essential part of the soil food web—rotting of plant debris is one important step in the chain that releases nutrients in a form that plant roots can absorb. In a balanced soil, the rot-causing fungi and bacteria behave themselves, feeding almost exclusively on dead plant tissue.

However, when the soil stays wet too long, especially during cool weather, rot-causing organisms go into overdrive and infect living roots and leaves as well as debris. Root rot fungi turn roots into a dark, mushy mess, and fungi and bacteria easily invade plant parts that touch the wet soil surface. For example, lower leaves of lettuce plants and tomato fruits resting on the soil are easy targets for rot organisms. Providing well-drained soil is a universal recommendation for successful vegetable gardening because rot is less likely when soil drains well.

RANGE: United States and Canada

CROPS AT RISK: All vegetable crops

DESCRIPTION: Soft rots caused by *Erwinia* bacteria typically show up first as water-soaked spots on roots or leaves. These spots enlarge and turn dark on the surface, while the tissue beneath becomes mushy. A thick bacteria-laden liquid sometimes oozes from cracks in root crops such as carrots. Secondary rot bacteria then invade, and the infected plants deteriorate into a nasty-smelling mess.

Black rot is a serious disease of cabbage and related crops. If the bacteria infect seedlings, the young plants turn sickly and die. Larger plants develop pale green, wilted leaves that then die and drop off. Leaf veins sometimes turn black. Secondary rots also set in as the infection spreads.

Crown rot of squash-family plants is caused by a *Fusarium* fungus and begins as water-soaked tissue at the base of the plants that then turns dark. Infected plants will break off at soil level because of the rotting crown.

Yellow leaves and slow growth are a clue that plants may be suffering from fungal root rot. Look for dark spots near the crown of the plant, or uproot an affected plant. If the side roots are dark and soft, root rot is the problem.

FIGHTING INFECTION: Remove rotted plant material from the garden whenever you find it. Reduce watering, especially overhead watering. Try drenching the soil with compost tea to introduce beneficial microbes (apply the tea in the morning so the soil surface will dry quickly).

If you find rot in stored produce, remove all the rotted vegetables and any healthy ones that were in contact with them. If produce was stored in damp sawdust or sand, move the remaining healthy produce to a fresh container with new filler material.

GARDEN CLEANUP: Remove infected plant material from the garden and compost it. At the end of the growing season, remove all crop debris from the garden and compost it.

NEXT TIME YOU PLANT: Promoting good drainage is the most important step you can take to reduce rot problems. If you have heavy, slow-draining soil, build up raised beds at least 6 inches high. Plant deep-rooted cover crops, such as rape, to break through compacted subsurface layers that prevent water from draining through the soil. Add mature compost to your soil.

Choose resistant varieties when possible and buy disease-free seed and healthy transplants.

Remove wild mustard and other cabbage-family weeds from the garden, because they can host black rot bacteria. Pay attention while you're hoeing or hand-weeding around crops. Cultivating too close to the base of your crops or too deeply in the soil can injure roots, leaving openings for rot organisms to invade.

CROP ROTATION: Traditional rotation isn't too effective against rot problems because the bacteria and fungi that cause rot are so ubiquitous in soil. If you're having trouble with a rot disease that infects a particular plant family, such as black rot of crucifers, setting up a crop rotation is helpful in your overall effort to reduce disease problems. The bacteria that cause black rot can't survive more than 2 months living free in the soil.

ROW COVERS

Fabric row covers help prevent pest problems, enhance seed germination, shelter young plants from chilly spring nights, and allow gardeners to extend the gardening season well into fall. Plastic row covers aren't as versatile because they don't let water and air through. If you use plastic row covers, you'll need to install a drip irrigation system and monitor temperature closely under the cover so that plants don't overheat.

The downside of using row covers is that they are petroleum-based products with a limited life span. Eventually, you'll end up throwing them away and buying more. Another drawback of using row cover fabric is the time required to put it in place and then to periodically remove it so you can pull weeds and check your crop. Experiment with row covers in your own garden to decide whether and when their problem-solving potential pays off.

CHOOSING ROW COVERS

Floating row covers are lengths of synthetic fabric that allow water, air, and light to pass through. They can be made of polypropylene, polyester, or polyvinyl, and they're often labeled as "garden blankets." Depending on the weight of the row cover, it provides up to 8°F of protection from the cold. Heavyweight covers keep plants warmer, but they let through only 50 percent of available light, compared to 85 percent for lightweight row covers. Super lightweight row covers don't provide any frost protection and are used only to screen out insect pests.

You can buy row covers in pieces as small as 5 × 25 feet for about $15. Larger pieces are cheaper per square foot, and you can cut them to the size you need or roll up the excess around a length of 2 × 4 lumber.

ROW COVER SUPPORTS

Draping floating row covers directly over plants works fine, but many gardeners like to prop up the fabric on supports. Supported row covers stay cleaner and look neater than unsupported covers. The supports can be hoops or something less formal.

- Fill several 1- or 2-liter soda bottles three-quarters full of water and screw on the caps tightly. Upend the bottles and push them a few inches into the soil along the length of the bed between plants. The bottles act like tent poles for the row cover, plus the water in the bottles will absorb heat during the day and radiate it at night.
- Cut a section of welded wire fence as an arching support. A section 6 to 7 feet long will cover a 3-foot-wide bed. Cut away the cross wires along both edges to leave

Row cover fabric held in place by pairs of wire hoops is easy to push up along one side when you want to reach the plants underneath.

exposed wires. Push the exposed ends into the soil on each side of the bed and drape the fabric over the fencing.

- Cut lengths of 9- or 10-gauge wire to make hoops to support the covers. Position the hoops along the bed, spread the fabric over it, and then place additional hoops over the fabric to lock it in place. When you want to uncover the bed, just push up the fabric along one side.

EXTENDING THE GROWING SEASON

As a single layer, lightweight row cover fabric provides a few degrees of extra warmth. But for serious season extension, you can work with heavier weight row covers, or double or triple up on lightweight row covers. It's also fine to throw a blanket or tarp over a floating cover on a night when a sudden dip in temperatures is predicted.

Weight down the fabric well enough to prevent cold drafts from sneaking underneath, but not to seal the edges completely. Rocks are handy for this purpose, but sharp rocks can tear the fabric. Plastic water bottles or jugs partly filled with water work well as weights. Or you can wrap the long edges of the row cover around lengths of 2 × 4s (and unwind them as needed to offer more slack). Yet another option is to staple the edges of the cover to wooden molding strips.

SEAL THE EDGES FOR PEST CONTROL

Although fabric row covers let air and water pass through, they block the movement of even tiny insects such as thrips and whiteflies. The key to

Molding strips are a lightweight but sturdy edging for floating row covers. The strips hold the covers in place and provide a convenient handle when you want to lift the cover and check the plants underneath.

and put back than loose soil, and they'll be less likely to tear the fabric.

If you plan to leave floating covers on for pest control up until harvest, monitor plants carefully during summer—temperatures may rise too high, and you'll be doing the plants more harm than you're trying to prevent from pests.

EXTENDING THE LIFE OF YOUR ROW COVERS

Over time, floating row covers will pick up some dirt and debris, but it's not worthwhile to wash them. Most of the dirt will shake off when the covers are dry, which they should be before you store them for winter. Fold or roll up the fabric and put it in a dry place out of the sun. Ideally, store it in your garage or garden shed. If you don't have the space, store the fabric outdoors, but keep it out of the sun.

pest protection with row covers is to seal the edges completely so pests cannot crawl under the covers. Once in the shelter of the row cover, chewing and sucking insects can ruin a stand of young plants in only a few days' time.

The standard recommendation for sealing row cover edges used to be to bury the edges in soil, but this is difficult to do and causes the fabric to break down more quickly. Instead, weight the edges of row covers with 1 × 2s or lengths of rebar. Another option is to shovel soil into double plastic bags (like grocery-store bags) or old pillowcases, and then place the bags on the row cover edges, pushing the bags into long, flat weights. Old socks filled with pebbles work well, too. The weights will be easier to take off

Scraps saved from torn row covers work well for wrapping a container-grown crop to provide heat and keep out pests.

Check the condition of row cover fabric before you store it at the end of the season. Cut off ripped and ragged edge pieces, but don't throw them away. You can use them as patches to repair small holes in large pieces of otherwise usable cover. Use waxed dental floss to sew the patches onto the cover. If you run out of row cover patches, you can use pieces of cast-off panty hose instead.

If a row cover has developed too many holes to be worth patching, cut it up into small intact sections, and tuck those into your planting supplies box or basket. They're handy for covering small patches of newly sown seeds to hold in surface moisture and boost germination.

Irregular pieces are useful for covering a frame around a container vegetable plant that you want to protect. For example, if you have a couple of potted eggplants, cover them with row cover until temperatures are well into the 70s. They'll like the extra heat and the chance to grow for a few weeks untroubled by flea beetles. Use coat hangers or other lightweight wire to mold a domelike frame. Drape the row cover fabric around the frame and clip it to the rim of the container with clothespins.

RUST

Rub the leaf of a crop infected with a rust fungus and you'll find a rusty coating of spores on your fingers. Different species cause rust in individual garden crops, and some species require a secondary host to complete their life cycle. For example, the species that causes rust in corn spends part of its life cycle in corn plants and the other part in sorrel (Oxalis spp.), a common weed.

However, moist winds carry rust spores long distances, allowing infections to occur even in areas where the alternate host is not present. A rust infection can take hold quickly during warm, humid weather.

RANGE: United States; southern Canada

CROPS AT RISK: Asparagus, beans (particularly pole beans), peas, corn, eggplant, Jerusalem artichokes, okra, onions, sweet potatoes

DESCRIPTION: Specific symptoms vary from crop to crop, but a rusty appearance is fairly common in infected plants. Asparagus plants show orange-red blisters on leaves and stems, while bean leaves develop reddish brown blisters on the undersides. Blisters also appear on bean pods. Blisters appear on the top surface of corn leaves and on stalks of infected corn plants. When an infection is severe, leaves turn yellow and drop off; plants become stunted and may die.

FIGHTING INFECTION: Pick off infected leaves and compost them. Switch to hand watering or drip irrigation rather than using sprinklers. If plants become seriously infected, pull them out and compost them.

GARDEN CLEANUP: Rust fungi can overwinter in crop debris. Clean up the garden thoroughly after harvest and compost the debris. If asparagus has been infected, cut all ferns to the ground at the end of the growing season and compost them or dispose of them away from the garden.

NEXT TIME YOU PLANT: For future crops, choose resistant or tolerant varieties when possible. Trellis pole beans and stake eggplant and okra. If past infections were severe, you can apply sulfur sprays to prevent the disease, but keep in mind that sulfur may burn crop leaves.

CROP ROTATION: A 2-year crop rotation by family may lessen rust problems by preventing rust fungi from building up in the soil (see Crop Rotation).

SALAD GREENS

Colorful, flavorful greens add originality and zest to homegrown salads, and they're so easy to grow that every gardener should try some. Most salad greens are trouble-free as long as you harvest them young, before pests or diseases have much chance to attack the plants. To ensure a healthy harvest, provide rich, well-drained soil and cover seedbeds with row cover fabric to prevent leaf-feeding insects such as aphids and flea beetles from bothering the crop.

MAKING MESCLUN

Mesclun is the French term for a mix of baby lettuces and other salad greens. Many North American seed companies now sell mesclun seed mixes—packets that include several types of seed mixed together. Trying one or more of these blends is a fun way to sample the wide variety of greens available. The trick, though, is to sow mesclun mixes so that each species included has a chance to germinate and grow.

Some salad greens (including lettuce) germinate very quickly in the right conditions, while others, such as corn salad, may take 2 weeks or longer to poke through the soil. If you sow a mesclun mix in a dense row, the fast germinators and strong growers, such as mustard, will outpace the weaker crops, resulting in a mix that's heavy on certain greens and lacking others.

One way to overcome the problem is to prepare a seedbed and use the stale seedbed technique (described in detail on page 436) to eliminate the first flush of weeds. Then you can sow the seeds lightly across the clear, well-moistened seedbed, spacing the seeds at least 1 inch apart. Gently cover the seeds with seed-starting mix (don't disturb the soil if you can help it),

firm it with your palm, and gently mist the area to moisten the top ½ inch of soil. Observe the bed closely and you'll see that some seedlings emerge within a few days. Protect the seedbed from heavy rain or strong sun until many of the seedlings have produced their first set of true leaves (see the Weather entry for techniques). By that time, even the slow germinators will have

If you learn to tell your salad greens apart at the seedlings stage, you can thin your stand of mesclun mix to favor the greens you like best.

emerged. If you can, learn to recognize the seedlings of the various components of your mesclun mix. If you see one or two types starting to dominate, thin them out to leave more room for other types to catch up and fill in.

SEPARATE YOUR GREENS

Another technique is to grow components of mesclun in individual wide or narrow rows. This method allows you to provide custom care for each type of green and also to vary the composition of your salads from day to day by selective harvesting. The drawback of this method is that it costs more to buy the seed than purchasing a salad mix, and it requires more space.

You can economize by placing a group order with friends and splitting up the seed. To solve the space problem for fall greens, sow the seeds in small patches in the shade of summer crops. As you harvest and pull out the warm-season crops, transplant the salad greens seedlings into the spaces that open up.

CHOOSING SALAD GREENS

Lettuce and spinach are the mainstays of green salads, but there's no need to stop there. Add endive, Asian cabbages, kale, and Swiss chard, too (you'll find complete coverage of these crops in their respective entries). Other salad greens that are gaining in popularity are arugula, corn salad (mache), cress, mustard, and radicchio. Radicchio is related to lettuce, while arugula, cress, and mustard belong to the cabbage family. If you discover pest or disease problems on these salad greens, refer to the Lettuce or Cabbage entries to diagnose the cause of the problem.

ARUGULA

FAMILY: Cabbage family

SECRETS OF SUCCESS: Arugula likes full sun but cool temperatures. If temperatures will exceed 70°F, plant arugula in the shade of other crops, or even in dappled shade under trees.

Sow seeds as soon as soil can be worked in spring. Sow small amounts of seeds weekly. In Zone 5 and colder, also sow a fall crop in September in a coldframe. In Zones 6 and 7, sow fall crops in the garden and provide some winter protection to extend the harvest. South of Zone 7, sow in fall when summer heat is past—repeat sowings as desired for harvest throughout winter. In the Gulf Region and the Southwest, try arugula as a winter crop, and shade the seedbed to cool it (see Weather for information on shading methods).

Provide a steady supply of water to encourage fast growth, which helps prevent bitterness. In arid climates, use a sprinkler to water arugula because the cooling spray of water also helps preserve the tenderness of the foliage. Begin harvesting arugula leaves when plants are as little as 2 inches tall.

CORN SALAD (MACHE)

FAMILY: Valerian family

SECRETS OF SUCCESS: Patience is the first requirement for growing corn salad. It is a mild salad green that's slower to germinate and grow than many other leafy crops. Mache does best in cool weather and overwinters well for a spring crop. Thin the stand until plants are 2 inches apart (you can eat the thinnings). Allow remaining plants to grow about 3 inches tall and then cut the whole rosette of leaves at one time.

CRESS

FAMILY: Cabbage family

SECRETS OF SUCCESS: Curly cress and upland cress are easier to grow than watercress and offer similar flavor (which is peppery but sweet). Curly cress grows so quickly that you can plant it in a large pot and harvest less than 2 weeks later. Cut the whole plant and let it regrow. Upland cress thrives in a garden bed and is fairly resistant to bolting. You can pick individual leaves or harvest the entire plant at once.

MUSTARD GREENS

FAMILY: Cabbage family

SECRETS OF SUCCESS: Mustards add a peppery bite to salads that ranges from mild to sharp depending on which type you choose. Some mustards sport red or almost purple leaves, while others are bright green. Baby leaves have milder flavor and are ready to pick about 3 weeks after sowing. Mizuna and mibuna are mild Asian mustards. Mizuna leaves are frilly, and mibuna has broad leaves with smooth edges.

Mustard tolerates heat and cold better than other greens and is quite vigorous, regrowing after multiple cuttings. Hot weather sharpens the flavor of mustards, so they are best used before summer heat sets in. You can cut mizuna and let it regrow, but pick individual young leaves of other mustards for salad use.

RADICCHIO

FAMILY: Lettuce family

SECRETS OF SUCCESS: In most areas, time the planting of radicchio so you can harvest it during cool fall weather, when the bitterness of the leaves becomes more moderate. Spring plantings will succeed in areas that have cool nights during the summer. Radicchio is quite hardy. If you harvest the heads at ground level in fall and provide a heavy row cover or other winter protection, the roots will survive and send up new growth in spring. Sow a bigger stand of radicchio than you wish to harvest. This provides a cushion against loss due to lack of uniformity— some plants may never form a proper head—and also against loss due to rot during a long period of cool and wet weather. If you choose heirloom varieties, check the growing instructions on the seed packet. You may need to cut topgrowth back in fall to stimulate heading.

SCAB

Scab-infected potatoes, beets, and other root crops don't look pretty, but they're still good to eat. Scab is a disease caused by a bacteria-like organism that also is somewhat like a fungus. It's only a problem for home gardeners if it becomes so severe that the entire surface of tubers or roots end up covered with scab. In healthy, moist soil, beneficial soil bacteria and fungi compete with scab and prevent it from becoming a serious problem.

An unrelated bacterium causes a disease of cucumbers that's also commonly called scab. Infected cucumbers develop water-soaked spots on leaves and fruits, and the infected areas on fruits dry to form a scab. It's usually not a problem in home gardens, though, because most cucumber varieties are scab resistant.

RANGE: United States; southern Canada

CROPS AT RISK: Potatoes, beets, carrots, parsnips, radishes, turnips

DESCRIPTION: Warty or corky blotches form on potato tubers; on other crops, scab appears as brown sunken areas on roots.

FIGHTING INFECTION: Scab usually isn't discovered until you harvest the crop. Cut away scabby areas and eat the rest; the disease doesn't affect flavor. Don't try to store infected roots or tubers because they'll be prone to rot by other bacteria or fungi.

GARDEN CLEANUP: Remove infected tubers and roots from the garden. It's okay to add peelings from scabby vegetables to your compost pile.

NEXT TIME YOU PLANT: Choose resistant varieties of potatoes. If possible, adjust soil pH in the area where you plan to plant potatoes or root crops to below 5.5. Maintain organic matter and even soil moisture—scab tends to be worse in dry soils. It's possible that applying *Trichoderma harzianum* to the soil may help with scab problems—watch for further research reports on this use of this biocontrol product (also see Disease Control).

CROP ROTATION: The microorganism that causes scab can survive indefinitely in soil, so rotating crops won't eliminate scab from your garden. However, setting up a long rotation of susceptible crops will help prevent scab from becoming worse from year to year. Including cover crops of rye and oats in a rotation is helpful.

SEASON EXTENSION

Cloches, covers, and coldframes are the most common devices for home gardeners who want to stretch the gardening season. In fact, season extension becomes a passion for some, especially those who garden in cold areas such as the Northeast states and provinces, high elevations of the Rocky Mountains, and the northern prairies. Greenhouses, hoop houses, and other equipment allow gardeners even in these extreme climates to garden long after frost strikes.

From a problem-solving standpoint, the first step in season extension is to find low-cost ways to experiment, allowing you to learn firsthand whether you enjoy stretching your growing season enough to invest money in materials that go beyond the basics. After that, if you want to try more sophisticated techniques, refer to the books on season extension listed in Recommended Reading on page 458.

SIMPLE TRICKS

The basic starting point for season extension is to plant crops a little earlier than usual in spring and then provide some kind of insulation from temperatures that would stress or kill them. Materials you have on hand around the house and garage often come in handy for these efforts.

Anchor a jug cloche by placing a rock on a fold-out flap or on the cap of the jug.

- For small plantings of cold-tolerant seeds when your soil is too wet to work in spring, make mini raised beds using bricks or scrap wood to build 2-inch-high frames, and fill them with compost.
- If you have lots of rocks, use them to surround your raised beds in spring. The rocks absorb heat during the day and release some of it to the soil during the night.
- To protect plants fast when low temperatures are forecast overnight, keep a stash of old sheets, blankets, curtains, and quilts to throw over the plants (you can buy these very cheaply at yard sales). If you store your covers in a garage or shed, stash them in a plastic tub with a tight lid, because mice will find them an attractive nesting site.
- The best-known homemade cloche is a translucent plastic jug with the bottom cut out. Leave the cap on for maximum protection, or remove it to vent the cloche. A common problem with jug cloches is that they blow away on windy days. To hold your jugs in place, anchor them either by putting a flat rock or piece of paver block on the cap, or by leaving the bottom of the jug attached by a small, uncut section, so the bottom can fold out to the side of the jug. Rest a rock or concrete paver on the flap.

PLANT LATE, HARVEST EARLY

One low-tech way to extend the growing season is to sow seeds of cold-tolerant crops such as spinach or lettuce 4 to 6 weeks before your first

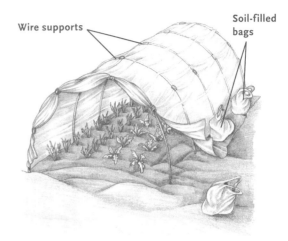

Wire supports

Soil-filled bags

A tunnel of clear plastic protects fall crops from hard frost.

fall frost and then protect them through winter. The plants will establish roots and a small rosette of leaves. Then they will have a head start in spring that beats what you can accomplish by starting seedlings indoors.

When your first hard frost is forecast, weed the area and water it (but don't saturate the soil). Pile on about 6 inches of shredded leaves. Spread a tarp over the leaves (to prevent them from getting wet, which would ruin their insulating value) and secure the tarp with bricks, rocks, or pieces of wood at the corners.

If you'd like to extend your fall harvest, plant the crops a little earlier, and at the time of hard frost, cover the plants with a clear plastic tunnel. You'll need 1½-mil plastic supported by hoops. In Zones 5 and north, a double layer of plastic provides better protection. To weight the edges and ends of the plastic, double up plastic bags (like the ones from a grocery store) and put some soil or sand in them. That way it's easy to open the row cover when you want to harvest. The plants may continue growing (slowly) until Thanksgiving or even Christmas. During January and February, though, cold temperatures and short days will result in very little growth, so just leave the plants covered and be patient. In late winter or early spring, remove the plastic and replace it with row cover fabric. The plants should resprout, offering a spring harvest that lasts until rising temperatures and longer days trigger bolting.

For a 3-foot-wide bed, you'll need lengths of plastic and row covers that are 8 to 9 feet wide. If you cut your own hoops from a roll of wire, make them about 7 feet long.

ROW COVERS

To extend your harvest on a larger scale, the next step is to buy enough row cover fabric to cover one (or more) of your garden beds. Row covers are easy to use, and if you use them in layers, you can extend the harvest season by weeks or even months. For details, see Row Covers.

COLDFRAMES

For many of us, row covers and low plastic tunnels are all we need to extend the season. But to explore even further, you can improvise a low-cost temporary coldframe. With six bales of straw and an old window or a wooden frame covered with heavy plastic, you've got the makings of a temporary coldframe. Set up the frame around a bed of existing plants in fall to extend the harvest, or use it in spring to harden off seedlings.

Keep a thermometer inside the coldframe to monitor the temperature. For cool-season crops, prop up the lid when the temperature rises above

Six bales of straw and an old window or a wooden frame covered with plastic serve well as a temporary coldframe. Use blocks of wood to vent the frame.

50°F. For warm-season crops, vent at 65°F. Venting the frame is especially important during humid weather to prevent disease problems. Use bricks or small pieces of scrap wood to prop open the lid. If the temperature inside the frame reaches 70°F, fully open or remove the lid. Be sure to recheck the frame in the early to mid-afternoon. Once the temperature inside drops below about 45°F, close the lid to capture heat to protect the plants overnight. To insulate the frame on a particularly cold night, cover it with an old blanket or quilt.

SEED STARTING

Planting seeds indoors when the weather is too cold (or too hot) for outdoor sowing opens up new gardening opportunities—plus it's lots of fun! Starting seeds indoors is easy, too, if you have the right seed-starting setup and know how to avoid common problems.

OVERCOMING POOR GERMINATION

Some seeds germinate virtually overnight in a wide range of conditions, but others are finicky. To avoid the disappointment of a poor start, investigate seed viability, check the quality and temperature of the soil mix, avoid damping-off, and try presprouting.

TEST SEED VIABILITY

One reason seeds don't sprout is that they're no longer viable—the embryo inside the seed has died. This can happen if seeds overheat or get wet in storage. The only way to assess seed viability is to test a sample of seed, and it's a good idea to test whenever you're working with leftover seed from a previous year.

To test seeds, dampen a few layers of paper

Seeds nestled in moist paper towels germinate quickly if they're viable. If the germination rate is less than 50 percent, the seed probably won't produce a good crop.

towels and spread 10 seeds over the surface. Roll up the paper towel and stand it on end in a mason jar, with the lid off (this allows substances in the seeds that could inhibit germination to seep down and away from the seeds). Put the jar in a warm place. Check the paper towels daily to be sure they're still moist. When you're reached the typical number of days to germination (you'll find this information on the seed packet), unroll the towels and see how many seeds have germinated. If they haven't sprouted, make sure the towels are still damp, roll them back up, and wait another couple of days.

Any seed that shows 50 percent or higher germination is worth using. If the germination rate is 25 to 50 percent, it indicates that the seed is weak and probably suffered from stress. If you have a lot of weak seed and hate to throw it away, try sowing it thickly as a cover crop. Seeds with a germination rate of less than 25 percent aren't worth sowing at all; discard the seed.

PRESOAK FOR SUCCESS

Some vegetable seeds have especially hard seed coats or seed coats that contain germination inhibitors. It can take these seeds more than 2 weeks to germinate. Try soaking the seeds of these crops in warm water for 24 hours before you sow them. Presoaking works well for beets, carrots, parsley, parsnips, and spinach seed.

Soaking bean or corn seeds can be too much of a shock, however, and can cause the cotyledons to crack. Instead, set up these seeds in a damp paper towel overnight, as if you were testing viability (see above). It's okay to put more than 10 seeds on a towel in this case—space them about 1 inch apart.

KEEP THE MIX WARM

Cold seed-starting mix is a common reason for slow germination. Check seed packets or other guidelines to find the ideal germination temperature for the seeds you want to start.

Start out warm by adding heated water (not boiling, but warmed on the stove or in an electric kettle) to your mix in a bucket. Stir the water and seed-starting mix well, using your hands or a long-handled trowel or large mixing spoon. Your goal is a mix that is moist, not sopping wet. Then transfer the mix to your containers and sow the seeds.

Place your seeded containers under seed-starting lights and leave the lights on continuously also to help add warmth. (A combination of warm white and cool white fluorescent tubes works well for seed starting.) Place a piece of black paper or fabric over the flats to block light but transmit heat. Check daily, and as soon as some seedlings poke through the surface of the mix, take off the cover, and change your lights to provide 16 hours of light per day.

The ideal way to warm the soil is to use a seed propagation mat especially designed to be safe to use in the moist environment of a seed-starting area.

STOP DAMPING-OFF

Fungal diseases can kill seedlings both in indoor flats and in garden beds, sometimes even before they emerge from the soil. Suspect damping-off if the problem is limited to a few containers here and there, especially if seedlings emerge and then keel over. Refer to Damping-Off to learn how to avoid problems.

NO MORE SPINDLY SEEDLINGS

Lack of light, stress, and mistakes in timing are the common causes for spindly seedlings, and they're easy to overcome.

IMPROVE THE LIGHTING

The most common mistake when it comes to lighting seedlings is giving them too little light. Perhaps you occasionally forget to switch on the lights—one day in darkness is enough to make seedlings stretch. The solution to a faulty memory is a timer. Set the timer so the lights will be on 16 hours a day. This is better than simply leaving the lights on all the time, because some plants actually suffer foliar damage and die when they're exposed to continuous light.

Ideally, seedlings should be close to the light source, with the top leaves only 2 inches away from the fluorescent bulbs. You can accomplish this by hanging your light fixtures from chains that you can adjust in length, and raising the lights as the seedlings grow.

Or, if your lights are fixed, use props such as bricks, old phone books, or stacks of newspaper to raise seedling flats up to the light. Newspaper is a great choice because you can decrease the size of the stack as the seedlings grow.

AVOID RICH POTTING MIXES

Seedlings are adapted to relying on the stored reserves in a seed for the first few weeks of life. If you plant them in a fertilizer-enriched potting mix, the food supply can be overwhelming, causing a rush of stem growth—and then your seedlings may flop over. Instead, plant seeds in a sterile germination mix especially designed for seed-starting. Once the seedlings have two pairs of true leaves, you can fertilize them with a half-strength solution of fish fertilizer or compost tea if you're worried about lack of nutrients.

CHOOSE CAPACIOUS CONTAINERS

Seedlings will grow in a wide variety of containers, from eggshells and egg cartons to yogurt cups and newspaper pots. But since your goal is to reap a bountiful harvest, it's worth your time and a little bit of money to buy standard equipment rather than relying on a hodgepodge of

Keep lights 2 inches above the tops of seedlings as they grow by adjusting the height of the fixture or the elevation of the seedlings.

cast-off materials. Pick what you prefer using: plastic six-packs, peat pots, or 3- and 4-inch plastic pots. Yogurt containers or other food containers also work, as long as you can collect enough of one size to meet your needs.

With standardized containers, it's easier to keep seedlings evenly watered and fertilized, and that will produce uniform, healthy seedlings.

CHANGE THE SCHEDULE

Seedlings sometimes become floppy or spindly from being pampered too long. They like their warm seed-starting area so much that they grow rapidly and then flop over. The simple solution is to shorten the period of time the seedlings spend in your seed-starting area; just recalculate seed-starting times. Just start your seeds 1 to 2 weeks later than whatever schedule you've used in the past. The seedlings will be smaller when you move them outside to start hardening off, but slightly undersize seedlings often survive the transition to the garden better than overgrown seedlings.

BRUSH YOUR SEEDLINGS

Use an old broom handle or the cardboard tube from gift wrapping paper to gently brush the foliage of a flat of seedlings—this will produce stockier growth. Start brushing seedlings as soon as the cotyledons are full size and the first set of true leaves is unfolding. Ideally, brush them for a minute and a half twice a day. (If you run a fan in your seed-starting area, it may substitute for brushing for the seedlings closest to the fan.)

HARDENING OFF

The last step in raising seedlings indoors is preparing them for conditions in the garden. Outdoors, seedlings must adapt to much stronger light, changes in soil moisture, and daily fluctuations in temperature. If you move them outside abruptly, the change in environment will be so great that the plants will go into shock and stop growing.

To avoid transplant shock, harden off seedlings by gradually introducing them to the outdoors. The ideal method is to move the seedlings outside for about 2 hours and then bring them back inside for the rest of the day. Increase the length of exposure by 2 hours a day until the plants are outside both day and night.

If you're at home during the day, this routine is easy. But if you're away from home during the day, you'll have to put the seedlings outdoors in the morning and take them back in at night. Provide some protection from wind and direct sun for the first week, and check each morning and evening to be sure the soil mix is moist.

One option is to put the seedlings in a coldframe for the day. Drape burlap or row cover over the coldframe cover to filter sunlight for the first few days. Prop open the lid partway on sunny days so it doesn't get too hot.

A screened porch makes a reasonable substitute for a coldframe. Find a spot on the porch that's protected from wind but receives direct sun for part of the day. If possible, change the location of the seedlings each day so they'll receive more direct sun.

If you don't have a coldframe, you can improvise one from a cardboard box. Cut the sides of the box on an angle like a coldframe. Set rocks or plastic bottles filled with water on the box flaps to prevent it from blowing away. Position the box where it will get sun for part of the day (you can move the box easily each day to increase the duration of sun). Drape burlap or row cover over the box to filter sunlight at first.

No matter which hardening-off method you use, remember to bring the seedlings back indoors overnight for the first week. After that, as long as temperatures aren't going to drop too low, you can leave the seedlings out overnight, too (minimum temperature varies by crop). As they adjust, you can begin to leave

A cardboard box with weighted flaps protects seedlings from wind and pests as you harden them off.

the coldframe cover off during the day and night, too, until you are ready to plant the seedlings in the garden.

SHALLOTS

Pleasantly mild onion flavor in a perennial package awaits those who grow shallots. Cooks prize shallots for their subtle flavor, which some describe as a mix of onion and garlic. Gardeners will appreciate how easy it is to grow shallots. Planted as sets in fall, these close cousins of onions produce foliage in fall, go dormant during winter, and then finish off in spring by producing a cluster of small, delicious bulbs.

If you like growing shallots, you may also want to try other kinds of multiplier onions, such as potato onions and Egyptian onions. In general, multiplier onions suffer few pest problems, and once some clumps are growing in your garden, you'll have a steady supply of onions for eating plus extras to replant for the next crop.

CROP BASICS

FAMILY: Onion family; relatives are asparagus, chives, garlic, leeks, and onions.

SITE: Shallots and other multiplier onions grow best in full sun but they can tolerate light shade.

SOIL: Shallots grow best in light, fertile, well-drained soil. The soil preparation instructions for onions also work well for shallots and other multiplier onions (see page 280).

TIMING OF PLANTING: Plant sets in fall or in early spring as soon as you can work in your garden. Plant shallots from seed in early spring.

PLANTING METHOD: All types of multiplier onions are commonly planted as sets (or bulblets for Egyptian onions). It's also possible to buy and plant seeds of some varieties of shallots.

CRUCIAL CARE: Mulch fall-planted shallots with several inches of straw in areas where winter temperatures drop below 0°F. Remove the mulch in early spring. Water plants with liquid fish or seaweed fertilizer as they resume growth in spring. Cut back on watering when the tops start to turn yellow.

GROWING IN CONTAINERS: Shallots and potato onions are a great winter windowsill crop, as explained below.

HARVESTING CUES: Snip individual leaves occasionally while plants are growing to use as flavoring or garnishes but keep in mind that this could reduce the growth of the bulbs. You can pull whole plants when young to use as green onions (scallions). To harvest mature bulbs for storage, allow the tops to turn yellow and fall over before digging the clusters. Carefully separate the bulbs and cure them in a warm dry place for about 1 week. High heat will cause flavor to intensify, so be sure to harvest the bulbs before temperatures climb if you prefer mild-flavored onions.

SECRETS OF SUCCESS

REPLANT AS YOU HARVEST. When you first order shallots or other multiplier onions from a supplier, you'll receive about 30 sets if you order 1 pound of sets (for Egyptian onions you'll receive bulblets). This is more than enough to start your own crop and share with a friend. After that first investment, you can use extra bulbs from your own harvest as the seed crop for the following year. When you harvest, sort

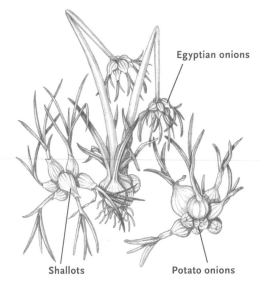

Egyptian onions

Shallots

Potato onions

Shallots and potato onions form clusters of pear-shaped or angular bulbs ranging in size from 1 to 3 inches. Egyptian onions, also called walking onions, produce a cluster of small bulblets at the top of the stalk as well as a group of bulbs at the base. Propagate by planting the bulblets, or allow the top-heavy stems to flop over and plant themselves.

potato onions and shallots by size. With potato onions, you'll keep the small bulbs for planting and eat the large ones. With shallots, it's a good idea to reserve the largest bulbs for replanting.

At the appropriate time for replanting, simply push the bulbs into the soil 1 to 2 inches deep (it's fine for the tip of the bulb to poke out through the soil). Planting time ranges from September to November, depending on where you live. Aim for the plants to produce three to five leaves before they go dormant for winter. Plants that grow too large in fall may produce a seedstalk prematurely in spring. If the plants do bolt, simply break off the seed stalk; the plants should still produce a harvestable cluster of bulbs.

Egyptian onions tend to propagate themselves, but if necessary, plant bulblets where you want them to grow.

POT THEM FOR WINTER GREENS. Chives grow well in containers outdoors but tend to languish indoors on a windowsill. Instead, use shallots and multiplier onions as a source of onion greens during winter. Plant a couple of shallot or potato onion bulbs into a light soil mix in a pot at the same time you plant the bulbs outdoors in the garden. When the weather turns too cold later in fall, bring the pot indoors to a sunny location for winter. Water lightly as needed, but let the soil surface dry between waterings. Snip leaves as needed for cooking and salads.

TROUBLESHOOTING PROBLEMS

Shallots and other multiplier onions rarely suffer pest or disease problems. If your crop develops a problem, refer to Onions to figure out the cause.

SLUGS AND SNAILS

Spring showers help your garden grow, but they also create perfect conditions for slug and snail feasts. These slimy pests stay hidden when the weather is dry or chilly, waiting for warm, wet nights to emerge and devour tender foliage. A severe slug attack can wipe out a stand of small seedlings overnight; older plants will show large, ragged holes in the leaves. Occasionally you'll see these creatures during the day as well, especially a rainy one.

Most garden slugs are dark colored and fairly small, but one brightly colored species with a reputation for especially large size and appetite is the banana slug of the Pacific Northwest. Generally, native snails aren't garden pests— it's the imported brown garden snail that is troublesome for gardeners. Slugs and snails belong to the mollusk family; they are relatives of lobsters, crabs, and other shellfish. Like many kinds of insects, though, slugs and snails overwinter as adults in a dormant state. (In mild-winter areas, they may remain active year-round.) The creatures emerge from dormancy at about the time soil thaws in the spring. At first, they will be very sluggish and cause little damage. As the weather warms, slug and snail feeding increases, and it can be devastating during a wet spring. In areas with hot, dry summers, damage tapers off as slugs seek out deep cover in the soil or under mulch where they can remain moist. Snails retreat inside their shells and seal themselves to plant stems, shaded walls, or protected spots such as the underside of wood siding. Slug and snail damage resurges in late fall in the humid South, too.

Spring and early summer are the prime seasons for these pests to lay eggs in most regions. Snails lay eggs in holes in the soil, while slugs lay eggs in soil cracks, under mulch or rocks, or even under outdoor flowerpots. When conditions are favorable, slug eggs will hatch in less than 3 weeks. In the North, one generation is common; two generations occur during years with lots of warm, wet weather. Snails can take up to 2 years to reach maturity.

There's no single control method that will eliminate slugs and snails, but using a combination of methods can save your garden from excessive damage. If slugs or snails are a serious problem in your area, try a combination of four tactics: changing the environment to make it less appealing to slugs and snails; providing hab-itat for ground beetles and other slug predators; handpicking frequently; and killing the pests using traps and/or baits.

PEST PROFILE

Many genera and species

RANGE: Throughout the United States and Canada

HOW TO SPOT THEM: Shiny trails on foliage are a sign that slugs or snails have been active in your garden at night; these trails are the dried slime left behind as the creatures glide along. Look underneath rocks or mulch or in other damp, dark spots during the day to find slugs and snails at rest. On a warm, moist night, go out in the garden well after dark and shine a flashlight on plants to catch these pests while they feed.

ADULTS: Black, brown, gray, or yellow creatures with soft bodies covered with sticky mucus.

Slug

Snail

Slug and feeding damage

Some species are as small as ¼ inch, while others reach as large as 7 inches long. Snails have a coiled shell; slugs do not. Baby slugs and snails are similar in appearance to adults but are smaller and may not have the same coloring.

EGGS: Translucent round balls filled with a watery liquid; laid in clusters in the soil or under plant debris

CROPS AT RISK: Nearly all crops, especially young seedlings, leafy crops, beans, tomato fruits, and strawberry fruits

DAMAGE: Slugs and snails rasp at foliage and can consume seedlings and small plants entirely. On larger plants, their feeding produces large, irregular holes in leaves or reduces leaves to nearly bare midribs. They also chew holes into strawberry fruits and tomato fruits that are near ground level.

CONTROL METHODS

PROVIDING HABITAT FOR PREDATORS. Ground beetles and rove beetles eat slugs and snails. Provide permanent shelter for these good bugs in and near your garden by planting cover crops and leaving some pathways or other areas with a permanent cover of stones or wood chip mulch. Birds eat slugs, too; see Birds for tips on encouraging pest-eating birds.

MANAGING MULCH. If slugs and snails are a serious problem in your garden, try raking up all surface mulch on garden beds at the start of the growing season. Wait until the warmth of early summer to renew the mulch. Limit mulch depth to 1 inch; this will protect the soil surface and conserve moisture, but the mulch itself will stay fairly dry and thus less welcoming to slugs and snails. Let grass clippings dry before using them as mulch.

HANDPICKING. Handpicking slugs and snails can be a successful control method, but only if you are dedicated. When slugs and snails are plentiful, handpick nightly, waiting until after 10:00 p.m. to pick. Use a flashlight to find the creatures on plants (be sure to lift leaves to check the undersides and areas around the base of plants). Wear rubber or latex gloves to avoid getting slug slime on your skin, or use tongs, chopsticks, or tweezers to pick off the pests and dump them in soapy water. Some gardeners stab the pests with shish kebab skewers. Try leaving a few dead snails or slugs in the garden; their presence will attract more slugs and snails to that spot.

If your composting area is located close to your vegetable garden, scout there, too. The moist conditions and plentiful supply of decaying plant material are very attractive to slugs and snails. You'll also find them under the foliage of dense groundcovers such as ivy.

Some gardeners kill slugs and snails by sprinkling salt on them, but doing so can cause harmful salt buildup in your soil. If you're curious to see how salt works, try it once, but then switch to collecting the slimy critters and dumping them in soapy water to drown.

To dispose of a large quantity of dead slugs or snails, bury them deeply in an active compost pile or about 4 inches deep in garden soil. A large pile of dead slugs or snails dumped on the soil surface may attract undesirable flies.

If you find egg masses around your garden, spread them out to dry in the sun or put them in a sealed plastic bag and throw them in the trash.

RAIDING SHELTERS. To supplement nighttime handpicking, set out some shelters for slugs and snails around your garden, such as boards

Set out boards propped up on small stones or sticks as lairs for snails and slugs. Check the refuges every morning; crush the pests or drown them in soapy water, and bury them in soil or a compost pile.

resting on small stones, sections of damp newspaper, or overturned flowerpots and grapefruit or melon rinds. First thing in the morning, check the undersides of these shelters for groups of slugs or snails that have spent the night there. Scrape them off the boards into a container of soapy water to drown. With pots and rinds, just rap the slug-filled shelter against the rim of the water container to dislodge the critters.

GARDEN CLEANUP. Periodically remove damp plant debris, boards, buckets, and empty plant pots and flats from around the garden except for those materials you're using as slug shelters. Don't leave pulled weeds in the garden; take them immediately to your compost pile.

CHANGING YOUR WATERING STYLE. If slugs and snails are a serious problem in your garden, switch to drip irrigation. Using a drip system allows the garden environment to remain much drier than sprinkler systems (or hand-watering) do. Whatever watering method you use, water in the morning so the soil surface dries quickly.

REPELLENT BARRIERS. Scratchy materials spread on the soil around plants may keep slugs and snails away. Try materials such as pine needles or crushed eggshells. Wood ashes, sawdust, sand, and lime are not as effective. Some gardeners claim that sweet gum balls work well as a slug barrier.

Whatever material you use, it's important to apply a scratchy barrier as a continuous band around the plants you want to protect. Scratchy barriers are less effective when they get wet, and that's a serious flaw, because wet conditions are when slugs and snails are most active. If you're relying on these barriers, be sure to handpick as well on rainy nights. Reapply barriers after each rainfall.

A commercial repellent with a soap-oil base repels slugs also. It needs to be applied in a continuous band according to label directions.

BEER TRAPS. Slugs and snails are attracted to fermenting mixtures such as beer or a solution of water, yeast, and sugar. If they become intoxicated by the fermenting liquid, they usually end up drowning in it. Take advantage of this by putting out beer traps made from an empty yogurt container, coffee can, or other container with vertical sides, as shown in the illustration below. The container should be at least 4 inches deep, with about 1 inch of liquid. Be sure 1 inch of the rim protrudes above soil level; otherwise, ground beetles in search of a slug meal may fall into the containers and drown. Commercial versions of these traps

A straight-sided container sunk into the soil and containing beer or a yeast solution will attract slugs.

are available, too, and most of them include a cover. Empty the traps daily, and change the liquid every few days. Keep in mind that these traps will attract slugs only from a distance of a few feet away, so they won't provide a garden-wide solution to slug problems.

COPPER BARRIERS. When slugs and snails come in contact with copper, the copper and slug slime interact chemically, causing the creature to receive an electric shock. This won't kill the slugs or snails, but it will repel them very effectively. You can protect framed raised beds by surrounding them with a strip of copper sheeting. Buy products especially designed for this purpose, or use copper flashing (available at hardware stores and home centers), as shown on page 370. Wear gloves when you handle copper flashing, and use tin snips to cut it. Tacks work well to fasten the flashing to the frame of the bed. Zinc-galvanized flashing will work, too, and it's usually less expensive than copper flashing. To attach copper or zinc barriers to a large plastic or clay container, use silicone caulk.

Miniature "fences" of copper or galvanized metal screening protect small plantings from slug and snail attack. Use an old pair of shears to cut the screening to the length and width you need. The barrier should extend 6 inches above soil level and 3 inches below ground, as shown in the illustration on page 371.

After you install a barrier, set cut baits in the protected area or check it at night for several nights to find slugs and snails that are already inside the area; handpick and kill any you find. Once you rid the protected area of the pests, the area should remain slug-free all season. The

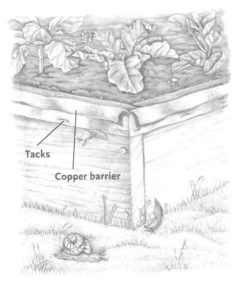

Tacks

Copper barrier

Tack copper flashing or a commercial copper barrier to the top outside edge of a raised bed frame to create a barrier that slugs and snails won't cross.

copper sometimes turns bluish green as it weathers, but this will not make it less effective.

COFFEE AND COFFEE GROUNDS. Research shows that a 1 percent solution of caffeine can kill small slugs and snails, apparently by a neurotoxic effect. Surrounding plants with a barrier of coffee grounds spread on the soil may protect them from slugs and snails; you can experiment to see how this works in your garden. Follow-up research still needs to be done to determine whether spraying coffee on plants is an effective slug repellent (coffee contains less than 1 percent caffeine). Also, there's been no research on the effect of caffeine or coffee sprays on beneficial insects.

IRON PHOSPHATE. Iron phosphate is a mined material that is toxic to slugs and snails. Once the pests ingest iron phosphate, they stop feeding and die within a few days. You may not see the dead pests around your garden because they usually crawl to a sheltered spot to die. Iron phosphate is available in commercial bait products (the bait is a nontoxic substance such as wheat gluten). Be sure to choose baits containing iron phosphate and not metaldehyde (which is toxic to people and pets). It's best to apply the bait on a warm, humid evening. If the soil surface is dry, water it first.

Iron phosphate is not toxic to children, pets, or wildlife, but it can cause eye irritation. If you spread the bait by hand, wear gloves and wash your hands after you apply it. Any bait that the slugs and snails don't eat will naturally degrade.

Note that iron phosphate products may contain a chelating agent that is not considered acceptable for use by certified organic growers.

CONTROL CALENDAR

BEFORE PLANTING

- Rake up surface mulch and leave soil uncovered until weather conditions are such that slugs and snails will be less active or until plants are large enough to withstand some damage.
- Maintain a portion of your garden in cover crops and provide refuges to shelter predatory ground and rove beetles.

AT PLANTING TIME

- Surround newly seeded areas with scratchy repellents to keep slugs and snails out.
- Affix copper barriers to the edge of raised beds, or create a low "fence" and attach a copper barrier to it.
- Surround newly seeded areas with a barrier of copper or galvanized flashing.

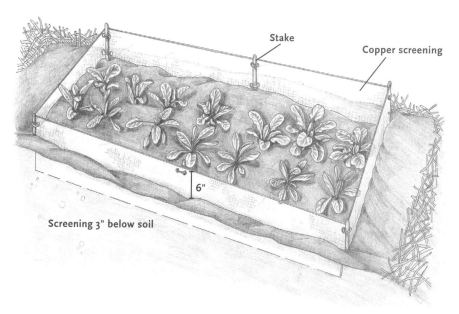

Stake

Copper screening

6"

Screening 3" below soil

To protect a newly seeded area from slug devastation when seedlings emerge, surround the bed with a fence of copper screening that extends both above and belowground. Small stakes or sturdy plant labels provide extra support.

WHILE CROP DEVELOPS

- At night, handpick slugs and snails in your garden and nearby groundcover beds and composting areas.
- Spread egg masses to dry in the sun or dispose of them in the trash.
- Put out "shelters" around the garden for slugs and snails. Check them first thing each morning and dispose of all the pests that have congregated there.
- In hot weather, seek out and kill snails hiding on walls or under building trim.
- Clear plant debris and other random items out of the garden periodically.

- Switch to drip irrigation.
- Put out traps baited with beer or a yeasty solution and collect trapped pests daily.
- Spread coffee grounds as a barrier around plants.
- Spread baits containing iron phosphate around susceptible plants.

AFTER HARVEST

- Clear debris out of the garden to remove potential overwintering sites.

❧ SOIL

Soil is as alive as the green plants that grow out of it, and your garden soil needs the same tender treatment that you bestow on your plants. You wouldn't think of trampling on your crops, hacking them up with sharp tools, or leaving them without water during a drought. And yet we sometimes inflict those punishments on the soil, not remembering the traumatic impact on the animals, insects, microorganisms, and plant roots that live under the soil surface.

If you're new to gardening, read "Healthy Soil, Healthy Plants, Healthy Harvest" on page 1 to learn about why soil is so important to gardening success. If you already understand how the soil food web feeds your crops, read on to learn how to solve soil-related problems and build up the soil food web in your garden.

EVALUATING YOUR SOIL'S PROBLEMS AND POSITIVES

How can you tell whether your soil is rich or poor? One way is to plant some crops and see how they do, and that's how many of us start out. If the plants grow well, it's safe to assume your soil is in good shape. If all the plants struggle, you can be pretty sure there's something wrong with the soil. If that's the case, the first step is to learn more about your soil and what it's lacking.

Soil characteristics such as texture, salt content, and pH vary both by region and within local areas. It's easy to find out what the dominant soil types are in your area by contacting your local Cooperative Extension office or joining a local gardening group. You can find out some specific qualities of the soil in your garden by trying a few do-it-yourself tests.

START WITH DRAINAGE

Drainage capacity varies with soil type. Sandy soils tend to drain very quickly, while dense clay soils hold water longer and can sometimes drain too slowly. Since almost all vegetable crops need well-drained soil, start by checking how quickly water drains through the soil in your garden. For this test you will need a large metal can (such as a juice can or coffee can) with both ends cut off, a stopwatch, a notepad and pencil, and a large jug of water. At one end of the can, use a permanent marker to mark intervals at 1, 2, and 3 inches from the rim.

In your garden, push the can into the soil until the rim is 3 inches above the soil line. Fill the can with water and watch it drain. It will drain pretty quickly at first, unless your soil is already saturated by rain. Continue filling the can and watching it drain, and begin timing the rate of water infiltration and writing it on your notepad. As you repeat the process, you see that the rate of drainage reaches a steady state. In well-drained soil, that rate will be ½ to 1 inch of water per hour. If your soil is draining more slowly, it's a cause for concern. The soil may be slow-draining clay, or it may be compacted overall. Another possibility is a compacted layer several inches or more than 1 foot below the soil surface (called hardpan). If your soil seems loose and open but isn't draining well, try sticking a piece of stiff metal wire down into the soil. If you meet resistance somewhere below the surface, that indicates a compaction layer.

If you have a large garden, repeat the drainage tests in a few different spots. It's possible that only a portion of your garden has a compaction problem.

DO SOME DIGGING

A little digging can tell you a lot about your soil. First, make sure the soil isn't too wet or too dry to work. Simply squeeze a handful of soil in your palm. If the soil forms a tight ball, it's too wet to work. If the soil won't form a ball at all but seems dusty and dry, it's too dry to dig. If the soil forms a loose ball that crumbles a bit when you open your hand, it's fine to dig.

Dig a hole about 12 inches deep and 12 inches wide, piling the soil on a tarp or old sheet. Notice whether it's easy to dig the hole or a strenuous effort. Does the soil come up as big clods, as a powder, or as small clumps? Soil that's easy to dig and consists of lots of small clumps is a sign that there's lots of biological activity underground. Large clumps that don't break apart aren't good for plants because roots will have trouble moving through the soil. Powdery soil is a sign that there's little organic matter, and therefore fertility is probably low.

Once the soil is out of the hole, sit down and watch the hole for a few minutes. You should see at least a few insects, centipedes, or spiders emerge into the hole and then disappear. You can also sift through some of the soil on the sides of the hole to find insects. If there are no insects or other crawly creatures in your soil, it's a sign that there's too little organic matter and biological activity in your soil.

Put the soil back into the hole a handful at a time, and count earthworms as you go. If you find more than five earthworms, there's adequate biological activity in your soil. Ten or more earthworms indicate an organically rich soil. Fewer than four is another sign that your soil has too little organic matter or that the soil pH is too low or too high.

TEST SOIL PH

pH is a measure of the relative acidity or alkalinity of the environment in the soil around plant roots. It's important because it influences the molecular form of nutrients such as nitrogen, potassium, phosphorus, magnesium, and boron. Plant roots can only absorb certain forms of these nutrients. In general, nutrients take the form that plants can "digest" easily when the soil pH is near neutral (7.0).

Garden centers and mail-order suppliers sell do-it-yourself pH tests. If you use one of these test kits, read the directions thoroughly. Small mistakes in your testing method can throw off results. You can also have a soil sample tested by your state Cooperative Extension service. The procedure for a Cooperative Extension test varies from state to state. Call your local office to find out the procedure where you live. Be sure to closely follow the directions given for taking a soil sample for the test.

IMPROVING DRAINAGE

Vegetable gardens need full sun, and that's often the deciding factor in where to set up a garden site. Many of us discover that the sunny site in our yard doesn't have good drainage. Poorly drained soil isn't good for vegetable crops because the water in the soil ends up filling all the spaces that would otherwise be filled with air. Most vegetable crops are fast-growing annuals, and their active roots can't grow and function without plenty of air in the soil. Plus, pathogens that cause root rot and other root diseases seem to thrive when the soil is wet and air levels are low.

Thus, if you have a choice of sites for a vegetable garden, pick the one with good drainage. If you have no choice except a poorly drained spot, there are three ways to deal with the problem. You can put in some type of drainage system, try to break up the compacted layer by planting a deep-rooted crop, or raise your garden above ground level.

DRAINAGE SYSTEMS

If your garden is on a slope, you may be able to lessen the drainage problem by funneling the flow of water around your garden. You could build up a berm on the uphill side of the garden or dig a swale. If you do this, though, be sure that you're directing the water to an appropriate endpoint such as a storm sewer and not simply onto your neighbor's property.

Installing drainage pipe is a sophisticated project that involves trenching, laying a protective fabric, and then wrapping drainage pipe with the fabric and refilling the trench. Either hire a professional to install the drainage tile or seek an experienced helper and thorough step-by-step instructions to follow if you decide to install drainage tile yourself. Before you begin, check to see if your town requires permits for this kind of work.

SOIL-OPENING CROPS

If you've got time and patience, you can avoid the hassle of installing a drainage system and instead plant a crop with large, deep roots that can break up compacted soil. This approach won't always solve a drainage problem, but it can be helpful for breaking through a hardpan layer. See "Cover Crops as Problem Solvers," on page 140, for suggestions of crops that can loosen soil and instructions on how to plant them. This is a long-term project because you need to let the crop grow long enough to produce the large, soil-opening roots and then allow those roots to break down. Once that process is complete, you'll be able to plant your vegetable crops.

RAISED BEDS

Most vegetable crops will provide a fine harvest as long as the top 12 inches of soil drains well because there will be enough rooting space (in fertile soil) for the plants to support their foliage, flowers, and fruits. Building foot-high raised beds can give you that 12 inches of good soil. Raised beds also warm up more quickly in spring.

Making raised beds can be a fancy project or a very simple one. If you have an existing garden,

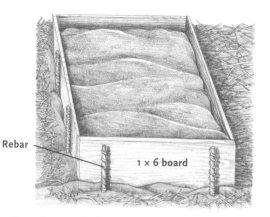

Rebar

1 × 6 board

To make a simple frame for a low raised bed, hammer sections of rebar into the soil to support 1 × 6 untreated wooden boards. The frame will last several years.

one way to make raised beds is simply to mark out permanent pathways and shovel the top few inches of soil from those pathways onto the beds. Then add about a 3-inch layer of compost and/or shredded leaves and peat moss, and you'll create fluffy beds that are about 8 inches higher than the pathways. If you like, you can use untreated 1 × 6 boards to make frames for the beds, or simply leave the sides open. Northern gardeners may want to create beds with a slight slope toward the south to capture more of the sun's light and heat, as shown below. Spread straw or wood chips in the pathways to prevent them from becoming too muddy or overly compacted. Orient the pathways to encourage surface water to drain out of the garden. If the land slopes, for example, run the paths up and down the slope.

BUILDING ORGANIC MATTER

Organic matter is the food source for the microorganisms that play so many beneficial roles in soil, and it's ultimately a major source of food for your plants, too, once the microbes digest it. Most home garden soils aren't naturally rich in organic matter. You need to keep adding it every year because the crops you grow keep consuming it. In turn, they transform it into the fruits, roots, stems, and leaves that you harvest and use to satisfy your appetite and fuel your health. You'll supply organic matter by adding compost and other amendments to the soil, by growing cover crops (see Cover Crops), and by using organic mulches (see Mulch).

HOW MUCH, HOW OFTEN?

How much organic matter do you need to add each year? That depends on what type of soil you have and where you live. As a general rule, sandy soil is lower in organic matter than loam soil, and organic matter breaks down faster in hot, humid climates than it does in cold, dry climates.

How much organic matter you need to add also depends on how intensively you garden. Do you use season extension devices so you can grow food almost year-round? Do you plant crops close together in wide rows? Or do you use standard spacing and limit your gardening to the frost-free months of the year? Naturally, the more you do to maximize your harvest, the more you'll need to feed your soil to replenish the food source for your crops.

One rule of thumb for temperate zone gardens (Zones 4 to 6) is to add ½ to 1 inch of compost per 4-month growing season. Gardeners in the humid Southeast need to add two or three times as much.

You may wonder how to quantify or measure a layer of compost ½ to 1 inch thick. Here's an equivalent that's easy to work with: You would need to spread a minimum of six 5-gallon buckets of compost per 100 square feet of garden, and a maximum of 12. You can literally scoop up the compost by the bucketful, or you can dump 12 buckets of compost into your wheelbarrow, spread it evenly, and then use paint to mark that level on the side of the cart. After that, just fill the cart to that level, and you'll know you have approximately enough compost to cover one bed that's 4 feet × 25 feet or two beds that are 3 feet × 15 feet.

If you prefer to avoid doing the math to figure out square footage of garden beds, you can make estimates based on a mental image of the harvest. This is based on the idea that you need to supply at least the amount of organic matter that you're going to remove from the garden. For example, if you plan to plant four tomato plants in a bed, picture how many tomatoes you'll harvest in total from those four plants. Perhaps three bushel baskets? Or enough tomatoes to fill your garden cart? If so, that's how much compost you want to apply to the bed. It's not an exact comparison, but it serves as a minimum standard. You may want to boost the amount of compost a bit, rather than keeping a strict one-to-one comparison of compost to harvested produce. Also, if you use this method, keep in mind

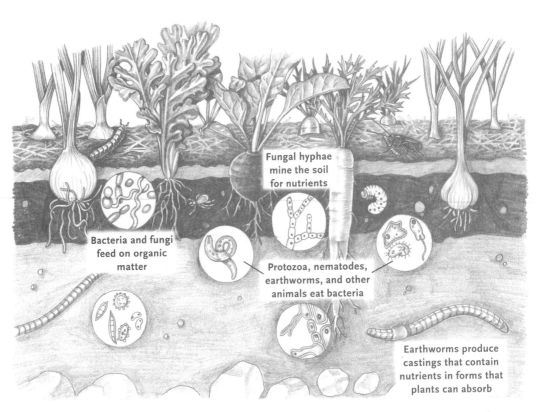

Fungal hyphae mine the soil for nutrients

Bacteria and fungi feed on organic matter

Protozoa, nematodes, earthworms, and other animals eat bacteria

Earthworms produce castings that contain nutrients in forms that plants can absorb

A complex network of insects, animals, and microorganisms is part of the natural balance in healthy garden soil.

COMBINE COVER CROPS AND COMPOST

One way to minimize the number of times you disturb the soil while maximizing your program of building soil organic matter is to coordinate your compost applications with your cover crops. After you harvest a vegetable crop, spread about ¹/₂ inch of compost over the soil surface. If you have small amounts of leftover bloodmeal, bonemeal, or other organic fertilizers, sprinkle them over the soil, too.

Next, sow cover crop seed. Rake the seed just enough to cover it. Tamp the area with the back of the rake. The cover crop will germinate rapidly in the compost. When the cover crop is ready to dig several weeks later (or after overwintering), you'll incorporate both the cover crop and the compost at once. If the cover crop has a lot of green tops, cut them off and use them as mulch somewhere in your garden or add them to your compost pile. That way you probably can avoid using a tiller and instead can hand-dig gently to incorporate the stubble.

---

that you'll want to use more organic matter if you have sandy soil (perhaps twice as much).

Some gardeners are even less scientific about how they add organic matter but are still very successful. For example, they sprinkle a blended organic fertilizer on the surface of their garden beds each fall and top that with a couple of inches of shredded leaves. In spring, they work the leaves lightly into the soil, and plant a few weeks later.

Another technique is to spread a light layer of compost on the soil surface each time before you apply a surface mulch. The mulch protects the compost from drying out and provides a sheltered yet airy environment in which soil animals and other organisms will quickly consume the compost and make it available to your plants.

WORKING THE SOIL WISELY

There's a myth that smooth, uniform soil is the gardener's goal. Based on the model of chemical-driven agriculture, this myth drives some gardeners to till the soil until it consists only of fine particles, with no rocks or bits of organic debris in sight. This model of the soil may work for large-scale plantings fed by synthetic chemical fertilizers and protected by chemical herbicides and pesticides, but it's not the right model for the home garden.

Messy soil is our goal. A soil that's rich in organic matter and that has an established structure provides the rich array of microenvironments needed for different forms of soil life. For the living things in the soil, tilling is like sending a hurricane through. The tiller blades kill them outright, it breaks apart their homes, and it changes the nature of the habitat so drastically that they can't find food afterward.

Does this mean you should never till? Perhaps not. If you have a large garden and limited time, tilling may be the only practical way to manage your garden. Whether to till or not is a personal choice, but if you do till, follow these practices to minimize damage.

- As with hand-digging the soil, always be sure the moisture content is right before you till. Tilling wet or dry soil can cause disastrous results.
- Run your tiller at low speed so the blades move slowly. This works the soil adequately but causes less damage to soil life.
- Whenever you till, incorporate some kind of organic matter at the same time. Till in a cover crop or the healthy crop residues that remain after harvest, or spread some compost or shredded leaves before you till.
- As soon as you finish tilling, cover the soil surface with an organic mulch. This will prevent the soil from losing as much moisture, and it will buffer the temperature within the soil. Moisture and moderate temperature are two important factors in encouraging soil life to rebound after disruption.
- Once you have planted a crop in the tilled area and the seedlings or transplants are settled in, drench the soil with compost tea to replenish the supply of beneficial bacteria and fungi. Why wait until after planting? Because the bacteria and fungi rely on the root exudates from your crop as a source of energy.

ᴥ SOUTHERN BLIGHT

Even northern gardens can suffer from a run-in with southern blight. The disease is aptly named, though, because it spreads rapidly in humid weather and temperatures higher than 85°F. A thick mat of white fungal growth carpets the soil surface around the base of infected plants. Many crops are vulnerable; tomatoes and beans are two of the most common victims. The fungus also feeds on decaying organic matter, so it's difficult to eradicate it from your garden once it's established. In the far North, though, cold winter temperatures may kill off the fungus in the soil.

The southern blight fungus is related to the fungus that causes white mold, and symptoms are somewhat similar. Consult White Mold for more information.

RANGE: Most common in the southern United States, but occasionally a problem as far north as Canada

CROPS AT RISK: Beans, peas, tomatoes, potatoes, peppers, sweet potatoes, cucumbers, melons, rhubarb, carrots, and other crops

DESCRIPTION: At first, leaves turn yellow and plants wilt. If you check the stem or crown near the soil surface, you'll find that it is water-soaked or dark. The white moldy growth follows soon after. Resting bodies (sclerotia) that look like tan or brown mustard seeds sometimes appear in the mold. Infected plants collapse and often die; carrot roots become mushy and rot.

FIGHTING INFECTION: Clear away fallen leaves and other organic debris on the soil surface near susceptible crops. Pull out infected plants and also gather up the top inch of soil from the area where the infected plant grew. Bury the plants and soil at least 8 inches deep, add them to a well-managed hot compost pile, or discard them. Carefully clean tools, shoes, gloves, and other items that come in contact with contaminated soil before using them elsewhere in the garden.

GARDEN CLEANUP: At the end of the gardening season, remove all surface crop residues by composting or by turning them into the soil several inches deep. Discard noticeably diseased crop residues.

NEXT TIME YOU PLANT: Solarize infected beds during the hottest part of summer. Feed the soil with high-quality mature compost to build populations of beneficial soil microorganisms. Applying *Trichoderma harzianum* to soil may help as well. Remove surface organic matter and dropped leaves, and/or cover beds with black plastic. Water the soil periodically with compost tea to continue adding beneficial bacteria and fungi. If you plant cover crops, be sure they have time to break down thoroughly before following them with a susceptible crop. As a practical measure, plant larger-than-usual amounts of susceptible crops to allow for some losses due to blight.

CROP ROTATION: Crop rotation won't eradicate southern blight, but growing a nonsusceptible crop such as corn in problem areas for a few years will cause the fungus to decline in the soil.

SPIDER MITES

Spider mites are a widespread pest in vegetable gardens, especially in arid climates. Too tiny to see without magnifications, these eight-legged pests bruise plant cells with their mouthparts, and the injured cells show up as stipples or speckles. A spider mite infestation can build up fast in hot weather, so it's important to try and limit the damage quickly. See Mites for prevention and control measures.

SPINACH

Spinach is a quintessential cool-weather crop. In rich, moist soil and moderate temperatures, spinach bursts into rosettes of tender, dark green leaves in a few weeks' time. However, you'll struggle with spinach if you try to coax it to produce during summer or persuade it to germinate in hot, dry soil. Unless you live in an area with cool summers (such as northern New England), stick with spinach in early spring, fall, and winter. For late spring and summer, plant a spinach substitute—there are several possibilities that will handle the heat just fine.

CROP BASICS

FAMILY: Beet family; relatives are beets and Swiss chard.

SITE: Sow spring crop in full sun and well-drained soil. Sow fall crop in the shadow of tall plants to provide shade early on, but open the crop to sun when the weather has cooled.

SOIL: Spinach does best in rich soil. Add a 2- to 4-inch layer of compost to poor-quality soil, and 1 inch to soil of average to good fertility. For a fall spinach crop in a garden that was enriched during spring, work in a ½-inch layer of compost before planting.

TIMING OF PLANTING: Sow seeds up to 8 weeks before last expected spring frost. Sow again 8 weeks before first expected hard frost for a fall crop. In cold-winter areas, sow again about 6 weeks before first expected hard frost to overwinter for early harvest the following spring. In mild-winter areas, repeat sowings every 2 to 3 weeks throughout fall as desired.

PLANTING METHOD: Sow seeds directly in garden beds, or start them indoors for an earlier spring harvest or to avoid the difficulty of sowing outdoors in hot soil for a fall crop.

CRUCIAL CARE: Keep spinach cool. Provide shade during unseasonable hot spells. Mulch the soil for fall crops to lower soil temperature.

GROWING IN CONTAINERS: Plant spinach as an edging in a large container designed for a patio tomato or pepper plant. The spinach will be ready to harvest shortly after you plant the centerpiece crop. Follow up the spinach with colorful annuals.

HARVESTING CUES: Start picking individual leaves when plants have formed six to eight leaves. Cut whole plants when leaves are 4 to 6 inches long.

SECRETS OF SUCCESS

START EARLY IN SPRING. The spring spinach season is short because increasing daylength triggers the plants' bolting mechanism and causes formation of flowerstalks. Once flowering begins, the quality and flavor of leaves decline. Temperatures above 70°F speed up flowerstalk formation, too, so a warm spell in spring can ruin a spinach crop fast.

To ensure that your spring crop will yield

tasty leaves for at least a few weeks before bolting, prepare a raised bed for planting in fall. Sow seed about 6 weeks before the first expected hard frost and protect the young seedlings over winter. Remove the cover in early spring and feed the plants with dilute fish emulsion and kelp to stimulate new growth.

If you don't want to fuss with overwintering spinach, sow seeds in a raised bed 8 weeks before your last expected spring frost. In the South, where temperatures climb quickly in spring, rig shade cloth over spring spinach to buffer it from heat; this can extend the harvest by a week or so.

TRY WIDE ROWS. Broadcast spinach seed lightly in bands 6 inches wide. As the seedlings grow, thin first to 2 inches apart, then 4, then 6. Eat all the thinnings—baby spinach is delicious!

GLORY IN FALL SPINACH. Fall is a great time for spinach in most areas. Southern gardeners can enjoy fresh spinach through New Year's Day, and in the Deep South, spinach may keep on producing well into winter.

In fall, daylength is decreasing. Thus, young plants won't be triggered to bolt, and spinach loves cool fall temperatures. The trick is to sow seeds early enough to allow plants to reach harvestable size before temperatures drop so low (near freezing) that growth stops. After that, they'll remain at a steady state until you want to harvest (spinach withstands temperatures as low as 20°F). In most areas, the time to plant a fall crop is mid- to late summer, about 8 weeks before the first expected hard frost of fall. However, daytime temperatures often exceed 80°F, and soil temperatures will be high as well. Because spinach germinates poorly when soil

temperature exceeds 75°F, it's important to cool the seedbed. Try to sow seeds in the shade of tall crops, and water the seeds daily. Once the seeds germinate, apply 2 to 3 inches of mulch.

Another way to improve germination when the weather is hot is to precondition your seed. Put the seed packet in a resealable plastic bag and store it in your freezer for 2 days. Then take out the seed you plan to sow. Roll it in a damp paper towel, set the towel on end in a plastic bag, and set it upright in your refrigerator for 5 to 7 days. After that, the seeds will be ready to sow outdoors. (While you're treating the seed, water your planting area well and mulch it—this will reduce the soil temperature. Remove the mulch when you're ready to plant.

GROW SAVOY AND SMOOTH. Spinach plants produce either savoyed (crinkled) leaves or smooth leaves. Smooth-leaved spinach is more

Smooth   Savoy

Some gardeners prefer the texture of smooth-leaved spinach for salads, but savoyed varieties are generally more disease-resistant.

SUMMER SUBSTITUTES FOR SPINACH

If spinach won't grow well in your garden during summer, don't struggle with it; switch to a substitute instead.

Swiss chard, a spinach relative that can withstand summer heat, is a fine spinach substitute. Its flavor and texture are different from that of regular spinach, but it's tasty, easy to cook, and just as vitamin-rich as spinach (see Swiss Chard).

Another spinach relative is New Zealand spinach, which forms bushy plants up to 2 feet tall. Sow New Zealand spinach seed outdoors after spring frosts are over and the soil is warm. It's not the easiest crop to persuade to germinate. Soak the seeds in warm water the night before you plan to plant, and sow heavily. Thin to 12 inches apart. When plants reach 1 foot tall, use scissors or a knife to cut the top 4 inches of stems at each harvest—young leaves are the best tasting and most tender. New Zealand spinach will continue producing new shoots until frost. This crop usually doesn't suffer from pest or disease problems, although it's potentially susceptible to many of the same problems as regular spinach.

Malabar spinach is not closely related to spinach or any of our common vegetable crops, but its shiny heart-shaped leaves taste quite a bit like spinach. Malabar spinach is a tender vine that grows 15 feet or longer in gardens. Because it needs a long growing season, start seeds indoors 6 weeks before the last spring frost and set out plants in your garden 2 weeks after the last frost. Provide a strong trellis to support the heavy vines, which will probably reach the top of the trellis and spill over the other side. Cut young stems and leaves as desired and eat them raw or steam them lightly. Seeds look like small purple berries and are easy to save for replanting the following season. Because it grows so vigorously, it's likely that Malabar spinach will produce more than you can eat. When fall comes and you're ready to change back to eating regular spinach, cut down the Malabar spinach vines and chop them up with a machete or other tool. Work them into a bed directly to boost organic matter, or combine them with shredded leaves to start a new compost pile.

New Zealand spinach

Malabar spinach

tender and easier to wash clean of grit, so it's preferable for eating raw in salads. However, varieties with savoyed leaves are generally more heat tolerant and disease resistant. Semi-savoyed varieties combine characteristics of both types. To be safe, choose at least one savoyed variety—that way, if wet weather sets in and disease becomes problematic, the disease-resistant savoy should still be harvestable. Sow each new variety you buy both in spring and fall—that way you'll find which varieties are best at tolerating heat stress and which are most productive during cool fall conditions.

REGIONAL NOTES

SOUTHEAST AND SOUTHERN PLAINS

White rust is a soilborne fungal disease that causes white blisters to form on leaf undersides. White spots also form on the topside of the leaf. Leaves turn yellow and severely infected plants may collapse. White rust spores spread in wind and splashing rain. Some varieties are partially resistant. If white rust appears in your garden, plant spinach in raised beds, switch to drip irrigation, and start a 3-year or longer rotation of beet-family crops.

PREVENTING PEST PROBLEMS

BEFORE PLANTING

- Prepare a raised planting bed in fall to provide good fertility and drainage for an early spring crop.

AT PLANTING

- Cover seeds or seedlings with row covers to keep out flea beetles, aphids, and leafhoppers.

AS CROP DEVELOPS

- Thin seedlings to maintain good air circulation around plants.
- Remove individual leaves that show leafminer tunnels or disease symptoms such as leaf spots.
- Handpick slugs, armyworms, and loopers.

AFTER HARVEST

- Dig in crop residues to reduce overwintering of pests and diseases.

TROUBLESHOOTING PROBLEMS

SEEDLINGS NEVER EMERGE OR NEW SEEDLINGS DIE. Damping-off is the probable cause in wet conditions; try replanting in a spot with better drainage (see Damping-Off for more information). In hot weather the problem is lack of germination due to high soil temperature. Sow spinach seeds again in the shade of tall crops, and water them daily to cool the soil.

NEW SEEDLINGS EATEN. A single cutworm can destroy several plants in one night. These pests hide just below soil level. See Cutworms for prevention and control information. Slugs and snails eat seedlings, too; turn to Slugs and Snails for control measures.

TUNNELS IN LEAVES. Leafminer flies lay eggs and their larvae tunnel through leaves. It's fine to harvest leaves that show tunneling; cut away the damaged areas and eat the rest.

To prevent future damage, cover seeded beds with row cover in spring. See Leafminers for more controls.

SMALL HOLES IN LEAVES. Flea beetles chew on leaves and ruin their appearance. A heavy infestation can destroy seedlings. See Flea Beetles for prevention and controls.

RAGGED HOLES IN LEAVES. Slugs, snails, and loopers are possible culprits; seek and find which pest is damaging your plants. Cut away damaged areas of leaves and use the rest. See Slugs and Snails and Cabbage Loopers for control information.

SKELETONIZED LEAVES. Armyworm feeding produces this damage pattern. Look for webbing on leaves and search inside it for armyworms. See Armyworms for control methods.

DISTORTED LEAVES. Aphids suck sap, causing leaves to curl and twist. Look for colonies on young leaves and undersides of larger leaves. Cut and discard damaged leaves. To protect regrowth from aphids, see Aphids.

WHITE PATCHES ON LEAVES. Suspect scorching or leafminers. Strong sun combined with heat causes tender leaf tissue to become scorched, especially if the plants are water stressed. To prevent scorching, add more mulch to retain soil moisture or use a soaker hose to ensure the soil stays moist during dry weather.

Leafminers tunnel through leaves and their feeding can be concentrated in one part of a leaf, resulting in damage that looks like white blotches rather than narrow tunnels. See Leafminers for control measures.

WHITE OR BROWN SPOTS ON LEAVES. Anthracnose or Cercospora leaf spot first appear as spots on leaves. These diseases usually become severe during long periods of wet, warm weather. Remove infected leaves and watch for new healthy growth. If the problem persists, see Anthracnose and Cercospora Leaf Spot for help.

YELLOW PATCHES ON LEAVES. Downy mildew (also called blue mold when it appears on spinach plants) first shows up as yellow spots on leaves. In cool, wet conditions, purplish or blue fuzz appears on leaf undersides. Remove infected plants. Many resistant varieties are available. See Downy Mildew for more information.

PALE OR YELLOWING LEAVES, STUNTED PLANTS. Virus diseases, root rot, and nitrogen deficiency are possible causes of yellowing leaves. Leafhoppers transmit curly top virus, and aphids transmit cucumber mosaic and beet western yellows viruses as they feed. Pull out diseased plants and destroy them. For more information see Beet Curly Top and Mosaic.

To check for root rot, uproot a couple of affected plants. If you find water-soaked brown roots, Fusarium rot is the problem. Replant in a new area with enriched soil. Water regularly (but don't overwater) to stimulate rapid growth.

If you suspect a nitrogen deficiency, water weekly with fish emulsion and side-dress with compost.

PLANTS FORM FLOWERING STALKS. Heat and increasing daylength in late spring trigger the bolting mechanism in spinach, causing flowerstalks to form. Uproot the plants and compost them. Wait until peak summer heat is past, and then sow a new crop.

SQUASH

Lack of space is a vexing problem for squash-loving gardeners. Staking and trellising can help solve the space problem, if you're willing to do the work to put up sturdy supports. To be successful with squash, you'll also need to anticipate and outwit the three top squash pests—cucumber beetles, squash bugs, and squash vine borers—and watch carefully for symptoms of disease.

Despite these potential problems, it is possible to grow great crops of summer squash (zucchini, yellow squash, and patty pan squash), winter squash (butternut, buttercup, acorn, spaghetti squash, and more), and pumpkins. Try some tricks of timing, use row covers when plants are young, help out with pollination when needed, and you'll be rewarded with a healthy harvest of delicious squash.

CROP BASICS

FAMILY: Squash family; relatives include cucumbers, gourds, and melons.

SITE: Plant squash in full sun.

SOIL: Squash will grow well in average soil as long as it's well drained. Wait until soil has warmed to about 60°F to plant.

TIMING OF PLANTING: In the North, wait to sow seeds or plant transplants about 2 weeks after your last expected spring frost. Try sowing summer squash again about 4 weeks later. These plants will extend the harvest if your first planting succumbs to disease. Southern gardeners can plant several succession plantings of summer squash if desired.

In the South, planting winter squash 2 weeks

after the frost date may be too early; time your planting so the crop will mature in fall when temperatures are declining but not yet dipping below freezing.

PLANTING METHOD: It's common to plant squash in hills, sowing several seeds in raised mounds of soil about 1 foot across. Put nutrients where they're needed by working a generous shovelful of compost into each planting hill before you sow seeds or set out transplants.

You can start seeds indoors in 4-inch peat pots, too; they will be ready for the garden about 3 weeks after sowing. Be sure to harden off the started seedlings by moving them outdoors to a coldframe or other protected spot for the last week before transplanting.

CRUCIAL CARE: Cover seedlings or transplants with row cover immediately at planting time to prevent early attack by cucumber beetles and other insect pests. Leave on the covers until the plants outgrow them or start to flower. By then, they should be better able to withstand some pest damage.

GROWING IN CONTAINERS: Plant summer squash in a 5-gallon container that's at least 10 inches deep. Winter squash don't grow well in containers—one exception is baby pumpkins, which will fill an 18-inch wide container and spill over the sides, too. Sow seeds or plant transplants, but thin to one sturdy plant per container.

HARVESTING CUES: Keep tabs on the flowers on summer squash plants even before fruits appear. Once the plants start to produce female flowers (the first flowers that form will be male), check plants daily. Harvest summer squash when they are 3 to 6 inches long.

Allow winter squash and pumpkins to reach mature size before harvesting. Watch for the rind to become a little dull, and lightly scratch the rind with your fingernail. If it doesn't leave a mark, that's a sign that the squash is ready to harvest.

SECRETS OF SUCCESS

EXPERIMENT WITH TIMING AND SUCCESSION SOWING. The best way to ensure that some of your squash escape pest problems is to sow seeds or plant transplants more than once. Your goal is to find a window of opportunity when pests are less active in your area (between the time of the first and second generations of cucumber beetle adults, for example). Northern gardeners have less flexibility with winter squash because the plants need warm temperatures but also a long growing season. One way around this is to sow seeds indoors 1 week before the last expected spring frost. Transplants will be ready about 2 weeks after the frost date. Plant your transplants and sow seeds in the garden at the same time.

GIVE THEM SPACE. The range of spacings for different kinds of squash is impressive. Summer squash plants will thrive when planted 1 to 2 feet apart in a 3-foot-wide bed. Vigorous winter squash and pumpkin plants, in contrast, should be planted up to 6 feet apart in rows 6 feet apart. Check the planting instructions for the specific varieties of squash you've purchased because even within a particular type, there are compact varieties and more sprawling types.

If you must squeeze plants more closely than the recommended spacing, pinch the growing tips after each vine has set a few fruits.

HARVEST RIGHT FOR QUALITY AND PRODUCTION. Pick summer squash when they are young and tender. Large summer squash develop a woody texture and reduce the plant's productivity. Use a knife or garden shears to cut the fruits with a little stem attached. If you try to pull them off, the fruit or the vine may break. Let winter squashes mature before you pick, but don't wait too long. Some authorities say that winter squash and pumpkins should be harvested before temperatures drop below 40°F; others say that light frost improves flavor. Inspect the squash frequently, including checking the undersides. If you find any spots that are discoloring or rotting, pick those fruits immediately and cook them (discarding unpalatable sections). Also harvest the rest of the undamaged fruit and prepare it for storage. If you leave it in the field, it may develop disease also.

Use pruners or loppers to cut winter squash and pumpkins from the vine, leaving a 2- to 3-inch-long stem on each one. Avoid using the stem as a handle to pick up the squash. Handle winter squash and pumpkins gently—gashes in the rind provide ideal spots for rot to invade when the fruits are in storage.

RAISING THE STAKES FOR SUMMER SQUASH

We think of summer squash as bushy plants, but they are actually short vines. Training those vines up short sturdy stakes is space-efficient, improves air circulation around plants, and keeps the fruit clean (and less prone to rot). At planting time, drive 4- to 5-foot stakes about 18 inches deep at each planting spot. Sow three seeds in a hill at each stake, popping the seeds into the soil about 2 inches from the stake. Once the seedlings have their first true leaves, thin to the sturdiest seedling. When the vine is about 12 inches long, fasten it to the stake with cloth ties or Velcro plant ties, supporting the vine every 6 inches. Add ties as the vine grows.

To keep your vines safe from insect pests, set up a tepee of stakes around each plant and wrap the tepee with row cover, tying it closed at the top and burying the cover in the soil at the base.

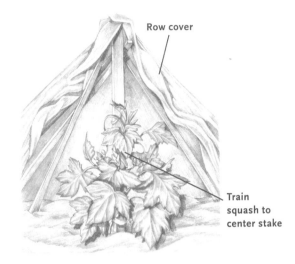

Row cover

Train squash to center stake

Winter squash also benefit from supports, but a single stake won't do the job. For trellis ideas for winter squash, see Melons.

REGIONAL NOTES

HEARTLAND

Cucurbit yellow vine disease is a relatively new problem on squash and pumpkin crops in the United States. It first appeared in Texas and Oklahoma in the late 1980s. Since then, it has spread to nearby states, including Colorado and Missouri, and has shown up in Connecticut and Massachusetts as well. The disease will probably continue to spread over time. Squash bugs spread the bacterium that causes yellow vine disease. Infected plants may wilt very suddenly and not recover. Sometimes leaves turn bright yellow before wilting sets in; yellowing often occurs just as vines start to set fruit. The only defense against this disease is to prevent squash bugs from infecting the vines. Melons and watermelons have been infected as well as squash and pumpkins.

EASTERN STATES

Pumpkins and zucchini crops from Georgia to New England are vulnerable to a fungus called Plectosporium blight. Oddly enough, the fungus has existed in the United States for a long time, but it infected only decaying plant matter. Beginning in the late 1980s though, it has been infecting living plants, too. The disease starts out subtly, causing small white spots on leaf veins and stems. As the spots grow together, the leaf veins may turn entirely white; leaves may drop prematurely. White pimples also appear on the surface of the fruits and sometimes cover the "handles" of pumpkins. Symptoms develop during wet weather and may disappear if dry conditions follow the rain. The fungus does survive over winter, so starting a 2-year or longer rotation of squash-family crops can help if this disease appears in your garden.

PREVENTING PROBLEMS

BEFORE PLANTING

- Sow a trap crop to attract cucumber beetles.

AT PLANTING

- Cover squash hills with row cover and seal the edges to keep out insect pests.

WHILE CROP DEVELOPS

- Spray exposed plants with kaolin clay to deter cucumber beetles.
- Remove row covers when plants start to flower.
- Cover nodes of winter squash vines with moist soil to encourage secondary roots.
- Spray vines with plant health boosters or compost tea to help prevent disease.
- Weed out squash-family weeds and volunteer squash seedlings.

AFTER HARVEST

- Remove and destroy any noticeably diseased crop residues, including damaged fruits. Turn under or compost all other crop debris.
- After removing winter squash vines, work the soil a few times to expose pests and pupae to predators and cold, then cover the soil for winter.

TROUBLESHOOTING PROBLEMS

SEEDLINGS CHEWED. Cucumber beetles can vigorously attack squash seedlings, sometimes eating the plants right down to the ground. Don't try to save young seedlings; replant instead and cover the seeded areas with row cover. See Cucumber Beetles for more controls.

PALE AREAS OR YELLOW FLECKS ON LEAVES. Squash bugs or spider mites may be the culprits. Examine leaves closely to find the pests. Look on leaf undersides and on young shoots for the small greenish squash bug nymphs or shield-shaped gray adults. Spider mites look like tiny specks, and there will be fine webs on the leaf undersides.

Handpick squash bugs. For more controls and to prevent future problems, see Squash Bugs.

If a spider mite problem is severe, spray insecticidal soap. For other controls, see Mites.

YELLOW AND/OR WATER-SOAKED SPOTS ON LEAVES. Angular leaf spot, downy mildew, and scab are possible causes. As angular leaf spot develops, the spots become clearly geometric and the diseased tissue sometimes drops out, leaving holes in the leaves. For controls, see Angular Leaf Spot.

To confirm that the problem is downy mildew, check leaf undersides, when purplish mold will develop on the spots. Remove infected leaves. Applying a copper spray may stop the problem from spreading. To avoid future problems, consult Downy Mildew.

As scab worsens, the leaves turn brown and die. The fruit will be infected, too. If young plants are already infected, it's best to uproot them and replant, using resistant varieties. See Scab for other controls.

SPOTS ON OLDER LEAVES. Yellow or brown spots, sometimes with a pattern of concentric rings, are typical of Alternaria blight. Remove infected leaves and take steps to improve air circulation and decrease moisture around the plants. See Alternaria Blight for more controls.

EDGES OF LEAVES DIE. Gummy stem blight is a fungus that infects pumpkins and sometimes other squashes. Water-soaked streaks also form on leaf stems and main stems. Drops of dark gummy ooze sometimes appear along the streaks. If plants become infected, chop them up well and turn them under or put them in a compost pile. The fungus persists in the soil, so start a 2-year or longer rotation of cucurbits. Cucumber beetle and aphid damage leaves plants more susceptible to gummy stem blight, as does powdery mildew. The fungus can infect seeds, too. To avoid bringing the problem into your garden this way, buy clean seed and keep in mind that commercially grown transplants may be infected but not show symptoms. Also, grow powdery mildew–resistant cultivars and cover young plants with row covers to prevent insect damage.

POWDERY WHITE COATING ON LEAVES. Powdery mildew is a widespread fungus. The coating will cover stems and fruits too if not checked. Handpick infected leaves and start other control measures as described in Powdery Mildew.

MOTTLED LEAVES. Cucumber mosaic virus causes mottling. Leaves may turn under. There is no treatment, and the virus will infect and ruin the fruits, too. Uproot and destroy the infected vines. To prevent future mosaic infections, consult Mosaic.

CURLED AND TWISTED LEAVES. Melon aphids suck sap, causing leaves to curl and twist. Look for colonies of yellow to dark green aphids on leaf undersides. Spray leaves with a strong stream of water to dislodge aphids; if the problem persists, see Aphids for more controls.

VINES WILT. Squash vine borer injury, nematode injury, or bacterial wilt can cause vines to wilt. The plants may recover overnight but then wilt again during the heat of the day. Check the base of the plant for borer holes with yellowish debris around them. If the problem is borers, there's some hope of reviving the vine—see Squash Vine Borer for details.

If you don't see evidence of borers, try cutting through a stem. If thick ooze is present, the problem is bacterial wilt, and there is no cure. Uproot and destroy the diseased vines. Cucumber beetles spread bacterial wilt as they feed; to prevent future infections, consult Cucumber Beetles.

Also uproot a wilted vine. If you find swellings on the roots, the problem is root-knot nematodes. Consult Nematodes to prevent future problems.

In the high heat of the desert or Deep South, afternoon wilting of vines may be a normal response. If you don't find any evidence of pests or diseases, check soil moisture. Water if the soil is dry more than 3 inches below the surface.

FLOWERS DON'T FORM FRUITS. The first flowers that form on squash vines are male flowers, which produce pollen, but no fruit.

Inspect the vines; you may just need to wait until female flowers form. However, if you find female flowers, and they aren't forming fruit, the problem is a lack of pollination. Insects pollinate squash flowers, and if the weather is cloudy and cool, the insects may not be working the blossoms. You can improve pollination by hand-pollinating, as shown below. One male flower has enough pollen to hand-pollinate several female flowers. The best time to pollinate is in the morning, after the dew has dried; the flowers will close up in the late afternoon.

MOLD ON BLOSSOMS OR SMALL FRUITS. Some fungal organisms can invade squash blossoms and then grow into the fruits as they develop. If you see small black dots sticking up out of the mold, the problems is a disease called wet rot. It is most likely to develop during wet weather. Remove and destroy the infected blossoms and fruits. Consult Rots to learn how to discourage rot problems in future plantings.

Another reason why young squash turn moldy is poor pollination. Try hand-pollinating new female flowers to improve pollination, and the problem may disappear.

HOLES IN BLOSSOMS AND FRUITS. Pickleworms feed on blossoms and fruits of summer squash (rarely on winter squash and pumpkins). Their tunneling in fruits allows rot to set in, so fruits usually aren't salvageable. For controls and to prevent future problems, see Pickleworms.

MOLDY OR DRY, ROTTEN SPOTS ON FRUITS. Many of the disease organisms that cause symptoms on leaves and stems of squash plants can also infect the fruits. These disease organisms can cause rotting or allow infection by secondary bacteria that produce wet or dry rots.

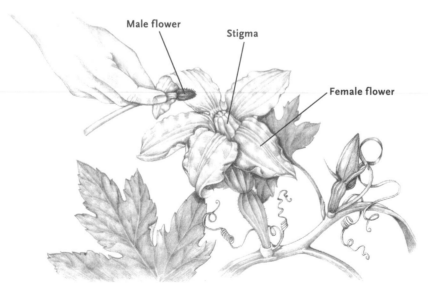

Use a cotton swab to transfer pollen from male flowers to female blossoms. Or pick the male flower, pull off the petals, and rub the pollen directly on the stigma of the female flower.

The fruits probably won't be usable. Remove diseased fruits to a hot compost pile or bury them deeply so that the seeds inside won't sprout in the future as volunteer seedlings (the seedlings could be infected with disease organisms). To prevent fruit rots in the future, improve air circulation around vines, use a support structure to keep vines and fruit off the ground, or place cans or other small supports under individual fruits.

BLOSSOM END OF SUMMER SQUASH REMAINS OPEN. This type of development is normal for winter squashes such as buttercup squash, but not summer squash. The problem results from a hormone imbalance due to stress—most likely cool temperatures or cloudy weather. When the weather changes, the fruits should form normally.

DRY ROT AT BLOSSOM END OF SQUASH. Blossom-end rot results from a lack of calcium while fruits are forming. It's usually due to water stress rather than calcium deficiency in the soil. Damaged fruit may be usable but will not keep in storage. To prevent this problem in the future, see Blossom-End Rot.

SUNKEN PITS IN SKIN OF SUMMER SQUASHES. Cold weather is the cause of chilling injury to summer squash fruits. Temperatures below 40°F can cause this. If it was a fluke cold spell, newly forming fruit should be normal. But if the growing season is ending, and you want to extend the harvest, cover plants with a blanket overnight to keep them warm. Remove the blanket during the day when temperatures rise.

SQUASH BUGS

About the time that squash vines poke through the soil, sleepy squash bugs are stretching their legs and peeking out from under nearby rock piles, leaf litter, and building foundations. Females settle in on young plants and begin laying eggs, a process that will continue until midsummer. Adults also begin feeding, sucking juices from plants and injecting a toxin that kills the tissue around the feeding site. As nymphs hatch, they feed, too, reaching the adult stage in 4 to 6 weeks.

In the North, there is only one generation per year, but it spans most of summer. In the South, the bugs may complete two or even three generations per year.

An invasion of squash bugs can kill young plants outright. On larger plants, heavy squash bug feeding may prevent plants from setting fruit. Vines sometimes wilt. If plants do produce fruit, the squash bugs will feed there, too, scarring the fruits.

Squash bugs are the worst pest of squash-family crops, and some types of squash are more vulnerable than others. Yellow straightneck and crookneck often suffer heavy damage, as do pumpkins and Cocozelle squash. Acorn squash and butternut squash seem more tolerant of squash bug damage. In general, melons and cucumbers are less likely targets than squash.

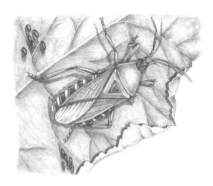

Squash bug

PEST PROFILE

Anasa tristis

RANGE: United States; some parts of southern Canada

HOW TO SPOT THEM: Check leaf undersides for rows of coppery eggs laid along the leaf veins. Also look for small yellow flecks on leaves, an early sign of squash bug feeding. Lay a board on the soil near squash plants and check underneath it in the morning to find bugs and nymphs.

ADULTS: Elongated shield-shaped, ⅝-inch, dark brown or gray bugs with a hard shell and orange and brown stripes on the abdomen. Adults give off an unpleasant odor when you crush them or when they congregate in groups.

NYMPHS: Wingless greenish insects with red heads and legs; as they age, bodies turn gray or grayish white and heads and legs darken.

EGGS: Shiny, copper-colored eggs laid in rows alongside veins on leaf undersides

CROPS AT RISK: Squash, pumpkins, cucumbers, melons, gourds

DAMAGE: Yellow flecks on leaves; leaves eventually dry up and die; young plants may be killed; vines wilt; pitted and scarred fruits, with areas of soft rot.

Squash bug nymphs

CONTROL METHODS

TIMED PLANTING. Plant as late as possible (still allowing time for the crop to mature). Your goal is to allow time for squash bugs in your area to emerge and seek out egg-laying sites elsewhere in your neighborhood, or starve to death, before your crop emerges.

GARDEN CLEANUP. After harvest, turn under all crop debris or move it to the center of an active compost pile. Also uproot squash-family weeds and remove boards or other items that could shelter overwintering adult squash bugs.

MULCHING. Squash bugs will hide under black plastic mulch. Instead use newspapers topped with straw, or no mulch at all.

HANDPICKING. Cut off and destroy leaves that have eggs on them. Handpick adults and nymphs or cut off severely infested leaves. Lay boards on the soil near squash plants and check the undersides every morning. Dump adults and nymphs into soapy water.

ROW COVERS. Cover seedbeds and newly planted transplants with row cover and seal the edges as described in Row Covers. Use row covers only for beds where no squash-family crops grew the previous year. Remove the covers when female flowers appear.

SPRAY. As a last resort, spray insecticidal soap or pyrethrins to kill young squash bug nymphs. Neem may also be effective. See Pest Control for more on these sprays.

CONTROL CALENDAR

AT PLANTING TIME

- Delay planting to avoid egg-laying adults.
- Mulch around plants with newspaper topped by straw.
- Cover seedbeds and transplants with row covers and seal edges.

WHILE CROP DEVELOPS

- Handpick adults, nymphs, and eggs.
- Spray nymphs with insecticidal soap.
- As a last resort, spray pyrethrins or neem.

AFTER HARVEST

- Turn under or remove all vine crop debris. Remove squash-family weeds.

SQUASH VINE BORERS

Zucchini is known for prolific production, but one small caterpillar can sever the lifeline of a zucchini or other squash vine and end the harvest before it even starts. The squash vine borer is the culprit, and summer squashes and pumpkins are its favorite host plants.

Squash vine borer moths—which many gardeners mistake for wasps—emerge and take flight in early summer, seeking out the bases of young squash vines as egg-laying sites. When the eggs hatch, the larvae immediately chew their way inside the vine. They feed inside for up to 6 weeks, disrupting the vascular tissue that allows water and nutrients absorbed by the roots to travel throughout the vine.

When the internal damage reaches a critical point the infested vine will suddenly collapse. If you can encourage rerooting, the vines may recover, but it's better to be proactive and encourage secondary rooting before vines wilt. Sometimes a borer will leave the vine and search out another to invade.

Two generations of borers complete their life cycle yearly in the South. There's only one generation in the North, but because the moths lay eggs over a period of about 2 months, the pest remains a problem all summer.

PEST PROFILE

Melittia spp.

RANGE: Eastern United States as far west as the Dakotas and Texas; the southwestern squash vine borer is found from Texas westward to southern Arizona.

HOW TO SPOT THEM: Set a yellow pan filled with water near your squash vines. The color attracts the moths and they will drown. This won't prevent a borer problem from developing but will alert you to their presence early. Remove the traps as soon as feasible because they also attract and drown insect-parasitic wasps. Inspect each vine (to about 3 feet out from the base) frequently for holes with frass sticking out or on the ground; these are entry sites of borers.

ADULTS: Wasplike moths with red or orange abdomens with black markings; metallic forewings and clear hindwings; 1½-inch wingspan (or smaller in the North)

LARVAE: Wrinkled, cream-colored caterpillars with brown heads; up to 2 inches long

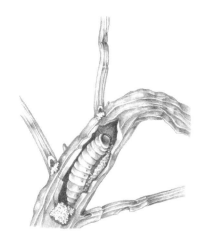

Squash vine borer inside vine

EGGS: Flat, reddish brown eggs laid near the base of the plant or on leaf undersides

CROPS AT RISK: Summer squash, zucchini, pumpkins, Hubbard squashes, and some other winter squashes; rarely a problem on butternut squash, cucumbers, or melons

DAMAGE: Larvae tunnel into vines and feed inside, disrupting the flow of water and nutrients. Diseases may invade damaged vines. Vines wilt suddenly and may not bear fruit.

CONTROL METHODS

TIMED PLANTING. Planting summer squash on the late side may allow the crop to escape injury. Or try planting a new mound every few weeks in the hope that some will escape invasion.

ROW COVERS. Cover seedbeds and young transplants with row covers to prevent moths from laying eggs. The critical area is the basal 3 feet of vine, which is where borers tend to enter. Once vines start to flower, remove the

Squash vine borer moth

covers or pull them back periodically and hand-pollinate.

HANDPICKING. Try to catch vine borer moths with an insect net. Crush egg clusters. Search for entry holes in the vines; slit the vine near the hole and carefully pull back the stem tissue to expose the interior. Remove the borer and crush it. Then close up the tissue and heap about 1 inch of moist soil over the surgical site to encourage healing and root formation. Baby this area by watering it with compost tea or a dilute organic fertilizer solution.

Another way to kill the larvae is to use a flashlight at night to illuminate the silhouette of borers inside squash vines. Stick a straight pin through the vine to impale the borer. Repeat every 3 days until you don't find any more borers.

ENCOURAGING SECONDARY ROOTING. Squash vines can form roots at points along the stem where a leaf stem emerges (nodes). Heap soil over some nodes along the length of every vine. That way, if a borer invades the base of the vine, the secondary roots can take over supplying the vine with water and nutrients.

PROVIDING EXTRA FERTILIZER. The faster a squash crop develops, the greater the chance it can set and mature fruit before borer damage causes growth to slow down or stop.

GARDEN CLEANUP. If a vine wilts because of borer injury, use a steel-tined rake or other implement to shred the vine as well as possible before removing it. This should expose or kill any larvae hidden in the vine, preventing them from pupating or moving on to another vine.

FALL CULTIVATION. Work the soil weekly for a few weeks in fall to expose pupae to predators, and then cover the soil for winter.

BIOCONTROLS. Prepare a solution of *Bacillus thuringiensis* or predatory nematodes according to the directions that come with the product. Use a garden syringe (available from some garden suppliers) to inject the solution into vines in the area near entry holes. If the borer ingests some of the solution as it feeds, it will sicken and die. For details on using Bt, see Pest Control; for information on predatory nematodes, see Nematodes.

CONTROL CALENDAR

AT PLANTING TIME

* Plant late or make several plantings to avoid egg-laying moths.
* Cover seedbeds or young plants with row covers.

WHILE CROP DEVELOPS

* Hand-pollinate crops that are protected by row covers.
* Handpick adults and eggs.
* Cover some nodes with moist soil to encourage secondary root formation.
* Fertilize vines to stimulate rapid growth.
* Slit stems and remove borers, or inject infested stems with Bt or predatory nematodes.

AFTER HARVEST

* Destroy infested crop residues to kill borers.
* Work the soil weekly for a few weeks in fall.

STINK BUGS

Stink bugs do smell bad, but it's the damage they cause to ripening tomatoes, pea pods, and other crops that worries gardeners. The southern green stink bug is one of the worst, especially in gardens in the Southeast.

In spring, stink bugs become active and lay eggs on leaf undersides. The nymphs feed on crops by sucking plant juices. Some species feed on leaves; others feed on flower buds and blossoms, which will shrivel and die back. Feeding injuries on fruits produces catfacing (distorted areas) or cloudy areas below the skin surface. Bean and pea seeds will be shriveled and unusable.

There are two or more generations of stink bugs each year, and the bugs can be serious pests of commercial cotton and soybean fields. If you live near fields of these crops, you may find that the stink bugs suddenly migrate to your garden in late summer or early fall.

The harlequin bug is another pest in the stink bug family (see Harlequin Bugs). Other types of stink bugs are beneficials that prey on plant pests. See Beneficial Insects for information on the beneficial spined soldier bug.

PEST PROFILE

Many genera and species

RANGE: Various species throughout the United States and Canada

HOW TO SPOT THEM: Check for clusters of barrel-shaped eggs on leaf undersides. Look for bugs on foliage (but wear heavy gloves if you handpick them, or your hands will smell terrible).

ADULTS: Shield-shaped, dull green or brown insects up to ½-inch long; wings folded across the body when at rest

LARVAE: Wingless oval insects that resemble adults; may have spots or stripes on their bodies

EGGS: Barrel-shaped eggs laid on plants

CROPS AT RISK: Tomatoes, peppers, beans, peas, corn, okra, melons, squash, cabbage, and other crops

DAMAGE: Flowers may wilt and die; fruits show catfacing or cloudy splotches; bean and pea

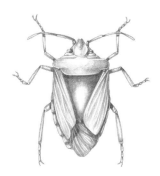

Stink bug

Stink bug nymph

pods may develop water-soaked areas and the seeds inside may shrivel or turn brown; okra pods are twisted and distorted; disease may ruin pods.

CONTROL METHODS

TIMED PLANTING. Plant crops as early as possible in spring in hopes of harvesting before a mass migration of stink bugs arrives in your area.

HANDPICKING. In the morning, knock the slow-moving bugs off plants into a container of soapy water. Crush egg clusters. Identify the bugs before you kill them, though, because some species are beneficial.

GARDEN CLEANUP. After harvest, compost or turn under crop residues to remove overwintering sites.

OIL SPRAYS. Spray lightweight oil or insecticidal soap to kill nymphs, but not when it's too hot, or the sprays may damage flowers or foliage. See Pest Control for information on oil sprays.

PYRETHRINS: As a last resort, spray pyrethrins to combat a sudden infestation of adult stink bugs in late summer or fall.

CONTROL CALENDAR

AT PLANTING TIME

• Plant early to avoid damage.

WHILE CROP DEVELOPS

• Handpick bugs and crush egg masses.
• Spray oil or insecticidal soap.
• As a last resort, spray pyrethrins.

AFTER HARVEST

• Compost or turn under crop residues.

SWEET POTATOES

Heat makes sweet potatoes happy, and they're a favorite crop for southern gardens. Northern gardeners can succeed with sweet potatoes, too, if they protect plants from cold at the beginning and end of the season. Planting sweet potatoes from slips—rooted sprouts—is simple. Purchased sprouts sometimes look limp and nearly dead, but they grow quickly in warm soil.

CROP BASICS

FAMILY: Morning glory family

SITE: Full sun is essential in the North; in very warm climates, sweet potatoes will still produce well in a site that receives some afternoon shade.

SOIL: Sweet potatoes will grow well in average soil. In sandy or poor soil, work in 2 to 4 inches of mature compost or cottonseed meal at 5 pounds per 100 square feet. Avoid unfinished compost because it may lead to tuber disease problems.

TIMING OF PLANTING: Plant slips outdoors after all danger of frost when soil is warm—55°F is the minimum soil temperature for outdoor planting; 65°F or warmer is ideal.

PLANTING METHOD: Planting rooted slips is the standard method for sweet potatoes.

CRUCIAL CARE: Weed and water regularly while plants are getting established. After that, sweet potatoes are a low-maintenance crop. Water established plants during dry spells.

GROWING IN CONTAINERS: Sweet potatoes will grow well in a half-barrel or similar size container in full sun.

HARVESTING CUES: Sweet potato plants don't change color or stop growing when the tubers are ready to harvest. Time your harvest according to the weather or the expected days to maturity, as described below.

SECRETS OF SUCCESS

SOUTHERN-STYLE SWEET POTATOES. In Zones 7 and warmer, most varieties of sweet potatoes thrive. When you choose varieties, be sure to select a nematode-resistant one if you've had problems with nematodes on other crops. Be sure the vines have room to spread because they'll grow up to 10 feet long. There's no need to rush the harvest, but keep in mind that the vines will continue growing and tubers enlarging as long as the air and soil are warm enough—sweet potatoes love soil temperatures in the 80s but will

stop growing when temperatures drop down close to 60°F. If you wait too long after your variety's expected days to harvest, tubers can grow more than a foot long. It is wise to harvest before nighttime temperatures start dropping into the 50s; otherwise, foliage may suffer cold damage, which in turn can cause tuber injury.

GROWING SWEETS IN THE NORTH. Choose a dwarf or early-maturing variety. Plant slips in raised mounds or ridges of soil, which will warm up quickly. Speed soil warming even more by covering the mounds with infrared transmitting (IRT) plastic mulch and planting the slips through slits in the mulch. It's wise to cover newly planted slips with cloches or row cover, both to add warmth and keep out insects. You'll have to remove cloches as the vines grow longer. Near the end of the growing period, cover the

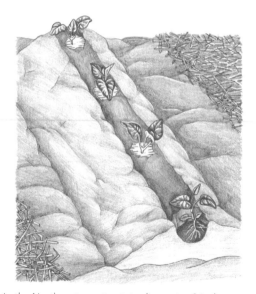

In the North, set sweet potato slips 12 to 18 inches apart in a raised planting ridge. If you're not planting through plastic mulch, shape a shallow basin in the soil around each slip to hold water.

plants again with row covers to add warmth because foliage can suffer cold injury at temperatures as high as 50°F. Watch the weather forecast and dig your crop before overnight temperatures drop below 40°F. Check the foliage daily. If any of it has blackened from cold, cut it off immediately and dig the tubers as soon as possible.

TREAT TUBERS WITH CARE. Start by cutting off most of the topgrowth so you can see what you're doing as you harvest. Sweet potato tubers seem tough and solid, but they have delicate skins. When you dig sweet potatoes, insert a spading fork gently into the soil about 1 foot away from the crown of the vine. If you sense that the tines are striking resistance or a solid object, pull out the fork and try again several inches farther out. Once you can insert the tines deeply, rock the tool back and forth to loosen the soil. Grasp the remaining bit of vines near the crown and shake them as well. Alternate shaking and rocking the tool until the tubers come free and you can lift them from the soil. Move them out of the sun right away to a warm, humid place to cure for about 10 days. Ideal curing temperature is 80° to 85°F. In the South, perhaps there's a spot on your porch out of direct sun. In the North, the closest you may be able to come to proper conditions is your kitchen counter.

PREVENTING PROBLEM

AT PLANTING TIME

- Cover plants with row cover to keep out flea beetles and sweet potato weevils.
- Put out wireworm traps if you suspect that wireworms are present in your garden.

WHILE CROP DEVELOPS

- Apply a repellent to deter flea beetles.
- Mound up soil around stems to deter sweet potato weevils from laying eggs.
- Cut back on watering a few weeks before harvest to avoid cracking of tubers.

AT HARVEST TIME

- Remove tubers from the garden immediately so they won't be injured by sun exposure.

AFTER HARVEST

- Remove all crop debris from the garden. Compost healthy material and discard noticeably diseased vines and tubers.

TROUBLESHOOTING PROBLEMS

SMALL HOLES IN LEAVES. Flea beetles chew thin grooves in the surface of sweet potato leaves, but usually they don't cause serious damage. The larvae eat tunnels through the roots. If your plants are heavily infested with adults, take steps to control them (see Flea Beetles).

VINES TURN BLACK OVERNIGHT. Exposure to cold temperatures can injure the vines. This is serious because the injury can translate into damaged tubers and poor flavor as well. Cut the vines at soil level immediately. Harvest the tubers the same day, or as soon as possible.

YOUNG LEAVES TURN YELLOW. This is a sign of Fusarium wilt or southern blight. Look for wilting and for dark internal areas in the stem near the base of the vines. Pull out infected plants to prevent the disease from spreading (see Fusarium Wilt).

If southern blight is the cause, stems may look water-soaked at the base, and a white mold often appears on the soil around the plants. Pull out infected plants and discard them or bury them deeply (see Southern Blight).

STUNTED PLANTS. Stunting can be due to root-knot nematodes, Fusarium wilt, or black rot. If you find swollen areas along sweet potato roots or cracks in the tubers, those are signs of root-knot nematodes. See Nematodes for more information.

If the stems are discolored at the base of the plants, the problem is probably Fusarium wilt. Pull out infected plants to prevent the disease from spreading (see Fusarium Wilt).

CRACKS IN POTATOES. Cracks may be due to root-knot nematodes (see Nematodes). If you rule out root-knot nematode problems, chances are the cracks are a reaction to a sudden uptake of water, either from heavy rains or irrigation, after a dry period. Salvage what you can from the tubers for use right away. Don't try to store cracked tubers.

BLACK BLOTCHES ON TUBERS. Scurf is a fungal disease that causes these black areas. They're limited to the surface, so you can cut them away and use the tubers. Tubers infected with scurf won't store well, though, so cook them right away. If it's too much to eat all at once, freeze the extra. To prevent recurrence, rotate your next planting to a new area, choose a resistant variety, and clear out rough organic matter from the bed before planting.

STRINGY FLESH IN TUBERS. Stringiness can be the result of wet weather, overwatering, poor soil drainage, or your choice of variety. For your next crop, choose a different variety, plant in well-drained soil, and cut back on watering for the last few weeks before harvest.

TUNNELS OR HOLES IN TUBERS. Wireworms and the larvae of sweet potato flea beetles and sweet potato weevils feed in sweet potato tubers. Wireworms are shiny, segmented larvae and are common in garden areas where lawn grass grew previously. See Wireworms for preventive measures.

Flea beetle larvae tunnel just under the surface of tubers. You can cut away the damaged areas and use the rest; the tubers won't keep well in storage.

Sweet potato weevils are southern pests. The weevils resemble blue-black ants; the area below their heads is reddish in color. They lay eggs in stems and roots, and the white, wormlike larvae tunnel their way through the plants into the tubers. Damaged roots will taste bitter, and rot often sets in. There are many generations per year. The larvae can spread through stored roots, too, so if you find any damaged tubers, check all stored tubers frequently. To avoid future problems, buy certified weevil-free planting stock, plant in a different part of your garden, and mound up soil around the base of vines.

SWISS CHARD

New gardeners and experienced gardeners love growing Swiss chard. This beautiful, tasty, nutritious crop produces well over a long season, and it rarely suffers from significant pest problems as long as it is well fed and receives enough moisture. Many gardeners have learned that Swiss chard makes a serviceable spinach substitute when conditions are too hot for spinach to grow well. Swiss chard leaves are thicker than spinach leaves, and their flavor is stronger. But served steamed or stir-fried, Swiss chard is tender and delicious. For the closest match to regular spinach flavor, try the chard variety called 'Perpetual Spinach'.

CROP BASICS

FAMILY: Beet family; relatives are beets and spinach.

SITE: Plant in full sun; will tolerate light shade.

SOIL: Rich soil is critical for fast growth and a long harvest. Add a 2- to 4-inch layer of compost before planting to poor-quality soil, or 1 inch of compost to soil of average to good fertility.

TIMING OF PLANTING: Sow seed as early as 2 weeks before last expected spring frost in most areas; or sow as desired anytime in spring. In cool-summer areas, summer sowings will do well also. In the Deep South, desert Southwest, and southern California, sow from late summer into fall, after high temperatures have passed.

PLANTING METHOD: Sow seeds directly in garden beds. Starting seedlings in containers is possible but not usually necessary.

CRUCIAL CARE: Check soil moisture regularly, and maintain even soil moisture from planting throughout the harvest.

GROWING IN CONTAINERS: An excellent container crop; see "Swiss Chard Centerpiece," on page 402.

HARVESTING CUES: Six is the cue: Begin cutting leaves about 6 weeks after sowing (in ideal growing conditions) or when plants reach about 6 inches tall or when plants have at least six well-developed leaves. Yellowing leaves are a sign of overmaturity.

SECRETS OF SUCCESS

PLANT WHEN YOU CAN. Spring is the traditional time to plant Swiss chard in most parts of the country, but if you miss out on planting Swiss chard in spring, try planting it later on. Even as far north as Zone 4, Swiss chard sown at the end of July will yield a nice harvest in fall. Plants can survive until temperatures drop below about 20°F. They tolerate heat well, too. Spring-planted Swiss chard may succeed as far south as Zone 8, especially if it's planted in a spot where tall crops will provide some shade. If you find your spring-planted crop going into decline or becoming tough because of high temperatures, plant a new crop for vigorous fall production.

SWISS CHARD CENTERPIECE

Finding space for Swiss chard in the garden—especially a small garden—can be a challenge because it grows for such a long season. A space-saving solution to the dilemma is to plant Swiss chard as the centerpiece in a half-barrel or other large container. Swiss chard has a beautiful vaselike form, and some varieties offer a rainbow of color. Since you harvest individual leaves, the plants will always look good once they're established. Plant three or four Swiss chard plants at the center of the container; surround them with low-growing annuals in matching colors, such as trailing nasturtiums, dwarf dahlias, or marigolds. Add parsley, too, for contrasting texture.

This kind of container planting is perfect for decorating a sunny deck or patio.

Here are some tips for success with container Swiss chard:

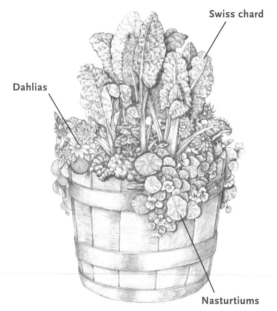

Swiss chard

Dahlias

Nasturtiums

- Start seedlings in a small flat indoors or in a protected spot outdoors. Transplant when the seedlings are 2 to 3 inches tall. This allows you to place the plants in exactly the formation you want in the container. The floppy transplants will become sturdier as they grow.
- Be diligent about watering. The plants will need more frequent watering than they would in a garden bed. Moisture stress can ruin their appearance or lead them to bolt.
- To support a healthy harvest, fertilize every 3 weeks.

If your goal is simply productivity, not a showpiece, you can use a 14-inch plastic pot to grow three Swiss chard plants. To prevent leafminer damage, rig a row-cover tent for the container, as shown on page 349.

THIN SEEDLING CLUMPS. Swiss chard seeds are actually dried husks that contain several tiny seeds. For each "seed" that you sow, a cluster of about five seedlings will emerge. After the first true leaves appear, thin the clusters to one plant each, using nail scissors to snip the excess seedlings at soil level. When the plants reach about 4 inches tall, thin them again to at least 6 inches apart, using the plants you pull for salads.

FEED FOR CONTINUED PRODUCTION. Side-dress the plants every 4 to 6 weeks. Use alfalfa meal, spreading 1 pound per 10 feet of row. Or water plants with a fish emulsion solution or apply a balanced organic fertilizer.

PAMPER PLANTS WHEN IT'S HOT. Hot, dry conditions can cause Swiss chard to decline. If a hot spell is predicted, mulch around the plants to keep the soil cooler and help retain moisture. Water more frequently. If the plants still go downhill, harvest what you can, and try sowing a new planting when the worst of the heat has passed.

HARVEST REGULARLY. Some sources recommend twisting the stalks free of the plant, but it's better to use a sharp knife to cut stems at the base. Be careful not to damage other stalks or the plant's growing point when you harvest.

REGIONAL NOTES

MIDSOUTH

In Zones 7 and 8, try setting up a plastic-covered tunnel over Swiss chard plants in fall. You'll extend the harvest well into winter, and plants may survive through to spring. Tender spring growth is delicious. But as daylength increases, the plants will go to seed; remove them and plant a new crop.

DESERT SOUTHWEST

Whiteflies may be troublesome; they will transmit viruses, and this can be a problem especially for commercial growers. It's the nymphs that suck the sap, and their feeding drains the plants of energy, leaving them more susceptible to the viruses. For control measures, refer to the Whitefly entry.

SOUTH

Infestation by root-knot nematodes can reduce the vigor of Swiss chard, resulting in a disappointing harvest, especially in southern areas where the nematodes survive over winter. In a home garden setting, it's hard to control nematodes with crop rotation because so many crops are susceptible. One better option is to solarize the soil before planting. Refer to page 186 for soil solarization directions; see the Nematodes entry for complete information on controlling root-knot nematodes.

PREVENTING PROBLEMS

AT PLANTING TIME

- Cover seeded areas with row covers and seal the edges (see Row Covers).
- Put out iron phosphate bait or a slug barrier to prevent slug and snail damage to seedlings.

AS CROP DEVELOPS

- Check under row covers once a week for signs of pests or disease
- Handpick slugs and snails as needed.
- Keep notes on productivity once you start cutting leafstalks. If you notice a decline in production, feed with compost tea and follow up by side-dressing with balanced organic fertilizer.

AFTER HARVEST

- If your crop suffered any disease problems, discard noticeably diseased material.

TROUBLESHOOTING PROBLEMS

SEEDLINGS NEVER EMERGE OR NEW SEEDLINGS DIE. Damping-off kills young seedlings before or immediately after they break through the soil. Replant in a spot with better drainage. Or pull seedlings, work compost and/or builder's sand into the soil to improve drainage, then replant. For more information on this disease, see Damping-Off.

NEW SEEDLINGS CUT OFF NEAR GROUND LEVEL. A single cutworm can destroy several plants in one night. These pests hide just below soil level. See Cutworms for prevention and control information.

TUNNELS IN LEAVES. Leafminer flies lay eggs and their larvae tunnel through leaves. It's fine to harvest leaves that show tunneling; cut away the damaged areas and eat the rest.

To prevent damage to future plantings, cover seeded beds with row cover in spring. For other control techniques, refer to Leafminers.

SMALL HOLES IN LEAVES. Flea beetles chew on leaves and ruin their appearance. A heavy infestation can destroy seedlings. For prevention and control measures, turn to Flea Beetles.

RAGGED HOLES IN LEAVES. Slugs, snails, cabbage loopers, and alfalfa loopers are possible culprits; seek and find which pest is damaging your plants. Cut away damaged areas of leaves and use the rest. Refer to Slugs and Snails and Cabbage Loopers for control information.

DEFOLIATED PLANTS. Armyworms or blister beetles can devastate a planting. Refer to Armyworms and Blister Beetles for control methods.

DISTORTED LEAVES. Aphids suck sap, causing leaves to become curled, twisted, or puckered. Look for colonies on young leaves and undersides of larger leaves. Cut and discard damaged leaves. To protect regrowth from aphids, refer to Aphids.

BRONZED LEAVES. Leaves develop a bronze coloring from exposure to cold. There's no need to treat this condition. If temperatures rebound, new leaves will be normal. If winter has arrived, it's a sign that harvest is over. Uproot the plants and compost them. Or try overwintering them; cut back the tops and cover the crowns with shredded leaves. Plants may survive and resprout in spring. Harvest young spring growth; plants will bolt in late spring or early summer.

STUNTED PLANTS; PALE LEAVES. Leafhoppers transmit beet curly top virus as they feed. The disease tends to be worse during sunny, hot, dry conditions. Pull out diseased plants and destroy them. The best way to prevent curly top is to stop leafhoppers from feeding; refer to Leafhoppers for controls.

These symptoms may also be a sign of root rot. Uproot a couple of affected plants. If you find soggy, rotted roots, Fusarium rot is the problem. Replant in a new area with enriched soil. Water regularly (but don't overwater) so that plants grow rapidly.

REDDISH OR BROWN SPOTS ON LEAVES AND STALKS. Cercospora leaf spot is a fungal disease, and it will show up when drainage is poor or if the plant is struggling due to lack of soil fertility and/or moisture. Improve conditions and cut off affected stalks; new growth often will be symptom-free. See page 109 for management techniques for this disease.

PRODUCTION OF NEW LEAVES DECREASES. Sometimes the simple reason for a decline in production is that you haven't harvested enough. When old leaves are left on the plant, they may suppress production of new young growth, so harvest Swiss chard regularly. If you see older leaves yellowing, cut them off and compost them.

PLANTS FORM FLOWERING STALKS. Swiss chard is a biennial: in its first year of growth, it forms stems and leaves only. After overwintering, it flowers, sets seed, and dies. Exposure to a period of cold temperatures can "trick" the plant into thinking it has overwintered, and it will bolt early. Cutting off the flowerstalks may trick the plant yet again, leading it to produce more harvestable stems and leaves. Experiment to see whether this works for you. If not, uproot the plants and plant new ones. Don't worry if it's not the ideal planting time in your area. Plants sown in late spring or summer won't grow as large as the early-spring crop, but as long as 2 months of frost-free weather remain in your growing season, the plants will produce some harvestable leaves. Try planting more than one variety, also. White-stemmed varieties seem less susceptible to bolting than red-stemmed varieties.

TARNISHED PLANT BUGS

Tarnished plant bugs work by subterfuge. In spring, these small insects seek out vulnerable buds, young fruits, and tender growing tips and lay eggs there, inserting the eggs directly into plant tissues. When nymphs hatch 5 to 7 days later, they feed in these sheltered areas, inserting their stylet (sucking mouthpart) into plants and injecting saliva that contains an enzyme to break down plant tissue. The injured tissue liquefies, and the nymphs suck the sap. Adults feed in the same way. The tissue around the feeding sites dies, and new growth is distorted.

When tarnished plant bugs feed on strawberry or tomato fruits, the fruits develop a distorted, dimpled appearance called catfacing. Tomato fruits also develop discolored areas under the skin. Feeding on flower buds can cause blossom drop, especially of peppers and eggplant.

After feeding for a few weeks, nymphs reach the adult stage. The new adults lay eggs, starting the cycle again. There are two or more generations per year, and adults overwinter under leaf litter, in cracks in tree bark, in alfalfa and clover fields, and in sheltered areas such as landscape borders and hedgerows.

These little bugs feed on an astonishing range of plants—more than 380 species—including many vegetable and fruit crops. Often, we mistake the damage they cause for some other type of problem, since it's not easy to see the small bugs and nymphs on the plants. Tarnished plant bugs also eat aphids and other soft-bodied insects, but this benefit doesn't outweigh the damage they can inflict on a vegetable garden.

No single control method works well against tarnished plant bugs. Row covers are impractical for large fruiting crops, and sprays often can't reach nymphs hidden in crevices of buds and young shoots. Because tarnished plant bugs are also serious pests of farm crops, researchers are working to improve biological controls. Plenty of native predators and parasitoids attack tarnished plant bugs, but they seek the bugs only on certain host plants, such as weeds and alfalfa. However, one imported parasitic wasp has become established in the Northeast and into eastern Canada; its presence there seems to be lessening overall problems with tarnished plant bugs. With luck, researchers will also find biocontrols that can successfully adapt and spread in the rest of the United States and Canada.

PEST PROFILE

Lygus lineolaris

RANGE: Throughout the United States and Canada

HOW TO SPOT THEM: Try sweeping an insect net through foliage of flower borders in early spring to find overwintering adults. If you find them, be prepared to protect your vegetable garden against them. Set up white sticky traps to monitor the activity of the bugs. Use a hand lens to look for adult bugs on crops in spring, especially crops that are producing flower buds and/or young fruit.

ADULTS: ¼-inch true bugs with brown bodies mottled with bronze, yellow, or reddish marks. The bugs have long, segmented antennae and a characteristic white triangular area on their back.

NYMPHS: Yellowish green, wingless, long-legged nymphs, up to ¼ inch long, some with black spots on their backs

EGGS: Laid in plant tissue

CROPS AT RISK: Lettuce, celery, tomatoes, potatoes, eggplant, peppers, cauliflower, cabbage, broccoli, turnips, Asian greens, cucum-bers, beans, carrots, Swiss chard, beets, asparagus, and other crops

DAMAGE: Catfacing of young fruits; cloud-spots on tomato fruits; yellowed, distorted growth; necrotic spots on lettuce and other crops; death of growing tips followed by sprouting of bushy sideshoots

Tarnished plant bug

Tarnished plant bug nymph

CONTROL METHODS

MANAGE WEEDS. Tarnished plant bugs will feed on a wide variety of weeds, including dandelion, chickweed, lamb's-quarters, smartweed, wild mustard, curly dock, and pigweed. Keep your spring garden free of these weeds to avoid attracting the bugs early in the season.

ROW COVERS. Cover small plants with row covers and leave the covers in place as long as possible. Seal row cover edges as described in Row Covers.

TRAP CROPS. Try planting patches of buckwheat around your garden, timed so that the buckwheat will flower just before the vulnerable crop. The tarnished plant bugs may be attracted to the buckwheat first and lay their eggs there instead of on your vegetables. Destroy the buckwheat before the next generation of nymphs matures.

GARDEN CLEANUP. After harvest, remove all crop debris from the garden in case nymphs or adults are hiding there.

KAOLIN. Spray vulnerable plants with kaolin clay before tarnished plant bugs become active, and reapply as needed. See Pest Control for information on kaolin.

GARLIC SPRAYS. Spray plants before tarnished plant bugs appear in your garden to deter feeding and egg laying. For information on garlic repellents, see Pest Control.

BIOCONTROL. Spray young nymphs with *Beauveria bassiana*; see Pest Control for details.

SPRAYS. As a last resort, spray infested plants with insecticidal soap or pyrethrins, but keep in mind that the spray will not kill nymphs that are deeply hidden in buds or growing tips of plants. Neem may be effective against young nymphs. See Pest Control for more on these sprays.

CONTROL CALENDAR

BEFORE PLANTING

- Clear all host weeds out of the garden.
- Plant buckwheat as a trap crop.

AT PLANTING TIME

- Cover plants and seedbeds with row cover.

WHILE CROP DEVELOPS

- Spray plants with garlic to repel bugs.
- Spray plants with kaolin clay to discourage feeding and egg laying.
- Spray young nymphs with *B. bassiana*.
- As a last resort, spray with insecticidal soap or pyrethrum. Neem may be effective.

AFTER HARVEST

- Remove infested crop debris from the garden.

THRIPS

Tiny thrips cause trouble for gardeners in several ways. When onion thrips or bean thrips gather in large numbers on the leaves of young plants, they can injure leaves badly enough to reduce yields (damage to older plants may look bad but shouldn't hurt the harvest). Western flower thrips feed on flowers rather than leaves, and injury by even a small number of these thrips can ruin the appearance of tomato fruits that develop from the injured flowers.

Thrips feed by rasping at the flower or leaf surface and slurping up the plant sap released; the feeding wounds are ideal sites for disease organisms to gain entry. Thrips present a triple threat to some crops because as they feed they can transmit certain viral diseases, including tomato spotted wilt.

Both adult thrips and their nymphs feed on plants, and they thrive in hot, dry conditions.

This is one pest that tends to be more serious in arid climates than in humid ones. The adults survive winter in plant debris or bark cracks, and they produce many generations per year. Western flower thrips overwinter in the southern half of the United States, the West Coast states, and British Columbia but are shipped to all regions of the United States and Canada on greenhouse-raised vegetable transplants.

Although there are many treatments for thrips, the pests can be hard to control overall because they are so small and can hide in nooks and crannies or foliage and flowers. In commercial and home greenhouses, releasing predatory mites can provide very effective control (see Mites).

Thrips

Thrips damage

PEST PROFILE

Many genera and species

RANGE: Throughout the United States and Canada

HOW TO SPOT THEM: Hold a white cloth or card near crop flowers and shake the flowers. The thrips will fall out and will show up as specks on the white surface.

ADULTS: Yellow to brown insects with fringed wings (visible through hand lens); so small that they appear only as specks on foliage

LARVAE: Yellow, wingless insects similar to adults

EGGS: Laid in leaves, buds, and flowers

CROPS AT RISK: Onions, cabbage, beans, tomatoes, and wide range of other crops

DAMAGE: Leaves of cabbage-family crops develop bronzed appearance. White or silvery flecks or patches and black spots form on leaves of onions and other crops. Tomato fruits form with blemishes or scars.

CONTROL METHODS

RESISTANT VARIETIES. Some varieties of cabbage are tolerant of thrips damage. Onion varieties with round leaves and an open-growth habit tend to suffer less damage because there are fewer places for thrips to hide among the foliage.

GROWING ONIONS FROM SEED. Commercial onion sets may be infested with thrips. To avoid the possibility of introducing thrips by purchasing sets, buy onion seed instead and start seeds indoors early.

ENCOURAGING BENEFICIALS. Plant flowering plants to attract lady beetles, lacewings, and predatory bugs, such as minute pirate bugs. These beneficials and others all help to suppress thrips outbreaks (see Beneficial Insects).

WATER REGULARLY. Plants suffering from moisture stress are more vulnerable to attack by thrips. Check soil moisture regularly, and don't allow the soil to dry out.

REFLECTIVE MULCH. Spread reflective mulch or aluminum foil over the soil surface around susceptible crops to confuse thrips, as described on page 308.

WATER SPRAYS. Direct a strong spray of water at foliage to kill leaf-feeding types.

REPELLENTS. Sprays containing garlic and/or hot pepper help to repel thrips. See Pest Control for details on repellents.

KAOLIN CLAY. Coat onion foliage with kaolin clay before thrips become active to deter the pests from feeding and laying eggs. See Pest Control for directions for using kaolin clay.

BIOCONTROL. If you catch an infestation early, apply *Beauveria bassiana* or spinosad to infested plants (but keep in mind that these biocontrols may have some negative effects on beneficial insects). See Pest Control for information.

OTHER SPRAYS. As a last resort, apply essential oil sprays, neem, or pyrethrins to kill thrips. For directions for using these materials, see Pest Control.

CONTROL CALENDAR

BEFORE PLANTING

- Select tolerant varieties of onions, cabbage, and other crops (if available).
- Start onion plants from seeds indoors rather than buying sets.
- Provide insectary plants and shelter for beneficial insects.

AT PLANTING TIME

- Surround susceptible crops with reflective mulch to confuse thrips.

WHILE CROP DEVELOPS

- Water plants regularly to prevent moisture stress.
- Coat onion plants with kaolin clay to deter thrips.
- Spray plants with garlic or hot pepper sprays to repel thrips.
- Spray plants with a strong stream of water to dislodge and kill thrips.
- Spray plants with *B. bassiana* or spinosad.
- As a last resort, apply essential oil sprays, neem, or pyrethrins.

TOMATOES

Every gardener can succeed with tomatoes. No matter what climate challenges you face—short growing season, high humidity, cool summers, or blazing sun—your tomato plants will produce ripe fruit if you choose varieties wisely and know how to work with the weather.

Always check a tomato variety's disease resistance and growth habit before you buy. Fusarium wilt, late blight, root-knot nematodes, and mosaic are just some of the serious diseases that can wipe out tomato plants, but fortunately there are plenty of resistant varieties available (including many with multiple-disease resistance). As for growth habit, tomatoes can be determinate, semideterminate, or indeterminate. Determinate plants reach only about 3 feet tall. They set their fruit in a rush—over a period of 2 weeks or so—then stop growing after the fruits ripen. These plants are a good choice if you want to can or freeze tomatoes. Indeterminate tomatoes are the classic rangy vines that keep producing foliage and fruits over a long season. With proper support and pruning, they're a gardener's delight. Semideterminate varieties fall between the other two types in size and fruit production.

CROP BASICS

FAMILY: Tomato family; relatives are eggplant, peppers, potatoes, tomatillos.

SITE: Plant tomatoes in full sun, but provide shelter from high winds by planting them downwind of other tall crops or even by setting up a section of slatted lattice in the path of prevailing winds.

SOIL: Tomatoes grow best in moderately fertile, well-drained soil. Be sure the soil is loosened 12 inches deep to encourage root system development. Work two shovelfuls of mature compost into the soil at the site of each planting hole.

TIMING OF PLANTING: Plant tomato transplants outdoors when the soil reaches 60°F, about 2 weeks after your last expected spring frost. For earlier planting, take steps to prewarm the soil and protect plants as described in "Protection Pays" below.

PLANTING METHOD: Sow seeds indoors 6 to 8 weeks before planting-out date. In the South, start a second crop using suckers, as described in "Pruning Tomatoes" on page 413.

CRUCIAL CARE: Provide some type of support for all tomatoes to reduce the risk of disease problems and ensure a better harvest. If your soil is less than optimal, water plants weekly with 2 cups of half-strength fish emulsion until they start to produce flowers.

GROWING IN CONTAINERS: Patio tomatoes will grow in a container as small as 6 inches wide and deep, but bigger is better. Indeterminate types need a 5-gallon container or larger; stake or cage them, and prune to prevent them from getting out of control. You can plant several different varieties in a large half-barrel planter—yields may not be optimum, but you'll enjoy the diversity of the harvest. Add plenty of compost to the potting mix for moisture retention. Monitor soil moisture carefully; large plants may need daily watering.

TOMATILLOS

Heat brings out the best in tomatillos. The rangy vines resemble tomato plants with heart-shaped or oval leaves. The fruits look like green or yellow cherry tomatoes, and they form inside a papery husk. Tomatillos are generally easy to grow, and one or two plants are all that most gardeners need to supply a plentiful harvest.

There's an interesting assortment of tomatillo varieties available with a range of fruit color, intensity of flavor, and length of time to maturity. They're good to eat fresh or to use in salsa and other dishes, and you can wash excess fruit and freeze it whole in plastic bags.

Water tomatillo plants regularly while they become established; then mulch the soil. After that, plants will be relatively drought tolerant. The vines tend to sprawl, so provide support or mulch the ground around plants well to prevent fruits from coming in contact with the soil.

Pick tomatillos when they are full size but less than mature to capture their tanginess, which is famous for adding that special flavor to green salsas and other Mexican dishes. For fresh eating you may prefer to let some of the fruits ripen further, which will allow their flavor to become sweeter and milder.

If you live where the frost-free season is short, place your container tomatoes on a child's wagon. That way you can easily move them indoors to a garage or shed anytime overnight temperatures are predicted to drop below 50°F.

HARVESTING CUES: Don't rush the harvest. Tomatoes ripen from the inside out. Let the skin turn red (if the variety is a red one), and you'll know the fruits are fully ripe. If temperatures exceed 85°F, red pigment won't form and instead, the fruits become yellow-orange. So when the weather is very hot, you may want to pick tomatoes while they're still pink and let them finish ripening indoors. Keep picked tomatoes out of the sun; they'll ripen best when the temperature is between 65° and 70°F.

SECRETS OF SUCCESS

SEED-STARTING SAVVY. If you start plants from seed indoors, sow seeds three times over a 2-week period. Begin sowing about 7 weeks before your last frost date. This will expand the length of your harvest by allowing you to set out some plants early (with protection). Choose early-maturing varieties for the first seeding to speed the early harvest even more.

Provide supplemental light for tomato seedlings, but don't light them for more than 18 hours per day. Although some kinds of plants can tolerate continuous light, tomato foliage will become mottled or bleached and withered under continuous light. Eighteen hours provides strong growth without triggering an abnormal growth reaction.

PRUNING TOMATOES

You don't have to prune tomatoes—you'll get a good crop even from unpruned plants. Pruning properly can increase yields and reduce disease problems, but pruning the wrong way can reduce yields and create problems. Pruning technique varies by type of tomato.

INDETERMINATES. When indeterminate plants reach about 1 foot tall, they'll begin producing side shoots from the main stem. Each side shoot has the potential to produce flowers, fruits, and even more side shoots. With caged tomatoes, you can let side shoots grow, and just nip off the growing tips if they begin to extend

beyond the boundaries of the cage. You may also want to prune off some side shoots when they're small to prevent the cage from becoming a dense mass of foliage that holds moisture and blocks air flow (prime conditions for disease). To remove a small side shoot, grip the sucker with your thumb and index finger and gently rock it back and forth. The sucker will break off on its own, and the wound left behind will quickly seal itself.

DETERMINATES. Minimal pruning is best for determinates. If you prune too many suckers, the plants won't produce enough foliage, increasing the chances of poorly flavored fruit and sunscald.

SUCKERS FOR A SECOND CROP. In the South and Southwest, a variation on pruning can provide you with a second set of young vigorous plants to grow for fall harvest, when your main crop may have succumbed to stress or disease. To start new plants, let some suckers grow to about 5 inches long. Use a sharp knife to cut these suckers off the plant in midsummer. Pinch off the lowest leaf on each stem and place the suckers in a container of water or moist vermiculite. Set them in a sheltered spot to root. Once the suckers sprout some vigorous white roots, transplant them to pots or to a garden bed. Treat them carefully for the first few days, providing shade from hot afternoon sun and watering daily.

Pinching small tomato suckers

PLANTING POINTERS. You don't need to pinch off flowers from transplants at planting time; the flowers won't interfere with the plant's adjustment to the garden. However, you should pinch off baby fruits. Otherwise, the plants will

pour all their energy into those fruits at the expense of root and leaf growth, resulting in stunted, unproductive plants.

One exception: If you have enough garden space to allow for a little inefficiency, allow one plant to retain its small fruits. Although the plant won't amount to much, those baby fruits will enlarge and ripen 1 to 2 weeks sooner than those on your other plants. You may decide it's worth the space to enjoy a few extra-early ripe tomatoes.

Tomato plants will sprout roots along buried sections of stem. You can bury the stem up to the first set of true leaves to encourage some supplemental rooting. Don't prune off healthy leaves, though, to expose more stem for burying—the shock of losing the leaves will make it harder for the plant to adjust to life outdoors, and that will slow growth and fruit set. Water in each transplant with about 2 cups of half-strength fish emulsion.

PROTECTION PAYS. Planting tomatoes outdoors early (up to 3 weeks before the standard recommended planting time) will speed the harvest, but only if you protect the plants properly. Prewarm the soil by covering it with black or red plastic mulch or by shaping a raised planting bed—or both (you can even put the plastic in place the preceding fall). Shield the plants from cold winds and chilly air by wrapping cages in clear plastic or surrounding the plant with a ring of roofing paper. Commercial devices such as Wall O' Water provide the ultimate in protection, and if you clean them and store them properly between uses, they will last for several years. Watering plants with warm water will help spur early growth, too.

Be sure to remove the protective coverings once temperatures reach the normal range for tomato growth; otherwise, you can damage the plants by overheating them. Once the soil is thoroughly warm (if you aren't using black plastic), cover the soil with newspaper topped with organic mulch or with a 3-inch layer of straw or chopped leaves. The mulch creates a barrier to prevent disease spores from splashing up from the soil onto tomato foliage. Several serious diseases of tomatoes, including Septoria leaf spot and late blight, can spread from soil to foliage.

PICK A SUITABLE SUPPORT. Supports for tomato vines save space, ease harvesting, and help prevent disease problems. For determinate tomatoes, standard commercial tomato cages work well. Or set up an arch of sturdy wire mesh fencing over a row of determinate tomatoes.

Indeterminate tomatoes need taller supports. Cages made from 6- to 8-foot-tall welded wire fencing or concrete reinforcing wire are popular

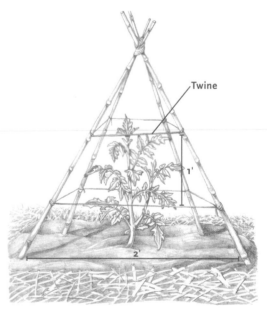

Twine

1'

2'

Tepee-style bamboo cages are easy to set up, manage, and store during the off-season. Tie the tomato vines loosely to the stakes as they grow.

because they are sturdy and long-lasting. If you make cages like these, plan on shaping them in three or four different diameters so you can stack them during the off-season. Eighteen inches is the minimum diameter; for that you'll need a 5-foot-wide section of fencing. A section that's 6¼ feet wide will make a cage 24 inches in diameter.

Bamboo tepees are cheaper and don't require as much storage space. Use four 8-foot, ¾-inch-diameter bamboo poles for each tepee. Lash the

BEYOND THE BASICS

TRELLIS-TRAINED TOMATOES TO PREVENT DISEASE

In the humid Southeast, caged tomato vines may be a target for disease problems. If disease has been a problem for your tomatoes, support them on a vertical trellis to improve air circulation. You can train tomatoes in a fan pattern on a vertical trellis with horizontal wooden slats or use a trellis with vertical twine tied off on a rigid frame made from electrical conduit, wooden posts and boards, or PVC pipe.

Attach vines to the fan-style trellis with pieces of soft cloth or old panty hose. Pass the tie loosely around the stem, then around the support, and tie off the ends at the support (don't tie a knot around a plant stem). With a vertical trellis, you'll use strong twine, tying it first at the base of the tomato plant (using a non-slip knot). Feed the twine to the top of the trellis, and then cut it off with about 2 feet of extra slack. As the vine grows, wind it around the twine. Untie the top end of the twine to adjust the pressure and use up the slack as needed.

Pruning is critical with trellised tomatoes. For the fan-style trellis, you want to allow two or three side shoots to develop (as needed for the number of fans), and then pinch out all other side shoots when they are small. For a vertical trellis, let each side shoot produce two or three leaves, then pinch the growing tip. This will provide plenty of foliage to feed developing fruits and protect them from sunscald, while still keeping the plants a manageable size. Whenever a stem reaches the top of a trellis, pinch the growing tip.

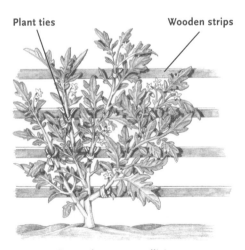

Plant ties **Wooden strips**

Fan-style tomato trellising

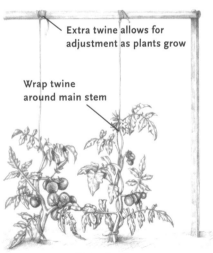

Extra twine allows for adjustment as plants grow

Wrap twine around main stem

Twine trellis

poles together at the top and push the bases into the ground a few inches. The poles should be 2 feet apart at the base. Set a tepee over each plant, and use strong twine for horizontal bracing around the tepee, 1 foot apart. As the vines grow, you can also fasten them loosely to the canes. These tepees are wind-resistant. For early-season protection, wrap the bottom foot of the tepee with clear plastic or roofing paper. Or you should have enough space to enclose transplants in Wall O' Water shelters for the first few weeks of growth, and then remove them. At the end of the growing season, store the bamboo canes upright in the corner of a garage or shed.

Set caged plants 3 feet apart (leaving 12 to 18 inches between cages) to allow good air circulation. Crowding cages may reduce yields because plants may suffer more symptoms of disease and or moisture stress.

REGIONAL REPORT

SOUTH AND WEST

Uneven ripening is a relatively new tomato problem caused by feeding by the silverleaf whitefly, which is an introduced pest in the Southeast and parts of the West. Tomatoes injured by these whiteflies don't ripen properly—parts of the fruit remain yellow or white while other parts of the fruit ripen. If you have only one or two tomato plants, you can try to prevent whitefly damage by tying a small piece of row cover around each cluster of fruit. If whiteflies move into a large planting of tomatoes, though, there are other steps you can take to combat them; see Whiteflies.

SOUTHEAST

If your tomato plants are growing normally on one side but have stunted growing tips on the other, the problem may be tomato spotted wilt virus. Other symptoms are thick, curled, yellow leaves. Sometimes small dark spots appear on leaves and brown streaks show up on stems. The fruits may be mottled with unusual ringed patterns of red, orange, or yellow.

Thrips spread this virus as they feed, and it's a widespread problem in commercial greenhouses, where it also can infect a wide variety of ornamental crops. If this disease suddenly appears in your garden, but you can't find any evidence of thrips, it's possible that it arrived in infected tomato transplants.

This disease is occasionally showing up in home gardens in other areas of the United States as well, from California and Colorado to New York. The virus can't survive long outside of living plant tissue or an insect host. If you find infected plants in your garden, remove and discard or destroy them. Don't buy any additional plants from a source that sold you infected plants.

FLORIDA, TEXAS, SOUTHWEST, AND CALIFORNIA

Folded leaves pinned together by webbing are a sign of tomato pinworms, small leafmining caterpillars that also tunnel into tomato fruits. The entry hole is very small, but it's enough to allow disease organisms entry, and that's what causes the real problem. Cut open a fruit that appears okay on the outside, and you may find a messy rot inside, or a tunnel that extends all the way to the center of the fruit. These pests can't survive winter except in the warmest parts of the United States, but there may be as many as eight generations a year in favorable climates. Pinworms may also appear in gardens in other areas if they are present on purchased transplants.

Inspect plants frequently for folded leaves or leaves with tunneling damage. Remove and destroy these leaves. Once a worm tunnels into a fruit, there's no way to undo the damage. If you live in an area where pinworms can overwinter, remove all crop residues from the garden and shred them before composting them. Also uproot and shred tomato-family weeds such as black nightshade and horse nettle.

PREVENTING PROBLEMS

BEFORE PLANTING

- Clear out weeds that can host tarnished plant bugs or other tomato pests.
- Plant insectary plants to attract and shelter beneficials that prey on aphids, thrips, and whiteflies (see Beneficial Insects).

AT PLANTING TIME

- Encircle each transplant with a cutworm collar.
- Cover plants with row cover to protect them from insects.

WHILE CROP DEVELOPS

- Remove row covers when plants outgrow them or temperatures rise.
- Spray plant health boosters to promote natural disease resistance (see Disease Control).
- Prune and train indeterminate tomatoes to maintain good air circulation around plants.
- Pick off individual leaves that show symptoms of disease and destroy them.

AT HARVEST

- Inspect fruit for any signs of disease and separate it from healthy fruit.

AFTER HARVEST

- Remove noticeably diseased crop residues and discard or destroy them.
- Cultivate the soil where tomatoes grew to expose pupae of hornworms and other pests to predators and cold.

TROUBLESHOOTING PROBLEMS

NEW SEEDLINGS EATEN. Cutworms or slugs and snails are responsible. Cutworms often cut transplants off cleanly near soil level; see Cutworms for preventive measures. Slugs and snails leave behind a slime trail; see Slugs and Snails for controls.

SMALL HOLES IN LEAVES. Flea beetles feed on tomato foliage and can be serious if they infest young transplants (see Flea Beetles).

DEFOLIATION. Tomato hornworms feast on leaves, sometimes leaving behind a bare midrib.

They are easy to handpick; for other controls, see Tomato Hornworms.

HOLES IN LEAVES. Colorado potato beetles and blister beetles sometimes feed on tomatoes. See Colorado Potato Beetles and Blister Beetles for controls.

SPOTS ON LEAVES. Early blight and late blight both begin as spots on leaves. Both can become very serious, and infected plants may die. Bacterial speck, bacterial spot, Septoria leaf spot, and gray mold also start out as small spots on leaves.

Early blight starts on lower leaves as target-like spots. Late blight begins as purplish or black water-soaked spots on lower leaves. White mildew may form on undersides of infected leaves. Late blight is a very serious problem both for home gardeners and commercial farmers, so it's important to diagnose which blight has attacked your tomatoes and take appropriate action. See Early Blight and/or Late Blight.

Black spots surrounded by a yellow halo are typical of bacterial speck; spots also show up on fruits. With bacterial spot, the spots are water-soaked, with an irregular shape.

To control bacterial diseases, pick off infected leaves. Switch to drip irrigation or soaker hoses for watering. Don't touch plants when wet. If the problem has been severe in the past, use copper sprays to prevent the spread of the disease. For the future, start a 3-year rotation of tomato-family crops, be sure to buy clean seed or healthy transplants, and disinfect all stakes and cages if you plan to reuse them (use a solution of 1 part household bleach to 12 parts water.)

Septoria leaf spot begins on lower leaves as gray spots that have a dark margin. Black dots may appear at the center of the leaf spots. When conditions are right, infected leaves rapidly die and fall off. Because this fungus does not cause symptoms on fruits, it's not as much of a concern as early or late blight. However, don't ignore Septoria leaf spot because if leaves become badly infected, yield will suffer. Control measures are similar to those for other fungal blights: See Early Blight for control information.

If gray mold is the cause of spot, a brownish gray fuzz will appear on the spots. See Gray Mold for controls.

WHITE POWDERY COATING ON LEAVES. Powdery mildew is a fungal disease named for the white coating that covers leaves and stems. Handpick infected leaves and start other control measures as described in the Powdery Mildew entry.

LEAF EDGES TURN BROWN. Bacterial canker is a bacterial disease that affects only tomatoes, but it can be a serious problem. Infected leaves curl and wilt. Cankers form on the stems, and the stems crack open. Dark spots with white edges appear on fruits. This bacterial disease usually reaches gardens through infected seeds or transplants, but once it is present, the spores can live for years in decomposing crop debris. If you're concerned that some of your plants are infected with bacterial canker, have the diagnosis confirmed by consulting an expert. Uproot and discard or destroy infected plants. Don't replant tomatoes in the spot where infected plants grew for at least 2 years.

MOTTLED LEAVES. Mosaic viruses cause mottling as well as leaf crinkling and abnormal growth. The viruses are transmitted by aphids and cucumber beetles. For preventive measures, see Mosaic.

LEAVES ROLL UP. Aphids or environmental conditions may be to blame; in the West, beet curly top virus is a possibility. Look for colonies of aphids on leaf undersides. Aphid-infested tomato plants usually produce good yields. Watch for natural enemies to come on the scene and control the problem for you.

If you can't find any other cause, the problems could have been caused by heavy rain or very strong sun (you may find that the rolled leaves are on the sunny side of the plant). The leaves may not recover, but plants should still produce fruit.

If plants are infected with beet curly top virus, the rolled leaves will become stiff and yellowed. Check leaf undersides—veins may turn purple (see Beet Curly Top).

CURLED AND TWISTED LEAVES. Herbicide drift onto tomato foliage can result in these symptoms. If you've sprayed any herbicides in other parts of your yard (or if your neighbor uses herbicides), that could be the cause. After several weeks, plants should resume normal growth.

LOWER LEAVES TURN YELLOW. Deterioration of older leaves can be a sign of lack of nutrients or of bacterial wilt, Fusarium wilt, or Verticillium wilt.

If bacterial wilt is the cause, young leaves may wilt also. Cut into a lower stem and you'll find brown discoloration. Uproot and discard infected plants. The bacteria can survive in soil for several years. To reduce future problems, enrich the soil with compost, and start a rotation of tomato-family crops. Be careful not to move infected soil from one bed to another in your garden. Some tomato varieties are tolerant of bacterial wilt.

Fusarium and Verticillium are fungal diseases that can persist in the soil (see Fusarium Wilt and Verticillium Wilt).

LOWER LEAVES APPEAR BRONZED. Spider mites and tomato russet mites can infest tomato foliage. Look closely at leaves for small dark specks and webbing on undersides. With russet mites, the leaves have a greasy shine but then dry up and fall off (see Mites for controls).

WHITE MOLD AT PLANT BASE. This is a symptom of Southern blight, a serious fungal disease in the southeast. Carefully uproot plants without knocking loose any of the fungal growth and discard or destroy them (see Southern Blight).

STUNTED PLANTS. Root-knot nematodes invade plant roots, causing yellowing leaves and poor growth. Uproot a plant and check the roots. Swelling and distortion of roots is a confirming sign. (Other types of nematodes also attack tomato roots in some parts of the country, but do not produce root swelling.) For controls, see Nematodes.

PLANTS WILT. Walnut wilt, white mold, or root rot are possible causes. Walnut tree roots secrete a substance called juglone that is toxic to tomatoes and many other kinds of plants. If there are black walnut trees near your garden, this could be the problem. In the future, try to plant tomatoes as far as possible from the trees.

White mold is a fungal disease that starts out as water-soaked spots on stems. White mold appears and then the branches wilt. To prevent future problems, see White Mold.

If roots are wet and brown, fungal root rot is to blame. To prevent this in the future, see Rots.

DRY BROWN SPOT ON BLOSSOM END OF FRUIT. Blossom-end rot is the result of a lack of calcium while fruit is forming, usually due to moisture stress. Fruits may be usable if you cut out the brown area. For prevention measures, see Blossom-End Rot.

FLOWERS DROP; NO FRUIT FORMS. Blossom drop has many potential causes: tarnished plant bug feeding, temperature extremes, too much nitrogen, moisture stress, or smog damage. Check first to see whether there are tarnished plant bugs present. These pests can be tough to control; see Tarnished Plant Bugs.

If weather or smog are to blame, the problem should resolve itself when the weather changes. For the future, cover flowering tomato plants if night temperatures below 60°F are predicted. There's no way to protect plants from high heat, but be sure that plants are well watered if night temperatures higher than 70°F or daytime temperatures higher than 85°F are predicted. As for smog damage, there's no remedy except to fight for pollution reduction in your region.

PUCKERED, SCARRED AREAS ON THE BOTTOM OF FRUITS. Catfacing can be caused by low overnight temperatures or tarnished plant bug feeding.

When nighttime temperatures drop below 60°F, it affects the flowers, and that results in catfaced fruit. The fruit will probably still be usable if you cut away the scarred parts. Don't try to set out large transplants too early, or blossoms can form when nights are still cold. During late summer, if a freak cold spell is predicted in your area, cover tomato plants with tarps or blankets overnight to keep the air around the plants warm.

If tarnished plant bugs are present, you may also see cloudy spots on the fruits, and fruit may be malformed. Rot often sets in. See Tarnished Plant Bugs for controls.

CRACKS IN FRUIT. Fruits crack when soil moisture level fluctuates, such as when a heavy rain comes after a dry spell. Harvest any cracked fruits immediately or rot will invade the openings. To prevent more cracking, mulch the soil. If heavy rain is predicted, pick as much fruit as possible before the storm hits, even fruit that is slightly underripe (it will ripen off the vine). If this is a common problem in your garden, choose cracking-resistant varieties.

WATER-SOAKED SPOTS ON FRUITS. Anthracnose or bacterial speck or spot are likely causes. Anthracnose is a fungal disease that shows up on ripe tomatoes. Centers turn black or start to rot. Infected fruits are usable if you cut out the bad spots, but they will go bad quickly after picking. See Anthracnose for prevention tips.

With the bacterial diseases, the spots may be sunken or scaly; on green fruits the spots may be ringed by a green halo. For the future, spraying copper-based fungicides when plants are young may help prevent the disease from spreading. Spray in the evening to avoid damaging leaves.

WHITE PATCHES ON FRUITS. If there is no other sign of disease except white or yellow patches on the fruit, the problem is sunscald. The areas may become water-soaked or dry and papery. Sunscald occurs when strong sun burns the skin of exposed tomatoes. This can happen if plants have lost leaves due to disease or if plants are overpruned. To prevent sunscald, don't prune plants late in the season, side-dress plants to boost foliage production, and take steps to reduce disease.

CLOUDY SPOTS ON FRUITS. The flesh underlying the spots will be hard and white, as a result of feeding by stink bugs (see Stink Bugs).

YELLOW AREAS ON RIPE FRUITS. Tomatoes grown in areas with lots of hot, bright sunshine are prone to this problem, which is called solar yellowing. It happens when tomatoes receive lots of bright sunshine and temperatures are higher than 85°F. Fruits are still edible, but flavor may be affected. To prevent the problem, reduce pruning in the future, and be sure plants don't suffer from moisture stress, which can reduce leaf growth.

GREEN SHOULDERS ON FRUIT. High temperature can cause this problem. The fruits will still be usable. Cut out portions that are too firm or unripe for eating. To prevent green shoulders, reduce pruning so there's plenty of foliage to shield developing fruit from the sun.

HOLLOW AREAS INSIDE FRUITS. Hollow spots result from poor pollination. Pollination may falter on cloudy days, during heavy rains, or when temperatures exceed 90°F or fall below 60°F. If the problem doesn't resolve itself when the weather changes, excess nitrogen in the soil may also be a contributing factor.

LIGHT-COLORED SPOTS ON FRUITS. Stink bug feeding leads to these cloudy spots, which stay green or white as the fruit turns red (see Stink Bugs).

If the spots are white with a yellow halo, they could be a symptom of gray mold. Infected fruits later turn brown and mushy (see Gray Mold).

HOLES IN FRUITS. Tomato fruitworms, tomato hornworms, and slugs and snails are possible culprits. Tomato fruitworms are the same pest as corn earworms; see Corn Earworms for controls. For hornworm control measures, see Tomato Hornworms. The best way to prevent slugs and snails from eating tomatoes is to stake or cage the plants to keep the fruit off the ground. For more controls, see Slugs and Snails.

TOMATO HORNWORM

The biggest caterpillar in the vegetable garden isn't necessarily the worst one. At 5 inches long, a full-size tomato hornworm is a startling sight, but it's rarely the cause of a crop failure. Hand-picking is often all that's needed to control hornworms.

At the start of the gardening season, the hornworm's parent emerges from the soil. This moth, called the five-spotted hawk moth, is also a large and striking insect. It feeds on nectar and lays eggs on the undersides of tomato-family crops and weeds. Eggs hatch as small green caterpillars that feed on foliage and grow alarmingly. After feeding for about a month, the caterpillars pupate in the soil. In some areas, there is only one generation per year, but in many regions, a second generation emerges to repeat the cycle. In fall, remaining caterpillars pupate, and the pupae rest in the soil over winter.

Watch for hornworms that have ricelike projections sticking out of their backs. These hornworms are the victims of parasitic wasps. The wasps laid eggs on these hornworms. Larvae hatched and fed inside the hornworm's body cavity. The white projections are the pupae formed by the wasp larvae. These hornworms won't live long, and the pupae will give rise to more parasitic wasps.

PEST PROFILE

Manduca quinquemaculata

RANGE: Throughout the United States and southern Canada

HOW TO SPOT THEM: Hornworms are very inconspicuous when small. Look for their dark green droppings on leaves or on the ground around plants. Try spraying plants with water to make the worms wriggle, and they may be easier to find.

ADULTS: Mottled gray moth has large, furry body with orange spots on sides of abdomen; wingspan up to 5 inches.

LARVAE: Green caterpillars up to 5 inches long with V-shaped white marks down the sides of the body and a black horn at the tail end; the related tobacco hornworm is similar but has seven white stripes down the side and a red horn. Occasionally, you may find a black tomato hornworm; this is just a color variant.

EGGS: Pearl-like yellowish green eggs laid singly on leaf undersides

Five-spotted hawk moth

Tomato hornworm

Hornworm parasitized by braconid wasp

CROPS AT RISK: Tomatoes, potatoes, eggplant, peppers

DAMAGE: Larvae chew on leaves; a large hornworm can consume several leaves in one day, leaving only the midribs behind; hornworms also chew into green fruits.

CONTROL METHODS

FALL CULTIVATION. In fall, turn the soil where tomato-family plants were growing to expose pupae to cold and birds.

HANDPICKING. Remove caterpillars from plants and squash them underfoot or dump them in soapy water. The horns are not harmful, and it's safe to handle these caterpillars. Leave parasitized caterpillars in the garden so that the natural population of braconid wasps can increase.

BIOCONTROL. Apply *Bacillus thuringiensis* when worms are 2 inches long or smaller. If worms are larger than this, handpicking will be more effective. Spinosad may be effective for hornworm control as well. See Pest Control for information on biological controls.

CONTROL CALENDAR

WHILE CROP DEVELOPS

- Check plants frequently for larvae and handpick those you find.
- Leave parasitized hornworms in the garden.
- Apply Bt or spinosad when caterpillars are small.

AFTER HARVEST

- Cultivate to expose pupae to cold and birds.

TURNIPS

Baby turnips are a great cool-weather crop for beginning gardeners. Turnips are easy to grow, and baby turnips mature quickly, so they work well as both spring and fall crops. Plus, baby turnips are surprisingly sweet, sure to delight those who've never eaten freshly harvested homegrown turnips. Try picking and cooking nutritious turnip greens, too. Turnips thrive in temperatures ranging from 40° to 60°F.

CROP BASICS

FAMILY: Cabbage family; relatives are arugula, broccoli, broccoli raab, cabbage, cauliflower, collards, horseradish, kale, kohlrabi, mustard, and radishes.

SITE: Plant turnips in full sun. For summer-sown seeds, shade the seedbed and new seedlings on hot afternoons.

SOIL: Turnips grow best in well-drained, rich soil that's been enriched with organic matter over time. If soil fertility is average to low, spread 2 inches of compost over the planting area and work it into the top 6 inches before planting.

TIMING OF PLANTING: Sow seeds for a spring crop 1 month before the last expected frost. In mild-summer areas, make succession sowings every 2 to 3 weeks throughout summer. In hot-summer areas, take a break. Sow your fall crop 8 weeks before the first fall frost (turnip seed germinates well even in hot soil). In mild-winter areas, you can sow seeds every 2 to 3 weeks throughout winter, ending in early spring.

PLANTING METHOD: Sow seeds directly in the garden.

CRUCIAL CARE: Once your summer-sown seedlings are a couple of inches tall, spread a thick layer of organic mulch to cool the soil.

GROWING IN CONTAINERS: Turnips will grow well in compost-enriched planting mix in containers at least 12 inches deep. Put the containers in full sun and monitor soil moisture. Water stress will harm the quality of the roots.

HARVESTING CUES: From aboveground, turnip plants don't offer any clues as to what's happening underground. To ensure you don't wait too long to harvest, on planting day make note of the expected maturity date on your garden calendar. When that day arrives, uproot one or two plants to check root size. Salad turnips are best when they're less than 2 inches in diameter. Storage types grow as large as 6 inches across.

SECRETS OF SUCCESS

SOW MORE THAN ONCE, MORE THAN ONE KIND. Turnip varieties range from crisp, fruity salad types that are delightful for eating raw to traditional storage turnips that hold in the soil like carrots for harvest throughout fall and into winter. Try more than one type to find your favorite. Choose varieties with different maturity dates to spread out the harvest. Also, be sure you sow more than once because it's good insurance against problems. Occasionally, pests such as flea beetles and cabbage maggots seriously damage turnips, or a late cold spell in spring can cause plants to bolt prematurely. If you make succession sowings, chances are that at least one batch of your turnips will escape pest damage and bolting. In general, fall plantings are less prone to pest damage than spring crops.

PICK PLENTY OF GREENS. Growing turnip greens is a cinch. It's a natural part of raising turnips because it's important not to crowd turnip plants. You'll need to thin when the young plants are about 4 inches tall, leaving them 2 to 4 inches apart (depending on the mature size of the roots). Use the thinnings as greens. After that, simply pick individual leaves as needed. Plants need to retain at least four leaves to provide a continuous source of energy for growth—

you can take the rest each time you pick. If you like lots of turnip greens, plant a patch especially for harvesting leaves. Broadcast seed lightly over a wide row, and don't thin the plants. Instead, use a cut-and-come-again technique to harvest the greens. In spring, stop cutting when the weather turns hot because heat affects the flavor of the greens. Pull out the plants and sow a cover crop or a warm-season vegetable in their place.

HANDLE THE HARVEST ACCORDING TO THE HEAT. In most areas, it's crucial to harvest spring turnips young, before summer heat ruins their flavor. If possible, keep your spring crop in the ground until it reaches the expected days to maturity. However, if temperatures rise above 80°F, dig your spring crop, even if the roots are very small. Yield will be lower, but flavor will be fine. If you delay, you may end up with "skunky" turnips that no one will want to eat.

Some storage turnip varieties require up to 80 days to reach mature size, which means you must plant your fall crop in midsummer. Luckily, while maturing roots need cool conditions, turnip seed germinates well in hot soil. The seedlings will withstand summer heat as long as you water them well. When fall arrives, turnip roots will hold well in cool soil, allowing you to spread the harvest over time. In fact, the sugar content of the roots rises as temperatures drop, improving the flavor.

PREVENTING PROBLEMS

BEFORE PLANTING

- Plant annuals and perennials with small flowers that attract beneficial insects that

Turnip   **Rutabaga**

Growing needs of both crops are very similar, but rutabagas take a little longer to mature. Plan to plant rutabagas about 3 months before your first fall frost date.

prey on aphids and other pests (see Beneficial Insects).

AT PLANTING TIME

- Cover seedbeds of spring crops with row cover to prevent flea beetle damage and block cabbage maggot flies.
- Insert wireworm traps into the soil near seeded rows of turnips.
- Apply parasitic nematodes to reduce wireworm populations.

WHILE CROP DEVELOPS

- Spray foliage that isn't protected by row covers with a garlic/hot pepper spray or kaolin clay to deter pests (but not if you plan to harvest the greens for eating).

TROUBLESHOOTING PROBLEMS

When planted in good soil, turnips rarely suffer from disease problems. A few pests are problematic, especially in spring. The most common turnip problems are listed below. If you see symptoms that aren't described here, refer to Cabbage for a full listing of pest, disease, and cultural problems.

LEAVES CURL. Check leaf undersides and around the crown for colonies of small round aphids. Use a strong spray of water to knock aphids off plants. See Aphids for more controls.

SMALL HOLES IN LEAVES. Flea beetles will congregate on young turnip plants in spring, riddling the leaves with small holes (see Flea Beetles).

HOLES IN LEAVES. Leaf-eating caterpillars are responsible; handpick caterpillars from the plants. For other controls, see Cabbage Loopers, Armyworms, and Imported Cabbageworms.

PLANTS FORM FLOWERSTALKS. Turnips are biennials. In spring, if a late cold spell comes along, the growing plants may think they've survived winter, and their natural response is to flower. Plants that bolt won't produce harvestable roots, so pull them out and compost them. In the future, start your spring crop later, or make succession sowings in spring.

LEAVES WILT. If healthy-looking leaves wilt during the day, pull up a wilted plant and check the root. If the root is misshapen and small, it's infected with clubroot fungus. Infected roots won't be palatable; dig and destroy them. To prevent future problems, see Club Root.

If the roots are normal, the problem may be harlequin bug feeding, especially in hot weather. Check the leaves for orange-and-black bugs. See Harlequin Bugs for controls.

PITHY ROOTS. Turnips turn pithy inside if they're left in the soil too long in spring. Exposure to heat also leads to pithiness. In the future, plant a little earlier and harvest roots on the small side (1 to 2 inches diameter).

TUNNELS IN ROOTS. Wireworms can be a serious problem in turnips, especially in new gardens converted from lawns. Look for leathery yellowish brown worms in the tunnels; these are wireworms (see Wireworms for controls).

Cabbage maggots also feed in turnip roots; the maggots are white and about ¼ inch long (see Cabbage Maggots).

VERTICILLIUM WILT

Verticillium is the V in tomato varieties labeled VFN resistant (F is Fusarium and N is nematodes). This fungus also is a serious problem in eggplant; less so for peppers, potatoes, and some other crops. It's widespread in garden soil, and in cool, wet conditions, the spores invade crop roots through small wounds. The fungus sends its hyphae through the vascular system, blocking water and nutrients from reaching plant tops. Thus, temporary wilting is a common early symptom. The fungus also produces a toxin that injures plant tissue.

Verticillium spores persist in the soil for up to 10 years, so if it's present in your garden, it's nearly impossible to eradicate. Instead, follow a strategy of choosing resistant varieties when they're available, promoting vigorous crop growth, and building populations of beneficial soil fungi.

RANGE: United States and Canada

CROPS AT RISK: Tomatoes, eggplant, peppers, potatoes, okra, cucumbers, melons, and occasionally other crops

DESCRIPTION: One or more branches wilt in the middle of the day but recover later on. In the following days, the branch or branches wilt and don't recover, even if you water. On tomatoes the first symptoms may be slow growth and yellow patches on lower leaves. Leaves may turn yellow and drop off. Eggplant plants infected while young may eventually die. Cut a stem (from the base of the plant) in half down the middle, and you'll see dark streaks where the fungus has blocked the vascular system.

Fusarium wilt produces similar symptoms, but it's usually a problem in hot weather rather than cool (see Fusarium Wilt).

FIGHTING INFECTION: Feed and water plants to avoid stress, but be careful not to saturate the soil with water. Mulch with compost or drench soil with compost tea to encourage beneficial soil fungi. If plants are infected before fruit set, they probably won't produce a worthwhile harvest. Uproot and dispose of them right away. Tomato plants sometimes don't develop symptoms until they've formed green or even ripe fruit. Harvest all fruits right away so you can destroy the plants and limit the spread of the fungus. The fruits will be edible (green fruit can ripen after picking).

GARDEN CLEANUP: Dispose of infected crop residues or bury them deeply in the soil.

NEXT TIME YOU PLANT: Before planting, add compost to the soil to boost populations of beneficial soil fungi. Take steps to control nematodes because the feeding damage they cause leaves roots more vulnerable to Verticillium infection (see Nematodes). Choose resistant varieties of tomatoes, and plant certified disease-free seed potatoes. Solarize infected beds during the hottest part of summer (this helps by killing nematodes as well as fungal spores).

CROP ROTATION: Plan a 3- to 4-year rotation of all Verticillium-susceptible crops if possible to reduce the concentration of the fungus in the soil.

WATERING

Understanding how soil and plants use—and lose—water reveals the answers to many watering problems. Once you know how to interpret soil moisture levels, you'll know when to water. Then you can tackle problems such as what to do if your water supply is limited, how to avoid watering small plants too heavily, and how to keep your garden watered when you are away.

WHEN AND HOW MUCH TO WATER

It takes experience to know how much water is required to wet the soil in your vegetable garden to a particular depth, such as 6 inches or 1 foot. The amount of water that soil can absorb depends on its texture, its structure, the amount of organic matter it contains, the volume of roots in the soil, and other factors. Water passes through sandy soil quickly, so it needs less water at a time but needs it more often. It may take more water to wet clay soil, but the soil will hold the water longer.

If you know what type of soil you have, perform a squeeze test to determine when to water. Pick up a handful of soil and gently squeeze it in your palm.

- If you have sandy soil, it's time to water when a handful of soil forms a clump when squeezed but falls apart when lightly touched. Apply ½ inch of water. If the soil is so dry and crumbly that it won't form a clump at all, your plants are probably suffering water stress. Apply 1 inch of water.
- For a loamy soil, if the soil forms a clump that falls apart when touched lightly, it's time to water. Apply 1 inch of water. If the soil is powdery and barely forms a clump, it's past time to water. Apply 1½ inches.

- When it's time to water clay soil, the clump will crumble a bit, but you should be able to spread a ribbon of soil between your thumb and forefinger. Apply 1½ inches of water. If clay soil has dried out excessively, it will be hard, possibly with surface cracks, and you won't even be able to scoop up a handful. Apply 2 inches of water.

SUIT WATERING TO PLANT SIZE

Your choice of whether and how much to water also depends on the type and size of crop. Shallow-rooted crops, such as lettuce, usually

HOW MUCH IS 1 INCH OF WATER?

For 100 square feet of garden (a bed 4 feet wide and 25 feet long), it's about 60 gallons. If you're using a sprinkler, you can measure how much water you've applied by setting out several empty cans (such as cat food cans) and watching to see how much water accumulates in the cans. Don't use just one can because your sprinkler probably doesn't apply equal amounts of water at all spots.

If you use a soaker hose, check the information on the product label; it may tell you about how much water per hour the hose emits.

need watering more often, while deep-rooted crops like tomatoes (which have roots that extend 2 feet below the soil surface) can go longer.

Another way to check soil moisture quickly is to dig down or just poke your finger a few inches into the soil. For seedlings, you should not allow the soil to dry out more than ½ inch deep. But once plants pass the seedling stage, allow the top 2 inches of soil to dry out before you do the squeeze test to decide how much to water.

NOT ENOUGH WATER

In arid climates, planting vegetables in sunken beds helps to conserve moisture, both by efficiently capturing rain and irrigation water and by protecting plants from drying winds.

Arid-climate gardeners also can capture the rain that falls during winter, when your garden is dormant, by spreading black plastic across a raised bed that has two small trenches along its length. Slits in the plastic allow water to drip through all winter and be absorbed by the soil. In spring, cut planting holes in the higher surface of the bed and set transplants in place.

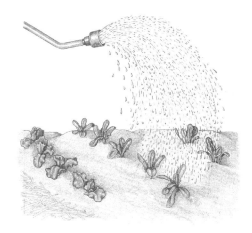

When you water young seedlings, mimic light rainfall by directing the spray in a gentle arc up and out over them.

WATERING WITH A HOSE

Many of us water our gardens with a handheld hose, and when municipal water restrictions are in place during droughts, it may be the only watering method permitted.

To avoid problems when you hand-water with a hose, take off the trigger grip nozzle on your hose—many a watering mistake is due to an overenthusiastic grip. Instead, designate a nozzle with a fan spray for the vegetable garden, and if you own one, use an extension handle. Many

DON'T RECYCLE WATER FOR VEGETABLES

Some municipalities allow use of gray water (water recycled from household uses other than toilets) in gardens, but don't use it in the vegetable garden. Gray water may contain bacteria that could be harmful if ingested. As with other sources of exposure to bacteria (from manure or compost tea), the risk is probably low, but unlike manure and compost tea, there are no guidelines to follow to ensure safe use.

Limit gray water use to ornamentals, and be sure you are complying with any local ordinances regarding it.

Be cautious as well about using water from your aquarium to water vegetable plants. A recent study in Australia traced cases of fever and diarrhea in children to water in home fish tanks that contained drug-resistant strains of salmonella bacteria.

Homemade hose guides protect your plants as you drag your hose into the garden.

watering wands have a setting for "shower" or "gentle shower," and those would be best to use.

Use an adjustable breaker (a small lever that enables you to turn the flow of water on and off without going to the faucet) so you can ease the setting to just the right pressure to imitate gentle rainfall. Position the head so it's a few inches above the soil and direct the water up in an arching stream. The goal when watering is to deliver 1 inch of water to the plants (see "How Much Is 1 Inch of Water?" on page 428).

CONTROL WHERE YOUR HOSE GOES

A hose dragged across a vegetable bed can flatten or break the foliage of young vegetable plants. Prevent catastrophes by fitting your garden with hose guides at strategic spots. Commercially made guides are available at garden centers, but it's easy to devise your own:

- Pound in a stake and set a clay pot over it.
- Place empty plastic pots at the critical corners and fill them with rocks (be sure the pots have drainage holes in the bottom).
- Pound in two stakes close together and insert the hose between them.

WATERING RESERVOIRS

If you will be away for only a few days, you can implant watering reservoirs around your garden, especially in crucial spots such as beside tomato plants and at the center of squash hills, so you don't need anyone to come in and water for you. Set up a soda bottle reservoir as shown below. Other good containers for this purpose include large, wide-mouthed metal cans (such as juice cans or institutional-size food cans), plastic gallon jugs, or plastic 1-gallon nursery pots sunken into the soil so only the top few inches stick out. The key is to plan ahead and dig holes for the reservoirs at planting time. Once the plants are growing, digging deeply to install a reservoir will damage roots.

Watering reservoirs are also a time-saver for watering container plants when you are home.

Punch holes in a soda bottle and bury it upside down near plants at planting time. Cut an opening in the exposed part to fill it.

WEATHER

Weather often outranks pests and diseases as the most dramatic problem for vegetable garden-ers. Pest and disease problems follow predictable patterns from year to year, but the weather breaks out of its expected course all too often. Learning to watch the forecast and plan your re-sponse strategy in advance will help you protect your garden from weather problems ranging from hot spells to strong wind, drought, and hail.

UNDERSTANDING WEATHER FACTORS

When it comes to weather and climate, it's important to understand some basic terminol-ogy and to learn about the "normal" weather patterns in your region.

FROST

Frost happens when water vapor in the air crystal-lizes as a thin dusting of ice on surfaces. Frost is related to both temperature and humidity. Light frost happens when the overnight low temperature drops to 32°F or just below, and is enough to kill "frost-tender" plants such as beans. A hard or kill-ing frost happens when the low temperature stays below 32°F for many hours, or when the mini-mum temperature reaches 24°F. Hard frost destroys most annual vegetable crops, but some tough crops can survive killing frosts under pro-tection of row cover, and a few—such as Brussels sprouts—withstand even lower temperatures.

FROST DATES

Government agencies in the United States and Canada track temperature data at thousands of locations. It's easy to find out the average date of the last expected spring frost and the first expected fall frost for your local area. Check with your Cooperative Extension Agent or use a Web browser to find the information (try typing in "average frost dates"). Keep in mind that these dates are averages. In any given year, your actual last spring frost may happen a few weeks earlier or later than the average date.

HARDINESS ZONES

The USDA tracks weather data over time, and from its long-term records on minimum winter temperature, it has compiled a map that divides North America into hardiness zones. This map is shown on page 472; Zone 1 has the lowest mini-mum temperature, Zone 11 the highest. Horti-culturists rank perennial plants (plants that live more than one year) by their winter hardiness. For example, asparagus is hardy to Zone 4, while artichokes are hardy only to Zone 7.

Knowing your hardiness zone is important even though you'll grow most vegetable plants as annuals. That's because, in general, the lower the Zone number, the cooler the weather is dur-ing summer, too. There are exceptions to this rule; for example, some areas along the West

Coast of the United States and Canada have a high hardiness ranking but also enjoy mild summer weather.

The USDA created the Hardiness Zone Map in 1965 and revised it in 1990. The revision changed the boundaries of each zone, generally toward the south. It's interesting to compare different versions of the maps. You may find that in 1965, your garden was considered Zone 5, but in the 1990 map, you're located in Zone 6. The Arbor Day Foundation released a map in 2004 that includes the 15 years of climate data since 1990, and their map also shows a southward shift in zone boundaries.

Many scientists attribute this warming trend to human activity, such as increased emission of carbon dioxide. Others say that the cause of the shifts are unknown, or are part of natural long-term patterns of climate change.

COMMON WEATHER PROBLEMS

Weather problems vary from place to place and year to year. The most common problems are lack of rain, high heat, wind, and unexpected cold and frost. In some cases, you can protect your garden from these problems, but sometimes the best defense is to take an experimental approach to planting. Add an extra-early or late planting of crops, so that if one planting succumbs during a weather crisis, another may survive. Follow the forecast daily, and stockpile materials that will help you shelter plants quickly when a heavy storm or early frost is forecast.

"Plant" a branch from a juniper or other evergreen in a garden bed as sunshade for a young transplant.

TOO LITTLE RAIN

Drought can strike anywhere in the United States and Canada. Building the organic matter content of your soil and covering the soil with mulch are two of the best defenses against drought, but inevitably, if there's no rain, the soil will dry out. See Soil and Mulch to learn how to build your soil's moisture-holding capacity. Consult Watering for smart watering strategies.

HOT SPELLS

Young transplants often need some help to adapt to life in the garden. Even if they've been hardened off, they still go through a transition while they're sending new roots into the soil. If the weather is unusually hot, cold, or dry during that time, the plants struggle. You can help them out by watering if it's dry, of course, but it's also important to shield them from cold and intense sunlight. Fortunately, you can do either with items you have on hand. To shelter plants from sun, try one of the following tricks.

- Pound in a wooden or metal stake near the plants and tie an old umbrella to the stake.
- Set laundry baskets (buy a few cheaply at thrift stores and dollar stores) upside down over the plants and weight the baskets with a rock or brick.
- Cut a branch from an evergreen shrub or tree and stick the cut end into the soil on the west side of the plant (see the opposite page).
- For small seedlings that are just emerging, use basket-weave-type plastic flats to provide filtered shade (see page 434). Simple upend the flats over the seedlings. You can still water the seedlings with the flats in place.
- To protect a whole bed of a summer planting of cool-season plants, pound in stakes at two corners of the bed and prop a section of picket fence against the stakes at a 45-degree angle (as shown on page 249).

- To create a strip of garden area with afternoon shade throughout late summer, plant a triple row of tall sunflowers along the western edge of your garden in late spring.
- Another way to shade a large group of plants is to buy shade cloth especially designed for garden use. It's available in different densities, and the lightest grade is probably what you'll want to buy. It comes in rolls that are 6 feet wide; a 20-foot roll costs about $20. Drape the fabric on some type of framework—don't let it rest directly on the plants.

To learn how to shelter young plants from a sudden cold snap, refer to Season Extension and Row Covers.

STRONG WIND

Air circulation in the garden is beneficial because it helps prevent disease. But gusting winds can sap soil and plants of moisture. In extreme cases, wind can topple trellises and other plant supports.

LOCAL CLIMATE QUIRKS

The more you learn about your local climate patterns, the better you'll be able to manage your garden. Each region of North America has distinctive climate features. For example, in the Southwest, hot winds in spring can quickly turn new transplants to toast unless you screen the plants from the drying air. Fall can be one of the best seasons for gardening in terms of temperatures, but it tends to be the driest time of year.

Gardeners in the Rocky Mountains will find that conditions differ dramatically over short distances depending on altitude. Gardeners at

high elevations will enjoy more rainfall but have to contend with earlier and later frost.

If you're new to gardening, ask fellow local gardeners what weather quirks you should be prepared for. Check the listings on page 458 for books, magazines, and newsletters that are customized to a particular region.

Every vegetable gardener should install a rain gauge and a thermometer in the garden, too. The readings you take in your own garden will be more relevant than those from your local radio or TV station.

Ideally, your vegetable garden is nestled into your yard in a spot with full sun and moderate air circulation, with landscape plantings at the boundary of your property that buffer the garden from the full force of wind. If you're still working on establishing that ideal situation, here is a tip for protecting your plants from too much wind: Add a rock wall along the appropriate side of your garden to block prevailing winds. Rock walls are better than solid walls because they allow some air to pass through, but they break the blasting force of the wind. Plus, rocks absorb heat during the day and release it at night, which can benefit heat-loving crops like peppers.

STORMS AND HAIL

A summer storm with high winds, drenching downpours, or hail can be a garden disaster. Heavy rain can wash out small seedlings and break or flatten delicate crops. Some parts of the country rarely experience hail, but it's not uncommon in the Midwest during thunderstorms. Hail can bruise plant tissues or tear through leaves, leaving the plants very vulnerable to disease.

Take as much advance action as you can when a serious storm is predicted at the height of gardening season. Harvest as much as you can. Cut back all your leafy greens to 1 inch above ground level. Pick ripening tomatoes, peppers, eggplant, and summer squash. Cover tomato cages with a "lid" of hardware cloth with ½-inch openings to deflect any large hailstones. Place some empty cardboard boxes over plants that most need protection, and weight the flaps with rocks. Don't use blankets or tarps as covers because heavy rain or hail could turn them into deadweights that will crush crops.

If your plants suffer storm or hail damage, hope that a spell of cool but dry weather will follow; it will give the plants the best chance to recover without disease problems.

For rows or beds of new seedlings, flip seedling flats upside down over the plants to offer filtered shade from hot sun.

WEEDS

Growing vegetables without growing weeds is one of the biggest challenges for organic gardeners. Because we carefully build rich soil, our gardens are a welcome environment for plants of many kinds—including weeds. The trick is to suppress the weeds without hurting our crops or the soil.

The secrets to success with weed control are simple: Stop weeds early, keep the soil covered as much as possible, learn how to fight persistent weeds, and never let weeds set seeds.

Do weeds have a positive aspect? Weeds do increase the diversity of a garden, and some produce deep roots that mine the subsoil for nutrients. However, the benefits of weeds usually don't outweigh the negatives: They compete with crops for water and nutrients, they host pest insects and diseases, and they restrict air movement. Perhaps the best way to view these plants as a benefit is to consider them an ever-present reason to enjoy a half hour in your garden, doing a little light weeding.

SIDESTEPPING WEED PROBLEMS

Almost all gardeners lose a crop sooner or later because they fail to weed during a critical stage, or they till a weedy bed that they shouldn't, or they let weeds go to seed. These are easy mistakes to make, especially when your garden is just one of the many activities that fill your life.

The good news is there are strategies that help busy gardeners cope with weed problems, so that you never face the depressing task of trying to uncover young vegetable plants hidden

under a jungle of weeds. You'll combine the following tactics:

- Limiting the food supply for weeds
- Pregerminating and killing weeds before you plant a crop
- Cultivating quickly and shallowly when weeds are small
- Mulching and planting cover crops to suppress weeds

CONTROL THE FOOD SOURCE

Standard soil preparation instructions for organic gardeners often read something like this: Spread 2 inches of compost over the surface of the bed and work it into the top few inches of soil. This technique creates a beautiful soil environment for extensive root growth. However, if your soil contains a large bank of weed seeds, it's also an invitation to weeds to sprout en masse because you're turning the soil—which brings weed seeds near the surface—and putting food (compost) within easy reach.

If you know that your soil is heavily stocked with weed seeds, try an alternate approach. Don't spread compost far and wide. Instead, disturb the soil surface as little as possible. If you plan to sow seeds, sow in narrow rows rather than wide rows or by broadcasting. Open a fairly deep furrow (2 or 3 inches), and fill it with compost to the depth at which you want to sow the seeds. Put your seeds in the furrow, and cover them with compost to soil level. This concentrates the nourishment right where the seedling roots can reach it and keeps it away from the potential weed population. If you're planting transplants, dig the planting hole extra deep and wide, and mix in compost as you plant and refill the hole. (If the plant ends up on a little hill, that's fine—just mold a low berm around it to retain water. This also helps prevent weed competition by keeping the bed surface dry—all the water soaks in right around your plants.)

With this approach, you'll need to side-dress every few weeks as the crops grow because the limited supply of compost won't last too long. Or, once the soil is warm enough, spread compost on the soil surface and top it with a weed-seed-free mulch such as chopped leaves.

Add compost to each planting site for transplants. A low berm around the transplant blocks water from running off, and that reduces weed seed germination.

WEED FIRST, SOW LATER

Once the weed seed bank is at a more reasonable level in your garden, you can return to the soil preparation method of spreading compost and working it in. Be smart, though, and outwit weeds with a trick called the stale seedbed method. After you've prepared the bed, water it well, then let it sit for 2 weeks. A hearty crop of weed seedlings will sprout. At that point, you can kill the weeds by hoeing gently just below the soil surface or by using a flame weeder (see "Flame Weeding" on page 438). Your goal is to disturb the soil as little as possible. After the hoed or flamed weed seedlings are dead, remove them from the soil surface and plant your crop seeds. Since you've already defeated the main flush of weed competition, your crop will have a much better chance to establish itself.

CULTIVATE WHEN IT COUNTS

In spring, to encourage the soil to warm up and dry out, we deliberately leave the garden uncovered. What happens when soil is bare? Weed seeds germinate and rhizomes sprout, and many of them establish themselves more quickly than crop plants will. There's a critical window of competition between crops and weeds that starts about a week after a crop emerges and lasts about 1 month. During that month, if you don't handicap the weeds, chances are good that they'll outpace your crop, and the crop will never attain its full size and productivity.

Fast and Easy Cultivation

If you love to hand-weed, you'll enjoy early season weed control. Invest in a thick kneeling pad

ABOUT ORGANIC WEED KILLERS

Insecticidal soap sprays that kill soft-bodied insects can also damage plant tissue, and garden product manufacturers have come up with herbicidal "soap sprays" as well— formulations of fatty acids that are strong enough to kill tender plants. Other organic herbicides have a vinegar base.

These herbicides work well for controlling weeds in the cracks along walkways and at the edges of patios, but they're not a good choice for the food garden. And of course, never use chemical herbicides in the vegetable garden. To safeguard your crops, your soil, and your health, stick with hoeing, mulching, and hand-weeding.

or a gardening stool to increase your comfort, and gently remove small weeds in crop rows and around seedlings. Be sure that the plants are dry before you weed (to avoid the risk of spreading disease spores), but to ease the job, weed when the soil is moist. If necessary, water the area the day before you weed.

If possible, weed in the morning on a sunny day, and leave the seedlings to wilt and die on the soil surface. Come back later or the following day and rake up the seedlings for your compost pile (seedlings left in the garden sometimes attract critters and diseases that feed on dying plant tissue, and then they may also spill over and attack your crop seedlings).

Take along an old screwdriver on your hand-weeding forays. It's handy for plunging into the soil to loosen up weeds with minimal disturbance to nearby crop seedlings.

USING A WEEDING TOOL. For weeding, try to find one long-handled hoe for clearing out weeds quickly, and also a short-handled tool for fine work closer to plant stems. Preferences vary when it comes to the design of a hoe blade, but it's worth it to buy a quality tool, whatever style you choose. And if back problems or arthritis in your hands bother you, investigate tool handle extensions that allow you to work without bending, and tools with padded handles that put less stress on your hands. For suppliers, turn to Resources for Gardeners on page 456.

When you use a hoe or hand weeder, keep the blade parallel to the soil surface. Mentally picture a line half an inch below the soil surface, and pull your hoe blade gently down the bed following that line. Your goal is to cut through weed stems or pull weed roots loose from the soil without flipping or tossing soil. That's because when you disturb the surface soil, you expose more weed seeds to light and stimulate them to germinate.

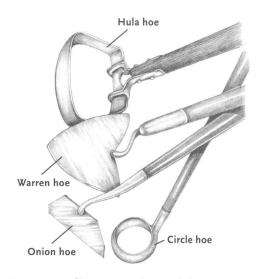

Try a variety of hoes to see which works best for you.

FLAME WEEDING. A flame weeder can be a useful part of your weed control efforts. You can use a flame weeder to kill weeds between rows or hills of seedlings, but it takes precise technique to avoid singeing crop seedlings. Safety is always the most important consideration when using a flame weeder. Your goal is to pass the flame very lightly over the bed—the flame needs to come in contact with a weed seedling for just one-tenth of a second to kill it. You shouldn't see the weed seedlings turn black as you work; if you do, you're applying too much heat. The flame could be heating the soil, too, which damages soil organisms and crop roots. If you've done the job right, the weeds will begin to droop and wither about 24 hours after being flamed, and a day after that they will be dead.

Observe these precautions when you use a flame weeder:

- Always station a 5-gallon bucket of water or a hose nearby (be sure the water supply is turned on—use a breaker at the nozzle so you can release the flow in an instant).
- Never use a flame weeder on a garden bed that is covered with mulch.
- Never use a flame weeder near any garden bed that is covered with dry mulch such as newspaper or wood chips.
- Never use a flame weeder near beds that contain dried-up plant residues.
- Never use a flame weeder on a windy day.

COVER THE SOIL

A good layer of mulch can very effectively prevent the return of weeds once you've prepared a garden bed (or after you've weeded). There are plenty of choices of mulch, from materials you have on hand—such as newspaper and grass clippings—to purchased plastic or organic mulches.

In some regions and circumstances, mulching the garden can lead to pest problems, providing hiding places for such undesirables as squash bugs and earwigs. And at the start of the season, leaving the soil uncovered to warm the soil is the most important consideration. But as soon as you can, put mulch in place. See the Mulch entry for details on mulch materials and how to manage mulch.

Another way to cover bare soil is to plant cover crops. These crops not only deny weeds the opportunity to grow but they also build soil structure, prevent erosion, and more. If you haven't been planting cover crops in your garden, start today. See Cover Crops to learn how.

Covering the soil with clear plastic to cause heat to build up and kill weed seeds is another way to fight weeds. This technique is called solarization and is discussed in detail on page 186.

KEEP AFTER SEED HEADS AND HOST PLANTS

Weed control is less critical at the height of the gardening season because most crops are large enough to withstand some weed competition. However, don't forget about weeds. Throughout the season, be on the watch for weeds that are going to seed and weeds that host serious vegetable crop diseases (particularly weeds in the cabbage and tomato families).

Weeds are prolific seed producers. A single pigweed plant can bear up to 250,000 seeds.

Stopping deposits to the seed bank in the soil is one of the most effective ways to reduce future weed problems. Weeds also wreak havoc if they break out in a nasty disease that can spread to crop plants. For example, nightshade can suffer from late blight, a devastating fungal disease that can wipe out tomato and potato crops.

Summer is not always a fun time to weed, but at the least, tour your garden every few days with a garbage bag and sharp garden clippers in hand. Cut flowerheads of weeds directly into the garbage bag (don't risk composting them, in case some seeds are already maturing in the flowerheads). You can come back later in the season when it's cooler to dig out the foliage and roots. If you spot weeds that are disease hosts, cut them off at ground level and stick them in the bag, too. (Never work around disease hosts when the garden is wet because disease spores spread easily in wet conditions.)

BATTLING BIGGER WEEDS

When weed seedlings escape your notice, they grow up to become full-size weeds. In most, but not all, cases it's a good idea to pull out such weeds, roots and all. But getting rid of big weeds can be tricky, depending on the type of roots they have. If a particular type of weed is fairly widespread in your garden, first pull up one to see what the roots are like.

TAPROOTS. For weeds with a long, sturdy central root, such as dandelion, insert a dandelion fork or a narrow trowel vertically into the soil alongside the weed. Pull gently on the top

and jab with the tool. You should feel the release as the root breaks or is cut well below the soil surface, and the root will pull freely from the soil. It's easiest to pull taprooted weeds a day or two after a soaking rain or watering.

FIBROUS ROOTS. When the soil is moist, it can be fairly easy to pull out fibrous-rooted weeds simply by grasping the plant at the crown, where roots and topgrowth meet, and pulling upward. Use a hand fork to loosen the clump first if it resists—don't injure your hands or wrists by overpulling.

BULBOUS ROOTS. Uproot these (wild onion, or onion grass, is an example) as you would fibrous-rooted weeds, but don't try to shake the soil free. The clump may include small bulblets that would fall free and then resprout in the bed. Carefully put the weeds, roots, and clinging soil into a hot compost pile to kill weed seeds. The heat will kill the bulbs, too. Otherwise, put these weeds into a discard pile or throw them out with household trash.

RHIZOMES. If your problem weed has lots of long, somewhat fleshy roots that run horizontally at or below the soil surface, like those of many grasses, proceed with caution! These roots are specialized stems called rhizomes. Even a small piece of rhizome left behind in the soil when you pull the plant has the ability to resprout roots and leaves. Sometimes the best course of action with rhizomatous weeds is to repeatedly cut back the tops at ground level, leaving the roots undisturbed. It's a long process, but eventually you can starve out the roots and rhizomes by removing the weed tops.

WEED EMERGENCIES

Sometimes weeds take over the garden much more quickly than you expect. Perhaps you had an extra busy week at work, or came down with a summer cold that kept you out of the garden entirely. If that particular week happens to be warm and wet, by the time you return to the garden, the weeds have taken over—everywhere. When a weed emergency like this happens, how do you get your garden back?

DECIDE WHAT TO SAVE

First, decide which crops are most worth saving. These may be crops such as melons or peppers that take the longest to produce. Or you may choose the vegetables you most like to eat. Weed out those beds (refer to the guidelines above to see which roots to uproot and which to cut off at ground level). Next, put down mulch without delay. An ideal choice is newspaper topped by straw or chopped leaves, to ensure that you smother any weeds that resprout from roots left in the soil.

SACRIFICE TO SAVE

Next, take stock of how much time and energy you have to deal with the rest of your garden. If you can't save the whole garden, don't try. Choose a bed or two for sacrifice to the garden gods. In these beds, you'll quickly cut everything to ground level—weeds and crops—and then cover the bed (as described below) while the plant material breaks down.

Before you bring in the lawn mower or string trimmer, though, check the beds for weeds that are going to seed. Use garden clippers to snip off all seed heads and discard them in plastic bags with your trash.

As you work, if you notice that a particular weed seems to dominate in a bed or throughout your garden, take time to identify it precisely. Refer to "Ten Weeds Worth Recognizing," below, for specific recommendations for bringing that weed under control, because you're bound to see it again in your garden in the future.

With weed seeds removed, cut all plants down to ground level. Cover the bed with black plastic, heavy cardboard, or an old wool (not synthetic) blanket, and weight down the edges. Later in the season, when you have time, you can uncover the bed and turn under the plant debris. Then wait a couple of weeks and if there's time left in the season, you can plant a fall crop.

TEN WEEDS WORTH RECOGNIZING

In many gardens, one or two species of weeds are predominant. It's worth it to identify these dominant weeds to the species level so that you can plan the most effective approach for bringing them under control.

BERMUDAGRASS (*Cynodon dactylon*)

RANGE: Southern two-thirds of the United States, occasionally a problem farther north

DESCRIPTION: Also called wiregrass and devilgrass, bermudagrass grows fast and is a particular problem in Southeastern gardens. This perennial grass has a network of creeping stems with thin, gray-green or bluish green leaves.

Bermudagrass

CONTROL STRATEGY: Dig out bermudagrass early and late in the season when you can search out and pull up its extensive roots without damaging crop roots. Let the plants dry in strong sun and then compost them. Solarizing a bed infested with bermudagrass will weaken the root system. Follow up by planting a cover crop to overwinter, and that will further reduce, although not eliminate, the problem. Mulching around crops helps, but bermudagrass often finds a way to pop up through mulch and then spread across the mulch surface.

BINDWEEDS (*Convolvulus* spp.)

RANGE: United States and Canada

DESCRIPTION: The pretty trumpetlike flowers and heart-shaped dark green leaves of bindweed, or wild morning glory, might tempt you to allow it to ramble in your garden, but don't be fooled. With roots that plunge down 10 feet and spreading vines, bindweed can be a nightmare in the garden.

CONTROL STRATEGY: Pulling bindweed seedlings is okay because the full root will come up when you pull it. However, once a plant has more than five leaves, the brittle roots will probably break as you pull them out, and pieces left behind will regrow. Mulching will help to weaken bindweed but won't control it fully. Solarizing and tilling aren't effective; flaming may weaken a stand over time.

If only one bed or area of your garden is infested with bindweed, you can eradicate the weed by covering the area tightly with black plastic for 2 years.

Another approach is to weaken the stand first, then mulch to suppress regrowth. Let the bindweed grow until it's about to flower, then use a string trimmer to cut it back to ground level. New shoots will sprout. Let these grow for 3 weeks, then cut them back to the ground. After that, dig in the bed and pull out as many roots as you can. Dispose of these roots in the trash or burn them.

Next, cover the bed with cardboard or several thicknesses of wet newspaper, overlapping pieces

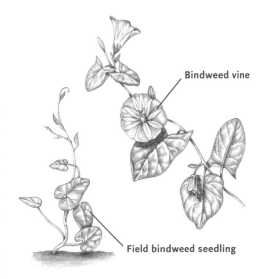

Bindweed vine

Field bindweed seedling

well. Top that with a few inches of chopped leaves or other dense organic mulch. Check it occasionally. Bindweed shoots will find their way up through the mulch covering. Cut them off.

The following season, you can try cutting holes through this covering to set transplants in place. Renew the top mulch covering. Water the transplants directly at the base, and watch carefully for any bindweed shoots that try to grow out through these openings.

By the following season, the bindweed should have died out. This is more work-intensive than black plastic, but it's kinder to the soil.

CANADA THISTLE (*Cirsium arvense*)

RANGE: Northern half of United States, Canada

DESCRIPTION: Sharp spines along the edges of the crinkled, silvery leaves make this tough perennial weed hard to handle. Upright stems up to 5 feet tall sport pink to purple flowers from midsummer through fall.

CONTROL STRATEGY: Wear thick gloves to cut back plant tops. Use a shovel or other tool to loosen roots before trying to pull out thistles. Let the roots dry thoroughly in the sun before composting. Never till this weed. Cutting back plant tops repeatedly will weaken the stand enough to allow successful gardening (but stay on the lookout for occasional shoots). When an area is seriously infested, though, try the mulching technique described for bindweed. If you don't take steps to weaken a strong thistle population, it could take over your entire garden.

CRABGRASSES (*Digitaria* spp.)

RANGE: United States and southern Canada

DESCRIPTION: This spreading annual grass sends out many thick, round stems from the base. Two common weed species are large crabgrass and smooth crabgrass. Large crabgrass leaves have white hairs along the edges. Smooth crabgrass leaves are smaller, with a purplish hue.

Flowering Canada thistle

Young Canada thistle shoots

Crabgrass

CONTROL STRATEGY: Pull crabgrass by hand, removing all roots as well as stems. Don't compost plants if seeds have started to form. After clearing an area, mulch it well to prevent regrowth.

LAMB'S-QUARTERS (*Chenopodium album*)

RANGE: United States, Canada

DESCRIPTION: Lamb's-quarters is a pretty, bushy weed with slightly toothed, medium green leaves. Later in the season the stems become woody, and dead stems last into winter.

CONTROL STRATEGY: It's easy to kill lamb's-quarters seedlings by hoeing and hand weeding, but watch for each new wave of plants. Young plants are edible, so you can add young, tender leaves to salads. Mulch beds once the soil is warm to prevent seeds from germinating. Never allow plants to set seed, and be mindful that plants produce multiple seed stalks; it doesn't help to cut back only the top of the plant. Don't compost any plants that have gone to flower.

NIGHTSHADES (*Solanum spp.*)

RANGE: Various species across the United States and Canada

DESCRIPTION: Nightshade rarely takes over a garden, but it's an important annual weed to control because it can host diseases and pests of the tomato family, including Colorado potato beetles. Nightshade has broad, wavy-edged leaves with narrow tips. The leaves are purplish underneath. White star-shaped flowers (they look very similar to pepper flowers) appear from midsummer to late summer. Later, flowers turn into green berries that mature to black.

CONTROL STRATEGY: Hoe and hand-pull seedlings and young plants to minimize the chance for the weeds to become hosts for insect pests and diseases. Never allow nightshade plants to reach the stage of producing seeds. Mulch areas infested with nightshade seeds to suppress germination.

Seedlings

Lamb's-quarters

Nightshade

Seedling

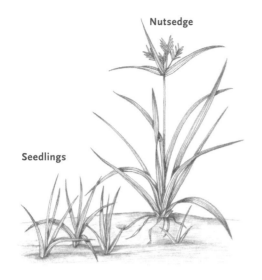

Nutsedge

Seedlings

NUTSEDGES (*Cyperus* spp.)

RANGE: United States, Canada

DESCRIPTION: Sedges are grasslike plants with th ick, yellowish green or dark green leaves that are V-shaped in cross section. The flower-heads are tufts of light-colored flower spikes at the top of the stems.

CONTROL STRATEGY: The best strategy for controlling nutsedge is to pull out small plants, but only small plants, by hand. Once plants have more than six leaves, stop pulling them. Other-wise, as you pull the plants, you will knock loose tubers that have formed in the soil, and each tuber will produce a new plant. You can cut the plants back to soil level (they will resprout), and be sure to cut off tops if flowerstalks begin to appear.

Mulching and solarization won't provide thorough control (but they won't hurt). If nut-sedge is limited to one part of your garden, try to plant a tall crop there because nutsedge won't grow as quickly when it's shaded.

PIGWEEDS (*Amaranthus* spp.)

RANGE: United States, southern Canada

DESCRIPTION: These annuals begin to appear in gardens in late winter and continue germinating through summer. Redroot pigweed is the most common type, with oval leaves on upright, rough stems. Prostrate pigweed grows as dense, ground-hugging mats. It has a fleshy root that is hard to pull—pieces left behind in the soil will resprout. The flowers are hard to spot; they form in small clusters in the leaf axils.

CONTROL STRATEGY: If pigweed grows past the seedling stage, uproot the plant as soon as you spot any flowers because these plants are highly prolific seed producers. Large pigweed plants develop a deep taproot, and you'll need to sever the taproot with a tool before you can pull out the plant. Don't compost plants that are in flower or bear seeds. Dry smaller plants well in the sun before composting, or they may rejuve-

Seedling

Redroot pigweed

nate in the pile and keep growing. Dense cover crops do a good job of smothering pigweed. Solarizing the soil will kill pigweed seeds in the top few inches of soil, which should give you a year's reprieve from battling this weed.

QUACKGRASS (*Elytrigia repens*)

RANGE: United States, except for some areas of the Deep South and Southwest; Canada

DESCRIPTION: This perennial grass has upright stems with long, narrow leaves. Leaves feel rough on the upper surface but smooth on the underside. Also called couch grass and wheat grass, quackgrass produces long rhizomes that sprout fibrous roots at the nodes.

CONTROL STRATEGY: Pull out individual plants, uprooting as much of the rhizomes as you can find. Bits of root will resprout, so work carefully. Let roots and rhizomes dry well before putting them in a compost pile. Never till a garden bed that's infested with quackgrass. Instead, follow the buckwheat weed-eradication plan described on page 138.

SHEPHERD'S PURSE
(*Capsella bursa-pastoris*)

RANGE: United States; Canada

DESCRIPTION: Seedlings of shepherd's purse resemble those of radishes and turnips, which isn't surprising because this annual weed belongs to the cabbage family. Shepherd's purse won't outcompete crops, but it's a weed to know and remove because it hosts some of the diseases that can infect cabbage and related crops, as well as the beet leafhopper, which transmits curly top virus.

CONTROL STRATEGY: Cultivate to uproot young plants; when they dry, add them to the compost pile. When the soil warms, mulch the soil to prevent more plants from germinating.

Quackgrass

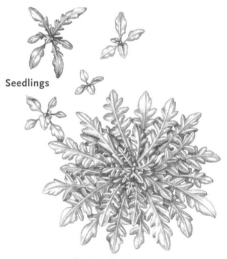

Seedlings

Shepherd's purse

WHITEFLIES

Common greenhouse pests, whiteflies also become garden problems at times, especially in late summer and fall in hot climates. In most areas, they die out over winter, but they will re-infest gardens when introduced there on the foliage of purchased transplants.

The tiny white adults lay their eggs on leaves of host crops. After the eggs hatch, the wingless nymphs crawl short distances across leaf and stem surfaces to find a suitable feeding site, where they will settle in and remain until they are ready to pupate. There are several generations per year in gardens until frost kills off these tender pests. Whiteflies can remain active year-round in greenhouses, and whitefly-infected transplants are often the source of a whitefly problem in home gardens.

In home gardens with plenty of biological diversity, natural predators often keep whiteflies from reaching damaging levels. But during hot, dry weather, or in arid climates, a whitefly problem can flare up suddenly and require treatment.

Silverleaf whitefly is an increasingly serious pest, not only because of the damage it causes directly, but also because it transmits disease to tomatoes in the Southeast and to lettuce and melon crops in the Southwest. Silverleaf whitefly feeding causes a disorder known as squash silverleaf in vine crops. Leaves turn yellow, and then the tissues between leaf veins become silvery and eventually bleached. Feeding by silverleaf whitefly can also cause yellowing of lettuce leaves and loss of color in broccoli stems and carrot roots.

PEST PROFILE

Trialeurodes spp., *Bemisia* spp.

RANGE: In gardens in the southern United States and along the West Coast; in greenhouses throughout the United States and Canada

HOW TO SPOT THEM: Whiteflies congregate on the undersides of leaves. Brush plants lightly with your hand or shake them gently. If whiteflies are present, they will fly up in a fluttery white cloud and then settle back on the plants. Also look for honeydew on foliage.

ADULTS: $\frac{1}{10}$-inch flies with white powdery scales covering the body and wings

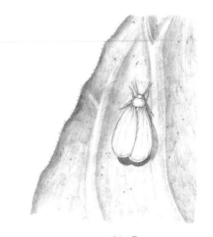

Whitefly

Whiteflies on tomato

LARVAE: Flat, wingless insects that feed in place

EGGS: Laid on leaves

CROPS AT RISK: Tomatoes, lettuce, melons, cucumbers, beans, and many other crops

DAMAGE: Plants are weakened; leaves turn yellow and drop off; young plants may become stunted; black sooty mold grows in honeydew coating foliage; plants may develop viral diseases.

CONTROL METHODS

ATTRACTING BENEFICIAL INSECTS. Lacewing larvae and lady beetle larvae eat whiteflies; plant small-flowered annuals and perennials to attract them. Also try spraying plants with a sugar-water solution to draw them to your garden. For details see Beneficial Insects.

TIMED PLANTING. In mild-winter areas, hold off on fall planting of cool-season crops until whitefly populations are declining after their summer peak.

INSPECTING TRANSPLANTS. Before you buy plants at a greenhouse or garden center, inspect them carefully. Brush your hand over a flat of plants here and there and watch for whiteflies to swarm. Check undersides of leaves for flies or honeydew. Don't buy plants from a supplier if you find whiteflies on any of their offerings because the plants you choose may have eggs or tiny nymphs present, even if you don't see them.

ROW COVERS. Cover seedbeds with row covers and keep them tightly covered as long as possible. Do not cover transplants unless you are certain they are free of whiteflies, nymphs, and eggs.

GARDEN CLEANUP. Remove infested crop debris after harvest and bury it deeply in a compost pile.

VACUUMING. Use a handheld vacuum cleaner to suck up whitefly adults (you'll need to follow up with another type of control to kill nymphs, too).

BIOCONTROLS. Spray infested plants with *Beauveria bassiana*. In greenhouses, release purchased parasitic wasps or predatory beetles to control whiteflies. Follow directions that come with the wasps for proper release.

NEEM. As a last resort, spray neem, which will kill young nymphs, prevent older nymphs from finishing their development, and deter adults from laying eggs on crop plants. Essential oil sprays also may be effective. See Pest Control for information.

CONTROL CALENDAR

BEFORE PLANTING

- Plant flowering plants and maintain permanent mulched or sheltered areas to provide habitat for beneficial insects that help control whiteflies.
- Inspect the stock of transplants at suppliers carefully and don't buy from suppliers if any plants show signs of whiteflies.

AT PLANTING TIME

- Plant cool-season crops as whitefly populations are declining.
- Cover seedbeds with row cover; cover transplants only if certain they are whitefly-free.

WHILE CROP DEVELOPS

- Use a handheld vacuum to remove adult whiteflies from plants.
- Spray *B. bassiana* on infested plants.
- In greenhouses, release purchased beneficials.
- As a last resort, spray neem or essential oil sprays.

AFTER HARVEST

- Remove and compost infested plant materials.

⤸ WHITE MOLD

Fluffy white mold on stems, fruits, or whole plants is the end result of an attack by the Sclerotinia sclerotiorum fungus. Resting bodies (called sclerotia) of this fungus last in the soil for years, awaiting a weeklong period of wet, warm weather. These conditions prompt the sclerotia to activate and send up mushroomlike structures that release active spores. Borne on moist breezes, these spores land on vulnerable crop stems and flowers and invade the tissue. White mold is also called stem canker, soft rot, and crown rot, and it can infect a very wide range of crops. Once a white mold infection starts, it spreads fast, often causing plants to collapse in a soggy heap.

Southern blight is caused by a related species and produces similar symptoms; see Southern Blight for details.

RANGE: United States; southern Canada

DESCRIPTION: Water-soaked spots on stems are an early symptom of white mold, but the spots may not catch your notice until they lead to girdling and wilting of leaves above the infected area. At other times, symptoms start near soil level as a wet rot—leaves and flowers turn brown and fall off. White mold sometimes appears on plant parts after they have fallen.

Sometimes black seedlike objects (sclerotia) appear in the white mold.

If the weather turns dry after an infection starts, the disease causes a firm, black rotting of the plant instead of the white mold and wet rot.

CROPS AT RISK: Most vegetable crops

FIGHTING INFECTION: Remove all infected plant parts from the garden as soon as you notice them. Handle the material carefully to avoid dropping sclerotia. Put it in bags for disposal or bury it deeply. Stop using sprinklers to water.

GARDEN CLEANUP: Sclerotia survive winter in the soil, but removing all crop residues at the end of the season is still helpful. Dispose of the residue or bury it deeply.

NEXT TIME YOU PLANT: The most important step is to improve soil drainage so that the soil surface won't stay wet for long periods of time. Plant resistant varieties when possible. Space plants widely to promote air circulation, and stake and trellis tall and vining crops.

Remove flowering weeds (such as ragweed and dandelion) because flowers on weeds are an ideal place for white mold spores to germinate. Biocontrol products that fight against *Sclerotinia* fungi are available for commercial farmers and may become available for home garden use in the future.

CROP ROTATION: Not effective. If white mold is a persistent problem in part of your garden, you'll need to improve soil drainage or else stop growing susceptible crops there for several years.

WIREWORMS

Lawn areas converted into vegetable garden beds are prime territory for wireworms. These beetle larvae feed on roots of lawn grasses, and they live as long as 6 years. Thus, wireworms are often ready and waiting in a new garden bed to attack crop roots.

Wireworms will nibble on the roots of most vegetables. When they feed on root crops such as carrots and potatoes, their tunneling ruins the crop's appearance and destroys its storage potential. When wireworms attack crops such as corn, the plants may be stunted or die.

If you till or dig a section of lawn to create a vegetable garden, it's wise to assume that wireworms are present. The adult form of this pest is a beetle commonly called a click beetle. You've probably seen click beetles around your yard and garden: they're the skinny brown beetles that will flip themselves over with a clicking sound when they accidentally get turned onto their back. The beetles eat leaves and flowers, but they don't do much damage. The problem results when they lay eggs around plant roots, introducing their long-lived larvae into the soil.

During winter, wireworms burrow deep into the soil so they won't freeze, returning near the surface in spring. They also "dive" deeper during hot summer weather. When they sense that their larval feasting years are at an end, wireworms form pupae in late summer. Adults, larvae, and eggs all are capable of surviving over winter.

Some wireworm species are pest predators rather than plant pests. These wireworms feed on larvae of wood borers in stumps and logs. Don't make a habit of squashing click beetles—some of them may be friends, not your foes.

PEST PROFILE

Limonius spp., *Agriotes* spp.

RANGE: Throughout North America

HOW TO SPOT THEM: To find out whether wireworms are present in your soil, put out wireworm traps (as described on the opposite page) and check them several days later. If you find wireworms in the traps, be prepared to protect your crops, especially root crops.

Click beetle

Wireworm larva

Wireworm tunneling

ADULTS: Click beetles are ⅓ to ¾ inch long, reddish brown or black with grooved wing covers.

LARVAE: Wireworms are brown or yellow, shiny, leathery worms up to 1½ inches long. Their bodies appear jointed.

EGGS: Laid in the soil

CROPS AT RISK: Most vegetables; damage is worst on carrots, potatoes, and corn.

DAMAGE: Seedlings weakened or killed, holes in tubers and storage roots. Feeding allows diseases to gain a foothold.

CONTROL METHODS

EXPOSING LARVAE TO PREDATORS AND FROST. Working the soil once a week in spring before planting and in fall after harvest brings wireworm larvae to the surface, where birds can eat them or exposure to freezing temperatures may kill them. If you keep chickens, turn them loose to feed in an infested bed just after you cultivate it.

TRAP CROPS. Wheat and lettuce are two fast-growing crops that attract wireworms. You can save a crop of carrots or potatoes from infestation by interplanting thickly planted bands of oats, wheat, or lettuce in the bed. See "Trap Cropping" on page 307 to learn more about this technique. With wireworms, it's best to remove the trap crop, roots and all, about 4 weeks after planting.

WIREWORM TRAPS. Pieces of carrot root or potato serve as the bait for these traps. One simple technique for making wireworm traps is to cut carrots or potatoes into 1-inch pieces (a good use for old, limp carrots or sprouted potatoes). Insert a wooden barbecue skewer into each piece. Stick these traps into the soil 2 to 4 inches deep (the end of the skewer will stick out of the soil) and several inches apart throughout your planting of susceptible crops. Check the traps twice a week; pick out and destroy any wireworms in them. When and if a trap becomes heavily infested, destroy the whole trap and replace it with a new one.

Another method of making a trap is to use a metal can, such as a coffee can. Use a large nail to poke holes in all sides of the can. Dig a hole between rows of susceptible crops and set the can in the hole. Put several pieces of cut-up carrot or potato in the can. Cover the can with a board, and top the board with mulch. Twice a week, open the trap and clean out the wireworms. If the bait is heavily infested, remove and destroy it and put in fresh bait. *Caution:* If you wish to discard used vegetable baits to your compost pile, first kill the wireworms by soaking the baits in soapy water.

REPELLENT CROPS. Plant a cabbage-family cover crop such as rapeseed in beds that you know or suspect are infested with wireworms. Cut down the crop when it is 6 inches tall and dig the plant material into the soil. As it decomposes, the crop will produce aromatic compounds that may repel the wireworms.

CONTROL CALENDAR

BEFORE PLANTING

- Cultivate weekly to expose larvae.
- Plant a trap crop between rows of susceptible crops.

AT PLANTING TIME

- In spring, delay planting if possible so that soil is warm (wireworms won't be so close to the surface).
- Put out wireworm traps in beds of susceptible crops.

WHILE CROP DEVELOPS

- Check traps and destroy larvae.
- Make a follow-up application of parasitic nematodes.

AFTER HARVEST

- Cultivate weekly to expose larvae and handpick them out of the soil.
- Remove all crop debris and weeds from the bed.
- Plant a repellent cover crop.

ADULT INSECTS AND RELATED ORGANISMS

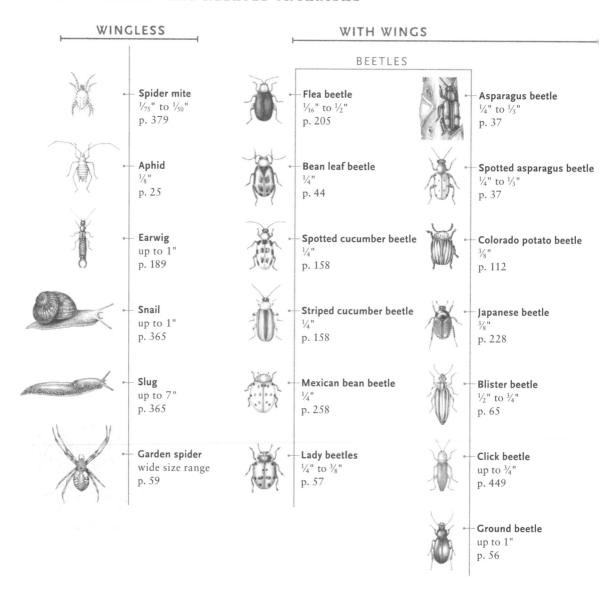

WINGLESS

WITH WINGS

BEETLES

Spider mite
$1/75$" to $1/50$"
p. 379

Aphid
$1/8$"
p. 25

Earwig
up to 1"
p. 189

Snail
up to 1"
p. 365

Slug
up to 7"
p. 365

Garden spider
wide size range
p. 59

Flea beetle
$1/16$" to $1/2$"
p. 205

Bean leaf beetle
$1/4$"
p. 44

Spotted cucumber beetle
$1/4$"
p. 158

Striped cucumber beetle
$1/4$"
p. 158

Mexican bean beetle
$1/4$"
p. 258

Lady beetles
$1/4$" to $3/8$"
p. 57

Asparagus beetle
$1/4$" to $1/3$"
p. 37

Spotted asparagus beetle
$1/4$" to $1/3$"
p. 37

Colorado potato beetle
$3/8$"
p. 112

Japanese beetle
$3/8$"
p. 228

Blister beetle
$1/2$" to $3/4$"
p. 65

Click beetle
up to $3/4$"
p. 449

Ground beetle
up to 1"
p. 56

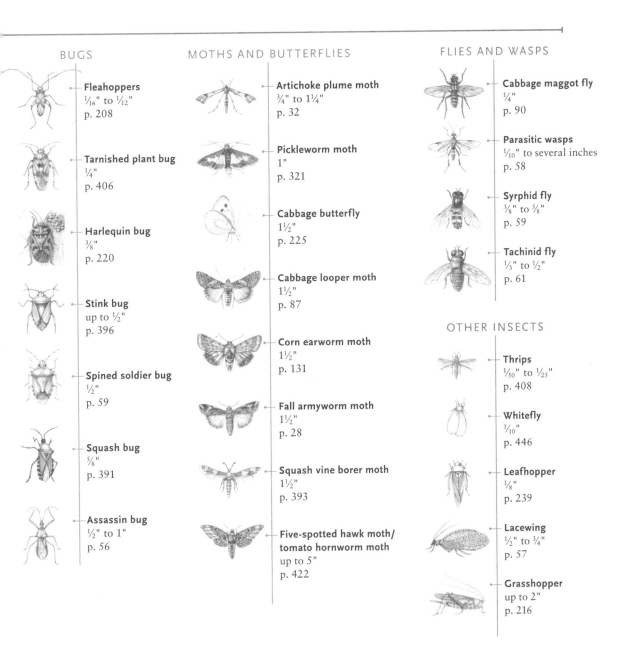

BUGS

Fleahoppers
$\frac{1}{16}$" to $\frac{1}{12}$"
p. 208

Tarnished plant bug
$\frac{1}{4}$"
p. 406

Harlequin bug
$\frac{3}{8}$"
p. 220

Stink bug
up to $\frac{1}{2}$"
p. 396

Spined soldier bug
$\frac{1}{2}$"
p. 59

Squash bug
$\frac{5}{8}$"
p. 391

Assassin bug
$\frac{1}{2}$" to 1"
p. 56

MOTHS AND BUTTERFLIES

Artichoke plume moth
$\frac{3}{4}$" to $1\frac{1}{4}$"
p. 32

Pickleworm moth
1"
p. 321

Cabbage butterfly
$1\frac{1}{2}$"
p. 225

Cabbage looper moth
$1\frac{1}{2}$"
p. 87

Corn earworm moth
$1\frac{1}{2}$"
p. 131

Fall armyworm moth
$1\frac{1}{2}$"
p. 28

Squash vine borer moth
$1\frac{1}{2}$"
p. 393

**Five-spotted hawk moth/
tomato hornworm moth**
up to 5"
p. 422

FLIES AND WASPS

Cabbage maggot fly
$\frac{1}{4}$"
p. 90

Parasitic wasps
$\frac{1}{10}$" to several inches
p. 58

Syrphid fly
$\frac{3}{8}$" to $\frac{5}{8}$"
p. 59

Tachinid fly
$\frac{1}{3}$" to $\frac{1}{2}$"
p. 61

OTHER INSECTS

Thrips
$\frac{1}{50}$" to $\frac{1}{25}$"
p. 408

Whitefly
$\frac{1}{10}$"
p. 446

Leafhopper
$\frac{1}{8}$"
p. 239

Lacewing
$\frac{1}{2}$" to $\frac{3}{4}$"
p. 57

Grasshopper
up to 2"
p. 216

LARVAE AND NYMPHS

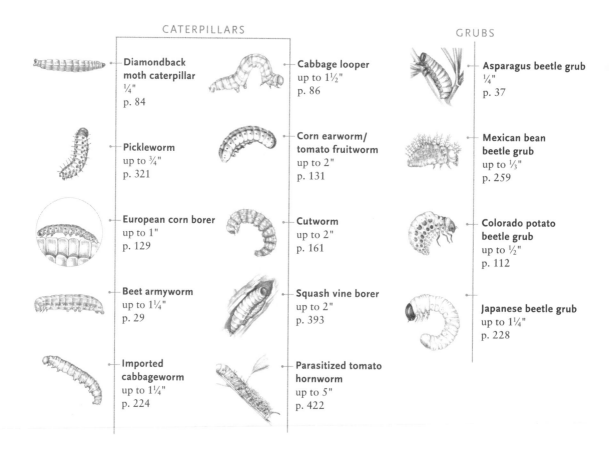

CATERPILLARS

Diamondback moth caterpillar
¼"
p. 84

Cabbage looper
up to 1½"
p. 86

Pickleworm
up to ¾"
p. 321

Corn earworm/ tomato fruitworm
up to 2"
p. 131

European corn borer
up to 1"
p. 129

Cutworm
up to 2"
p. 161

Beet armyworm
up to 1¼"
p. 29

Squash vine borer
up to 2"
p. 393

Imported cabbageworm
up to 1¼"
p. 224

Parasitized tomato hornworm
up to 5"
p. 422

GRUBS

Asparagus beetle grub
¼"
p. 37

Mexican bean beetle grub
up to ⅓"
p. 259

Colorado potato beetle grub
up to ½"
p. 112

Japanese beetle grub
up to 1¼"
p. 228

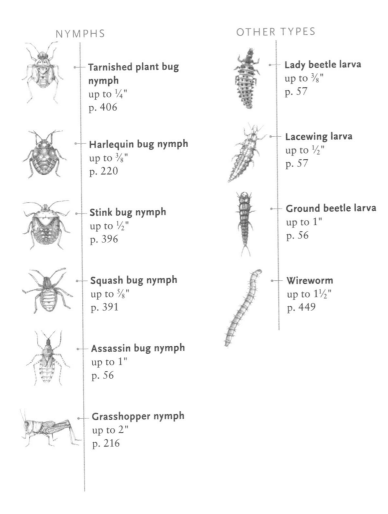

NYMPHS

Tarnished plant bug nymph
up to ¼"
p. 406

Harlequin bug nymph
up to ⅜"
p. 220

Stink bug nymph
up to ½"
p. 396

Squash bug nymph
up to ⅝"
p. 391

Assassin bug nymph
up to 1"
p. 56

Grasshopper nymph
up to 2"
p. 216

OTHER TYPES

Lady beetle larva
up to ⅜"
p. 57

Lacewing larva
up to ½"
p. 57

Ground beetle larva
up to 1"
p. 56

Wireworm
up to 1½"
p. 449

RESOURCES FOR GARDENERS

Shop by mail for an excellent selection of seeds, garden supplies, and helpful tools. Many of these mail-order companies offer organic, heirloom, and rare seeds.

MAIL-ORDER SEEDS AND PLANTS

Abundant Life Seeds
PO Box 157
Saginaw, OR 97472-0157
Phone: 541-767-9606
Fax: 866-514-7333
www.abundantlifeseeds.com

Baker Creek Heirloom Seeds
2278 Baker Creek Road
Mansfield, MO 65704
Phone: 417-924-8917
Fax: 417-924-8887
www.rareseeds.com

Bountiful Gardens
18001 Shafer Ranch Road
Willits, CA 95490
Phone: 707-459-6410
Fax: 707-459-1925
www.bountifulgardens.org

The Cook's Garden
PO Box C5030
Warminster, PA 18974
Phone: 800-457-9703
Fax: 800-457-9705
www.cooksgarden.com

Garden Trails
5730 W. Coal Mine Avenue
Littleton, CO 80123
Phone: 888-476-9123
Fax: 303-972-6950
www.gardentrails.com

Heirloom Seeds
PO Box 245
West Elizabeth, PA 15088-0245
Phone: 412-384-0852
www.heirloomseeds.com

High Mowing Seeds
76 Quarry Road
Wolcott, VT 05680
Phone: 802-472-6174
Fax: 802-472-3201
www.highmowingseeds.com

Ed Hume Seeds
PO Box 73160
Puyallup, WA 98373
Phone: 253-435-4414
Fax: 253-435-5144
www.humeseeds.com

Irish Eyes-Garden City Seeds
5045 Robinson Canyon Road
Ellensburg, WA 98926
Phone: 509-964-7000
Fax: 509-964-7001
www.gardencityseeds.net

Johnny's Selected Seeds
955 Benton Avenue
Winslow, ME 04901
Phone: 877-564-6697
Fax (US only): 800-738-6314
Fax (non-US): 207-861-8363
www.johnnyseeds.com

John Scheepers Kitchen Garden Seeds
23 Tulip Drive
PO Box 638
Bantam, CT 06750
Phone: 860-567-6086
Fax: 860-567-5323
www.kitchengardenseeds.com

Kitazawa Seed Company
PO Box 13220
Oakland, CA 94661-3220
Phone: 510-595-1188
Fax: 510-595-1860
www.kitazawaseed.com
Specializes in Asian vegetables

Native Seeds/SEARCH
526 N. 4th Avenue
Tucson, AZ 85705-8450
Phone: 866-622-5561
Fax: 520-622-5591
www.nativeseeds.org
Nonprofit organization dedicated to preserving and distributing seed of native crops of the Southwest

Nichols Garden Nursery
1190 Old Salem Road NE
Albany, OR 97321-4580
Phone: 800-422-3985.
Fax: 800-231-5306.
www.nicholsgardennursery.com

Pinetree Garden Seeds
PO Box 300
New Gloucester, ME 04260
Phone: 207-926-3400
Fax: 888-527-3337
www.superseeds.com

Renee's Garden Seeds
6116 Highway 9
Felton, CA 95018
Phone: 888-880-7228
reneesgarden.com

Salt Spring Seeds
Box 444, Ganges PO
Salt Spring Island, British
Columbia V8K 2W1
Canada
Phone: 250-537.5269
www.saltspringseeds.com
Untreated heirloom vegetable seeds for Canadian gardeners

Seeds for the South
410 Whaley Pond Road
Graniteville, SC 29829
Fax: 803-232-1119
www.seedsforthesouth.com

Seeds Trust
PO Box 596
Cornville, AZ 86325
Phone: 928-649-3315
Fax: 928-649.8181
www.seedstrust.com

Southern Exposure Seed Exchange
PO Box 460
Mineral, VA 23117
Phone: 540-894-9480
Fax: 540-894-9481
www.southernexposure.com

Territorial Seed Company
PO Box 158
Cottage Grove, OR 97424-0061
Phone: 800-626-0866
Fax: 888-657-3131
www.territorialseed.com

Tinyseeds.com
12165 Claude Court, Suite 104
Northglenn, CO 80241
Phone: 303-550-4679
www.tinyseeds.com

Turtle Tree Seed
Camphill Village
Copake, NY 12516
Phone: 800-620-7388
Fax: 678-202-1351
www.turtletreeseed.org
Biodynamic seed

Underwood Gardens
1414 Zimmerman Road
Woodstock, IL 60098
Phone: 815 338-6279
Fax: 888 382-7041
www.underwoodgardens.com
Heirloom and rare seeds

MAIL-ORDER TOOLS AND SUPPLIES

Extremely Green Gardening
Company
PO Box 2021
Abington, MA 02351
Phone: 781-878-5397
Fax: 781-878-5582
www.extremelygreen.com

Gardeners Supply Company
128 Intervale Road
Burlington, VT 05401
Phone: 888-833-1412
www.gardeners.com

Gardens Alive
5100 Schenley Place
Lawrenceburg, IN 47025
Phone: 513-354-1482
www.gardensalive.com

Peaceful Valley Farm and Garden
Supply
PO Box 2209
125 Clydesdale Road
Grass Valley, CA 95945
Phone: 888-784-1722
www.groworganic.com

Planet Natural
1612 Gold Avenue
Bozeman, MT 59715
Phone: 800-289-6656 (orders only)
Fax: 406-587-0223
www.planetnatural.com

Seeds of Change
3209 Richards Lane
Santa Fe, NM 87507
Phone: 888-762-7333
www.seedsofchange.com

RECOMMENDED READING

GARDENING HOW-TO

Ashton, Jeff. *The 12-Month Gardener.* New York: Sterling (Lark Books), 2001.

Bartholemew, Mel. *Square Foot Gardening.* Emmaus, PA: Rodale, 2005.

Bradley, Fern Marshall, and Jane Courtier. *Vegetable Gardening.* Pleasantville, NY: Readers Digest. 2006.

Cunningham, Sally Jean. *Great Garden Companions.* Emmaus, PA: Rodale, 1998.

Weaver, William Woys. *Heirloom Vegetable Gardening.* New York: Henry Holt, 1999.

Caldwell, Brian. *Vegetable Crop Health.* NOFA Handbook Series. Athol, MA: NOFA Interstate Council, 2004.

Coleman, Eliot. *The New Organic Grower's Four Season Harvest.* White River Junction, VT: Chelsea Green Publishing Company, 1999.

Creasy, Rosalind. *The Edible Salad Garden.* Boston: Periplus Editions, 1999.

Denckla, Tanya L. K. *The Gardener's A–Z Guide to Growing Organic Food.* North Adams, MA: Storey, 2003.

Flint, Mary Louise, and Steve H. Dreistadt. *Natural Enemies Handbook.* Berkeley: Univeristy of California Press. 1998.

Hart, Rhonda Massingham. *Deerproofing Your Yard & Garden.* North Adams, MA: Storey, 2005

Lanza, Pat. *Lasagna Gardening.* Emmaus, PA: Rodale, 1998.

Lovejoy, Sharon. *Trowel and Error.* New York: Workman, 2003.

McGee, Rose Marie Nichols, and Maggie Stuckey. *McGee & Stuckey's Bountiful Container.* New York: Workman, 2002.

Ogden, Shepherd, *Straight-Ahead Organic.* White River Junction, VT: Chelsea Green, 1999.

Roth, Sally. *The Gardener's Weather Bible.* Emmaus, PA: Rodale, 2003.

Shepherd, Renee, and Fran Raboff. *Recipes from a Kitchen Garden.* Ten Speed Press, 1993.

Smith, Edward C. *The Vegetable Gardener's Bible.* North Adams, MA: Storey, 2000.

Staub, Jack. *75 Exciting Vegetables for Your Garden.* Salt Lake City, UT: Gibbs Smith, 2005.

PEST, DISEASE, AND WEED ID GUIDES

Cebenko, Jill Jesiolowski, and Deborah L. Martin, eds. *Insect, Disease & Weed I.D. Guide.* Emmaus, PA: Rodale, 2001.

Cranshaw, Whitney. *Garden Insects of North America.* Princeton, NJ: Princeton University Press, 2004.

Flint, Mary Louise. *Pests of the Garden and Small Farm.* Berkeley: University of California Press, 1998.

Howard, Ronald J., J. Allan Garland, and W. Lloyd Seaman, eds. *Diseases and Pests of Vegetable Crops in Canada.* Ottowa, Ontario: Entomological Society of Canada, 1994.

Pleasant, Barbara. *The Gardener's Bug Book.* Pownal, VT: Storey, 1994.

———. *The Gardener's Guide to Plant Diseases.* Pownal, VT: Storey, 1995.

GARDENING ORGANIZATIONS AND WEB SITES

Thousands of Web sites offer information on vegetable gardening. Search for on-line answers to questions on specific topics at *Organic Gardening* magazine's Web site: www.organicgardening.com/links/. For reliable and up-to-date information about vegetable gardening and organic methods visit the home pages of these membership organizations.

You'll also find excellent region-specific information about gardening at the Web site of your state university system.

Canadian Organic Growers
www.cog.ca

A nonprofit educational and networking organization especially for Canadian farmers and gardeners. Members receive The Canadian Organic Grower *magazine.*

Maine Organic Farmers and Gardeners Association
www.mofga.org

Longstanding nonprofit organization that promotes organic food production. Sponsor of the Common Ground Country Fair, other educational events, and the Maine School Garden Network. Members receive the MOF&G newspaper and free admission to the fair.

***Mother Earth News* magazine**
www.motherearthnews.com

Web site offers searchable archive of articles from the magazine, covering natural homebuilding, livestock, alternative energy as well as organic gardening.

National Gardening Association
www.garden.org

Members qualify for on-line gardening courses, expert answers to gardening questions, quarterly newsletter. Web site offers articles on gardening, pest control library, message boards, seed swap, and more.

Northeast Organic Farming Association
www.nofa.org/index.php

Nonprofit organization that promotes healthy food and organic farming. Members receive quarterly newspaper and can join one of the seven state chapters. State and regional conferences offer workshops on a variety of topics for both organic farmers and gardeners.

Northwest Coalition for Alternatives to Pesticides
www.pesticide.org

Nonprofit organization with a goal of protecting people and the environment by advancing healthy solutions to pest problems. Offers free program with e-mail tips and a Q&A phone hotline for gardeners in the Northwest who are trying to convert to pesticide-free gardening.

Oregon Tilth
www.tilth.org

Nonprofit organization to promote organic farming. Members receive In Good Tilth *newsletter and discounts on admission to Oregon Tilth events.*

Organic Consumers Association
www.organicconsumers.org

Information on farm issues, food safety, environmental issues, organic standards, and more. Sponsoring organization for the Organic Farmers and Gardeners Union.

***Organic Gardening* magazine**
www.organicgardening.com

Articles on a wide range of organic gardening subjects, message boards, free online newsletter, submit questions to "Garden Girl" for expert advice. Subscribe to the magazine, which is the original source of information about organic home gardening in the United States.

Seattle Tilth
www.seattletilth.org

Sponsors demonstration gardens, community gardens, gardening classes, and a natural lawn and garden hotline. Web site offers information on organic gardening and composting and a listing of events in the Seattle region.

INDEX

Boldface *page references indicate illustrations.*

<u>Underscored</u> *references indicate boxed text and tables or charts.*

A

Acalymma spp., 158–60, **159, 160**
Alfalfa meal, as fertilizer, <u>202</u>
Alternaria blight (Alternaria leaf
    spot), 9–10, 181–82
Alternaria solani (Alternaria
    blight), 9–10
Amaranthus spp. (pigweeds), 444–
    45, **444**
Angular leaf spot, 10
Animal pests
    armadillos, 20
    assessing the problem, 11–12
    deer (*See* Deer)
    gophers, 20, **20**
    ground squirrels, 21
    hares, 21
    moles, <u>22</u>
    pets as, 22–23
    rabbits, 11
        fences for, 16–18, **17, 19,** 166,
          167, **171**
        instant defenses against, 13,
          356, 357
        scare devices for, 13–15, **14**
        traps and, 19–20
    raccoons, 11
        fences for, 16–18, **17, 18, 19,**
          166, 167, **171**
        instant defenses against, 13,
          356, 357
        scare devices for, 13–15, **14**
    skunks, 21
    snakes, 21
    solutions for
        barriers, 20, **20,** 22, **22,**
          165–66, **166**
        categories of, 12–13
        fencing, 12, 16–18, **17, 18,**
          19, 166, 167, 168, **171**
        instant defenses against, 13,
          356, 357
        repellents, 12–13, 15–16
        scare devices, 12, 13–15, **14**
        trapping, 19–20
    voles, 21–22, **22**
    wildlife pressure, <u>12</u>
    woodchucks (groundhogs)
        fences for, 16–18, **17, 19,** 166,
          167, 168, **171**
        traps and, 19–20
Anthracnose, 23–24
    Bacillus subtilis and, 180–81
    bicarbonate and, 181–82
    copper and, 183
    neem oil and, 182–83
Ants, honeydew and, 25
Aphids, 25–27, **26**
    control calendar, 27
    control methods, 26–27
    honeydew and, 25, 26
    profile of, 25–26, **26**
    viral diseases and, 174
Armadillos, 20
Armyworms, 28–30, **28, 29**
    beet, 28–30, **29**
    control calendar for, 30
    control methods for, 29–30
    fall, 28–30, **28, 29**
    profile of, 28–29, **28, 29**
Artichokes, 30–34, **31**
    crop basics, 30–31
    harvesting, 31, **31**
    problem prevention, 32
    secrets of success, 31–32
    troubleshooting, 32–34, **32**
        gray mold (botrytis blight)
          and, 219
Arugula, **352,** 353
    as trap crop, <u>307</u>
Asparagus, 34–38, <u>35</u>, **37**
    crop basics, 34–35
    motherstalk cultural method, <u>35</u>
    problem prevention, 36–37
    regional notes
        growing in the Deep South, 36
    secrets of success, 35–36
    troubleshooting, 37–38, **37**
        gray mold (botrytis blight)
          and, 219
        protecting from deer, <u>170</u>
Asparagus beetles, **37,** 38
    spotted, **37,** 38
Assassin bugs, 56, **56**
Aster yellows, 39, 239
Avena sativa (oats), as cover crop, <u>141</u>

B

Bacillus subtilis, as disease-control
    agent, 180–81

Bacillus thuringiensis (Bt), 311–12
Bacteria
  as beneficial organisms, 174
  as disease-causing organisms
    (*See* Bacterial diseases)
Bacterial diseases, 174
  biofumigation and, 185
  copper and, 183
  soil solarizing and, 186
Bacterial fruit blotch, 255
Bacterial spot, copper and, 183
Bactericides, copper as, 183
Baking soda, as fungicide, 182
Barriers. *See also* Fences
  as insect control method
    collars, 91, **91**, 162, 306
    kaolin clay, 306–8
    row covers, 306
    screening cones, 91, **91**
  for small animal pests, 20, **20,**
    22, **22,** 165–66, **166**
Beans, 40–45, **41, 42, 42, 44**
  bell as cover crop, 140
  crop basics, 40
  fava as cover crop, 140
  lima
    downy mildew and, 187
  problem prevention, 43
  regional notes
    growing in the far North,
      42–43
  secrets of success, 41–42, **41, 42,**
    42
  supports for, 41–42, **41, 42, 42**
  troubleshooting, 43–45, **44**
    Anthracnose symptoms on, 23
    beet curly top (curly top
      virus) and, 46
    corn earworm and, 131–33,
      131, 132
    gray mold (botrytis blight)
      and, 219
Beauveria bassiana, 312–13
Beet curly top (curly top virus), 46
  leafhoppers and, 239
Beet family, 147
Beets, 47–51, **48**
  crop basics, 47
  problem prevention, 49
  regional notes
    growing in the western
      United States and Canada,
      48–49
  secrets of success, 47–48

troubleshooting, 49–51
  armyworms and, 28
  beet curly top (curly top
    virus), 46, 239
  beet webworms, 48–49
  Cercospora leaf spot and, 109
  downy mildew and, 187
  Fusarium wilt and, 210
  leafminers and, 241–43, **243**
Beet webworms, 48–49
Bermudagrass (wiregrass,
    devilgrass), 440–41, **441**
Bicarbonate, as disease control
    agent, 181–82
Bindweeds, 441–42, **441**
Biofumigation, 185
Birds, 62–65, **63,** 64, **64**
  attracting, 62
  crows, 64–65, **64**
  jays, 64–65
  as pests, 11–12, 62–65
    barriers for, 63, **63, 64**
    scare tactics, 13, 14, **14,**
      63–64
Blackleg, cabbage family members
    and, 83–84
Black rot, cabbage family members
    and, 84, 346
Blights. *See* Alternaria blight
    (Alternaria leaf spot); Early
    blight; Late blight
Blister beetles, 65–66, **66**
  as beneficial insect, 65, **66**
  control calendar, 66
  control methods, 66
  profile of, 65–66, **66**
Bloodmeal, as fertilizer, 202
Blossom-end rot, 67–68
Bonemeal, as fertilizer, 202
Boron, deficiency symptoms, 201
Botrytis blight (gray mold), 218–19
  neem oil and, 182–83
Brassica napus (rape), as cover crop,
    141
Brassica oleracea Acephala Group
    (kale), as cover crop, 141
Broccoli, 68–74, **69, 70**
  crop basics, 68–69
  problem prevention, 71
    protecting with trap crops,
      307
  regional notes
    growing in the Northeast and
      eastern Canada, 70

secrets of success, 69–70
troubleshooting, 71–73
  Alternaria blight and, 9
  club root and, 110–11
  Swede midge and, 70–71
  temperature-induced failures,
    69–70, **69,** 72
Broccoli raab, 73, **73**
Brussels sprouts, 74–78, **75 76**
  crop basics, 74–75
  problem prevention, 77
  secrets of success, 75–77, **75, 76**
  troubleshooting, 77–78
    Alternaria blight and, 9
    club root and, 110–11
Bt *(Bacillus thuringiensis),* 311–12
Buckwheat, as cover crop, 138, 140

C

Cabbage, 79–86, **79, 81**
  crop basics, 80–81
  problem prevention, 82–83
  regional notes
    growing in the Deep South, 82
    growing in the Southeast, 82
  secrets of success, 81–82, **81**
  troubleshooting, 83–86
    cabbage yellows, 210, 211
    club root and, 110–11
    cross-striped cabbageworms,
      82
    yellowmargined leaf beetle, 82
Cabbage family members, 147
  biofumigation and, 185
  in crop rotation plan, 148, **149,**
    150
  troubleshooting
    Alternaria blight, 9
    blackleg, 83–84
    black rot, 84, 346
    cabbage yellows, 210, 211
    Fusarium wilt, 210
    harlequin bugs, 220–21, **220**
Cabbage loopers, 86–89, **87**
  control calendar, 88
  control methods, 88
  profile, 87–88, **87**
Cabbage maggots, 89–92, **90, 91**
  control calendar, 92
  control methods, 90–92, **91**
  profile, 90, **90**
  turnip damage by, **90**

Cabbageworms
  cross-striped, 82
  imported (*See* Imported
    cabbageworms)
Cabbage yellows, 210, 211
Calcium, deficiency symptoms, <u>201</u>
Canada thistle, 442, **442**
Cantaloupe. *See* Melons
Capsella bursa-pastoris (shepherd's
  purse), 445, **445**
Cardoons, <u>33</u>, **33**
Carrot family members, <u>147</u>
  in crop rotation plan, <u>148</u>, **149,**
    150
Carrots, 93–100, **94,** <u>95</u>, **95**
  crop basics, 93
  problem prevention, 96–97
  regional notes
    growing in the Far West, 96
    growing in the Southeast, 96
  secrets of success, 93–96, **94,**
    <u>95</u>, **95**
  troubleshooting, 97–100
    Alternaria blight and, 9
    aster yellows, 39, **240**
    Cercospora leaf spot and, 109
    motley dwarf disease, 96
Cauliflower, 100–104, <u>101</u>, **101**
  blanching, <u>101</u>, **101**, 103
  crop basics, 100–102, <u>101</u>, **101**
  problem prevention, 102–3
  secrets of success, 102
  troubleshooting, 103–4
    Alternaria blight and, 9
    club root and, 110–11
Celery, 104–8, <u>105</u>, **105**
  blanching, <u>105</u>, **105**
  crop basics, 104–5
  problem prevention, 106
  regional notes
    growing in Canada and
      northern United States, 106
  secrets of success, 105–6, <u>105</u>, **105**
  troubleshooting, 107–8
    aster yellow symptoms on, 39
Cercospora leaf spot (Cercospora
  blight, Cercospora leaf
  blight, frog-eye leaf spot),
  109
Chenopodium album (lamb's-
  quarters), 443, **443**
Chinese broccoli, <u>73</u>, **73**
Chives, <u>282</u>
Chrysopa spp. (lacewings), 57, **57**

Chrysoperla spp. (lacewings), 57, **57**
Cirsium arvense (Canada thistle),
  442, **442**
Cloches, **356**, 357
  as instant animal pest protector,
    13
Clover
  alsike, <u>140</u>
  Berseem, <u>140</u>
  Crimson, <u>140</u>
  red, <u>141</u>
  sweet, <u>141</u>
  white (white Dutch, New
    Zealand white), <u>141</u>
Club root, 110–11, 173
Coldframes, 358–59, **358**
Collards, **232.** *See also* Kale
Colorado potato beetles, 112–14,
  112, 113
  control calendar, 114
  control methods, 113–14
  profile of, 112–13, **112, 113**
Comfrey tea, as fertilizer, 203–4
Companion planting, 115
Compost
  buying, 121
  disease control and, 178
  making (*see* Composting)
  as mulch, 265
  soil improvement and, <u>377</u>
Composting, 116–24, **117,** <u>118</u>,
  119, 120, <u>122</u>, **123**
  bins, 117–18, **117**
    screening, **119**
  making hot, <u>122</u>
  materials for, 116–17
    avoiding contaminated, <u>118</u>
  sheet, 120–21
  trench, 120, **120**
  troubleshooting, 118–20, <u>122</u>
Compost tea, 121–24, **123**
  disease control and, 178, 179
  safety of, 123–24
Convolvulus spp. (bindweeds),
  441–42, **441**
Copper
  as bactericide, 183
  as fungicide, 183
Corn, 123–30, <u>126</u>, **126, 129**
  crop basics, 124–25
  pollination, <u>126</u>, **126**
  problem prevention, 127–28
    bird pests and seedlings,
      64–65, **64**

  regional notes
    growing in the East and
      Midwest, 127
  secrets of success, 125–26, <u>126</u>,
    126
  troubleshooting, 128–30, **129**
    corn earworms and, 131–33,
      131, 133
    fall armyworms and, 28, **29**
    Japanese beetles and, **229**
    raccoons and, 11 (*see also*
      Raccoons)
    Stewart's wilt and, 127
Corn borer, European
  corn and, 129, **129**, 130
Corn earworm (tomato fruitworm,
  cotton bollworm,
  vetchworm), 131–33, **131,**
  132
  control calendar, 133
  control methods, 132–33
  profile of, 131–32, **131, 132**
Corn leaf blights, 128
Corn salad (mache), 353–54
Cotton bollworm (corn earworm),
  131–33, **131, 132**
Cottonseed meal, as fertilizer, <u>202</u>
Couch grass (quackgrass, wheat
  grass), 445, **445**
Cover crops, 2–3, 4, 133–42, **135,**
  136, <u>137</u>, **137,** <u>138</u>, <u>139</u>,
  <u>140–141</u>
  beans
    bell, <u>140</u>
    fava, <u>140</u>
  buckwheat, <u>138</u>, <u>140</u>
  clover
    alsike, <u>140</u>
    Berseem, <u>140</u>
    crimson, <u>140</u>
    red, <u>141</u>
    sweet, <u>141</u>
    white (white Dutch, New
      Zealand white), <u>141</u>
  cowpeas, <u>140</u>
  disease prevention and, 134
  hairy vetch, <u>141</u>
  importance of using, 2–3, 4
  kale, <u>141</u>
  killing, 139, 142
  as living mulch, 266–67
  as nematode control, <u>275</u>, **275**
  oats, <u>141</u>
  oilseed radish, <u>141</u>

as pathways, <u>137</u>, **137**
peas
    Austrian winter, <u>140</u>
    field, <u>140</u>
    southern, <u>140</u>
planting through dead, <u>139</u>
rape, <u>141</u>
rye
    cereal, <u>140</u>
    winter, <u>140</u>
ryegrass, annual, <u>140</u>
soil improvement and, 134–35,
    374, <u>377</u>
soybean, <u>141</u>
sudangrass, <u>141</u>
troubleshooting, 139
weed suppression and, 134, <u>138</u>
when to plant, 135–38, **135,
    136**
Cowpeas, as cover crop, <u>140</u>
Crabgrass, 442–43, **442**
Crenshaw melons. *See* Melons
Cress, 354
Crioceris asparagi (asparagus
    beetle), **37,** 38
Crioceris duodecimpunctata (spotted
    asparagus beetle), **37,** 38
Crop rotation, 142–51, **144, 145,**
    <u>146</u>, <u>147</u>, <u>148</u>–<u>149</u>, **149,**
    <u>150</u>, **150**
    companion planting and, 115
    disease control and, 143, 185–
      86
    improvising, 147, 151
    as insect control method, 91
    insect pest control and, 303–4
    as nematode control, 274
    by plant family, 146, <u>147</u>, <u>148</u>–
      <u>150</u>, **149, 150**
    by plant feeding habit, <u>146</u>
    rules of, 144–46, **144, 145,** <u>146</u>
    usefulness of, 143
Crown rot, 346
Crows, seedlings and, 64–65, **64**
Cucumber beetles, 158–60, **159,
    160**
    control calendar, 160
    control methods, 159–60, **160**
    profile of, 158–59, **159**
Cucumber mosaic virus. *See*
    Mosaic
Cucumbers, 151–57, **152, 154**
    crop basics, 151–52
    problem prevention, 155

regional notes
    for dry western states, 154
    secrets of success, 152–54, **154**
    troubleshooting, 155–57
      Anthracnose symptoms on,
        23–24
Cucurbit yellow vine disease, 387
Curly top virus (beet curly top), 46
Cutworms, 161–62, **161, 162**
    control calendar, 162
    control methods, 162, **162**
    profile of, 161–62, **161**
Cynodon dactylon (bermudagrass,
    wiregrass, devilgrass),
    440–41, **441**
Cyperus spp. (nutsedge), 444, **444**

D

Damping-off, 163–64, 360–61
    biofumigation and, <u>185</u>
    Trichoderma harzianum and, 181
Deer, 11, 164–72, **166, 167,** 167,
    168, <u>169</u>, **169,** <u>170</u>, 171
    caging crops, <u>170</u>
    fences for
      electric, 168–72, **171**
      low, 165–66, **166**
      plastic, 167–68, **168**
      repellent, 165
      tall wire, 166–67, **169**
      wood, <u>167</u>, **167**
    population density and, <u>12</u>
    repellents, 165
      ultrasonic, <u>169</u>
    scare devices for, 14, 165
Devilgrass (bermudagrass,
    wiregrass), 440–41, **441**
Diabrotica undecimpunctata
    (cucumber beetle), 158–60,
    159, 160
Digitaria spp. (crabgrasses), 442–43,
    442
Disease control, 172–86, <u>176</u>, <u>178</u>,
    179, <u>183</u>, <u>185</u>, <u>186</u>. *See also*
    Diseases; Environmental
    conditions and disorders;
    specific crop; specific disease
    Bacillus subtilis and, 180–81
    bacterial diseases, 174
      controlling, 183, <u>185</u>, <u>186</u>
    baking soda spray as, 182
    bicarbonate and, 181–82

biofumigation and, <u>185</u>
breaking the cycle, 182–86
compost and, 178
compost tea and, 178, 179
copper and, 183
crop debris disposal and, 184
crop rotation and, 185–86
disinfectant solution for, 296
fungal diseases, 173–74 (*see also*
    Fungus diseases)
garlic and, <u>183</u>
harpin and, 178–80
insect pests and, 39, 174, 177
milk and, <u>183</u>
neem oil, 182–83
plant health boosters, 177–80
    compost, 178
    compost tea, 178–79
    harpin, 178–80
plant selection and, 177
plant self defense, 175
    boosting, 177–80
seeds and, 176
soil and, 175–76
    solarizing, <u>186</u>
sources of contamination, 173, 174,
    175, 176–77, **179,** 182, 264
systemic acquired resistance
    (SAR), 175
Trichoderma harzianum, 181
variety selection and, 177
viral diseases, 174–75
Diseases. *See also* Disease control;
    Nutritional deficiencies
    Alternaria blight (Alternaria leaf
      spot), 9–10, 181–82
    angular leaf spot, 10
    Anthracnose, 23–24
      controlling, 180–81, 182–83
    aster yellows, 39
    bacterial, 174
      biofumigation and, <u>185</u>
      copper and, 183
      soil solarizing and, <u>186</u>
    bacterial fruit blotch, 255
    beet curly top (curly top virus),
      46
      leafhoppers and, 239
    blackleg
      cabbage family members and,
        83–84
    black rot
      cabbage family members and,
        84, 346

Diseases *(cont.)*
blossom-end rot, 67–68
Botrytis blight (gray mold),
182–83, 218–19
cabbage yellows, 210, 211
Cercospora leaf spot (Cercospora
blight, Cercospora leaf
blight), 109
club root, 110–11, 173
corn leaf blights, 128
cover crops and, 134
crop rotation and, 143, 185–86
crown rot of squash family
members, 346
cucumber mosaic virus, 263–64
cucurbit yellow vine disease, 387
curly top virus (beet curly top), 46
damping-off, 163–64, 360–61
biofumigation and, 185
Trichoderma harzianum and,
181
early blight, 188–89
Bacillus subtilis and, 180–81
copper and, 183
frog-eye leaf spot (Cercospora
leaf spot), 109
fungal *(see* Fungal diseases)
fungal leaf blights, 183
Fusarium wilt, 210–11
gray mold (Botrytis blight),
182–83, 218–19
gummy stem blight, 156, 256,
389
late blight, 237–38, 238, **238**
Bacillus subtilis and, 180–81
copper and, 183
mildew, downy, 187
Bacillus subtilis and, 180–81
bicarbonate and, 181–82
copper and, 183
milk and, 183
neem oil and, 183
mildew, powdery, 337–38
Bacillus subtilis and, 180–81
bicarbonate and, 181–82
neem oil and, 182–83
mosaic, 263–64
motley dwarf disease of carrots,
96
Phomopsis blight of eggplant, 193
Plectosporium blight, 387–88
root rots, 185
rust, 182–83, 350–51
scab, 185, 355
Septoria leaf spot, 183

soft rot, 346
southern blight, 378–79
Stewart's wilt, 127
tomato spotted wilt virus, 416
Verticillium wilt, 185, 427
viral, 174–75
white heart of lettuce, 39
white mold, 448–49
white rust, 383
Disorders, defined, 173
Drainage, 372–73
improving, 374–75, **375**
Drought, dealing with, 432

E

Early blight, 188–89
Bacillus subtilis and, 180–81
copper and, 183
Earwigs, 189–91, **190**
control calendar, 191
control methods, 190–91, **190**
profile of, 189–90, **190**
Eggplant, 191–96, **193**
crop basics, 191–92
problem prevention, 194
regional notes
growing in the Southeast, 193
secrets of success, 192–93, **193**
troubleshooting, 194–96
blossom-end rot and, 67–68
early blight and, 188–89
Phomopsis blight, 193
Verticillium wilt, 427
Elytrigia repens (quackgrass, couch
grass, wheat grass), 445,
445
Encarsia formosa (parasitic wasps), 61
Endive, 196–98, **197**
Belgian, 198, **198**
crop basics, 196–97
secrets of success, 197, **197**
troubleshooting, 197–98
Environmental conditions and
disorders. *See also*
Nutritional deficiencies;
specific crop
blossom-end rot, 67–68
nutrient deficiencies, 175,
199–200, 201
weather-related disorders, 175
Epicauta spp. (blister beetles),
65–66, **66**
Epsom salts, as fertilizer, 202

F

*Fagopyrum esculentum (Polygonum
esculentum)* (buckwheat), 140
Feather meal, as fertilizer, 202
Fences
for deer
electric, 168–72, **171**
low, 165–66, **166**
plastic, 167–68, **168**
repellent, 165
tall wire, 166–67, **169**
wood, 167, **167**
for small animal pests, 12,
16–18, **17, 18, 19,** 20, **20,**
22, **22,** 166, 167, 168, **171**
Fertilizers, 199–205, 199, 201,
202–203, 204
about using, 199, 200
analyzing labels, 201
applying, 204–5
blended organic, 202
homemade, 204
choosing, 200–204, 202–203
comfrey tea as, 203–4
compost tea as, 121–24, **123**
manure tea as, 202–3
types of organic, 201–2, 202–203
Fish products, as fertilizer, 203
Flea beetles, 205–7, **206**
control calendar, 207
control methods, 206–7
profile of, 206, **206**
trap crops and, 307
Fleahoppers, 208–10, **208, 209**
control calendar, 209–10
control methods, 209
profile of, 208–9, **208, 209**
Forficula auricularia (earwigs),
189–91, **190**
Frog-eye leaf spot (Cercospora leaf
spot), 109
Frost, 431
Fungal diseases, 173–74
controls for
Bacillus subtilis and, 180–81
baking soda spray and, 182
bicarbonate and, 181–82
biofumigation and, 185
soil solarizing and, 186
Trichoderma harzianum and,
181
Fungal leaf blight, 183
Fungal leaf spot, *Bacillus subtilis*
and, 180–81

Fungi
  as beneficial organisms, 173
  as disease-causing organisms
    (*see* Fungus diseases)
Fungicides
  Bacillus subtilis as, 180–81
  baking soda as, 182
  bicarbonate as, 181–82
  copper as, 183
  garlic as, 183
  milk as, 183
  neem oil as, 182–83
Fusarium wilt, 210–11

G

Garden basics, 2–4
  cover crops, 2–3, 4, 133–42,
    **135, 136, 137, 137, 138,
    139, 140–141**
    as living mulch, 266–67
    as nematode control, 275, **275**
  interplanting, 3
  mulch, 2–3 (*see also* Mulch)
  planting (*see* Planting)
  plot size, 3–4
  sanitation
    as disease control method,
    182, 184, 264
    disinfection solution for, 296
    as insect pest control method,
    305–6
  soil, 3 (*see also* Soil)
  weeds (*see* Weeds)
Garlic, 212–15, **212, 213, 213**
  crop basics, 212–13, **212, 213,
    213**
  as fungicide, 183
  problem prevention, 215
  regional notes
    growing in the North, 213, **213**
    growing in the South, 213
    growing in the West, 214
  secrets of success, 213–14
  troubleshooting, 215
    leek moth and, 246
    white rot, 214
Garlic spray, as insect pest
  repellent, 308–10
Glycine max (soybeans), 141
Gophers, 20
  barriers for, 20, **20**
Gourds. *See* Squash
Grass clippings, as mulch, 265

Grass family, 147
Grasshoppers, 216–18, **217, 301**
  control calendar, 218
  control methods, 217–18
  profile of, 216–17, **217**
Gray mold (Botrytis blight),
  182–83, 218–19
Green manure. *See* Cover crops
Greensand, as fertilizer, 203
Ground beetles, 56–57, **56**
Groundhogs. *See* Woodchucks
  (groundhogs)
Ground squirrels, 21
Guano, as fertilizer, 203
Gummy stem blight, 156, 256, 389
Gypsum, as fertilizer, 203

H

Hail, dealing with, 434
Hairy vetch, as cover crop, 141
Halticus bractatus (fleahoppers),
  208–10, **208, 209**
Handpicking, as insect control
  method, 304
Hardening off, 362, **363**
Hares, 21
Harlequin bugs, 220–21, **220**
  control calendar, 221
  control methods, 221
  profile of, 220–21, **220**
Harpin, 178–80
Heat waves, dealing with, 432–33
Helicoverpa zea (corn earworm),
  131–33, **131, 132**
Hemerobius spp. (lacewings), 57, **57**
Herbicides, 437
Herbs, protecting from deer, 170
Heterorhabditis bacteriophora
  (beneficial nematodes),
  271–72
Honeydew
  ants and, 25
  aphids and, 25, 26
Honeydew melons. *See* Melons
"Hopperburn," leafhoppers and,
  239, **240**
Horseradish, 222–24, **223**
  crop basics, 222
  problem prevention, 223
  secrets of success, 222–23, **223**
  troubleshooting, 224
Hot pepper spray, as insect pest
  repellent, 308–10

I

Imported cabbageworms, 225–27,
  225, 226
  control calendar, 227
  control methods, 226–27
  profile of, 225–26, **225, 226**
Insecticidal soap, 316–17
Insecticides. *See* Pesticides
Insect pest control
  barriers, 306–8
    collars, 91, **91,** 162, 306
    kaolin clay, 306–8
    row covers, 13, 306, 348–49
    screening cones, 91, **91**
  biological controls, 310–15
    Bacillus thuringiensis (Bt),
    311–12
    Beauveria bassiana, 312–13
    homemade, 315
    milky spore disease, 230
    spinosad, 314
  commercial traps, 310
  cover crops and, 134
  crop rotation and, 91, 143
  evaluating the damage, 299
  hands-on control methods
    garden sanitation, 274, 305–6
    handpicking, 304
    vacuuming, 305
    water sprays, 305
  identifying pests, 301–3, **301,
    302, 303**
    illustrated key to adults,
    452–453, **452–453**
    illustrated key to larvae and
    nymphs, 454–455, **454–455**
  monitoring crops for pests,
    300–301
  newly emerging pests, 300
  pesticides, 315–20
    combination sprays, 317
    insecticidal soap, 316–17
    neem, 318–19
    oil sprays, 317–18
    OMRI approved, 6–7
    pyrethrum, 319–20
  preventing pests, 299, 303–4
  repellents
    foil mulches, 308
    garlic spray, 308–10
    hot pepper spray, 308–10
  row covers as, 13, 306, 348–49
  trap crops and, 307
  weather and, 314

Insect pests, 365–71, **366, 367, 368, 369, 370, 371.** *See also* Insect pest control; *specific crop; specific pest*
ants, 25
aphids, 25–27, **26**
  viral diseases and, 174
armyworms, 28–30, **28, 29**
asparagus beetles, **37,** 38
  spotted, **37,** 38
beet webworms, 48–49
blister beetles, 65–66, **66**
cabbage loopers, 86–89, **87**
cabbage maggots, 89–92, **90, 91**
cabbageworms
  cross-striped, 82
  imported, 225–27, **225, 226**
Colorado potato beetles, 112–14, **112, 113**
corn borers, European
  corn and, 129, **129,** 130
corn earworms, 131–33, **131, 132**
cotton bollworms (corn earworms), 131–33, **131, 132**
cucumber beetles, 158–60, **159, 160**
cutworms, 161–62, **161, 162**
  as disease vectors, 177
earwigs, 189–91, **190**
flea beetles, 205–7, **206,** 307
fleahoppers, 208–10, **208, 209**
grasshoppers, 216–18, **217,** 301
harlequin bugs, 220–21, **220**
illustrated key to adults, 452–453, **452–453**
illustrated key to larvae and nymphs, 454–455, **454–455**
imported cabbageworms, 225–27, **225, 226**
Japanese beetles, 228–30, **228, 229**
leafhoppers, 239–41, **240**
  disease transmission and, 39, 46
leafminers, 241–43, **242**
leek moths, 246
Mexican bean beetles, 258–60, **259, 302**
mites, 261–63, **262, 263,** 303
nematodes, 176, 272–74, **273,** 275, **275**
pepper weevils, 294–95
pickleworms, 321–23, **321**
potato psyllids, 333

saltmarsh caterpillars
  lettuce and, 251
silverleaf whiteflies, 416
slugs, 303, 365–71, **366, 367, 368, 369, 370, 371**
snails, 303, 365–71, **366, 367, 368, 369, 370, 371**
spider mites, 261–63, **262, 263,** 303
squash bugs, **303,** 391–93, **392**
squash vine borers, 393–95, **394**
stink bugs, 396–97, **396**
Swede midge, broccoli and, 70–71
tarnished plant bugs, 406–8, **407**
thrips, 408–10, **409**
tomato fruitworms (corn earworm), 131–33, **131, 132**
tomato hornworms, 422–23, **422, 423**
tomato pinworms, 416–17
vetchworms (corn earworms), 131–33, **131, 132**
whiteflies, 446–48, **446, 447**
wireworms, 449–51, **450**
yellowmargined leaf beetles, 82
Insects. *See* Insect pests; Insects, beneficial
Insects, beneficial, 51–61, 53, 54–55, **55, 56, 57, 58, 59, 60, 61**
assassin bugs, 56, **56**
blister beetles as, 65, **66**
encouraging, 3
flowers attractive to, 53, 54–55, **55**
garden habitat for, 52–53
ground beetles, 56–57, **56**
insectary garden for, 53–55, 54–55, **55**
lacewings, 52, 57, **57**
lady beetles, 52, 57–58, **57, 58**
parasitic wasps, 58, **58**
  Encarsia formosa, 61
  Pediobius foveolatus, 61, 258–59
predatory mites, 61, 261
releasing, 61
requirements of, 51–52, 55
spiders, 58, **59**
spined soldier bugs, 59, **59,** 60, **60**
syrphid flies, 59, **59,** 61
tachinid flies, 61, **61**
Interplanting, 3, 325–26, **326**
Iron, deficiency symptoms, 201
Italian turnip, 73, **73**

J
Japanese beetles, 228–30, **228, 229**
  control calendar, 230
  control methods, 229–30
  profile of, 228–29, **228, 229**
Jays, seedlings and, 64–65, **64**

K
Kale, 231–34, **232, 352**
  as cover crop, 141
  crop basics, 231
  problem prevention, 232–33
  secrets of success, 231–32
  troubleshooting, 233–34
    club root and, 110–11
Kaolin clay, as insect pest control, 306–8
Kelp (seaweed), as fertilizer, 203
Kohlrabi, 234–36, 236, **236**
  crop basics, 234–35
  secrets of success, 235, 236
  troubleshooting, 235–36
    club root and, 110–11

L
Lacewings, 52, 57, **57**
Lady beetles, 52, 57–58, **57, 58**
Lamb's-quarters, 443, **443**
Late blight, 237–38, 238, **238**
  Bacillus subtilis and, 180–81
  copper and, 183
Leafhoppers, 239–41, **240**
  aster yellows and, 39
  beet curly top (curly top virus) and, 46
  control calendar, 241
  control methods, 240–41
  profile of, 239, **240**
Leafminers, 241–43, **242**
  control calendar, 243
  control methods, 243
  profile of, 242, **242**
Leaves, as mulch, 266, **266**
Leek moth, 246
Leeks, 244–47, **245**
  crop basics, 244
  problem prevention, 246
  regional notes
    growing in Canada, 246
  secrets of success, 245–46, **245**

troubleshooting, 246
leek moth, 246
Legume family members, 147
in crop rotation plan, 148, **149,**
150
Leptinotarsa decemlineata
(Colorado potato beetle),
112–14, **112, 113**
Lettuce, 247–52, **249,** 250, **352**
crisphead ('Iceberg'), 250
crop basics, 247–48
problem prevention, 251
regional notes
growing in the Southwest, 251
secrets of success, 248–50, **249,**
250
troubleshooting, 251–52
armyworms and, 28
aster yellow symptoms on, 39
corn earworm and, 131–33,
131, 133
gray mold (Botrytis blight)
and, 219
saltmarsh caterpillar and, 251
white heart, 39
Lettuce family members, 147
in crop rotation plan, 148, **149,**
150
Liriomyza spp. (leafminers), 241–43,
242
Lolium multiflorum (annual
ryegrass), 140

M

Mache (corn salad), 353–54
Magnesium, deficiency symptoms,
201
Mallow family members, 147
Manure tea
as fertilizer, 202–3
safety of, 203
Marigolds
African
nematodes and, 275
French
nematodes and, 275, **275**
Melanoplus spp. (grasshoppers),
216–18, **217**
Melilotus officinalis (sweet clover),
141
Melons, 253–58, **254**
crop basics, 253–54
problem prevention, 255–56

regional notes
growing in high altitudes,
254–55
growing in the East, 255
growing in the Far North,
254–55
secrets of success, 254, **254**
troubleshooting, 256–58
Anthracnose symptoms on,
23–24
bacterial fruit blotch, 255
blossom-end rot and, 67–68
gray mold (Botrytis blight)
and, 219
raccoons and, 11 (*see also*
Raccoons)
Mesclun, 352–53, **352**
Mexican bean beetles, 258–60,
259, 302
control calendar, 260
control methods, 260
profile of, 259, **259**
Mildew
downy, 187
Bacillus subtilis and, 180–81
copper and, 183
milk and, 183
neem oil and, 182–83
powdery, 337–38
Bacillus subtilis and, 180–81
bicarbonate and, 181–82
neem oil and, 182–83
Milk, as fungicide, 183
Milky spore disease
Japanese beetles and, 230
Mites, 261–63, **262, 263,** 303
harmful, 261–63, **262, 263**
predatory (beneficial), 61, 261
Mizuna, **352**
Mold, 173
Moles, 22
Morning glory family members,
147
Mosaic, 263–64
Motley dwarf disease
carrots and, 96
Mulch, 265–70, **266, 267,**
268–269, **269**
allergies and, 270
foil as insect repellent, 308
importance of using, 2–3
materials useful as, 265–69,
266, 267, 268–269, **269**
compost, 265
grass clippings, 265

leaves, 266, **266**
living (cover crops), 266–67
newspaper, 267, **267**
pine needles, 267
plastic, 267–68, 268–269,
269
straw, 268–69
wood chips, 269
troubleshooting, 269–70
as weed control method, 438
Murgantia histrionica (harlequin
bug), 220–21, **220**
Muskmelons. *See* Melons
Mustard
as biofumigant, 185
as trap crop, 307
Mustard greens, 354
Mycorrhizae, 173

N

National Organic Program (NOP),
5–6, **6**
Neem, 318–19
as fungicide, 182–83
Nematodes, 176, 271–74, **273,**
275, **275**
beneficial, 271–72
harmful, 272–74, **273,** 275, **275**
Newspaper, as mulch, 267, **267**
Nightshade, 443, **443**
Nitrogen, deficiency symptoms,
201
NOP (National Organic Program),
5–6, **6**
Nutritional deficiencies, 175,
199–200
symptoms of, 201
Nutsedge, 444, **444**

O

Oats, as cover crop, 141
OFPA (Organic Foods Production
Act), 5–6
Oil sprays, 317–18
Okra, 276–78, **277**
crop basics, 276
problem prevention, 277
secrets of success, 276–77, **277**
troubleshooting, 277–78
OMRI (Organic Materials Review
Institute), 6–7, **6**

Onion family members, <u>147</u>
  in crop rotation plan, <u>148,</u> **149,**
    150
  troubleshooting
    leek moth and, 246
    onion maggots and, 246
    onion thrips and, 246
    white rot and, 214
Onions, 278–83, **281,** <u>282</u>
  crop basics, 279–80
  Egyptian (walking), **364**
  potato, **364**
  problem prevention, 282
  secrets of success, 280–82, **281**
  troubleshooting, 282–83
    aster yellow symptoms on, 39
Organic Foods Production Act
    (OFPA), 5–6
Organic gardening
  National Organic Program
    (NOP) and, 5–6, **6**
  Organic Foods Production Act
    (OFPA) and, 5–6
  Organic Materials Review Institute
    (OMRI) and, 6–7, **6**
  standards for, 5–7, <u>5,</u> **6**
Organic Materials Review Institute
    (OMRI), 6–7, **6**
Organic matter
  building in soil, 375–77, <u>377</u>

P

Parasitic wasps, 58, **58**
  Encarsia formosa, 61
  Pediobius foveolatus, 61, 258–59
Parsnips, 284–86, **284**
  crop basics, 284, **284**
  problem prevention, 285
  secrets of success, 285
  troubleshooting, 285–86
Pathways, cover crops as, <u>137,</u> **137**
Peas, 286–91, **287,** <u>288,</u> **288**
  Austrian winter, as cover crop,
    <u>140</u>
  crop basics, 286–87
  field, as cover crop, <u>140</u>
  problem prevention, 289
  secrets of success, 287–89, **287,**
    <u>288,</u> **288**
  southern, as cover crop, <u>140</u>
  troubleshooting, 289–90
Pediobius foveolatus, 61, 258–59

Peppers, 291–98, **293, 294**
  crop basics, 292
  problem prevention, 295
  regional notes
    growing in the Far South,
      294–95
    growing in the Far West, 295
  secrets of success, 293–94, **293,**
    294
  troubleshooting, 295–98
    blossom-end rot and, 67–68
    corn earworm and, 131–33,
      131, 133
    early blight and, 188–89
    pepper weevils, 294–95
Pepper weevils, 294–95
Pergomya spp. (leafminers),
    241–43, **242**
Pesticides, 315–20
  combination sprays, <u>317</u>
  insecticidal soap, 316–17
  neem, 318–19
  oil sprays, 317–18
  OMRI approved, 6–7
  pyrethrum, 319–20
Pests. *See* Animal pests; Birds;
    Insect pests; Pets
Pets, as pests, 22–23
pH, 373–74
Phomopsis blight of eggplant, 193
Phosphorus, deficiency symptoms,
    <u>201</u>
Pickleworms, 321–23, **321**
  control calendar, 323
  control methods, 322
  profile of, 321–22, **321**
Pieris rapae (imported
    cabbageworms), 225–27,
    225, 226
Pigweeds, 444–45, **444**
Pine needles, as mulch, 267
Pisum sativum var. *arvense*
    (Austrian winter pea, field
    pea), <u>140</u>
Plant families. *See also specific*
    family
  in crop rotation, 146, <u>147,</u>
    <u>148–150,</u> **149, 150**
Planting
  double-row spacing, 325, **325**
  interplanting, 3, 325–26, **326**
  quantity to plant, 324
  succession planting, 325, 326,
    326

timing, 323–24
    as insect pest control strategy,
      304
Plant supports
  for beans, 41–42, **41,** <u>42,</u> **42**
  for cucumbers, **154**
  for eggplants, **193**
  for peas, **287, 288**
  for tomatoes, 414–16, **414,** <u>415,</u>
    415
Plastic
  as deer fence, 167–68, **168**
  as mulch, 267–68, <u>268–269,</u> **269**
  as season extender, 293–94,
    294, 357–58, **357**
Plectosporium blight, 387–88
Podisus maculiventris (spined soldier
    bugs), 59, **59,** <u>60,</u> **60**
Polygonum esculentum (Fagopyrum
    esculentum) (buckwheat),
    <u>140</u>
Potassium, deficiency symptoms,
    <u>201</u>
Potatoes, 330–37, **331,** <u>332,</u> **332**
  crop basics, 330–31
  problem prevention, 333–34
  regional notes
    growing in the Southeast, 333
    growing in the western states,
      333
  secrets of success, 331–33, **331,**
    <u>332,</u> **332**
  troubleshooting, 334–38
    early blight and, 188–89, 205
    "hopperburn," 239, **240**
    late blight and, 237–38, <u>238,</u>
      238
    potato psyllids, 333
Potato psyllids, 333
Pumpkins. *See* Squash
Pyrethrum, 319–20

Q

Quackgrass (couch grass, wheat
    grass), 445, **445**

R

Rabbits, 11
  barriers for, 16–18
  fences for, 166, 167, **171**

scare devices for, 13–15, **14**
traps and, 19–20
Raccoons, 11
  barriers for, 16–18, **17, 18, 19**
  fences for, 166, 167, **171**
  scare devices for, 13–15, **14**
Radicchio, 354
Radishes, 339–42, <u>340</u>, **340**
  crop basics, 339
  kinds of, <u>340</u>, **340**
  oilseed
    as cover crop, <u>141</u>
  problem prevention, 341
  secrets of success, 339–40
  troubleshooting, 341–42
    club root and, 110–11
Raised beds, 374–75, **375**
  gopher-proof, 20, **20**
Rape, as cover crop, <u>141</u>
Rapeseed, as biofumigant, <u>185</u>
Raphanus sativus (oilseed radish), <u>141</u>
Rapini, <u>73</u>, **73**
Repellents
  for deer, 165, <u>169</u>
  for insect pests
    foil mulch as, 308
    garlic spray as, 308–10
    hot pepper spray as, 308–10
  for small animal pests, 12–13,
    15–16
Rhubarb, 342–245, **343**
  crop basics, 342–43
  problem prevention, 344
  regional notes
    growing in the Deep South, 344
  secrets of success, 343, **343**
  troubleshooting, 344–45
Rock phosphate, as fertilizer, <u>203</u>
Root rots
  biofumigation and, <u>185</u>
Rots, 345–46
Row covers, 347–50, **348, 349**
  choosing, 347
  extending life of, 349–50
  as insect pest control device, 13,
    306, 348–49
  as season extender, 348, 358
  supports for, 347–48, **348**
Rust, 350–51
  neem oil and, 182–83
Rutabaga, **425**. *See also* Turnips
Rye
  cereal, as cover crop, <u>140</u>
  winter, as cover crop, <u>140</u>

Ryegrass, annual
  as cover crop, <u>140</u>

S

Salad greens, 352–54, **352**
  arugula, **352**, 353
  corn salad (mache), 353–54
  cress, 354
  kale, 231–34, **232, 352**
  lettuce, 23, 28, 39, 219, 247–52,
    249, <u>250</u>, 251, **352**
  mesclun, 352–53, **352**
  mizuna, **352**
  mustard greens, 354
  radicchio, 354
  seedlings of, 352, **352**
Saltmarsh caterpillar
  lettuce and, 251
SAR (systemic acquired resistance),
  175
Scab, 355
  biofumigation and, <u>185</u>
Scallions, 282
Scare devices
  for deer, 14, 165
  for small animal pests, 12,
    13–15, **14**
Season extension, 356–59, **356,
  357, 358**
  cloches, **356,** 357
  coldframes for, 358–59, **358**
  planting date adjustment, 357
  plastic-covered cages as,
    293–94, **294**
  plastic tunnels as, 357–58, **357**
  row covers, 348, 358
  tricks for, 356–57, **356, 357**
Seaweed (kelp), as fertilizer, <u>203</u>
Secale cereale (cereal rye, winter
  rye), <u>140</u>
Seedlings
  Anthracnose symptoms on, 24
  caring for indoors, 361–62, **361**
  damping-off of, 163–64, 360–61
  hardening off, 362, **363**
  thinning, 328
  transplanting, 329, <u>329</u>, **432, 434**
Seeds, 359–60, **359**. *See also*
  Seedlings
  boosting germination, 328, **359,**
    360
  disease transmission and, 176

sowing indoors, 360–61
sowing outdoors, 327–28, **327**
testing viability of, 359–60, **359**
Septoria leaf spot
  copper and, 183
Shallots, 363–65, **364**
  crop basics, 364
  secrets of success, 364–65
  troubleshooting, 365
Shepherd's purse, 445, **445**
Silverleaf whitefly, 416
Skunks, 21
Slugs, <u>303</u>, 365–71, **366, 367,
  368, 369, 370, 371**
  control calendar, 370–71
  control methods, 367–70, **368,
    369, 370, 371**
  mulch and, 270
  profile of, 366–67, **366, 367**
Snails, <u>303</u>, 365–71, **366, 367,
  368, 369, 370, 371**
  control calendar, 370–71
  control methods, 367–70, **368,
    369, 370, 371**
  mulch and, 270
  profile of, 366–67, **366, 367**
Snakes, 21
Soft rot, 346
Soil, 372–78, **376, <u>377</u>**
  compost and, <u>377</u>
  cover crops and, 134, 374, <u>377</u>
  creating healthy, 2–3
  crop rotation and, 143
  disease control and, 175–76
    solarizing, <u>186</u>
  drainage, 372–73, 374–75, **375**
  evaluating, 372–74
  fertility of, 134, 143
  importance of healthy, 2
  natural balance in, **377**
  nutrient deficiencies in, 199–200
  organic matter
    adding, 375–77, <u>377</u>
  pH, 373–74
  solarizing
    as disease preventive, <u>186</u>
    nematodes and, 274
  testing, 199
  tilling, 377–78
Solanum spp. (nightshades), 443,
  443
Sooty mold, aphids and, 26
Sorghum bicolor var. *sudanense*
  (sudangrass), <u>141</u>

Southern blight, 378–79
Soybean meal, as fertilizer, 203
Soybeans, as cover crop, 141
Spider mites. *See* Mites
Spiders, 58, **59**
Spinach, 380–84, **381, 382,** 382
  crop basics, 380
  Malabar, 382, **382**
  New Zealand, 382, **382**
  problem prevention, 383
  regional notes
    growing in the Southeast, 383
    growing in the Southern
      Plains, 383
  secrets of success, 380–81, 383
  troubleshooting, 383–84
    leafminers and, 241–43, **243**
    white rust, 383
Spinach family members, 147
  in crop rotation plan, 148, **149,**
    150
Spined soldier bugs, 59, **59,** 60, **60**
Spinosad, 314
Spotted asparagus beetles, **37,** 38
Squash, 385–91, 387, **387,** 390
  crop basics, 385–86
  problem prevention, 388
  regional notes
    growing in the East, 387–88
    growing in the Heartland, 387
  secrets of success, 386, 387, **387**
  troubleshooting, 388–91
    blossom-end rot and, 67–68
    Fusarium wilt and, 210–11
Squash bugs, **303,** 391–93, **392**
  control calendar, 393
  control methods, 392–93
  profile of, 392, **392**
Squash family members, 147
  in crop rotation plan, 148, **149,**
    150
  troubleshooting
    Alternaria blight and, 9
    angular leaf spot and, 10
    beet curly top (curly top
      virus) and, 46
    Cercospora leaf spot and, 109
    crown rot, 346
    cucurbit yellow vine disease,
      387
    downy mildew and, 187
    gray mold (botrytis blight)
      and, 219
    Plectosporium blight, 387–88

squash bugs, **303,** 391–93, **392**
squash vine borer, 393–95,
  394
Squash vine borer, 393–95, **394**
  control calendar, 395
  control methods, 394–95
  profile of, 394, **394**
Steinernema carpocapsae (beneficial
  nematodes), 271–72
Steinernema feltiae (beneficial
  nematodes), 271–72
Stewart's wilt, 127
Stink bugs, 396–97, **396**
  control calendar, 397
  control methods, 397
  profile of, 396–97, **396**
Storms, dealing with, 434
Straw, as mulch, 268–69
Succession planting, 325, 326, **326**
Sudangrass, as cover crop, 141
Sunflower seedlings
  bird pests and, 64–65, **64**
Swede midge, broccoli and, 70–71
Sweet potatoes, 397–400, **398**
  crop basics, 397–98
  problem prevention, 399
  secrets of success, 398–99, **398**
  troubleshooting, 399–400
    Fusarium wilt and, 210–11
Swiss chard, 382, 401–5, 402, **402**
  crop basics, 401, 402, **402**
  problem prevention, 403
  regional notes
    growing in the Desert
      Southwest, 403
    growing in the Midsouth, 403
    growing in the South, 403
  secrets of success, 401–3
  troubleshooting, 403–5
    Fusarium wilt and, 210
    leafminers and, 241–43, **243**
Symphytum × *uplandicum* (Russian
  comfrey)
  tea from as fertilizer, 203–4
Syrphid flies, 59, **59,** 61
Systemic acquired resistance
  (SAR), 175

T

Tachinid flies, 61, **61**
Tagetes erecta (African marigold)
  nematodes and, 275

Tagetes patula (French marigold)
  nematodes and, 275, **275**
Tarnished plant bugs, 406–8, **407**
  control calendar, 408
  control methods, 407–8
  profile of, 406–7, **407**
Thrips, 408–10, **409**
  control calendar, 410
  control methods, 409–10
  profile of, 409, **409**
Toads, shelter for, 54
Tobacco mosaic virus. *See* Mosaic
Tomatillos, 412
Tomatoes, 411–21, **413,** 413, **414,**
    415, **415**
  crop basics, 411–12
  problem prevention, 417
  regional notes
    growing in California, 416–17
    growing in Florida, 416–17
    growing in Texas, 416–17
    growing in the South, 416
    growing in the Southeast, 416
    growing in the Southwest,
      416–17
    growing in the West, 416
  secrets of success, 412–16, 413,
    413, 414, 415, **415**
    pruning, 413, **413,** 415
  troubleshooting, 417–21
    Anthracnose symptoms on, 24
    armyworms and, 28
    beet curly top (curly top
      virus) and, 46
    blossom-end rot and, 67–68
    corn earworm (tomato
      fruitworm) and, 131–33,
      131, 132
    early blight and, 188–89
    Fusarium wilt and, 210
    gray mold (botrytis blight)
      and, 219
    late blight and, 237–38, 238,
      238
    silver whitefly, 416
    tomato pinworms, 416–17
    tomato spotted wilt virus, 416
    Verticillium wilt, 427
Tomato family members, 147
  in crop rotation plan, 148, **149,**
    150
Tomato fruitworm (corn
  earworm), 131–33, **131,**
  132

Tomato hornworm, 422–23, **422, 423**
  control calendar, 423
  control methods, 423, **423**
  profile of, 422–23, **422, 423**
Tomato pinworms, 416–17
Tomato spotted wilt virus, 416
Transplanting, 329, *329,* **432, 434**
Traps
  for insect pests
    commercial, 310
    crops as, 307
  small animal pests and, 19–20
Trichoderma harzianum, as disease control agent, 181
Trichoplusia ni (cabbage loopers), 86–89, **87**
Trifolium alexandrinum (Berseem clover), 140
Trifolium hybridum (alsike clover), 140
Trifolium incarnatum (crimson clover), 140
Trifolium pratense (red clover), 141
Trifolium repens (white clover, white Dutch clover, New Zealand white clover), 141
"Turnip broccoli," 73, **73**
Turnip mosaic virus. *See* Mosaic
Turnips, 424–26, **425**
  crop basics, 424
  Italian, 73, **73**
  problem prevention, 425–26
  secrets of success, 424–25
  troubleshooting, 426
    Alternaria blight and, 9
    cabbage maggots and, **90**
    club root and, 110–11

U

USDA hardiness zones, 431–32
  map, 472

V

Vacuuming, as insect control method, 305
Verticillium, 427
  biofumigation and, 185
Verticillium wilt, 427
Vetch, hairy
  as cover crop, 141

Vetchworm (corn earworm), 131–33, **131, 132**
Vicia faba (fava bean, bell bean), 140
Vicia villosa (hairy vetch), 141
Vigna unguiculata ssp. *unguiculata* (cowpea, southern pea), 140
Viral diseases, 174–75
Voles, 21–22
  barriers for, 22, **22**

W

Watering, 428–30, *428, 429,* **429, 430**
Watermelons. *See also* Melons
  troubleshooting
    Anthracnose symptoms on, 23–24
    blossom-end rot and, 67–68
Water sprays, as insect control method, 305
Weather, 431–34, *433,* **434**
  common problems with, 432–34
    drought, 432
    heat waves, 432–33
    storms and hail, 434
    wind, 433–34, **434**
  disorders related to, 175
  insect pest control and, 314
  local quirks, 433
  understanding, 431–32
    frost, 431
    frost dates, 431
    hardiness zones, 431–32
Web sites
  for National Organic Program (NOP) regulations, 6
  for *Organic Gardening* magazine database, 6
  for Organic Materials Review Institute information, 6
Weeds, 435–45, **436, 437, 441, 442, 443, 444, 445**
  avoiding problems with, 435–39
  control methods
    cultivation, 436–37
    flame weeding, 438
    herbicides and food crops, 437
    mulch, 438 (*see also* Mulch)
    sowing techniques and, 435–36
    tools for, 436–37, **437**
    vigilance, 438–39

dealing with large, 439
  salvaging overrun gardens, 440
  ten common
    bermudagrass (wiregrass, devilgrass), 440–41, **441**
    bindweeds, 441–42, **441**
    Canada thistle, 442, **442**
    crabgrasses, 442–43, **442**
    lamb's-quarters, 443, **443**
    nightshades, 443, **443**
    nutsedges, 444, **444**
    pigweeds, 444–45, **444**
    quackgrass (couch grass, wheat grass), 445, **445**
    shepherd's purse, 445, **445**
Wheat grass (quackgrass, couch grass), 445, **445**
Whiteflies, 446–48, **446, 447**
  control calendar, 448
  control methods, 447
  profile of, 446–47, **446, 447**
White heart of lettuce, 39
White mold, 448–49
White rot, onion-family members and, 214
White rust, 383
Wind, dealing with, 433–34, **434**
Wiregrass (bermudagrass, devilgrass), 440–41, **441**
Wireworms, 449–51, **450**
  control calendar, 451
  control methods, 450–51
  profile of, 450, **450**
Wood ashes, as fertilizer, 203
Wood chips, as mulch, 269
Woodchucks (groundhogs)
  barriers for, 16–18, **17, 19**
  fences for, 166, 167, 168, **171**
  traps and, 19–20

Y

Yellowmargined leaf beetle
  cabbage and, 82

Z

Zinc, deficiency symptoms, 201
Zucchini. *See also* Squash
  blossom-end rot and, 67–68

USDA PLANT HARDINESS ZONE MAP

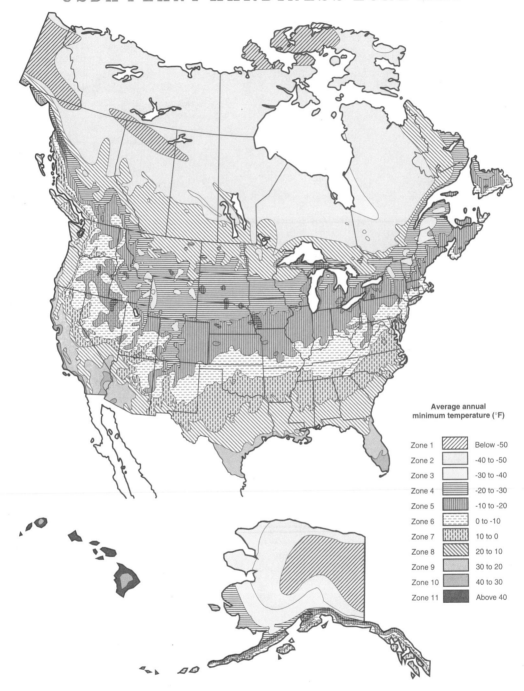

Average annual minimum temperature (°F)

Zone 1		Below -50
Zone 2		-40 to -50
Zone 3		-30 to -40
Zone 4		-20 to -30
Zone 5		-10 to -20
Zone 6		0 to -10
Zone 7		10 to 0
Zone 8		20 to 10
Zone 9		30 to 20
Zone 10		40 to 30
Zone 11		Above 40

This map is recognized as the best indicator of minimum temperatures available. Look at the map to find your area, then match its pattern to the key at right. When you've found your pattern, the key will tell you what hardiness zone you live in. Remember that the map is a general guide; your particular conditions may vary.